master electrician's workbook

BASED ON THE 1990 NEC®

James G. Stallcup

AMERICAN TECHNICAL PUBLISHERS, INC.
HOMEWOOD, ILLINOIS 60430

2 3 4 5 6 7 8 9 - 89 - 9 8 7 6 5 4 3

Printed in the United States of America

Library of Congress Cataloging-in-Publication Data

Stallcup, James G.
Master electrician's workbook.

1. Electric wiring—Examinations, questions, etc. 2. Electric apparatus and appliances—Installation—Examinations, questions, etc. 3. Electricians—Licenses—United States. I. National Fire Protection Association. National electrical code, 1990. II. Title.
TK3271.S73 1990 621.319'24'0218 89-17981
ISBN 0-8269-1715-1

Contents

Introduction

MASTER ELECTRICIAN'S WORKBOOK BASED ON THE 1990 NEC® is designed for journeyman electricians preparing for the Master's Exam. As it is comprehensive in its approach to a technical subject, it may be used by anyone wishing to learn more about the 1990 National Electrical Code®. Beginning with basic information, and progressing through calculations, step-by-step procedures in the illustrations and text "show and tell" how to apply provisions of the 1990 National Electrical Code® to answer electrical questions and solve calculations. Frequent reference to and study of the 1990 National Electrical Code® is required. Copies of the National Electrical Code® 1990 (NFPA No. 70) may be purchased from:

National Fire Protection Association
Batterymarch Park
Quincy, MA 02269

The Instructor's Guide for this book contains answers and Code® substantiation for all questions. Answers, *solutions,* and Code substantiation are given for all problems. Copies of the Instructor's Guide may be purchased from American Technical Publishers, Inc.

Blank copies of calculation forms, designed specifically for this book, may be reproduced for instructional use only. These forms, located at the back of this book, shall not be reproduced and sold to others. Calculation forms that may be reproduced for instructional use only are clearly marked,

© by American Technical Publishers, Inc.
This form may be reproduced for instructional use only.
It shall not be reproduced and sold.

The Publisher

CALCULATIONS

When total wattage, or VA, is to be divided by phase-to-phase (3ϕ) voltage times 1.732, the following values are substituted:

for 208 volts × 1.732, use 360
for 230 volts × 1.732, use 398
for 240 volts × 1.732, use 416
for 440 volts × 1.732, use 762
for 460 volts × 1.732, use 797
for 480 volts × 1.732, use 831

WATTS—VOLT-AMPERES

In general, within the National Electrical Code®, the term *watts* (w) has been superseded by the term volt-amperes (VA) for the computation of loads. However, references to nameplate ratings still reflect the term *watts* on certain loads.

CONDUCTORS

Unless otherwise specified, copper conductors are sized on THWN per Table 310-16. Derate by Column A or B of Note 8 to Tables 310-16–310-19 (consider as loads with diversity unless noted otherwise).

MANDATORY USE OF "SHALL"

Section 110-1 of the 1990 NEC® states that mandatory rules use the word "shall." Always refer to the NEC® for mandatory rules.

FRACTIONS OF AN AMP

Chapter 9, B. Examples of the 1990 NEC® states that fractions less than .5 may be dropped.

RECEPTACLE LOADS

All receptacle loads exceeding 10,000 VA in commercial and industrial locations are reduced per Table 220-13.

Branch Circuits

Chapter 1

Branch circuits extend between the last overcurrent protection device protecting the circuit conductors supplying *utilization equipment.* Any equipment that uses electrical energy is utilization equipment. This equipment is either cord- and plug-connected to an outlet or hard-wired. Overcurrent protection devices protecting branch circuits are fuses or circuit breakers located in the service equipment panelboard or subpanel. Branch circuits are characterized by the NEC® by their use in the wiring system, and rules are applied for either a general-purpose or direct branch circuit.

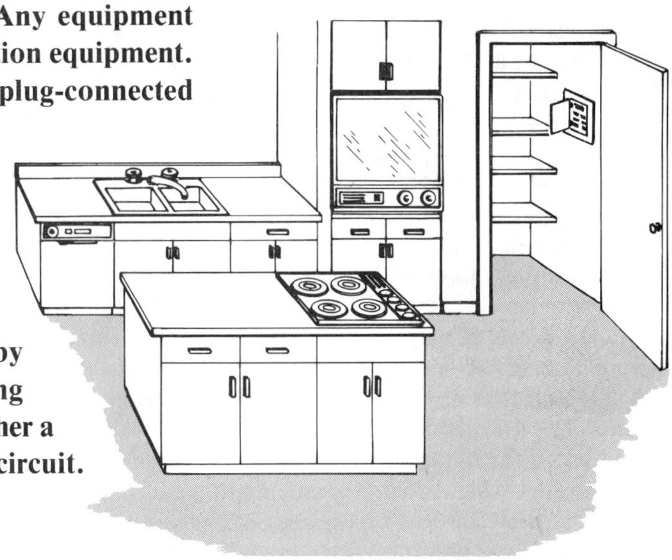

CLASSIFICATIONS—210-3

Branch circuits are classified by the rating or setting of the overcurrent protection device protecting the circuit. For example, if #14 THHN Cu conductors supply a branch circuit feeding receptacle outlets, and the circuit is protected by a 15 A overcurrent protection device, the branch circuit is classified at 15 A. However, if a 20 A overcurrent protection device is used to protect #10 THHN Cu conductors to reduce the voltage drop due to a long run between the service panel and the receptacle outlets, the branch circuit is classified at 20 A and not 30 A.

Overcurrent protection devices rated at 15, 20, 30, 40, and 50 A can be used to supply two or more outlets. A 20 A branch circuit with six lighting and seven receptacle outlets can only be protected by a 20 A overcurrent protection device (either fuses or CB) with #12 Cu conductors.

A direct circuit can be any rating required to supply the load. For example, if a piece of equipment requires a branch circuit to be rated at least 150 A, the size overcurrent protection device and conductors shall be rated at 150 A. See Figure 1-1. Note that branch circuit conductors shall have an ampacity rating not less than the overcurrent protection device and the maximum load served.

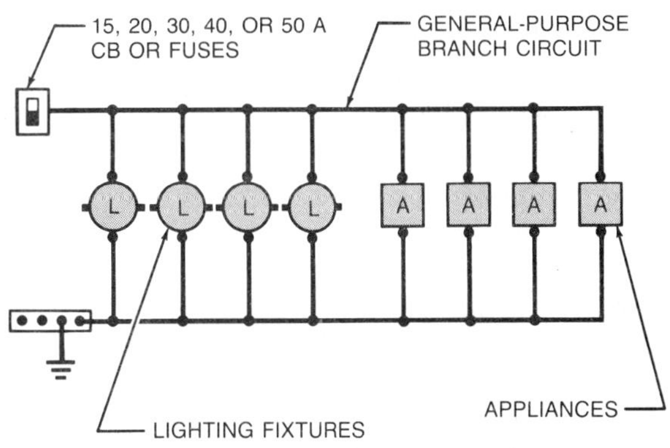

Figure 1-1. Any number of outlets can be used to supply power to lighting and/or appliances from a 15 A through 50 A general-purpose branch circuit.

PERMISSIBLE LOADS—210, PART B

The load served on an individual branch circuit shall not exceed the rating of the overcurrent protection device. Branch circuits with more than one outlet are grouped in ratings of 50 A or less. An individual branch circuit can supply any size load with no restrictions to ampere ratings. Branch circuits serving two or more outlets shall have the load divided to ensure enough circuits for the number of outlets required for lighting and appliances.

15 or 20 A Circuits—210-23(a)

Branch circuits rated at 15 or 20 A can serve lighting fixtures and/or appliances. For example, a 15 or 20 A, 120 V, 2-wire branch circuit can supply a maximum load of 1800 VA at 15 A (15 A × 120 V = 1800 VA) and 2400 VA at 20 A (20 A × 120 V = 2400 VA). A 15 or 20 A, 120/240 V, 3-wire branch circuit can supply a maximum load of 3600 VA at 15 A (15 A × 240 V = 3600 VA) and 4800 VA at 20 A (20 A × 240 V = 4800 VA). These computations are derived by multiplying amperage times voltage.

The amount of power that can be supplied to the load of a three-phase (3ϕ) branch circuit is found by multiplying amperage times voltage times $\sqrt{3}$ ($\sqrt{3}$ = 1.732). A 15 A, 208 V, 3ϕ system can supply a maximum load of 5400 VA (15 A × 208 V × $\sqrt{3}$ = 5400 VA). A 20 A, 208 V, 3ϕ system can supply a maximum load of 7200 VA (20 A × 208 V × $\sqrt{3}$ = 7200 VA).

The use of the following values eliminates one mathematical step when solving 3ϕ calculations:

for 208 V × 1.732, use 360
for 230 V × 1.732, use 398
for 240 V × 1.732, use 416
for 440 V × 1.732, use 762
for 460 V × 1.732, use 797
for 480 V × 1.732, use 831

The rating of any one cord- and plug-connected appliance shall not exceed 80% of the rating of the overcurrent protection device that protects the circuit.

What size overcurrent protection device is required for a 15 A, 120 V, 1ϕ dishwasher that is cord- and plug-connected per 422-8(d)(2)?

Step 1. *210-23(a):*
15 A × 80% = 12 A
20 A × 80% = 16 A

Answer. **OCPD = 20 A**

A 15 A overcurrent protection device is too small for the 15 A load of the dishwasher. It exceeds 12 A, which is the permissible loading range of the 15 A overcurrent protection device. The 20 A overcurrent protection device can be used. Its amperage rating, after the 80% derating, exceeds the 15 A full-load current rating of the dishwasher.

Appliances fastened in place can be connected to a branch circuit serving lighting units and/or cord- and plug-connected appliances, provided the rating of the fastened-in-place appliance does not exceed 50% of the branch circuit rating.

What is the maximum rating of a fastened-in-place appliance that can be connected to a 20 A circuit with two or more outlets? The branch circuit also supplies lighting outlets.

Step 1. *210-23(a):* 20 A × 50% = 10 A

Answer. **Maximum rating = 10 A**

A fastened-in-place appliance with a rating of 10 A or less can be connected to the 20 A circuit after applying the 50% rule. Notice that 10 A, or 50% of the branch circuit, is not exceeded. The remaining 50% of the branch circuit is used for lighting and/or cord- and plug-connected appliances. See Figure 1-2.

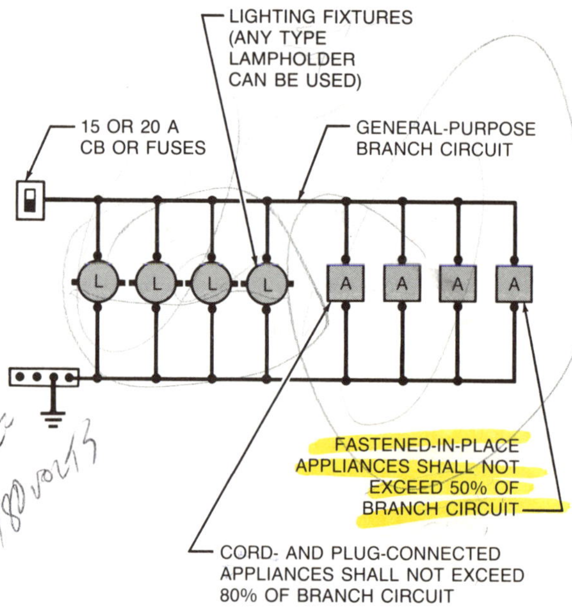

Figure 1-2. Lighting unit and fixed appliance loads are permitted on a 15 A or 20 A general-purpose branch circuit.

30 A Circuits—210-23(b)

Branch circuits rated at 30 A can serve fixed lighting units with heavy-duty lampholders in any type occupancy except dwelling units or appliances can be supplied in any occupancy. Individual cord- and plug-connected appliances shall not exceed 80% of the branch circuit rating. A cord- and plug-connected appliance drawing more than 24 A shall not be connected to this type of circuit. The 24 A rating is derived by the following procedure:

Step 1. *210-22(c):* 30 A × 80% = 24 A

Step 2. *210-23(b):* 30 A branch circuit

Answer. **Rating = 24 A**

What is the load for a fixed appliance drawing 22.5 A?

Step 1. *220-3(a):* 22.5 A × 125% = 28 A

Step 2. *210-23(b):* 30 A branch circuit

Answer. **Load = 28 A**

A single appliance, used at continuous-duty, in any type occupancy can be supplied by a 30 A branch circuit. See Figure 1-3.

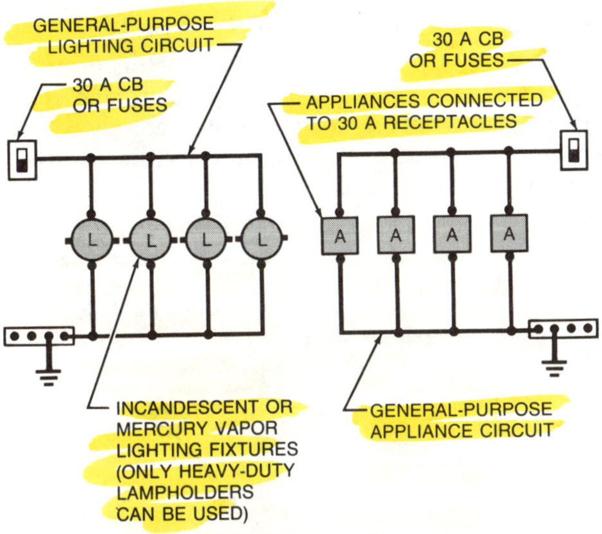

Figure 1-3. Either lighting or appliances, but not both, can be connected to multioutlet 30 A circuits.

40 or 50 A Circuits—210-23(c)

Branch circuits rated at 40 or 50 A can supply fixed lighting units with heavy-duty lampholders or infrared heating units in other than dwelling units or cooking appliances in any type occupancy. A 40 or 50 A circuit can be used to supply a single load for equipment such as a water heater, dryer, or heating unit. Note that 40 or 50 A branch circuits can be used to supply one outlet or circuits having more than one outlet. See Figure 1-4.

What is the load for a 38.5 A water heater?

Step 1. *422-14(b):* 38.5 A × 125% = 48 A

Answer. **Load = 48 A**

Circuits Larger Than 50 A—210-23(d)

Branch circuits larger than 50 A shall not supply outlets used for lighting loads. Multioutlet branch circuits for lighting are limited to a maximum load of 50 A. However, multioutlet branch circuits larger than 50 A can supply any combination of loads that are not connected to lighting units.

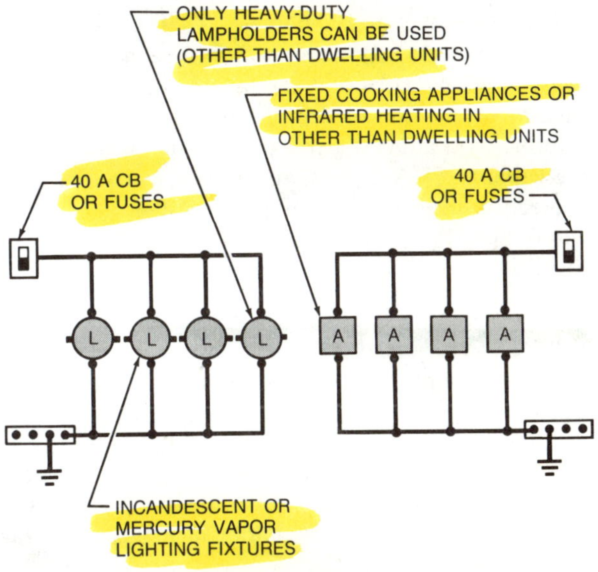

Figure 1-4. Multioutlet 40 A or 50 A circuits can be used to supply power to fixed lighting units, fixed appliances, or infrared heating units.

BRANCH CIRCUIT CONDUCTORS—310

Conductors used for general wiring are listed in Table 310-16 at 60°, 75°, or 90°C. Table 310-13 lists the temperature rating of conductors and their conditions of use. Table 310-16 is based on not more than three current-carrying conductors in a raceway or cable and an ambient temperature of not more than 30°C or 86°F. Table 310-16 lists both copper and aluminum conductors with the types of insulation available.

60°C—Table 310-16, Column 2

Column 2 of Table 310-16 lists TW and UF types of insulation for general wiring. Conductors with types of insulation bearing the W rating are listed for dry and wet locations per 310-8(a). The temperature rating of these conductors is 60°C per Table 310-13 and Table 310-16.

Conductors #14 through #1 are to be used with 60°C ampacities and terminated on overcurrent protection devices rated at 100 A or less. A #12 TW conductor shall be terminated on a 20 A overcurrent protection device and used at 60°C ampacities listed in Table 310-16.

A piece of electrical equipment requires a conductor rated at 80 A to supply its operating power at 480 V, 3ϕ. What size TW or THW Cu conductors is required to supply this load in a dry and wet location?

Step 1. *Table 310-13:* Load = 80 A

Step 2. *Table 310-16:* #3 TW or #3 THW

Answer. *240-3, Ex. 4:* **Conductors = #3 TW or #3 THW Cu**

If nonmetallic sheathed cable or UF cable is used per 336-26 or 339-5, the ampacity of each conductor is sized at 60 °C. For example, a #6 nonmetallic sheathed cable (Romex/rope) with THHN insulation shall be used at 60 °C per Table 310-16, which limits the cable to 55 A or less. A 50 or 60 A overcurrent protection device can be used per 240-3, Ex. 4 and 240-6. See 310-8(b) and Table 310-13 for dry and wet locations.

75 °C—Table 310-16, Column 3

The conductors most commonly used in Column 3 of Table 310-16 are THW and THWN. These conductors are approved for use in dry and wet locations per 310-8(a) and Table 310-13. Column 3 also lists underground service entrance (USE) as a wiring method for service entrance, feeders, and branch circuits to utilization equipment.

Conductors #1/0 and larger are to be used with 75°C ampacities and terminated on overcurrent protection devices rated over 100 A. Conductors with higher temperature ratings can be used if the ampacities are matched to the terminals of the overcurrent protection device and equipment. Terminals are marked either 60°, 75°, or 60°/75°C. A 60°C marking permits only 60°C ampacities to be applied. A 75°C marking allows only 75°C ampacitics to bc used except for con ductors #14 through #1, in which case a 60°C rating shall be applied. Note that 75°C terminals can be used with any size conductor listed in the 75°C column. For example, a #3 THWN Cu conductor can be used at 100 A at 75°C but only at 85 A at 60°C. The temperature rating of the conductors shall match the temperature rating of the device and equipment and ampacity values shall be applied accordingly.

90 °C—Table 310-16, Column 5

Conductors with 90°C insulation, such as XHHW and THHN, can be connected to terminals rated 60°, 75°, or 90°C. The ampacities shall be used according to the temperature rating of the terminals. Higher ampacity ratings of the 90°C conductors shall not be used unless they are connected to devices and equipment with 90°C terminals. For example,

A #8 THHN Cu conductor is terminated to a 60°C circuit breaker in the service panelboard. What is the ampacity used to supply the load?

Step 1. *Table 310-16:* #8 THHN Cu = 55 A

Step 2. *Table 310-16:*
60°C #8 conductor and terminal = 40 A

Answer. **Ampacity = 40 A**

Note: Conductors #14 through #1, which are to be 100 A devices or less, shall be rated at 60°C ampacities. Conductors #1/0 and larger, which are connected to devices over 100 A, shall be used at 75°C ampacities when the temperature ratings of conductors, devices, and equipment do not correspond.

The following ampacity values are based on the temperature ratings of terminals using a #1/0 THHN Cu conductor:

60°C. . . . . .125 A
75°C. . . . . .150 A
90°C. . . . . .170 A

A #1/0 THHN Cu conductor can serve a load of 125 A or less on 60°C terminals. It can serve a 150 A load or less on 75°C terminals, and a 170 A load or less on 90°C terminals. Terminals and equipment shall be matched.

Derating by Note 8 to Tables 310-16–310-19

Higher temperature ratings of conductors are beneficial where derating factors are applied for more than three current-carrying conductors in a raceway, or conductors routed through ambient temperatures exceeding 30°C or 86°F. For example,

Fifteen current-carrying conductors are pulled in the same conduit to supply a 19 A load on each conductor, and the load has been increased 125% for continuous duty. Will #12 THWN Cu conductors serve the load or will higher temperature rated conductors be required?

Step 1. *Table 310-16:* #12 THWN Cu = 25 A

Step 2. *Note 8 to Tables 310-16–310-19:*
15 conductors = 70%
25 A × 70% = 17.5 A

Step 3. *Table 310-16:* #12 THHN Cu = 30 A

Step 4. *Note 8 to Tables 310-16–310-19:*
15 conductors = 70%
30 A × 70% = 21 A

Answer. **#12 THWN Cu with 75 °C insulation cannot be used**
#12 THHN Cu with 90 °C insulation can be used

Note: The ampacity of the conductors after derating shall be equal to or higher than the 19 A load.

What is the ampacity of three #2/0 THHN Cu conductors routed through an ambient temperature of 128°F?

Step 1. *Table 310-16:* #2/0 THHN Cu = 195 A

Step 2. *Table 310-16, Ampacity Correction Factors:*
128°F = 76%
195 A × 76% = 148.2 A

Answer. **Ampacity = 148.2 A**

The ampacity of conductors may be derated at least three times. This occurs when there are more than three current-carrying conductors in a raceway. Table 310-16 and Note 8 to Tables 310-16–310-19 list the percentage by which each conductor is derated. If the conductors are routed through ambient temperatures exceeding 86°F, the ampacity of the conductors shall be derated according to the Ampacity Correction Factors of Table 310-16. The correction factors are based on the ambient temperature surrounding conductors.

For *continuous-duty loads* (loads that operate for three hours or more), the load is multiplied by 125% per 220-3(a), or the overcurrent protection device is derated by 80% per 384-16(c). The rules for derating the ampacity of conductors in branch circuits are

1. Derate for over three current-carrying conductors per Note 8 to Tables 310-16–310-19.

2. Derate by the Ampacity Correction Factors per Table 310-16.

3. Derate for continuous-duty loads per 220-3(a) or 384-16(c).

4. Derate by Column A or B of Note 8 to Tables 310-16–310-19. (Consider as loads with diversity unless noted otherwise.)

These rules may also be applied when derating conductors for feeder circuits.

What is the ampacity of six current-carrying #3 THHN Cu conductors used at continuous duty and routed through an ambient temperature of 102°F?

Step 1. *Table 310-16:* #3 THHN Cu = 110 A

Step 2. *Note 8 to Tables 310-16–310-19:*
Six #3 THHN Cu = 80%
110 A × 80% = 88 A

Step 3. *Table 310-16, Ampacity Correction Factors:*
102°F = 91%
88 A × 91% = 80 A

Answer. **Ampacity = 80 A**

What size overcurrent protection device is required to serve a load of 60.5 A using #3 THHN Cu conductors with a new ampacity of 80 A?

Step 1. *384-16(c):* Ampacity = 80 A
80 A × 80% = 64 A

Answer. *240-3, Ex. 4:* **OCPD = 80 A**

Note: 60.5 A × 125% = 75.6 A (76 A) and 240-3, Ex. 4 permits an 80 A overcurrent protection device. Article 384-16(c) requires the 80 A overcurrent protection device to be derated by 80% (80 A × 80% = 64 A). After derating the 80 A overcurrent protection device, the remaining ampacity of 64 A is larger than the 60.5 A load and complies with the NEC®. See Figure 1-5. Additional information is available in the *Electrical Construction Materials Directory,* published by Underwriters Laboratories (UL).

What is the ampacity of fourteen #3 THHN current-carrying copper conductors installed in a raceway with no load diversity?

Step 1. *Table 310-16:* #3 THHN Cu = 110 A

Step 2. *Note 8 to Tables 310-16–310-19:*
Fourteen #3 THHN Cu = 50%
110 A × 50% = 55 A

Answer. **Ampacity = 55 A**

VOLTAGE LIMITATIONS—210, PART A

Lighting fixtures mounted inside, outside, or on buildings are installed according to the voltage levels of the conductors. Voltage limitations are divided into three categories:

1. 120 V, or less, between conductors, or 120 V to ground;

2. Not over 277 V to ground;

3. Over 277 V to ground, but not over 600 V between conductors.

The maximum voltage permitted to supply lighting fixtures and plug-in appliances rated 1440 VA or less in dwelling units and guest rooms in hotels, motels, or similar occupancies is 120 V between conductors. This level of voltage specifically applies to dwelling units such as one-family dwellings, apartment units in multifamily dwellings, and units in townhouses and condominiums. Guests rooms in hotels, dormitories, etc. are also covered. In such occupancies, 120 V between conductors can supply lighting fixtures or receptacles used for cord- and plug-connected loads rated 1440 VA or less. See Figure 1-6.

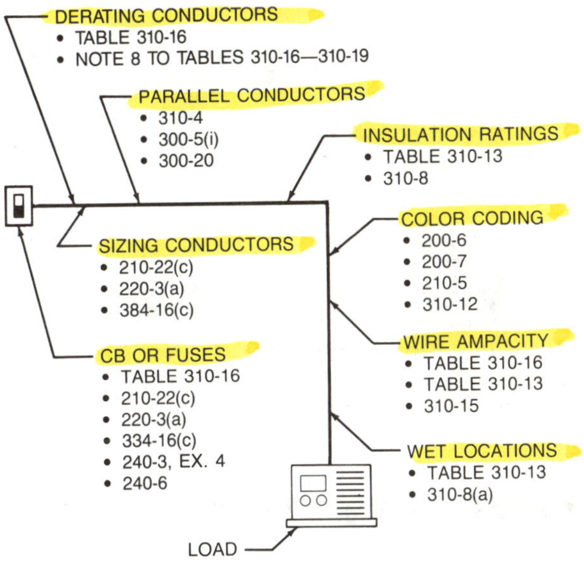

Figure 1-5. Branch circuit conductors shall be installed based upon the number in a raceway, temperature of surrounding air (ambient temperature), or the use of the load served.

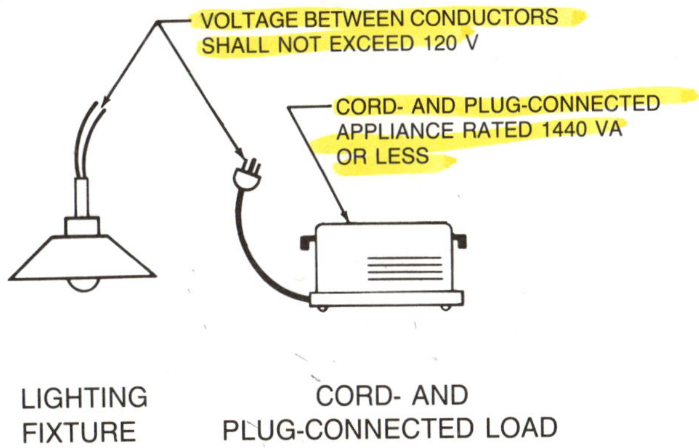

Figure 1-6. In dwelling units, apartments, townhouses, and condominiums, any lighting fixtures and receptacles used for plug-connected loads rated at 1440 VA or less shall be supplied at 120 V or less between conductors.

120 V Between Conductors—210-6(b)

In any type of occupancy, 120 V between conductors can supply loads to medium-base screw-shell lampholders, ballasts for fluorescent or high intensity discharge (HID) lighting fixtures, and cord- and plug-connected or permanently wired appliances. Cord- and plug-connected or permanently wired appliances rated over 1440 VA are usually supplied by direct circuits.

Such appliances include A/C units, heating units, ranges, dryers, water heaters, etc. Appliances rated 1440 VA or less are generally connected to general-purpose circuits that supply more than one outlet. See Figure 1-7.

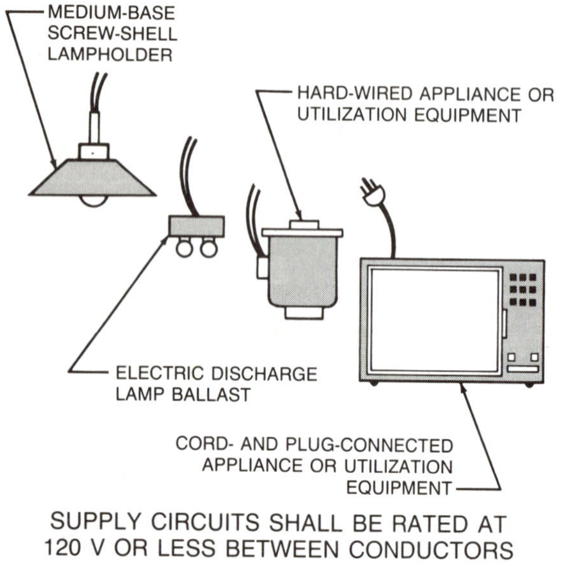

Figure 1-7. Branch circuits with 120 V or less between conductors can supply medium-base screw-shell lampholders, ballasts for fluorescent or HID lighting fixtures, hard-wired appliances or other utilization equipment, and cord- and plug-connected equipment.

277 V to Ground—210-6(c); 225-7(c)

Circuits with over 120 V between conductors are supply circuits rated at 208, 240, 277 or 480 V, but not over 600 V. See Figure 1-8. The voltage to ground is over 120 V but not over 277 V. Circuits with any of these voltage ratings are allowed to serve incandescent lighting fixtures with mogul-base screw-shell lampholders or ballasts for electric discharge lighting fixtures. These circuits can serve cord- and plug-connected or permanently wired appliances or other utilization equipment. Incandescent or electric discharge fixtures with mogul-base screw-shell lampholders supplied by 480/277 V, 4-wire systems are no longer required to be installed at a minimum height of 8′-0″ above finished grade.

Lighting fixtures supplied by 480/277 V circuits are mounted on the outside of buildings and structures, or on poles on industrial or commercial premises per 225-7(c). The fixture shall be located at least 3′-0″ from windows, platforms, fire escapes, etc.

600 V Between Conductors—210-6(d); 225-7(d)

Circuits rated over 277 V to ground, but not exceeding 600 V between conductors, can supply ballasts for electric discharge fixtures mounted not less than 22′-0″ above the finished grade. These fixtures are used for lighting of outdoor areas such as parking lots, athletic fields, highways, and bridges. Utilization equipment

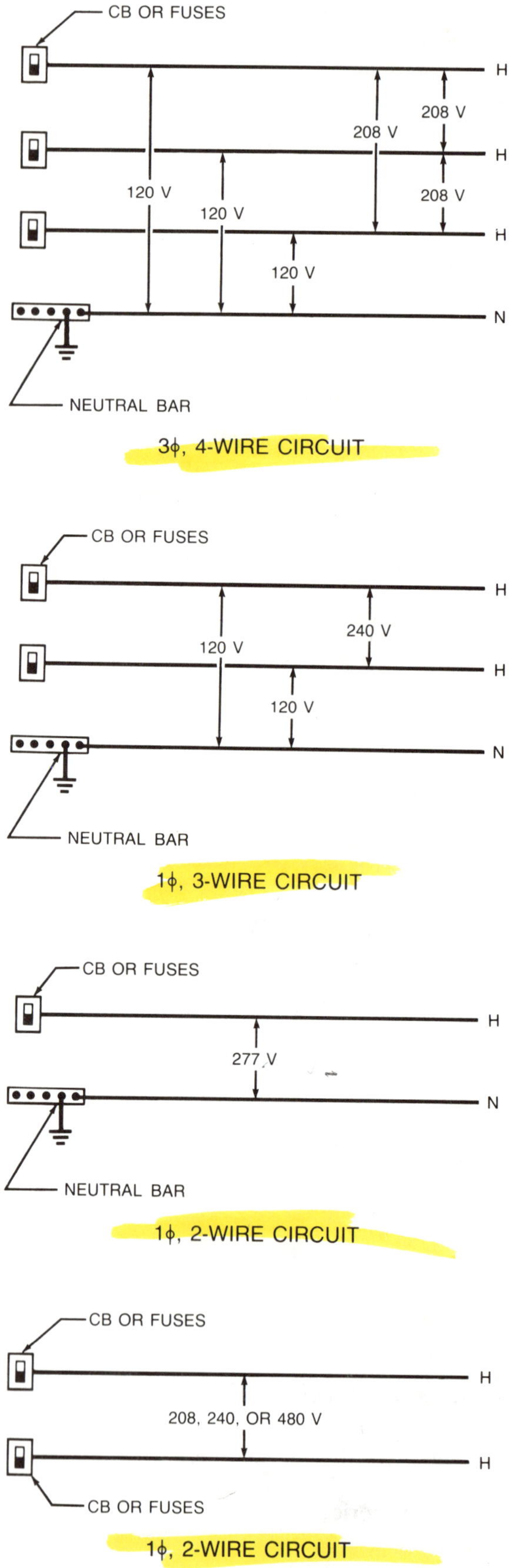

Figure 1-8. Many circuits can be utilized to supply incandescent lighting with mogul-base screw-shell lampholders, electric discharge ballasts, and utilization equipment.

that is either cord- and plug-connected to a receptacle or permanently wired can be supplied with circuits having these voltage levels. For structures such as tunnels, the mounting height shall be at least 18′-0″ above finished grade.

These circuits are usually derived from ungrounded 480 V, 3-wire systems or corner-grounded, 3-wire systems. The ungrounded systems will have 480 V to ground if the leg is accidentally grounded. The corner-grounded system has 480 V to ground because one leg is intentionally grounded. Circuits exceeding 277 V to ground, but not over 600 V between conductors, can supply the auxiliary equipment of electric discharge lamps per 225-7(d).

DETERMINING AMPERAGE

Amperage for 1ϕ circuits is determined by dividing VA by the supply voltage ($I = VA/V$). The same basic formula with the addition of $\sqrt{3}$ is used for 3ϕ circuits. The formula then reads, $I = VA/V \times \sqrt{3}$. Conductors and overcurrent protection devices for branch circuits are selected from the amperage rating.

Finding Amperage—1ϕ Circuits

Amperage for 1ϕ circuits is obtained by dividing the total VA rating of the electrical equipment by the supply voltage. The formula $I = VA/V$ is applied as follows:

Find the amperage of an 1800 VA load on a 120 V, 1ϕ branch circuit.

Step 1. $I = \dfrac{VA}{V}$

$$I = \frac{1{,}800 \text{ VA}}{120 \text{ V}} = 15 \text{ A}$$

Answer. **Amperage = 15 A**

Find the amperage of a 7680 VA load on a 240 V, 1ϕ branch circuit.

Step 1. $I = \dfrac{VA}{V}$

$$I = \frac{7680 \text{ VA}}{240 \text{ V}} = 32 \text{ A}$$

Answer. **Amperage = 32 A**

Finding Amperage—3ϕ Circuits

Amperage for 3ϕ circuits is obtained by dividing the total VA rating of the electrical equipment by the supply voltage times $\sqrt{3}$ ($\sqrt{3} = 1.732$). Dividing VA by V times the $\sqrt{3}$ evenly distributes the amperage on phases A, B, and C.

When solving 3ϕ problems, instead of multiplying the supply voltage by 1.732 and then dividing this figure into the volt-amperage, use the values shown in the 3ϕ voltage chart. This chart will eliminate one math step in the calculation. For example:

Find the amperage of an 8640 VA load on a 208 V, 3ϕ branch circuit.

Step 1. $I = \dfrac{VA}{V \times \sqrt{3}}$

$$I = \frac{8640 \text{ VA}}{360} = 24 \text{ A}$$

Answer. **Amperage = 24 A**

Find the amperage of a 20,800 VA load on a 240 V, 3ϕ branch circuit.

Step 1. $I = \dfrac{VA}{V \times \sqrt{3}}$

$$I = \frac{20{,}800 \text{ VA}}{416} = 50 \text{ A}$$

Answer. **Amperage = 50 A**

LIGHTING LOADS—220, PART B

Lighting is provided by incandescent or electric discharge lighting fixtures. Either one or both types of lighting can be used in a building. Branch circuits supplying lighting loads are computed either at *noncontinuous duty* or *continuous duty*. The loads on the circuits are calculated at noncontinuous duty for lighting used for less than three hours at any given time. The loads on the circuits are calculated at continuous duty for lighting used for three hours or more without an OFF period.

Noncontinuous Duty—220-3(a)

Branch circuits supplying noncontinuous lighting are computed at 100% of the total VA rating of the circuit. These loads include lighting fixtures and cord- and plug-connected table lamps or floor lamps used at various intervals of time. Where such lighting loads are properly used, they will never overload the circuit during their time of use. For example, 13 lighting outlets that are cord- and plug-connected or hardwired can utilize a 20 A circuit as long as the total VA rating of the loads does not exceed 20 A.

How many outlets are permitted on a 20 A branch circuit used at noncontinuous duty?

Step 1. *220 3(c)(5):* $I = \dfrac{VA \times 100\%}{V}$

$$I = \frac{180 \text{ VA} \times 100\%}{120 \text{ V}} = 1.5 \text{ A per outlet}$$

Step 2. *220-4(a): No. of outlets* $= \dfrac{device}{1.5 \text{ A}}$

$$\textit{No. of outlets} = \frac{20 \text{ A CB}}{1.5 \text{ A}} = 13$$

Answer. **Outlets = 13**

Note: A 15 A branch circuit can have up to 10 outlets connected (15 A CB/1.5 A = 10 outlets).

Continuous Duty—220-3(a)

Branch circuits supplying continuous lighting loads shall have their VA rating increased by 125%. Continuous lighting loads are incandescent or electric discharge lighting fixtures installed for illumination in commercial or industrial buildings. Only 10 outlets can be connected to a 20 A branch circuit.

How many outlets are permitted on a 20 A branch circuit used at continuous duty?

Step 1. *220-3(c)(5); 220-3(a); 210-22(c):* $I = \frac{VA \times 125\%}{V}$

$I = \frac{180\ VA \times 125\%}{120\ V} = 1.875\ A$ per outlet

Step 2. *220-4(a):* $No.\ of\ outlets = \frac{device}{1.875\ A}$

$No.\ of\ outlets = \frac{20\ A\ CB}{1.875\ A}$

Answer. **Outlets = 10**

Note: A 15 A branch circuit can have up to eight outlets connected (15 A CB/1.5 A × 125% = 8).

Resistive Loads—210-22(c); 384-16(c)

A *resistive load* has no inductive loads. Branch circuits that supply resistive loads such as incandescent lighting and resistive heating are figured at 100% times their VA rating for noncontinuous duty or 125% times their VA rating for continuous duty. See 424-3(b). If resistive loads are connected to a 20 A branch circuit and the loads are not used at continuous duty, each ungrounded phase conductor can be loaded up to 20 A. For resistive loads used at continuous duty, a 20 A branch circuit shall not be loaded to more than 80% of the 20 A rating.

What is the maximum resistive continuous-duty load that can be connected to a 20 A branch circuit?

Step 1. *210-22(c); 384-16(c):* Derate device 80%
20 A × 80% = 16 A

Answer. **Load = 16 A**

What size overcurrent protection device is required to serve a 16 A load?

Step 1. *220-3(a):* Increase load 125%
16 A × 125% = 20 A

Answer. **OCPD = 20 A**

Note: The neutral conductor is not considered a current-carrying conductor where it carries only the unbalanced load from resistive loads on the phase conductors.

Electric Discharge Loads—210-22(b)

Branch circuits supplying the ballast of electric discharge lighting loads such as fluorescent, mercury vapor, metal halide, high pressure sodium, or low pressure sodium are sized at 125% of the VA rating of each ballast. The wattage of each bulb or lamp shall not be used. Some electricians figure the total wattage of the lamps and increase this value by 125% to obtain the total lighting load. In most cases, this requires a larger lighting load than the VA rating of each ballast times 125%. This method should not be used because it does not comply with the NEC®.

What is the lighting load for a branch circuit supplying ten 1.5 A ballasts that serve twenty F40, CW lamps used at continuous duty?

Step 1. *210-22(b):* $I = ballast\ A \times no.\ of\ ballasts$
$I = 1.5\ A \times 10 = 15\ A$

Step 2. *220-3(a):* $Load = I \times 125\%$
$Load = 15\ A \times 125\% = 19\ A$

Answer. **Load = 19 A**

Note: The neutral conductor is considered a current-carrying conductor for electric discharge lighting loads.

Outlets—220-3(c)(3)(4)(5)

Where the VA rating of ballasts or the wattage of incandescent bulbs is unknown, branch circuits are computed at 180 VA per outlet, or 600 VA for outlets serving heavy-duty lampholders. Lampholders connected to a 20 A branch circuit shall have a wattage rating of less than 660 watts for admedium types and 750 watts for other types per 210-21(a). Outlets on branch circuits are calculated at 100% for noncontinuous duty and 125% for continuous duty. See Figure 1-9.

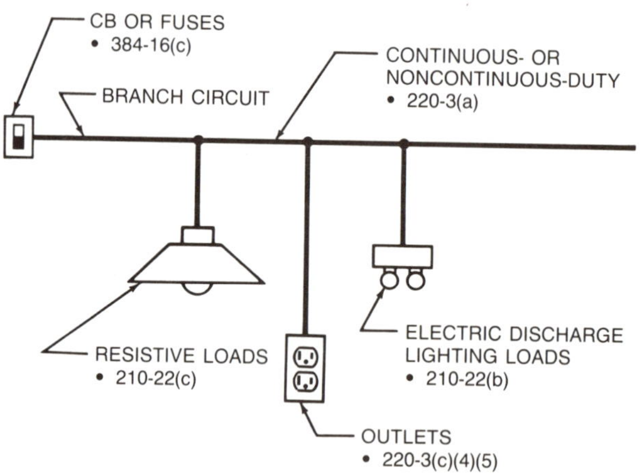

Figure 1-9. Branch circuit conductors supplying lighting loads are computed at 100% for noncontinuous-duty operation and at 125% for continuous-duty operation.

What size branch circuit conductors, using THHN Cu, and overcurrent protection device is required for 12 outlets used at noncontinuous duty to supply lighting fixtures?

Step 1. *220-3(a):*
$Load = no.\ of\ outlets \times VA \times 100\%$
$Load = 12 \times 180\ VA \times 100\% = 2160\ VA$

Step 2. *220-4(a):* $I = \frac{VA}{V}$

$$I = \frac{2160\ VA}{120\ V} = 18\ A$$

Step 3. *Table 310-16 and Footnotes; 240-3, Ex. 4; 240-6:* #12 THHN Cu = 30 A

Answer. **Conductors = #12 THHN Cu**
OCPD = 20 A

What size branch circuit conductors, using THHN Cu, and overcurrent protection device is required for 10 lighting outlets used at continuous duty in a commercial building?

Step 1. *220-3(a): Load = no. of outlets × VA × 125%*
Load = 10 × 180 VA × 125% = 2250 VA

Step 2. *220-4(a):* $I = \frac{VA}{V}$

$$I = \frac{2250\ VA}{120\ V} = 19\ A$$

Step 3. *Table 310-16 and Footnotes; 240-3, Ex. 4; 240-6:* #12 THHN Cu = 30 A

Answer. **Conductors = #12 THWN Cu**
OCPD = 20 A

SHOW WINDOW LOADS—220-3(c)(5)

Branch circuits supplying show window lighting loads are calculated at a minimum of 180 VA per outlet. Since show window lighting can operate for three hours or more to provide illumination for advertisement, the total VA rating is multiplied by 125% for each branch circuit. *Note:* See page 45 for linear footage method.

How many outlets can be installed on a 20 A branch circuit supplying lighting loads in a show window?

Step 1. *220-3(c)(5); 220-3(a):*

$$\text{No. of outlets} = \frac{\text{device} \times 120\ V}{VA \times 125\%}$$

$$\text{No. of outlets} = \frac{20\ A \times 120\ V}{180\ VA \times 125\%} = 10.6$$

Answer. **Outlets = 10**

LIGHTING TRACK LOADS—410-102

Branch circuits that supply lighting tracks are computed at 180 VA for each 2′ of track. For a multicircuit length of lighting track, the load is divided between the number of circuits.

What is the VA rating of 26′ of lighting track and the number of outlets permitted?

VA rating

Step 1. *410-102:* $VA = \frac{\text{track length}}{2'} \times 180\ VA$

$$VA = \frac{26'}{2'} \times 180\ VA = 2340\ VA$$

Answer. **Rating = 2340 VA**

Outlets

Step 1. $\text{No. of outlets} = \frac{VA}{180\ VA}$

$$\text{No. of outlets} = \frac{2340\ VA}{180\ VA} = 13$$

Answer. **Outlets = 13**

A multicircuit lighting track served by two circuits shall have the VA rating divided as follows: 180 VA for each 2′/2 circuits = 90 VA. Each 2′ of lighting track supplied by two circuits has 90 VA for each circuit. A multicircuit lighting track served by three circuits shall have the VA rating divided as follows: 180 VA for each 2′/3 circuits = 60 VA. Each 2′ of lighting track supplied by three circuits has 60 VA for each circuit.

SIGN LOADS—600, PART A

A sign circuit is required for each commercial occupancy having *ground floor access* (accessible to pedestrians). Ground floor access is ground floor footage with an entrance for pedestrians from a sidewalk, street, enclosed mall, etc. Signs used for commercial occupancies are considered to be used at continuous duty where they burn for three hours or more.

Rating—600-6(a)

Branch circuits supplying lamps, ballasts, and transformers, or combinations of these, are limited to 20 A or less. Branch circuits serving transformers for electric discharge lighting units are limited to 30 A or less. For example, branch circuits rated 20 A or less can serve transformers for channel letters and ballasts for electric discharge lamps. Where transformers only are connected to branch circuits to serve neon or channel letters, the rating of the branch circuit is 30 A or less.

Size Required—600-6(b)

At least one 20 A branch circuit shall be routed from the service equipment to the outside of the premise and placed in an accessible location. This 20 A branch circuit is used to supply an outlet for a sign or outline lighting. No other loads shall be connected to this branch circuit.

Minimum Load—600-6(c)

Branch circuits supplying the lighting load of signs are calculated at a minimum of 1200 VA. If the sign is used at noncontinuous duty, the circuit can be loaded up to 20 A. A sign used at continuous duty shall not be loaded to more than 80% of the branch circuit rating.

What is the amperage of a 20 A, 120 V branch circuit used at continuous duty to supply a wall sign?

Step 1. *600-6(b)(c): VA = A × V × 80%*
VA = 20 A × 120 V × 80% = 1920 VA

Step 2. *210-22(c); 220-3(a); 384-16(c):*
20 A × 80% = 16 A

Answer. **Amperage = 16 A**

Note: 1200 VA or 10 A (1200 VA/120 V = 10 A) is the minimum used for the branch circuit load.

RECEPTACLE LOADS—220, PART B

Receptacle outlets are installed in dwelling units, apartments, townhouses, condominiums, and commercial or industrial locations where there are cord- and plug-connected appliances, utilization equipment, and table lamps or floor lamps. Cord- and plug-connected loads to branch circuits are connected in an arrangement not to load up the circuit. Branch circuits supplying cord- and plug-connected loads are *general-purpose* (serving more than one outlet) or *direct* (serving one outlet). Receptacle outlets are located so that cord- and plug-connected items can be plugged in without the use of extension cords. See Figure 1-10.

General-purpose Branch Circuits—220-3(c)(5)

A general-purpose branch circuit supplies more than one outlet where cord- and plug-connected items are connected to the circuit. The outlets on a general-purpose branch circuit are computed at 180 VA each. The number of outlets times 180 VA times 100% is used to find the load of a branch circuit supplying outlets of noncontinuous duty. The number of outlets times 180 VA times 125% is used to find the load of a branch circuit supplying outlets of continuous duty.

What is the load of six receptacle outlets supplying cord- and plug-connected loads used at noncontinuous duty?

Step 1. *220-3(c)(5); 220-3(a):*
Load = no. of outlets × 180 VA × 100%
Load = 6 × 180 VA × 100% = 1080 VA

Answer. **Load = 1080 VA**

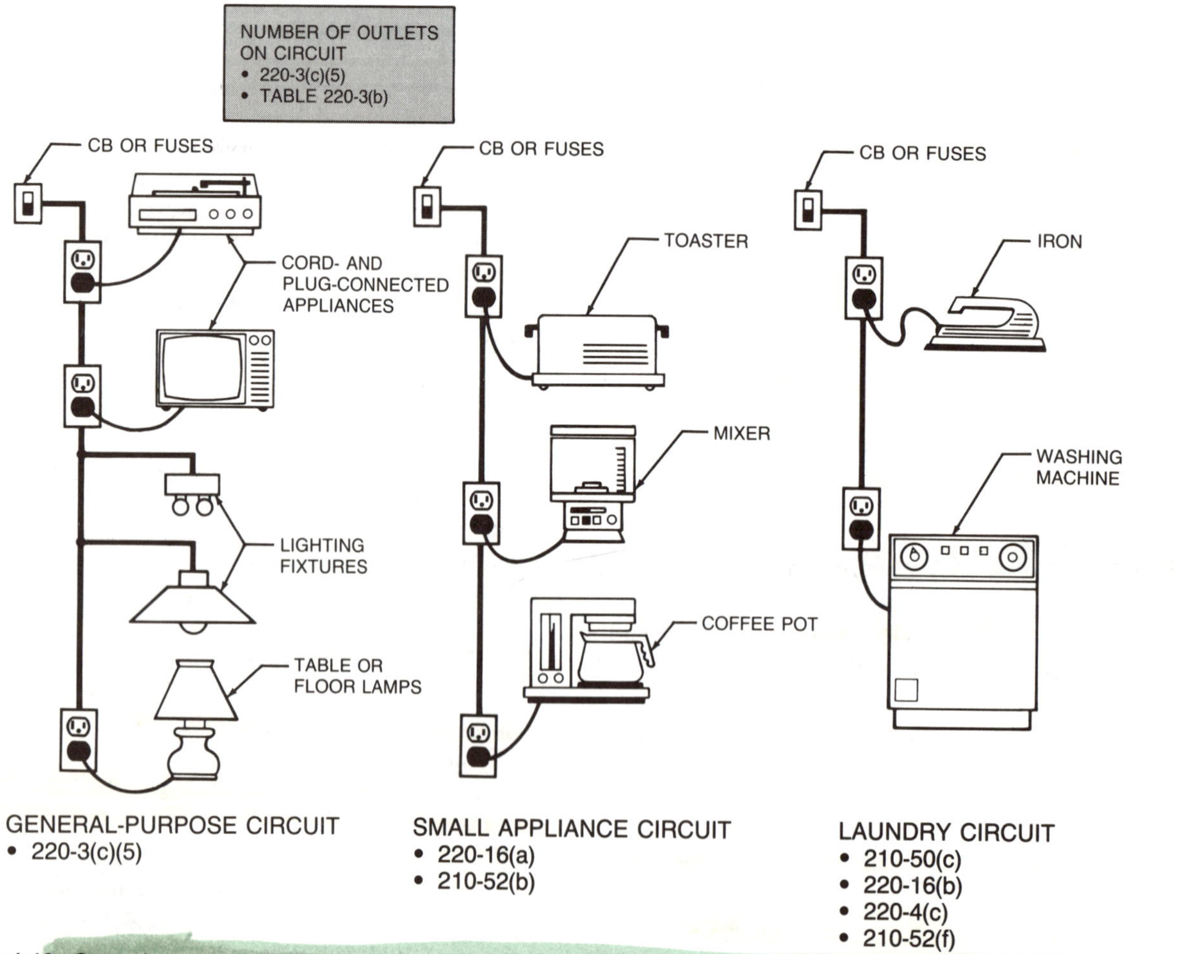

Figure 1-10. General-purpose branch circuits supplying receptacle outlets that serve cord- and plug-connected lights or appliances are figured at 100% for noncontinuous-duty and at 125% for continuous-duty operation.

What is the load of 10 receptacle outlets supplying cord- and plug-connected office equipment used at continuous duty?

Step 1. *220-3(c)(5); 220-3(a):*

$$Load = no.\ of\ outlets \times 180\text{ VA} \times 125\%$$
$$Load = 10 \times 180\text{ VA} \times 125\% = 2250\text{ VA}$$

Answer. **Load = 2250 VA**

Small Appliance Circuits—210-52(b); 220-16(a)

Small appliance circuits are circuits that supply receptacle outlets located in the kitchen, pantry, breakfast room, and dining room of a dwelling. These circuits shall not supply any other outlets except those located in these rooms. Small appliance circuits supply cord- and plug-connected appliances such as microwave ovens, toasters, coffee pots, and electric skillets. Refrigeration equipment can be cord- and plug-connected to the small appliance circuits. There shall be at least two 20 A branch circuits of #12 Cu conductors with grounding provisions to supply these outlets. The small appliance circuits shall not be used to supply outlets for lighting fixtures or fixed appliances. See 210-52(b).

Laundry Circuit—210-50(c); 220-16(b); 210-52(f)

A laundry circuit is one 20 A branch circuit installed exclusively for the laundry room. Only cord- and plug-connected appliances that pertain to laundry equipment shall be connected to this circuit. For example, a washing machine and electric iron can be cord- and plug-connected to the laundry circuit. Lighting outlets or other electrical items shall not be connected to this circuit.

Number on Circuit—220-3(c)(5); 220-4(d); Table 220-3(b)

The number of outlets permitted on a branch circuit is determined by dividing the ampacity of the circuit breaker or fuse by 1.5 A (180 VA/120 V = 1.5 A) per 220-3(c)(5). This method is used in commercial and industrial locations such as offices and stores. If the outlets supply noncontinuous-duty loads, 180 VA or 1.5 A is multiplied by 100%. Where the loads connected to outlets are used at continuous duty, 180 VA or 1.5 A is multiplied by 125%.

How many outlets used at noncontinuous duty are permitted to be connected to a 15 A CB or fuse on a 120 V branch circuit?

Step 1. *220-3(c)(5); 220-3(c):*

$$No.\ of\ outlets = \frac{device}{1.5\text{ A} \times 100\%}$$
$$No.\ of\ outlets = \frac{15\text{ A CB}}{1.5\text{ A} \times 100\%} = 10$$

Answer. **Outlets = 10**

How many outlets used at continuous duty are permitted to be connected to a 20 A CB or fuse on a 120 V branch circuit?

Step 1. *220-3(c)(5); 220-3(c):*

$$No.\ of\ outlets = \frac{device}{1.5\text{ A} \times 125\%}$$
$$No.\ of\ outlets = \frac{20\text{ A CB}}{1.5\text{ A} \times 125\%} = 10.7$$

Answer. **Outlets = 10**

In dwelling units, the number of outlets permitted is computed on the VA per square foot basis. The branch circuit capacity is multiplied by the voltage and divided by the VA per square foot per Table 220-3(b). All general-use receptacle outlets rated at 20 A or less and located in one-family, two-family, and multifamily dwellings and in guest rooms of hotels, motels, etc. are computed at 180 VA per outlet.

How many general-use receptacle outlets are permitted to be connected on a 15 A CB or fuse on a 120 V branch circuit in a dwelling unit?

Step 1. *220-4(d); Table 220-3(b):*

$$Circuit\ sq.\ ft. = \frac{V \times device}{VA\ per\ sq.\ ft.}$$
$$Circuit\ sq.\ ft. = \frac{120\text{ V} \times 15\text{ A CB}}{3\text{ VA}} = 600\text{ sq. ft.}$$

Step 2. *220-3(c)(5); 220-4(d):* Any number of outlets permitted in the 600 sq. ft. area

Answer. **Outlets = Any number**

■ *Master Electrician's Test Tip.* Most examination boards require this procedure to determine the number of outlets permitted on a general-purpose circuit in a dwelling unit.■

Individual Circuits—220-3(a)

The procedure used to determine the size branch circuit to supply power to an individual cord- and plug-connected load is to multiply the load by 100% for noncontinuous duty and 125% for continuous duty. The power formula is applied to determine the amount of amperage each cord- and plug-connected load will pull. In the power formula, I = amps; P = volt amps (or watts); and E = voltage. Refer to the inside back cover of this book.

To use the power formula, cover the value to be found. For example, when I is covered, the formula reads $I = P/E$. When P is covered, the formula reads $P = I \times E$. When E is covered, the formula reads $E = P/I$. *Note:* In this book, a variation of the power formula is used. This variation is $I = VA/V$. This variation is used because the NEC® recognizes volt amps (VA) for load calculations per the examples in Chapter 9.

What size THWN Cu conductors is required to serve a continuous-duty processing machine pulling 10,800 VA per leg? The supply voltage for the branch circuit is 240 V, 1ϕ.

Step 1. *Text:* $I = \frac{VA}{V}$

$$I = \frac{10{,}800\text{ VA}}{240\text{ V}} = 45\text{ A}$$

Step 2. *220-3(a):* 45 A × 125% = 56 A

Step 3. *Table 310-16:* #6 THWN Cu = 65 A

Answer. **Conductors = #6 THWN Cu**

An individual branch circuit can have any rating and supply any load for which it has been properly sized. Cord- and plug-connected loads are usually rated at 60 A or less.

Multiwire Branch Circuits—210-4

Multiwire branch circuits shall serve only line-to-neutral loads with more than one hot conductor sharing the neutral. Single-pole circuit breakers shall be used to supply multiwire circuits to lighting loads, receptacle loads, etc. In dwelling units, multiwire circuits are supplied by 2-pole or single-pole circuit breakers with approved handle ties where more than one device or piece of equipment is on the same yoke of a receptacle.

A fusible switch can be used where all ungrounded conductors of the multiwire circuit are disconnected simultaneously. This requirement provides safety for personnel working on or troubleshooting a multiwire circuit with two hot legs serving the bottom and top outlets of a receptacle. With both hots OFF, there is no shock hazard to a person getting between the neutral. Electricians and homeowners have been shocked by placing their bodies between the neutral and loads downstream when disconnecting the neutral from the terminals of a receptacle.

Section 300-13(b) requires the neutral of multiwire circuits to be pigtailed to prevent different voltage levels between the ungrounded conductors where the neutral is lost as a result of a bad or loose connection. The voltage on each phase conductor depends on the levels of current between each conductor. The phase conductor with the smaller load has a higher voltage than the phase conductor with the higher load because of the resistance of the circuits. The neutral conductors of a multiwire branch circuit are pigtailed when supplying terminals of receptacles with the connecting link broken. See Figure 1-11.

COOKING EQUIPMENT LOADS—220, PART B

Branch circuit loads for ranges, cooktops, and wall-mounted ovens are determined by the demand factors in Columns A, B, and C of Table 220-19. Cooking equipment in dwelling units is permitted to have a demand factor applied because the unit, when in use, will never operate at its full kW rating. Where the kW rating of a cooking unit exceeds the values listed in Column A, B, or C, the Notes to Table 220-19 are applied. See Figure 1-12.

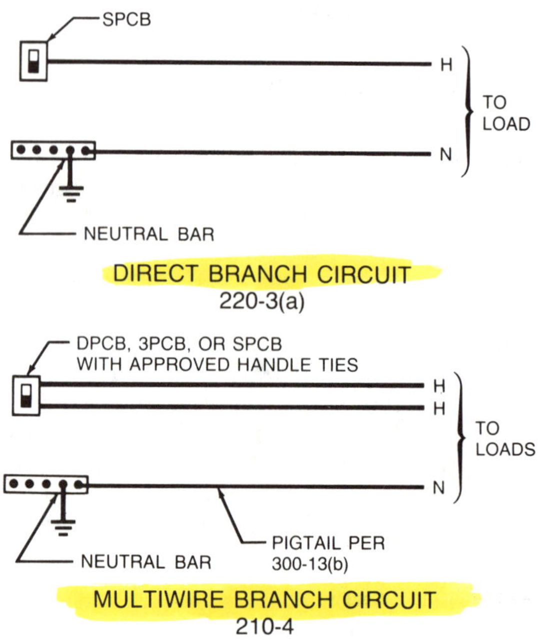

Figure 1-11. Direct or multiwire branch circuits supply receptacle outlets for cord- and plug-connected lighting or appliances.

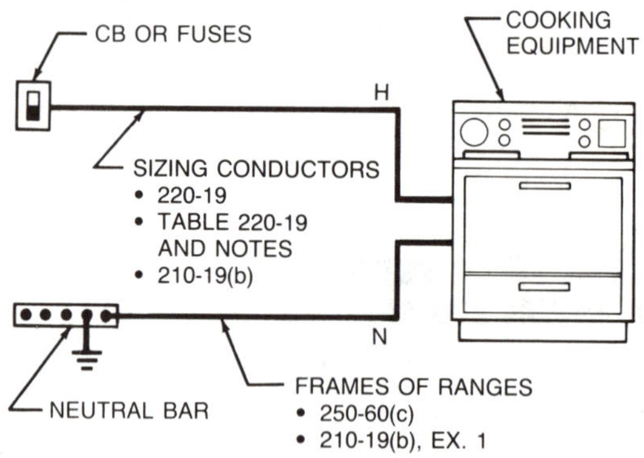

TABLE 220-19				
COOKING EQUIPMENT	COL. A	COL. B	COL. C	NOTES
9 to 12 kW	✓			
3½ to 8¾ kW			✓	
Less than 3½ kW		✓		
Over 12 to 27 kW				Note 1
Unequal ratings over 12 kW to 27 kW				Note 2
Cooktop with one or two ovens				Note 4

Figure 1-12. Branch circuit conductors for cooking equipment are sized from demand factors listed in Table 220-19 and accompanying notes.

Column A—Table 220-19

Column A of Table 220-19 is used for cooking equipment rated from over 8¾ kW to 12 kW. For example, a range with a 9.5 kW rating has a demand load of 8 kW per Table 220-19, Column A.

> What is the demand load for a 12 kW range supplied by a 240 V, 1ϕ branch circuit?
>
> Step 1. *Table 220-19, Column A:* 12 kW range = 8 kW
>
> Answer. **Load = 8 kW**

Column B—Table 220-19

Column B of Table 220-19 is used for cooking equipment rated less than 3½ kW. For example, a small range of 2.5 kW has a demand load of 2 kW (2.5 kW × 80% = 2 kW).

> What is the demand load for a 3 kW range supplied by a 240 V, 1ϕ branch circuit?
>
> Step 1. *Table 220-19, Column B:* 3 kW × 80% = 2.4 kW
>
> Answer. **Load = 2.4 kW**

Column C—Table 220-19

Column C of Table 220-19 is used for cooking equipment rated from 3½ kW to 8¾ kW. Based on one range, the demand load is 80% per Column C. For example, a range rated at 7.5 kW has a demand of 6 kW (7.5 kW × 80% = 6 kW).

> What is the demand load for a 8.75 kW range supplied by a 240 V, 1ϕ branch circuit?
>
> Step 1. *Table 220-19, Column C:*
> 8.75 kW × 80% = 7 kW
>
> Answer. **Load = 7 kW**

Note 1 to Table 220-19

Note 1 to Table 220-19 is applied where the kW rating of the cooking equipment is over 12 kW but not over 27 kW. All kW ratings that exceed 12 kW are multiplied by 5%. The kW rating of one range listed in Column A of Table 220-19 is multiplied by this total. For example, the demand load for an 18 kW range is determined by subtracting 12 kW from 18 kW, multiplying the difference by 5% (18 kW − 12 kW × 5% = 30%), and multiplying times the demand. The demand for one range in Column A is 8 kW. The load for the 18 kW range is 10.4 kW (8 kW × 130% = 10.4 kW). Overcurrent protection devices and branch circuit conductors are selected from the 10.4 kW demand load.

> What is the demand load for a 24 kW range supplied by a 240 V, 1ϕ branch circuit?
>
> Step 1. *Note 1 to Table 220-19:*
> 24 kW − 12 kW = 12 kW
> 12 kW × 5% = 60%
>
> Step 2. *Table 220-19, Column A:*
> 8 kW × 160% = 12.8 kW
>
> Answer. **Load = 12.8 kW**

Note 4 to Table 220-19

Note 4 to Table 220-19 can be applied to a counter-installed cooktop with one or two wall-mounted ovens. The total kW of each cooking unit is totaled and all kW exceeding 12 kW is multiplied by 5%. The kW rating of one range (not three) listed in Column A of Table 220-19 is multiplied by this total. For example, the demand load for a 9 kW cooktop and two ovens rated at 8 kW each is determined by adding the kW of each unit together (9 kW + 8 kW + 8 kW = 25 kW); subtracting 12 kW from the total (25 kW − 12 kW = 13 kW); and multiplying the difference by 5% (13 kW × 5% = 65%). The difference is 65%. The demand load for one range in Column A is 8 kW. Multiply 8 kW by 165% to find the demand load for the cooktop and ovens (8 kW × 165% = 13.2 kW). The demand load is 13.2 kW.

> What is the demand load for a 10 kW cooktop and two 6 kW ovens connected to a 240 V, 1ϕ branch circuit?
>
> Step 1. *Note 4 to Table 220-19:*
> 10 kW + 6 kW + 6 kW = 22 kW
> 22 kW − 12 kW = 10 kW
> 10 kW × 5% = 50%
>
> Step 2. *Table 220-19, Column A:*
> 8 kW × 150% = 12 kW
>
> Answer. **Load = 12 kW**

Taps—210-19(b), Ex. 2. Taps to connect a cooktop and ovens can be made from a 50 A branch circuit when the tap conductors are sized from the kW rating of each piece of cooking equipment per Note 4 to Table 220-19. The following example illustrates the procedure for sizing the taps to each piece of cooking equipment using nonmetallic sheathed cable (Romex).

> What size conductors is required to supply the taps from the branch circuit conductors for a 10 kW cooktop and two 6 kW ovens? The supply voltage is 240 V, 1ϕ.
>
> <u>*Cooktop*</u>
>
> Step 1. *Note 4 to Table 220-19:* $I = \frac{VA}{V}$
>
> $$I = \frac{10{,}000\ \text{VA}}{240\ \text{V}} = 41.7\ \text{A}$$
>
> Step 2. *Table 310-16; 336-26:* #6 60°C conductors = 55 A
>
> Answer. **Conductors = #6-2 with ground (nonmetallic sheathed cable)**

Ovens

Step 1. *Note 4 to Table 220-19:* $I = \frac{VA}{V}$

$$I = \frac{6000 \text{ VA}}{240 \text{ V}} = 25 \text{ A}$$

Step 2. *Table 310-16; 336-26:*
#10 60°C conductors = 30 A

Answer. **Conductors = #10-2 with ground (nonmetallic sheathed cable)**

Many inspectors require junction boxes for the taps to the 50 A branch circuit conductors to be located as near as possible to the cooking equipment, usually directly behind or underneath the equipment.

Commercial—220-3(a)

The branch circuit load to commercial cooking equipment is computed at continuous or noncontinuous duty. A commercial cooking load of 10 kW shall be calculated at 100% for noncontinuous duty. If a commercial cooking load operates at continuous duty, a load of 12 kW is calculated at 125%. A 12 kW cooking load times 125% equals 15 kW (12 kW × 125% = 15 kW).

The overcurrent protection device, conductors, and conduit or cable are sized from the 15 kW rating. For example, a commercial cooking load of 6.5 kW is used in a restaurant at noncontinuous duty. The load is determined by multiplying 6.5 kW by 100% (6.5 kW × 100% = 6.5 kW). For an 11 kW commercial cooking load used at continuous duty, the 11 kW load is multiplied by 125% (11 kW × 125% = 13.75 kW). See Table 220-20 for selecting demand factors for more than one piece of cooking equipment in a kitchen.

What is the load for a 12 kW cooking unit used at continuous duty and noncontinuous duty and supplied by a 240 V, 1ϕ branch circuit?

Continuous duty

Step 1. *220-3(a):* 12 kW × 125% = 15 kW

Answer. **Load = 15 kW**

Noncontinuous duty

Step 1. *220-3(a):* 12 kW × 100% = 12 kW

Answer. **Load = 12 kW**

DRYING EQUIPMENT LOADS—220-18

The branch circuit load for drying equipment is the larger of either the VA rating of the nameplate or 5000 VA. For a dryer with 6000 W listed on the nameplate, a 6000 VA rating is used to size the branch circuit. Four or fewer dryers are computed at 100%. Five or more dryers are computed at a reduction of the total VA rating, which is based on the number of dryers. A dryer with a nameplate rating of 4500 W is computed at 5000 VA for a branch circuit supplying the receptacle outlet used to cord- and plug-connect the unit. Section 220-3(c), Ex. 5 and 220-18 are used only for dryers in dwelling units. Commercial dryers used in laundry facilities shall be computed at 100% for noncontinuous duty and 125% for continuous duty per 220-3(a). See Figure 1-13.

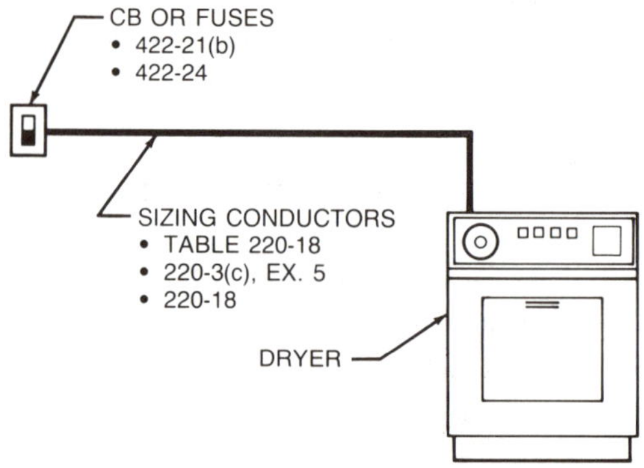

Figure 1-13. Branch circuit conductors and components are sized by the kW rating of the dryer, which shall be 5,000 VA or more for dwelling units.

What is the branch circuit load for a 6500 W dryer used in a dwelling unit?

Step 1. *220-3(c), Ex. 5; 220-18:*
6500 VA × 100% = 6500 VA

Answer. **Load = 6500 VA**

What is the load for an 8000 W commercial dryer used eight hours a day in a 24-hour laundry facility?

Step 1. *220-3(a):* 8000 VA × 125% = 10,000 VA

Answer. **Load = 10,000 VA**

WATER HEATER LOADS—422, PART C

Water heater units are equipped with elements that are connected by number to heat water at different stages. One group of elements is used to heat water in small amounts. When a larger amount of hot water is needed, additional elements are connected to the circuit to provide the proper amount of heat. Overcurrent protection devices and conductors shall be sized to carry the minimum or maximum load of the water heater during the cycle.

Conductors—422-14(b)

Conductors are selected at not less than 125% of the nameplate rating of the water heater. This requirement includes all fixed storage water heaters with a capacity of 120 gallons or less. Larger sized conductors act as a heat sink between the overcurrent protection devices and the point of connection to the terminals of the elements. For example, a 240 V, 1ϕ, 8500 W water heater requires a branch circuit equivalent to 10,625 VA (8500 VA × 125% = 10,625 VA).

Overcurrent Protection—422-14(b)

Overcurrent protection devices shall be selected and sized to prevent nuisance tripping when elements are connected or disconnected. Resistance heating elements pull approximately 10% of the water heater's connected load. Overcurrent protection devices are sized at not less than 125% of the total VA rating. For example, a water heater with a 37 A connected load requires a 50 A overcurrent protection device (37 A × 125% = 46 A) per 422-14(b) and 240-3, Ex. 4.

Disconnecting Means—422-21(b)

A disconnecting means is not required at a water heater when the overcurrent protection device of the branch circuit is readily accessible. When the service panel is readily accessible, the overcurrent protection device located in the service panel, outside or inside the premises, can serve as the disconnect. For cord- and plug-connected water heaters, an accessible cord and plug can serve as the disconnecting means per 422-22(a). See Figure 1-14.

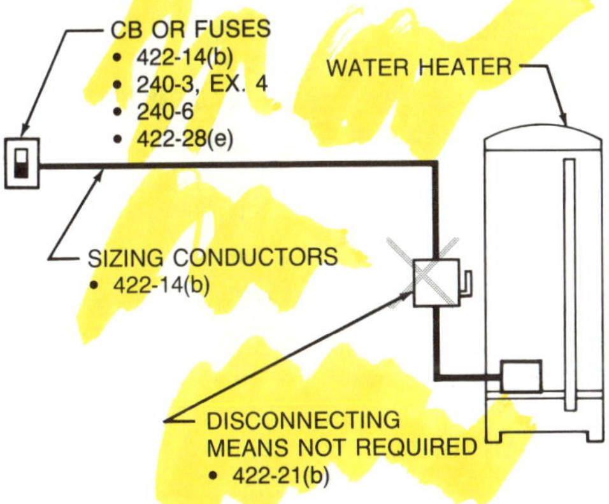

Figure 1-14. Branch circuit conductors and components are sized by the kW rating of the water heater.

What size conductors and DPCB is required to supply power to a 240 V, 1ϕ water heater pulling 6912 VA? Use nonmetallic sheathed cable (Romex).

Conductors

Step 1. Text: $I = \frac{VA}{V}$

$$I = \frac{6912 \text{ VA}}{240 \text{ V}} = 28.8 \text{ A}$$

Step 2. *422-14(b); 422-4(a), Ex. 2:*
28.8 A × 125% = 36 A

Step 3. *Table 310-16; 336-26:* #8 60°C conductors = 40 A

Answer. **Conductors = #8-2 with ground (nonmetallic sheathed cable)**

DPCB

Step 1. *422-14(b):* 28.8 A × 125% = 36 A

Step 2. *240-3, Ex. 4; 240-6:*
40 A is the next higher standard size

Answer. **DPCB = 40 A DPCB**

HEATING LOADS—424, PART A

Heating elements in heating units are rated at 5 kW. The elements are stacked in the heating unit to provide the rated kW for a particular size occupancy. Two stacked elements provide a 10 kW heating unit. Four stacked elements provide a 20 kW heating unit. Branch circuit conductors and overcurrent protection devices are sized and selected based on the kW rating of each heating unit.

Conductors—424-3(b)

Branch circuit conductors supplying power to a heating unit are computed at 125% of the heating element load plus 125% of the blower motor load when present. The ampacity of the conductors shall be at least 125% of the total VA rating of the heating elements and the blower motor load. For example, a 240 V, 1ϕ, 15 kW heating unit with a 3.5 A blower motor pulls 66 A (15 kW × 1000/240 V = 62.5 A). The branch circuit conductors are sized at 125% times 66 A, which equals 82.5 A (66 A × 125% = 82.5 A).

Overcurrent Protection—424-3(b)

The branch circuit overcurrent protection device shall not be less than 125% of the heating elements and blower motor load. If an overcurrent protection device does not correspond to this rating, the next higher size overcurrent device is selected per 240-3, Ex. 4. This requirement prevents the overcurrent protection device from tripping open when the heating unit requires all the 5 kW rated elements to satisfy the heating load called for by the thermostat. For example, a 10 kW heating unit pulling 42 A with a 2.5 A blower motor has a total connected load of 44.5 A (42 A + 2.5 A = 44.5 A). The overcurrent protection device shall be equivalent to 125% times 44.5 A (44.5 A × 125% = 56 A). The overcurrent protection device shall be rated no less than 60 A per 240-6. This is the next higher standard size. The overcurrent protection device can be sized on the ampacity of the conductor per 240-3, Ex. 4.

Disconnecting Means—424-19

A disconnecting means is required for a self-contained heating unit with a controller that energizes the circuits to the heating elements and blower motor. The disconnecting means shall be located within sight and within 50′ of the heating unit per 100, 424-19, and 430-102. The disconnecting means can be a fused or nonfused disconnect, an automatic breaker, or a nonautomatic circuit breaker. See Figure 1-15.

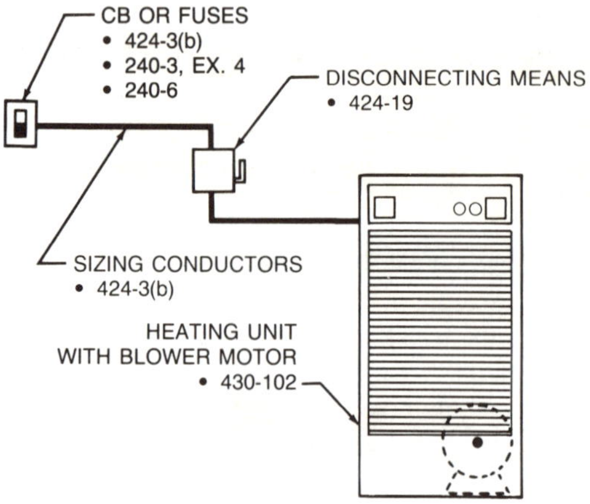

Figure 1-15. Branch circuit conductors and components are sized by the kW rating of the heating unit.

What size conductors and circuit breaker is required for a 25 kW, 240 V, 1ϕ heating unit with a 4 A blower motor? Use THWN Cu conductors.

Conductors

Step 1. *Text:* $I = \frac{VA}{V}$

$$I = \frac{25 \text{ kVA} \times 1000}{240 \text{ V}} = 104 \text{ A}$$

Step 2. *424-3(b):* 104 A + 4 A × 125% = 135 A

Step 3. *Table 310-16:* 150 A = #1/0 THWN

Answer. **Conductors = #1/0 THWN Cu**

CB

Step 1. *424-3(b):* 104 A + 4 A × 125% = 135 A

Step 2. *240-6; 240-3, Ex. 4:*
150 A is the next higher standard size

Answer. **CB = 150 A DPCB**

AIR CONDITIONING LOADS—440, PART C

Air conditioning units are rated in Btu or tons. One ton contains 12,000 Btu. A three-ton compressor in a central air conditioning unit has 36,000 Btu (3 × 12,000 Btu = 36,000 Btu). The air conditioner load is computed by the square footage and construction of the premises. The size air conditioner rated in Btu is selected from this calculation. The conductors shall be sized large enough to carry the load of the compressor, and overcurrent protection devices shall be selected to allow the compressor to start without tripping open due to high inrush current.

Conductors—440-32

Branch circuit conductors supplying power to air conditioning units shall be rated at least 125% of the full-load current rating of the compressor. If the branch circuit selection current is the higher full-load current rating, this rating shall be used instead of the compressor's full-load current rating.

The branch circuit selection current is determined by the manufacturer. It shall be higher or equal to the full-load current rating of the compressor per 440-4(c), FPN. For example, an air conditioning unit with a compressor that has a nameplate rating of 25 A shall have branch circuit conductors with a rating of 31 A per 440-32 (25 A × 125% = 31.25 A).

Conductors are sized large enough to prevent damage to insulation caused by an overload. Overload relays are usually adjusted to trip open on overloads exceeding 140% of the full-load current rating of the compressor. Conductors are sized and selected at 125% of the compressor's full-load current rating to compensate for such a condition. The load of the condenser motor is added at 100% when present.

Overcurrent Protection—440-22(a)

The overcurrent protection device is rated to allow the compressor to start and run. For the minimum rating, the overcurrent protection device shall not exceed 175% of the compressor's full-load current rating or the branch circuit selection current, whichever is greater. For example, the minimum rating of the overcurrent protection device for a central air conditioner with a full-load current rating of 44 A is 70 A (44 A × 175% = 77 A). Note that 175% times the full-load current rating shall not be exceeded.

If an air conditioner unit will not start using 175%, a maximum of 225% times the full-load current rating can be used. A full-load current rating of 44 A times 225% will permit a 90 A device to be used (44 A × 225% = 99 A). The condenser motor is added at 100% to this total for central A/C units when not figured with the nameplate rating.

An air conditioning unit installed on the roof or on the west side of the premises may have trouble starting and operating because of heat from the sun during hot summer months. Some units have the size of the overcurrent protection device listed on the nameplate of the unit. If the unit calls for a fuse, it shall be protected by that size fuse only per 110-3(b). Where the unit calls for fuses or a heating, air conditioning, and refrigeration (HACR) circuit breaker, they can be used to protect the circuit. See the Underwriters' Laboratory book, *Electrical Appliance & Utilization Equipment Directory*. The heading, Air Conditioning Equipment, gives further information.

Disconnecting Means—440-14

A disconnecting means shall be provided within 50′ and within sight of central air conditioning equipment per Article 100 of the NEC®. The disconnecting means cannot be locked open and out of sight as a disconnect for a motor can be. It must be within sight of the air conditioning unit. This allows service personnel to serve the equipment without the threat of an electrical shock from someone energizing the power circuit to the unit. The disconnecting means is usually mounted adjacent to the air conditioning unit, and a pigtail is used to connect the branch circuit from the disconnect to the unit. The overcurrent protection device installed in the service equipment located outside and beneath the meter can serve as the disconnecting means, if within 50′ of the unit. See Figure 1-16.

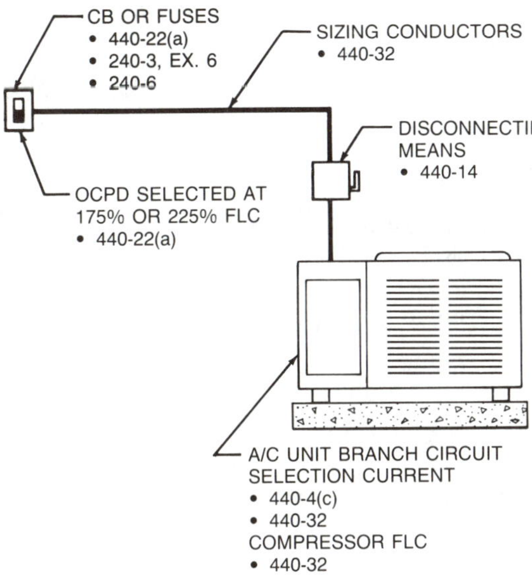

Figure 1-16. Branch circuit conductors and components are sized by the FLC of the compressor or the branch circuit selection current, whichever is greater.

What size conductors and overcurrent protection device is required to supply power to a 240 V, 1ϕ air conditioning unit with a nameplate rating of 5520 VA? Use nonmetallic sheathed cable and a DPCB.

Conductors

Step 1. *Text:* $I = \frac{VA}{V}$

$$I = \frac{5520\text{ VA}}{240\text{ V}} = 23\text{ A}$$

Step 2. *440-22(a):* 23 A × 125% = 28.75 A

Step 3. *Table 310-16; 336-26:* 29 A = #10 60°C

Answer. **Conductors = #10-2 with ground (nonmetallic sheathed cable)**

DPCB

Step 1. *440-22(a):* 23 A × 175% = 40.25 A

Step 2. *240-3, Ex. 6; 240-6:*
40 A is the next lower standard size

Answer. **DPCB = 40 A DPCB**

MOTOR LOADS—430, PARTS B AND D

Motors are rated in horsepower. The amount of work that a motor can perform depends upon its horsepower rating. A motor with a high horsepower rating can do more work than a motor with a low horsepower rating. The overcurrent protection devices and conductors used to supply power to motors are sized and based on the horsepower, voltage, and number of phases. See Figure 1-17.

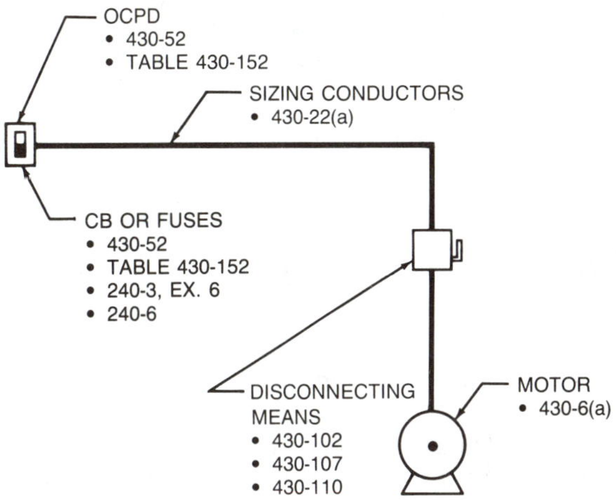

Figure 1-17. Branch circuit conductors and components are sized by the FLC of motors from Table 430-148 for 1ϕ and Table 430-150 for 3ϕ per 430-6(a).

Conductors—430-22(a)

Branch circuit conductors supplying power to motors are sized with enough copper or aluminum to withstand the starting and stopping or continuous operation of the driven load. Branch circuit conductors shall be able to pull an overload that motors may produce because of a stuck bearing in the motor or the driven load. An overload can be generated by the driven equipment, requiring more power to drive the load. To prevent the insulation of conductors from failing, resulting from excess heat from an overload, branch circuit conductors are increased by 125% of the motor's full-load current rating.

The full-load current ratings of 1ϕ motors are obtained from Table 430-148 per 430-6(a). Full-load current ratings for 3ϕ motors are selected from Table 430-150 per 430-6(a). The full-load current ratings for motors rated 110, 115, or 120 V are selected from the 115 V column for single phase. The full-load current ratings for motors rated 220, 230, or 240 V are selected from the 230 V column for 1ϕ or 3ϕ. For motors rated at 200 or 208 V, the full-load current rating in each table is increased by 15% for 200 V and 10% for 208 V.

The full-load current ratings for motors rated at 440, 460, or 480 V are selected from the 460 V column for 3ϕ. For 550, 575, or 600 V, 3ϕ motors, the 575 V column is used for selecting the full-load current rating based on the horsepower of each motor.

By applying Table 430-150, the following 20 HP motor has the full-load current rating of each motor listed in amps based on the voltage.

230 V—54 A
208 V—54 A × 110% = 59.4 A
200 V—54 A × 115% = 62.1 A

For example, a 15 HP, 230 V, 3ϕ motor has a full-load current rating of 42 A per Table 430-150. The branch circuit conductors are selected at 125% times 42 A, which equals 52.5 A (42 A × 125% = 52.5 A). The branch circuit conductors selected from Table 310-16 shall be equivalent to 53 A.

Overcurrent Protection—430-52; Table 430-152

Inrush current of a motor occurs when power is applied to the windings and the motor starts. *Running current* is the amount of current required to drive a load. Overcurrent protection devices are sized large enough to hold the inrush current of the motor and allow the motor to accelerate the driven load. The inrush current of most motors is four to six times the running current of the motor, based on its code letter per Table 430-7(b). A motor marked with code letter G and a full-load current rating of 20 A has an inrush current of 120 A (20 A × 6 = 120 A).

Section 430-52 requires that the overcurrent protection device not exceed the percentages listed in Table 430-152, based on the type device used in the columns. If the percentages in Table 430-152 do not correspond to a standard device listed in 240-6, the next higher standard size rating may be used per 430-52, Ex. 1. If the motor will not start and accelerate its driven load, 430-52, Ex. 2 is applied. The percentages that an overcurrent protection device can be increased by applying 430-52, Ex. 2 are as follows:

Nontime delay fuse (NTDF) shall not exceed 400% of the motor's FLC for fuses rated 600 A or less.

Time delay fuse (TDF) shall not exceed 225% of the motor's FLC.

Circuit breaker (CB) shall not exceed 400% of the motor's FLC for ratings of 100 A or less, or 300% of the motor's FLC for ratings over 100 A.

Instantaneous trip CB (ITCB) shall not exceed 1300% of the motor's FLC.

A 25 HP, 230 V, 3ϕ motor has a full-load current rating of 68 A per Table 430-150. Find the size overcurrent protection device required. (Apply 430-52, Ex. 1.)

NTDF

Step 1. *Table 430-152:* 300%
68 A × 300% = 204 A

Step 2. *430-52; 240-3, Ex. 6; 240-6:*
200 A is the next lower standard size

Step 3. *430-52, Ex. 1; 240-3, Ex. 6; 240-6:*
225 A is the next higher standard size

Answer. **NTDF = 225 A**

Note: 250 A NTDF can be used per 430-52, Ex. 2 (68 A × 400% = 272 A).

TDF

Step 1. *Table 430-152:* 175%
68 A × 175% = 119 A

Step 2. *430-52; 240-3, Ex. 6; 240-6:*
110 A is the next lower standard size

Step 3. *430-52, Ex. 1; 240-3, Ex. 6; 240-6:*
125 A is the next higher standard size

Answer. **TDF = 125 A**

Note: 150 A TDF can be used per 430-52, Ex. 2 (68 A × 225% = 153 A).

CB

Step 1. *Table 430-152:* 250%
68 A × 250% = 170 A

Step 2. *430-52; 240-3, Ex. 6; 240-6:*
150 A is the next lower standard size

Step 3. *430-52, Ex. 1; 240-3, Ex. 6; 240-6:*
175 A is the next higher standard size

Answer. **CB = 175 A**

Note: 250 A CB can be used per 430-52, Ex. 2 (68 A × 400% = 272 A).

ITCB

Step 1. *Table 430-152:* 700%
68 A × 700% = 476 A

Step 2. *430-52; 240-3, Ex. 6; 240-6:*
476 A = maximum setting

Answer. **ITCB = 476 A**

Note: 884 A setting can be used per 430-52, Ex. 2 (68 A × 1300% = 884 A).

■ *Master Electrician's Test Tip.* Some examining boards require that the next lower standard size overcurrent protection device be selected unless the problem states that the motor will not start and accelerate the load. ■

Disconnecting Means—430-102; 430-107

An approved disconnecting means is required to disconnect the power supply to a motor to allow personnel to service the motor without the danger of the branch circuit being accidentally energized. The disconnecting means to disconnect a motor circuit can be provided by the following methods:

1. A disconnecting means is installed adjacent to the controller with the motor located within 50′ and within sight.
2. A disconnecting means that can be locked in the open position is located adjacent to the controller. The motor does not have to be located within sight.
3. If the disconnecting means located by the controller cannot be locked open and the motor is located out of sight, an additional disconnect shall be provided within sight and within 50′ of the motor.

What size conductors and overcurrent protection device is required to supply power and allow a 30 HP, 208 V, 3ϕ motor to start and run? Use THWN Cu conductors and TDF. (Apply 430-52, Ex. 1.)

Conductors

Step 1. *Table 430-150:* 30 HP = 80 A
80 A × 110% = 88 A

Step 2. *430-22(a):* 88 A × 125% = 110 A

Step 3. *Table 310-16:* 115 A = #2 THWN

Answer. **Conductors = #2 THWN Cu**

TDF

Step 1. *Table 430-152:* 88 A × 175% = 154 A

Step 2. *430-52, Ex. 1; 240-3, Ex. 6; 240-6:*
175 A is the next higher standard size

Answer. **OCPD = 175 A TDF**

Note: 300 A NTDF (88 A × 300% = 264 A)
225 A CB (88 A × 250% = 220 A)
616 A ITCB (88 A × 700% = 616 A)

WELDER LOADS—630, PARTS B AND D

Transformer arc, motor-generator arc, and resistance welders are used in industry. Manual or automatic welding can be performed with all three types of welders. Individual branch circuit conductors and overcurrent protection devices for welders are sized by the duty cycle and demand factors of each welder. Duty cycles express the percentage of time that a welder is actually welding or the percentage of time current flows through the weld. Branch circuits are sized on a *demand factor* (multiplier) based and selected on the duty cycle of each type of welder. See Figure 1-18.

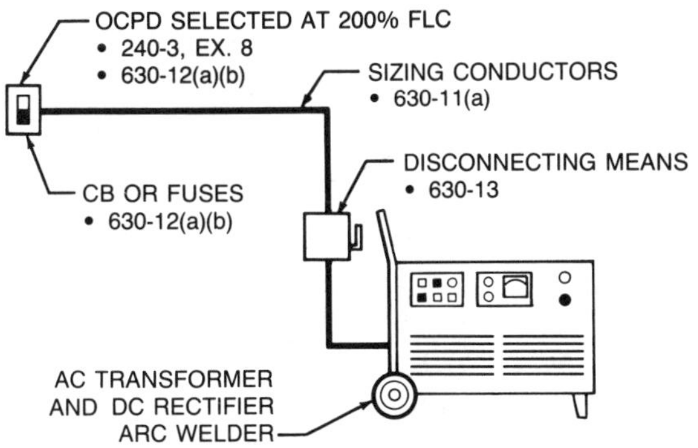

Figure 1-18. Branch circuit conductors and components of AC transformer and DC rectifier arc welders are sized by the primary FLC of each type of welder.

Duty Cycle—630-11(a); 630-21(a); 630-31(a)

Demand factors of welders are selected from the tables in 630-11(a), 630-21(a), and 630-31(a). For example, welders with a duty cycle of 50% have a demand factor of 71%. This multiplier is applied to determine the size branch circuit conductors required to supply power to the welder. If the duty cycle of a welder is known, the demand factor can be computed by taking the square root of the percentage listed on the nameplate of the welder. For example, a welder with a duty cycle of 50% has a multiplier of 71% ($\sqrt{50\%}$ = 71%). The multiplier is slightly higher for motor-generator arc welders because of the size of the motor.

When a welder welds for only two 15-cycle periods per second, the welder operates for a time of 30 seconds. Since all supplying voltage is usually 60 Hz, the duty cycle is equal to 50% (duty cycle = 30/60 × 100 – 50%). For a welder designed to operate for six 4-cycle welds per second, the duty cycle is equal to 40% (duty cycle = 6 × 4/60 = 40%). By taking the square root of 40%, the multiplier equals 63% ($\sqrt{40\%}$ = 63%). Branch circuit conductors that supply power to an individual welder are determined by multiplying the primary full-load current rating of the welder by the multiplier, which is based on the duty cycle.

AC Transformer and DC Rectifier Arc Welders—630, Part B

Arc welders perform welding operations through the medium of an arc drawn between the work and a metal rod. In arc welders, metal from the rod (electrode) is added to the weld.

Conductors—630-11(a). The current-carrying capacity of branch circuit conductors shall not be less than the rated primary current of the welder times a duty cycle factor listed in the table per 630-11(a). For example, an AC transformer and a DC rectifier arc welder has a nameplate rating of 42 A with a duty cycle of 20%. From the table in 630-11(a), the multiplier is 45%. The rating of the branch circuit conductors is equivalent to 19 A (42 A × 45% = 18.9 A). This rating is based on the fact that the operation of the welder occurs and recurs according to a prescribed cycle.

Overcurrent Protection—630-12(a). The overcurrent protection device is selected at not more than 200% of the welder's primary full-load current rating listed on the nameplate. There is an exception to 630-12(a) for conductors already protected at a rating not exceeding this value. If the provisions of 630-12(a) are not applied, the overcurrent protection device shall not exceed 200% of the branch circuit conductors supplying the welder. For example, an AC transformer and DC rectifier arc welder has a nameplate rating of 42 A. The rating of the overcurrent protection device shall not exceed 84 A (42 A × 200% = 84 A) when 630-12(a) is applied. Where 630-12(a) is applied, the overcurrent protection is selected at 90 A because 84 A does not correspond to a standard device per 240-6, or the device opens unnecessarily per 240-3, Ex. 8, because of the type of welding done.

What size THWN Cu conductors and overcurrent protection device is required to supply an AC transformer and DC rectifier arc welder that has a full-load current rating of 72 A with a 40% duty cycle?

Conductors

Step 1. *630-11(a):* Multiplier = 63%
72 A × 63% = 45 A

Step 2. *Table 310-16:* 50 A = #8 THWN

Answer. **Conductors = #8 THWN Cu**

OCPD

Step 1. *630-12(a):* 200%
72 A × 200% = 144 A

Step 2. *240-3, Ex. 8; 240-6; 630-12:*
125 A is the next lower standard size

Answer. **OCPD** = 125 A

Note: By applying 630-12, a 150 A overcurrent protection device can be used.

■ *Master Electrician's Test Tip.* Some examining boards require that 630-12(b) be applied when individual protection of conductors is not provided. The overcurrent protection device shall not exceed the ampacity of conductors used to supply power to the welder. ■

Motor-Generator Arc Welders—630, Part C

Motor-generator arc welders use the arc welding principle to perform work. A metal electrode is used to strike an arc and weld two metal parts together. The electrode is melted by the arc and supplies the metal needed to join the metal parts. See Figure 1-19.

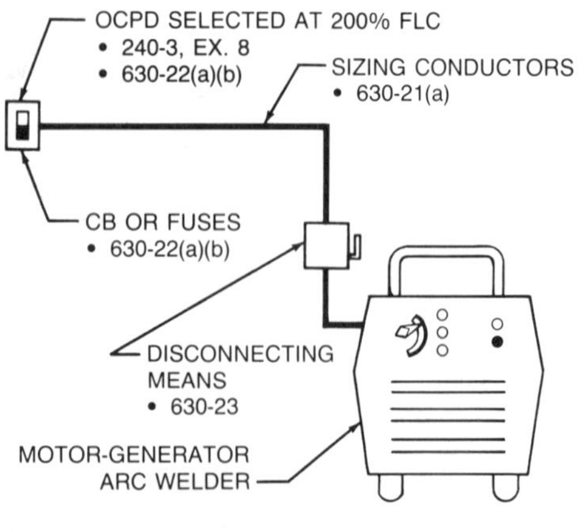

Figure 1-19. Branch circuit conductors and components of motor-generator arc welders are sized by the primary FLC of the welder.

Conductors—630-11(a). The current-carrying capacity of the branch circuit conductors shall not be less than the rated primary current of the welder times a duty cycle factor listed in the table per 630-21(a). For example, a motor-generator arc welder has a nameplate rating of 58 A with a duty cycle of 60%. From the table in 630-21(a), the multiplier is 81%. The rating of the branch circuit conductors is 47 A (58 A × 81% = 47 A).

Overcurrent Protection—630-12(a). The overcurrent protection device is selected at not more than 200% of the welder's primary full-load current rating listed on the nameplate. There is an exception to 630-22(a) for conductors already protected at a rating not exceeding this value. If the provisions of 630-22(a) are not applied, the overcurrent protection device shall not exceed 200% of the branch circuit conductors supplying the welder. For example, a motor-generator arc welder has a nameplate rating of 58 A. The rating of the overcurrent protection device shall not exceed 200% of 58 A, which equals 116 A (58 A × 200% = 116 A) per 630-22(a). Where the provisions of 630-22(a) are applied, a 125 A overcurrent protection device is used because 116 A does not correspond to a standard size device per 240-6 or the device opens unnecessarily because of the type of welding done.

What size THWN Cu conductors and overcurrent protection device is required to supply a resistance welder that has a full-load current rating of 85 A with an 80% duty cycle?

Conductors

Step 1. *630-21(a):* Multiplier = 91%
85 A × 91% = 77 A

Step 2. *Table 310-16:* 85 A = #4 THWN

Answer. **Conductors = #4 THWN Cu**

OCPD

Step 1. *630-22(a):* 200%
85 A × 200% = 170 A

Step 2. *240-3, Ex. 8; 240-6; 630-22:*
150 A is the next lower standard size

Answer. **OCPD = 150 A**

Note: By applying 630-22(a), a 175 A overcurrent protection device can be used. See 630-22(b).

Resistance Welders

Resistance welders weld by virtue of a heavy current that flows through the small area of materials in contact at a particular time. Resistance welders do not add metal, but welding is accomplished by pressing two metal parts together as they reach the molten state. See Figure 1-20.

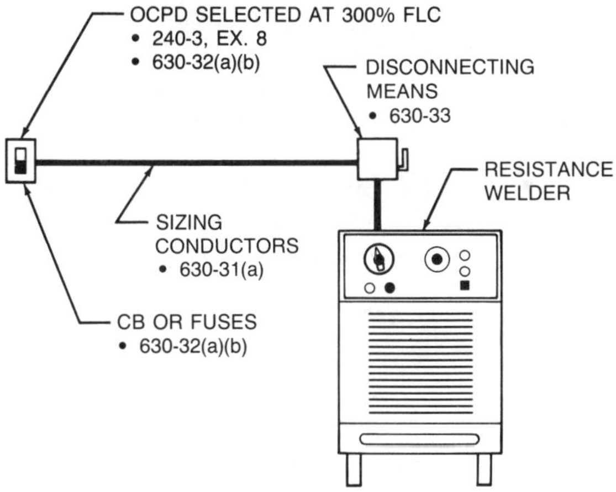

Figure 1-20. Branch circuit conductors and components of resistance welders are sized by the primary FLC of the welder.

Conductors—630-31(a). The current-carrying capacity of the branch circuit conductors shall not be less than the rated primary current of the welder times a duty cycle factor listed in the table per 630-31(a). For example, a resistance welder has a nameplate rating of 69 A with a duty cycle of 30%. From the table in 630-31(a), the multiplier is 55%. The rating of the branch circuit conductors is 38 A (69 A × 55% = 37.95 A). This rating is based on the fact that the operation of the welder occurs and recurs according to a set plan.

Overcurrent Protection—630-32(a). The overcurrent protection device is selected at not more than 300% of the welder's primary full-load current rating listed on the nameplate. There is an exception to 630-32(a) for conductors already protected at a rating not exceeding this value.

If the provisions of 630-32(a) are not applied, the overcurrent protection device shall not exceed 300% of the branch circuit conductors supplying the welder. For example, a resistance welder has a nameplate rating of 69 A. The rating of the overcurrent protection device shall not exceed 300% of 69 A, which is 207 A (69 A × 300% = 207 A) when 630-32(a) is applied. When 630-32(a) is applied, the overcurrent protection can be rated at 225 A because 207 A does not correspond to a standard device per 240-6, or the device opens unnecessarily because of the type of welding done.

What size THWN Cu conductors and overcurrent protection device is required to supply a resistance welder that has a full-load current rating of 95 A with a duty cycle of 50%? Apply 630-32 (a). Use FLC rating.

<u>Conductors</u>

Step 1. *630-31(a):* Multiplier = 71%
95 A × 71% = 67.5 A

Step 2. *Table 310-16:* 85 A = #4 THWN

Answer. **Conductors = #4 THWN Cu**

<u>OCPD</u>

Step 1. *630-32(a):* 300%
95 A × 300% = 285 A

Step 2. *240-3, Ex. 8; 240-6; 630-32(a):*
275 A is the next lower standard size

Answer. **OCPD = 275 A**

What size THWN Cu conductors and overcurrent protection device is required to supply a resistance welder that has a full-load current rating of 95 A with a duty cycle of 50%? Apply 630-32(b). Use conductor rating.

<u>Conductors</u>

Step 1. *630-31(a):* Multiplier = 71%
95 A × 71% = 67.5 A

Step 2. *Table 310-16:* 85 A = #4 THWN

Answer. **Conductors = #4 THWN Cu**

<u>OCPD</u>

Step 1. *630-32(b):* Multiplier = 300%
85 A × 300% = 255 A

Step 2. *240-3, Ex. 8; 240-6; 630-32(b):*
250 A is the next lower standard size

Answer. **OCPD = 250 A**

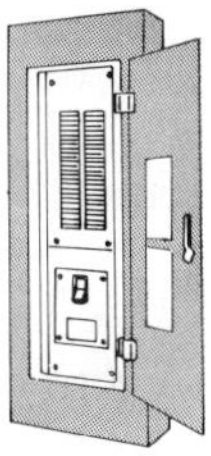

Branch Circuits

Date ____________

Chapter 1—TEST

Name ____________

1. What is the total VA rating of a 120 V, 1ϕ branch circuit supplied by a 20 A CB?
2. What is the total VA rating of a 208 V, 3ϕ branch circuit supplied by a 20 A 3PCB?
3. What is the maximum size lighting load that can be connected to a 20 A, 1ϕ branch circuit supplying a fixed appliance load of 7 A?
4. What size branch circuit is required to supply a 14 A, 120 V, 1ϕ cord- and plug-connected dishwasher?
5. What is the maximum size lighting load that can be connected to a 15 A, 120 V, 1ϕ branch circuit supplying a 6 A fixed equipment load?
6. Can a 26 A cord- and plug-connected piece of utilization equipment be connected to a 30 A branch circuit?
7. What size branch circuit is required to supply a fixed lighting load of 21.5 A used at continuous duty?
8. What size branch circuit is required to supply a commercial deep fryer that is used at continuous duty and pulls 30 A?
9. What size load can be connected to 60 °C rated terminals supplied by a #6 THWN Cu conductor?
10. What size nonmetallic sheathed cable (Romex) is required to supply a fixed appliance load of 49.6 A used at continuous duty?
11. Can #4 THWN Cu conductors be terminated to a 90 A CB rated at 60 °C and supply a load of 82 A at noncontinuous duty?
12. What is the maximum load permitted on three #4/0 THHN Cu conductors in an ambient temperature of 129 °F?
13. What is the load allowed for seven current-carrying #1 THWN Cu conductors routed in an ambient temperature of 135 °F?
14. What size circuit breakers is required for six current-carrying #3 THHN Cu conductors routed through an ambient temperature of 112 °F? The conductors supply a continuous load of 70 A.
15. What is the maximum voltage between conductors for lighting fixtures installed in dwelling units?
16. What is the maximum VA rating of cord- and plug-connected equipment in guest rooms of hotels or motels?
17. What is the maximum voltage to ground for pin-type electric discharge lighting fixtures installed in office buildings?
18. What is the mounting height for lighting fixtures installed on poles and supplied by a 480 V corner-grounded branch circuit?
19. What is the amperage for an industrial plant with a connected load of 441,000 VA? The supply voltage is a 120/208 V, 4-wire system.

______ **20.** How many 1.5 A lighting fixtures can be connected to a 20 A noncontinuous-duty branch circuit?

______ **21.** How many 1.5 A lighting fixtures can be connected to a 30 A continuous-duty branch circuit?

______ **22.** What is the load for a 15 A circuit breaker protecting a continuous load?

______ **23.** How many 2-wire branch circuits are required to supply thirty 1.5 A ballasts? Use 20 A CBs.

______ **24.** What size 120 V branch circuit, using THWN Cu conductors, is required to supply 12 lighting fixtures pulling 200 W each at noncontinuous duty?

______ **25.** How many 20 A, 120 V branch circuits are required to supply 76′-0″ of lighting track?

______ **26.** How many general-purpose outlets per circuit are permitted on three 20 A branch circuits supplying duplex receptacles (noncontinuous duty)?

______ **27.** What is the load for 52 duplex receptacles used at noncontinuous duty on a 120 V branch circuit?

______ **28.** What is the total square footage permitted to be served by a 20 A branch circuit in a dwelling unit?

______ **29.** What is the amperage for a 11.5 kW, 240 V electric household range?

______ **30.** What size nonmetallic sheathed cable is required to supply an 8000 VA, 240 V household oven tapped from a branch circuit?

______ **31.** What is the demand load for an 8 kW range in a dwelling unit?

______ **32.** What is the demand load for a 22 kW household range?

______ **33.** What is the demand load for a branch circuit supplying a 10 kW cooktop, an 8 kW oven, and a 6 kW oven installed in the kitchen of a dwelling unit?

______ **34.** What is the amperage for a 16 kW, 240 V, 1ϕ commercial cooker used at continuous duty?

______ **35.** How many amps does a 5500 VA, 240 V, 1ϕ household dryer pull?

______ **36.** What is the amperage of a 240 V, 1ϕ, 7200 VA water heater?

______ **37.** What is the ampacity used to select the branch circuit conductors supplying a 35 kW, 240 V, 1ϕ heating unit with an 840 VA blower motor?

______ **38.** What size THWN Cu conductors and overcurrent protection device is required to supply a 30 kW, 240 V, 1ϕ heating unit with a 4 A blower motor?

______ **39.** What is the minimum size circuit breaker required to allow a 42 A air conditioner to start and run?

______ **40.** What size THWN Cu conductors and overcurrent protection device is required for a central air conditioner that pulls 52 A?

______ **41.** What size THWN Cu conductors is required to supply a 60 HP, 200 V, 3ϕ induction motor?

______ **42.** What size TDFs is required to allow a 75 HP, 208 V, 3ϕ induction motor to start and run? The motor is rated code letter G and has a service factor of 1.15. Use next higher standard size.

______ **43.** What size THWN Cu conductors and circuit breaker is required to allow a 200 HP, 460 V, 3ϕ induction motor to start and run?

______ **44.** What is the duty cycle for a welder that employs three 4-cycle welds per second?

______ **45.** What size conductors and overcurrent protection device is required to supply power to an AC transformer arc welder with a primary current rating of 48.5 A and a duty cycle of 90%? Use THWN Cu conductors.

Feeder Circuits

Chapter

2

Feeder circuit conductors include all conductors between the service equipment or the source of a separately derived system, and the overcurrent protection devices protecting the branch circuit. Feeder circuits supply subpanels in dwelling units, apartment buildings, and commercial and industrial occupancies. They can be routed in commercial and industrial buildings where taps are made to the feeder conductors to supply panelboards, disconnects, control centers, and other types of electrical equipment.

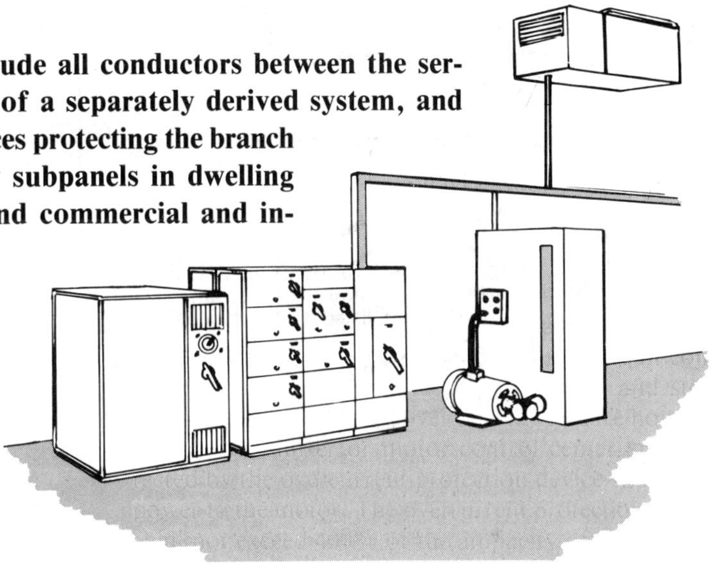

LOADS—220, PART B

Feeder circuit conductors and overcurrent protection devices are sized and selected per 220, Part B. Loads are grouped and percentages applied according to their conditions of use.

Voltage—220-2

Voltage ratings used to determine the ampacity of electrical components of feeder circuits are listed in 220-2. See Figure 2-1. Ampacity (I) is found by dividing the total volt-amp rating (VA) by voltage (V). The voltage rating to size the feeder circuit for voltages not listed is one of the following:

120 V
120/240 V
240 V
480/277 V
480 V
600 V

A feeder circuit load of 15,600 VA is supplied by a 120/240 V, 1ϕ system. What is the ampacity?

Step 1. *Text:* $I = \frac{VA}{A}$

$$I = \frac{15{,}600 \text{ VA}}{240 \text{ V}} = 65 \text{ A}$$

Answer. **I = 65 A**

A 120/208 V, 3ϕ feeder circuit has a total connected load of 28,800 VA. What is the ampacity?

Step 1. $I = \frac{VA}{V \times \sqrt{3}}$

$$I = \frac{28{,}800 \text{ VA}}{360} = 80 \text{ A}$$

Answer. **I = 80 A**

The ampacity used to select the feeder circuit components is 80 A per phase. Per phase denotes that 80 A is connected to phase A, phase B, and phase C, respectively. This is verified by dividing 28,800 VA by 120 V per phase and dividing the total by three (the number of 120 V circuits in a 120/208 V, 4-wire system). For example,

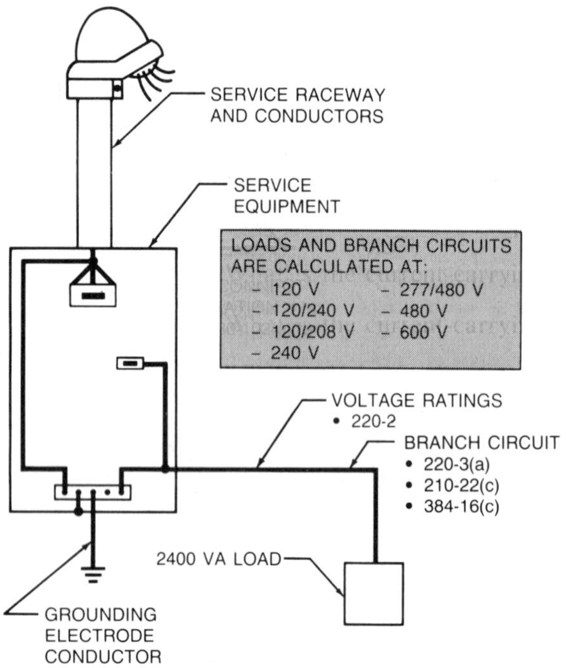

BRANCH-CIRCUIT RATINGS

NONCONTINUOUS DUTY

What is the amperage of a 2400 VA load used at noncontinuous duty?

Step 1. *220-3(a):* $I = \frac{VA \times 100\%}{V}$

$$I = \frac{2400\text{ VA} \times 100\%}{240\text{ V}} = 10\text{ A}$$

Answer. **Amperage = 10 A**

CONTINUOUS DUTY

Step 1. *220-3(a):* $I = \frac{VA \times 125\%}{V}$

$$I = \frac{2400\text{ VA} \times 125\%}{240\text{ V}} = 12.5\text{ A}$$

Answer. **Amperage = 13 A**

Figure 2-1. Branch circuits are sized from the amperage of loads, which are determined from VA divided by the voltage ratings per 220-2. For problems with the voltage listed, that voltage is used to calculate the load.

$$I = \frac{VA}{V}$$

$$I = \frac{28{,}800\text{ VA}}{120\text{ V}} = 240\text{ A}$$

$$I = \frac{240\text{ A}}{3} = 80\text{ A}$$

I = 80 A

Phase-to-phase voltage (208 V) times $\sqrt{3}$ (1.732) divided into the total VA rating (28,800 VA) equally distributes the amperage on each 120 V leg. When figuring 3ϕ problems, use the 3ϕ values chart to eliminate one mathematical step.

Noncontinuous Duty—220-10(b)

Noncontinuous-duty loads operate for a time period of less than three hours per Article 100. The VA ratings of noncontinuous-duty loads are calculated at 100%, or demand factors are applied to certain loads based upon their condition of use. For example, a connected load of 32,000 VA is calculated at 100% of 32,000 VA (32,000 VA × 100% = 32,000 VA), and the feeder circuit is sized on this VA rating.

Noncontinuous-duty loads can be portable, fixed cord- and plug-connected, or permanently connected (hard-wired) loads of all types. Regardless of the type of load supplied, noncontinuous-duty loads shall not operate for more than three hours.

Continuous Duty—220-10(b)

Continuous-duty loads operate for three hours or more. These loads do not operate at varying or intermittent duty. For example, a processing machine in an industrial plant cuts, forms, and glues cartons eight hours a day. This is a continuous-duty load. Power is supplied continuously to continuous-duty loads, and the loads operate for three hours or more without the circuit being disconnected.

The VA rating of continuous-duty loads is multiplied by 125% to obtain the total VA rating for sizing feeder circuits. For example, a processing machine pulls 25,920 VA and operates for six hours a day. The VA rating used to size the circuit is 32,400 VA (25,920 VA × 125% = 32,400 VA).

Demand Factors—220, Part B

Demand factor is the maximum VA demand of the total connected load in a system. Certain loads, based on their condition of use, are eligible to have a demand factor applied to their VA rating. The maximum demand for a feeder circuit is equal to the connected load times the demand load. Loads on which a demand factor is applied are listed in the NEC®. Requirements are precise and can only be applied under specific conditions of use. See Figure 2-2.

The total rating of a feeder circuit supplying certain loads under specific conditions of use can be reduced

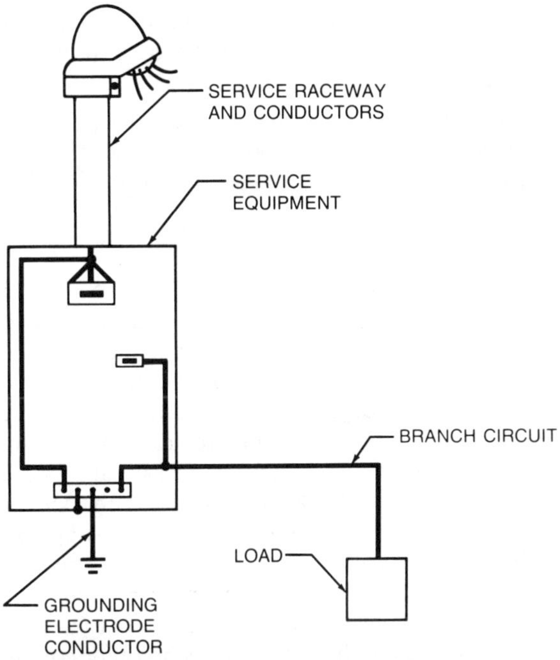

CALCULATION OF LOADS	
Continuous duty—220-3(a)	VA × 125%
Noncontinuous duty—220-3(a)	VA × 100%
Demand factors—220-10–220-22	VA × %

Figure 2-2. Branch circuits are determined by multiplying the VA ratings of the loads by 125% for continuous duty and 100% for noncontinuous duty. Demand factors are applied for certain loads.

according to demand factors. Some loads that can be reduced by applying a demand factor are

- Lighting loads . . . . Table 220-11
- Receptacle loads . . . . Table 220-13
- Dryer loads . . . . Table 220-18
- Range loads . . . . Table 220-19
- Kitchen equipment loads . . . . Table 220-20
- Dwelling unit loads . . . . Table 220-30
- Existing dwelling unit loads . . . . . Table 220-31
- Multifamily dwelling unit loads . . Table 220-32
- School loads . . . . Table 220-34
- Restaurants . . . . Table 220-36
- Farm loads (individual load) . . . . . Table 220-40
- Farm load (total of loads) . . . . Table 220-41

What is the load of a number of noncontinuous-duty receptacles supplying office machines at 22,400 VA?

Step 1. *Table 220-13:*

First 10,000 VA × 100% = 10,000 VA
Next 12,400 VA × 50% = 6,200 VA
16,200 VA

Answer. **Load = 16,200 VA**

The VA rating is calculated per Table 220-13. The first 10,000 VA are computed at 100% and the remaining VA are calculated at 50%. See 220-13 and Table 220-11. Receptacle outlets at 180 VA can be added to the lighting load and the demand factors of Table 220-11 applied.

■ *Master Electrician's Test Tip.* Demand factors of Table 220-13 can only be applied to receptacles used at noncontinuous duty. Demand factors are not applied to continuous-duty loads. ■

Loads that are thermostatically controlled, used at varying duty, or connected momentarily to the line are eligible to have a demand factor applied. For example, a demand factor of 65% is applied to the total VA rating of six or more pieces of thermostatically controlled electric cooking equipment. Power is only supplied to the equipment when the thermostat is calling and therefore the equipment never uses power for three hours or more.

Neutral—220, Part B; 310

The neutral is a grounded conductor that is intentionally grounded at the service equipment. It must be sized to carry the *maximum unbalanced current.* The maximum unbalanced current is the load between the neutral and any one phase (ungrounded) conductor. Under certain conditions, a demand factor is applied to the neutral.

Sizing—220-22. The neutral is sized to carry the calculated load or the maximum unbalanced load of the system. The calculated load, up to 200 A, is computed at 100%. All amperage above 200 A, except 3ϕ, 4-wire wye feeder circuits supplying power to electric discharge lighting, data processing, or similar equipment, is computed at 70%. An additional demand factor of 70% is calculated for household ranges, countertop cooking units, wall-mounted ovens, and dryers. For example, a load of 250 A is calculated for a feeder circuit supplying electric discharge lighting. The reduction of 70% is not permitted. The 250 A is computed at 100%. See Figure 2-3.

What is the neutral load for a feeder circuit supplying 275 A of electric discharge lighting?

Step 1. *220-22:* 275 A × 100% = 275 A

Answer. **Load = 275 A**

A feeder circuit has 265 A of electric discharge lighting and 285 A of incandescent lighting. What is the neutral load?

Step 1. *220-22:*

Electric discharge lighting =
265 A × 100% = 265 A
Incandescent lighting =
First 200 A × 100% = 200 A
Next 85 A × 70% = 59.5 A
524.5 A

Answer. **Load = 524.5 A**

If a 12 kW household range has a demand of 8 kW per Table 220-19, Column A, the demand factor is found by multiplying 8 kW by 70%.

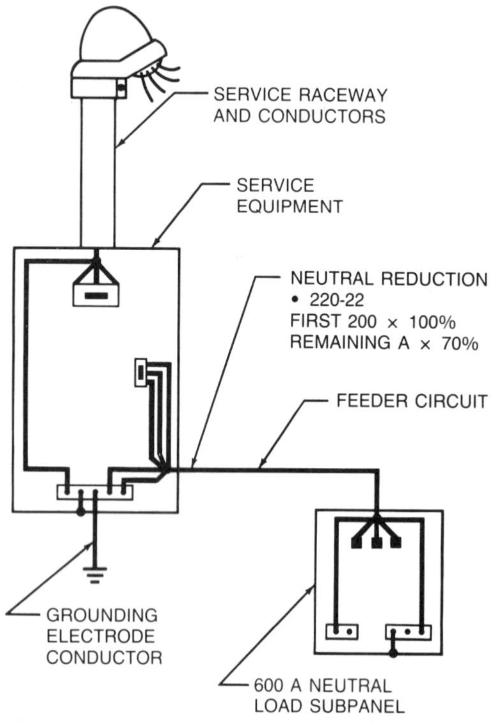

SIZING NEUTRALS

What is the amperage of the neutral with a 600 A load?

Step 1. *220-22:*

First 200 A × 100% =	200 A
Next 400 A × 70% =	280 A
Total =	480 A

Answer. **Amperage = 480 A**

Figure 2-3. The neutral conductors are sized by taking 70% of all amperage that exceed 200 A. The 70% reduction does not apply to loads with harmonic currents.

What is the demand load for the neutral having a 12 kW range on the phase conductors?

Step 1. *Table 220-19, Column A:* 12 kW = 8 kW demand

Step 2. *220-22:* 8 kW × 70% = 5.6 kW

Answer. **Load = 5.6 kW**

For a 5-wire, 2ϕ feeder circuit supplying a connected load of 250 A in each phase, the total load of the neutral is increased by 140% (250 A × 140% = 350 A). The current in the neutral conductor of a 3-wire, 2ϕ system is 141% greater than either of the other two phase conductors. The 141% is derived by taking the square root of two ($\sqrt{2}$ = 141%). The $\sqrt{2}$ is used in a 5-wire, 2ϕ system because the unbalanced current in the neutral is collected from all ungrounded phase conductors. The phase conductor with the higher ampacity is multiplied by 140% per 220-22 instead of 141%. The neutral is sized on this total calculation of 140% times the phase conductor with the higher ampacity rating. Sizing the neutral conductor at 140% times the phase conductor with the higher ampacity rating prevents it from being overloaded by 120 V loads switched in and out on the circuits.

What is the neutral load for a 5-wire, 2ϕ feeder circuit with each phase rated at 160 A?

Step 1. *220-22:* 160 A × 140% = 224 A

Answer. **Load = 224 A**

Conditions of Use—Note 10 to Tables 310-16–310-19. Note 10 is divided into three parts: (a), (b), and (c). Part (a) does not require that the neutral be considered a current-carrying conductor when carrying only unbalanced current from other ungrounded conductors. The neutral conductor carries very little load when circuits are balanced. See Figure 2-4.

For example, a 1ϕ, 2-wire circuit with one ungrounded and one grounded conductor supplies a 120 V load of 12 A. Because there is only one hot ungrounded conductor in this 2-wire circuit, the neutral carries the same amount of current (12 A) as the ungrounded conductor. There is no other conductor in the circuit to create an unbalanced load.

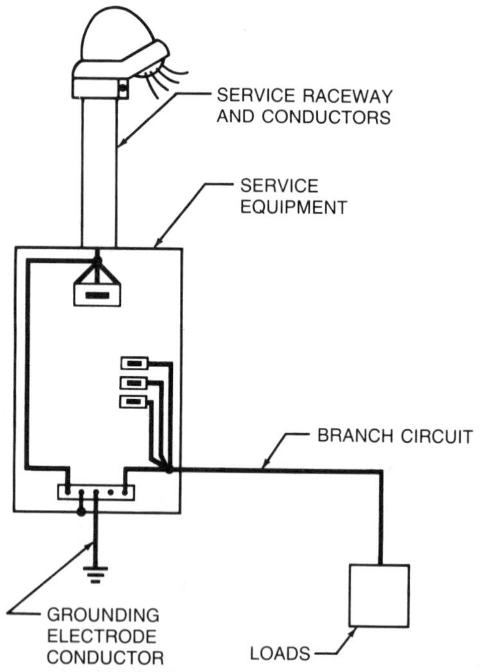

CONDITIONS OF USE—TABLES 310-16–310-19	
Part A	Unbalanced loads
Part B	1ϕ taps are same size
Part C	Neutral is same size as phase conductors for harmonic loads

Figure 2-4. The neutral is not considered a current-carrying conductor when carrying unbalanced current. It is considered a current-carrying conductor when supplying inductive or harmonic loads.

What is the neutral load of a 1ϕ, 120 V, 2-wire circuit serving a 12 A load?

Step 1. *220-22; Note 10 to Tables 310-16–310-19:*
Ungrounded conductor load = 12 A
Grounded conductor load = 12 A

Answer. **Load = 12 A**

A 3-wire circuit has two ungrounded conductors and one grounded conductor. The neutral carries the unbalanced load of the two ungrounded conductors. For example, a 3-wire circuit carries 38 A on one ungrounded conductor and 49 A on the other ungrounded conductor. Because the two ungrounded conductors are sharing the same neutral, the unbalanced load of the neutral is determined by subtracting 38 A from 49 A (49 A − 38 A = 11 A). The neutral carries an unbalanced load of 11 A. If both ungrounded conductors pull the same load there is no unbalanced load present and the neutral load is 0 A.

The ungrounded conductors of a 3-wire circuit carry 68 A and 56 A. Find the unbalanced neutral load.

Step 1. *220-22; Note 10 to Tables 310-16–310-19:*

Ungrounded conductor A =	68 A
Ungrounded conductor B =	56 A
	12 A

Answer. **Load = 12 A**

Part (b) requires that the neutral of a 3-wire, 120/208 V feeder circuit be the same size as the ungrounded conductors for a feeder circuit tapped from a 4-wire, 120/208 V system. The neutral conductor carries approximately the same current as the ungrounded conductors and is counted as a current-carrying conductor; therefore, the derating factors of Note 8 are applied. For example, a 120/208 V, 1ϕ tap with 165 A on one ungrounded conductor and 175 A on the other ungrounded conductor is made from a 4-wire, 120/208 V, 3ϕ feeder. The total 120 V load on the grounded neutral conductor is 85 A. Because the tap is 120/208 V, 1ϕ made from a 4-wire wye, 3ϕ system, the neutral is sized at 175 A and is considered a current-carrying conductor. The neutral is sized on the largest ungrounded conductor. It would be sized from the 120 V, 85 A load if it were not part of a 120/208 V, 1ϕ tap from a 4-wire wye system.

What is the neutral load for a 120/208 V, 1ϕ tap made from a 4-wire wye, 3ϕ system with 175 A on phase A and 165 A on phase B?

Step 1. *Note 10(a) to Tables 310-16–310-19:*
Ungrounded conductor A = 175 A
Ungrounded conductor B = 165 A
Grounded conductor N = 85 A

Answer. **Load = 175 A**

Part (c) requires the neutral of a 4-wire wye, 3ϕ system supplying loads with electric discharge lighting, data processing equipment, or similar loads to be the same size as the phase conductors. The neutral conductor collects the *harmonic currents* generated by these loads and is considered a current-carrying conductor. The harmonic currents generated by these types of loads in a 4-wire wye, 3ϕ system are approximately 25% per phase. The neutral collects all harmonic currents from the ungrounded conductors in a 4-wire wye system because they are all in phase. When adding 25% per phase for the ungrounded conductors, the total percentage of harmonic currents in the neutral is 75%. Section 220-22 does not permit the neutral to be reduced 70% for current from 120 V loads exceeding 200 A. Harmonic currents are created by flux variations in ballasts, transformers, motors, data processing equipment, and similar loads where the 120 V neutral loads in the feeder circuit are harmonic.

What is the total percentage of harmonic currents on phase A, B, and C on a 4-wire wye, 3ϕ system?

Step 1. *Note 10(c) to Tables 310-16–310-19:*

Ungrounded conductor A =	25%
Ungrounded conductor B =	25%
Ungrounded conductor C =	25%
	75%

Answer. **Percentage = 75%**

Demand Factors—220-22. A demand factor of 70% is applied to all neutral loads exceeding 200 A except for circuits supplying electric discharge lighting, data processing equipment, and similar equipment. On 120 V neutral loads with no harmonic currents present in the neutral, current exceeding 200 A is reduced 70%.

Ungrounded conductors for phases A, B, and C total 385 A. What is the neutral load exceeding 200 A without the flow of harmonic currents reduced?

Step 1. *Note 10(c) to Tables 310-16–310-19:*
Ungrounded conductors phases A, B, and C = 385 A of neutral loads

Step 2. *220-22:*

First 200 A × 100% =	200 A
Next 185 A × 70% =	129.5 A
	329.5 A

Answer. **Load = 330 A**

Because the neutral collects the harmonic currents generated from the ungrounded conductors in a 4-wire wye system, it is considered a current-carrying conductor. The neutral is connected as a current-carrying conductor where Note 8 to Tables 310-16–310-19 is applied for the total number of conductors in a conduit, cable, etc.

The 120 V loads are 300 A per phase in a feeder circuit supplying equipment with harmonic currents. What is the neutral load?

Step 1. *Note 10(c) to Tables 310-16–310-19:*
Ungrounded conductors = 300 A

Step 2. *220-22:* 300 A × 100% = 300 A

Answer. **Load = 300 A**

Where the number of current-carrying conductors in a raceway is 10 through 24, 25 through 42, or 43 and above, a load diversity of 50% is applied per Note 8 to Tables 310-16–310-19. If twelve #2/0 THWN Cu conductors are pulled in the same raceway between the panelboard and load, the 50% diversity is applied by dividing the ampacity on each conductor. A #2/0 THWN Cu conductor is rated at 175 A per Table 310-16. With twelve conductors in a conduit, Note 8 to Tables 310-16–310-19 requires that each conductor be derated 70% and a load diversity of 50% be applied.

What is the load for each of twelve #2/0 THWN Cu conductors in a raceway supplying a panelboard?

Step 1. *Table 310-16:* #2/0 THWN = 175 A

Step 2. *Note 8 to Tables 310-16–310-19:*
10 through 24 = 70%
175 A × 12 × 70% = 1470 A
1470 A ÷ 12 = 122.5 A

Step 3. *Note 8 to Tables 310-16–310-19:*
Applying 50% load diversity
175 A × 12 × 50% = 1050 A
1050 A ÷ 12 = 87.5 A

Answer. **Load = 87.5 A**

The load diversity of 50% can also be applied by dividing the load on each #2/0 conductor using the following procedure:

Step 1. *Note 8 to Tables 310-16–310-19:*
Applying 50% load diversity

Three #2/0 × 110 A =	330 A
Three #2/0 × 90 A =	270 A
Three #2/0 × 80 A =	240 A
Three #2/0 × 70 A =	210 A
	1050 A

Answer. **Load = 1050 A**

Note that the load on any one of the #2/0 THWN Cu conductors did not exceed 122.5 A, which is 175 A times 70% per Note 8 to Tables 310-16–310-19. The 70% derating factor is based on twelve #2/0 THWN Cu conductors in conduit. The load on each conductor can be any value at or below 122.5 A, as long as the total load on all conductors does not exceed 1050 A. A load diversity of 50% is applied to ensure safe loading of each conductor. See Figure 2-5.

NUMBER OF CONDUCTORS	DERATING FACTORS
4–6	80%
7–9	70%
*10–24	70%
*25–42	60%
*43 and above	50%

*Apply 50% load diversity.

Figure 2-5. Derating factors are based upon the number of conductors in a raceway. Apply 50% load diversity for raceways containing 10 or more conductors.

WIRING METHODS—310

The temperature rating of copper and aluminum conductors is determined by the insulation used. The ampacity of conductors is based on the type of insulation used per Table 310-16. A #3 TW conductor rated at 60°C has an ampacity of 85 A. A #3 THWN Cu conductor rated at 75°C has an ampacity of 100 A, and a 90°C THWN Cu conductor has an ampacity of 110 A. The number of conductors permitted in a conduit is determined by the insulation of the conductors. The higher ampacity ratings are used where current-carrying conductors in conduit and cables exceed three, or they are routed through high ambient temperature areas per the Ampacity Correction Factors to Table 310-16.

60°C Conductors—Table 310-16; Table 310-13

The ampacity of 60°C insulated conductors is found in Table 310-16. Their conditions of use are listed in Table 310-13. A #8 TW Cu conductor has an ampacity of 40 A per Table 310-16. Table 310-13 lists TW as having a moisture-resistant thermoplastic insulation that can be used in dry or wet locations. See 310-8(a) for the types of insulation permitted in wet locations.

What is the ampacity of six #4 TW Cu conductors in conduit? Only three are current-carrying. What size conduit is required to run the conductors underground between the panelboard and equipment?

Ampacity

Step 1. *Table 310-16:* #4 TW = 70 A

Step 2. *Note 8 to Tables 310-16–310-19:*
70 A × 100% = 70 A

Answer. **Ampacity = 70 A**

Conduit

Step 1. *310-8(a); Table 310-13:*
#4 TW is permitted in wet locations

Step 2. *Table 3A, Ch. 9:* Six #4 TW = 1½" conduit

Answer. **Conduit = 1½"**

Conductors rated with 60°C insulation can be terminated to overcurrent protection devices and equipment and used at the ampacities listed in Table 310-16 under the 60°C column. The 60°C ratings in Table 310-16 can be used for 60°C, 75°C, or 90° terminals for ampacities adjusted for such use. See Figure 2-6.

75°C Conductors—Table 310-16; Table 310-13

The ampacity of 75°C conductors is rated at a greater value than 60°C conductors. A #4 THWN Cu conductor has an ampacity of 85 A per Table 310-16 under the 75°C column. THWN insulation can be installed in dry or wet locations per 310-8(a) and Table 310-13.

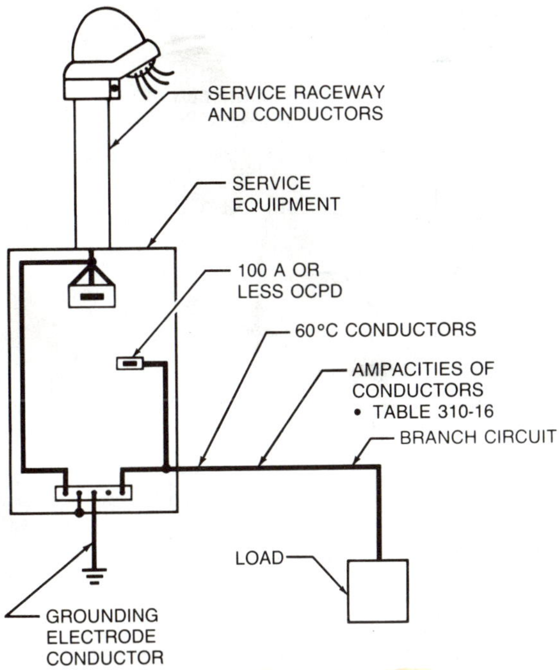

60°C RATINGS		
CONDUCTOR SIZE	AMPACITY RATING	TERMINAL RATING
#8 THWN Cu	40 A	40 A
#3 THWN Cu	85 A	85 A
#1/0 THWN Cu	125 A	125 A

Figure 2-6. Branch circuit conductors and their terminals require the same temperature rating based on the ampacities of Table 310-16.

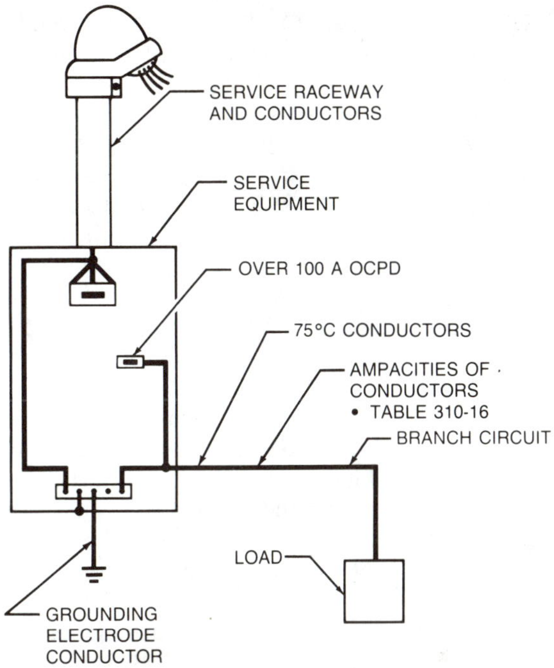

75°C RATINGS		
CONDUCTOR SIZE	AMPACITY RATING	TERMINAL RATING
#8 THWN Cu	50 A	50 A
#3 THWN Cu	100 A	100 A
#1/0 THWN Cu	150 A	150 A

Figure 2-7. Higher ampacity conductors of Table 310-16 are derated and used at lower temperature values when connected to terminals of overcurrent protection devices and equipment.

What are the ampacity and size conduit required for six #4 THWN Cu conductors installed in a dry and wet location? Only three conductors are current-carrying.

Ampacity

Step 1. *Table 310-16:* #4 THWN = 85 A

Step 2. *Note 8 to Tables 310-16–310-19:*
85 A × 100% = 85 A

Answer. **Ampacity = 85 A**

Conduit

Step 1. *310-8(a); Table 310-13:*
#4 THWN is permitted in dry and wet locations

Step 2. *Table 3B, Ch. 9:*
Six #4 THWN = 1¼″ conduit

Answer. **Conduit = 1¼″**

Conductors rated with 75°C insulation are used at 60°C ampacities per Table 310-16 when terminated to overcurrent protection devices and terminals of equipment rated at 100 A or less. Conductors rated at 75°C are used at ampacities of 75°C per Table 310-16 when terminated to overcurrent protection devices and terminals of equipment rated over 100 A. See Figure 2-7.

What is the ampacity of a #4 THWN Cu conductor connected to a 60°C terminial rated at 100 A or less?

Step 1. *Table 310-16:* #4 THWN = 85 A

Step 2. *110-3(b); Table 310-16:*
Terminated at 100 A or less
60°C = 70 A

Answer. **Ampacity = 70 A**

Additional information is available in the *Electrical Construction Materials Directory,* published by Underwriters' Laboratories (UL). See this book for ampacities of conductors terminated to overcurrent protection devices and equipment. The ampacities are selected for 60°C conductors terminating to terminials rated at 100 A or less and 75°C conductors terminating to terminials rated over 100 A.

90°C Conductors—Table 310-16; Table 310-13

The ampacities of 90°C conductors are greater in rating than either the 60°C or 75°C conductors. These greater ampacities can be used where too many current-carrying conductors are routed through an ambient temperature exceeding 30°C. For example, a circuit of few current-carrying conductors supplying a load of 36 A are installed in a conduit. What size TW or THWN Cu conductors are required to supply the 36 A load? A #8 TW Cu conductor has an ampacity of 40 A per Table 310-16, but Note 8 to Tables 310-16–310-19 requires the eight #8 TW Cu conductors to be derated 70%. After derating 40 A by 70%, the remaining ampacity of 28 A (40 × 70% = 28 A) is not sufficient to supply the 36 A load.

A #8 THHN Cu conductor has an ampacity of 55 A. After derating 55 A by 70% per Note 8, the new ampacity is 39 A (55 A × 70% = 38.5 A). The 39 A rating of the #8 THHN Cu conductor is greater than 36 A and can be used at 60°C to supply the load. The eight #8 THHN Cu conductors can be pulled in the same conduit to serve the load and terminated to 40 A overcurrent protection devices.

Three #1 THWN Cu conductors were selected to supply a load of 125 A, but after the installation, the conductors were found to be routed through an area with a constant temperature of 103°F. Because of the temperature, the #1 THWN Cu conductors are derated by 88% per the Ampacity Correction Factors to Table 310-16. Only 114 A are available after derating the 130 A of a #1 THWN Cu conductor (130 A × 88% = 114.4 A).

A #1/0 THWN Cu conductor is required to supply a load of 125 A after derating the 150 A rating per Table 310-16. After applying the derating factor of 88%, the new ampacity is 132 A (150 A × 88% = 132 A). However, a #1 THHN Cu conductor rated at 150 A has an ampacity of 137 A after derating by 91% per the Ampacity Correction Factors to Table 310-16 (150 A × 91% = 136.5 A). The new ampacity of the #1 THHN Cu conductor can be used to supply the 125 A load and terminated to a 125 A or 150 A overcurrent protection device. *Note:* Terminals shall match conductor rating.

Can #1/0 THWN Cu conductors be terminated to a feeder circuit with 90°C, 75°C, or 60°C terminals?

90°C

Step 1. *Table 310-16:* 90°C #1/0 THWN Cu = 150 A

Answer. **Yes**

75°C

Step 1. *Table 310-16:* 75°C #1/0 THWN Cu = 150 A

Answer. **Yes**

60°C

Step 1. *Table 310-16:* 60°C #1/0 THWN Cu = 125 A

Answer. **Yes**

Note that #1/0 THWN Cu conductors can be terminated to a 150 A overcurrent protection device with terminals rated at 90° C and 75° C and can supply a load of 150 A. A load of 125 A terminated to 60° C terminals can be served and protected by a 125 A overcurrent protection device. See Figure 2-8.

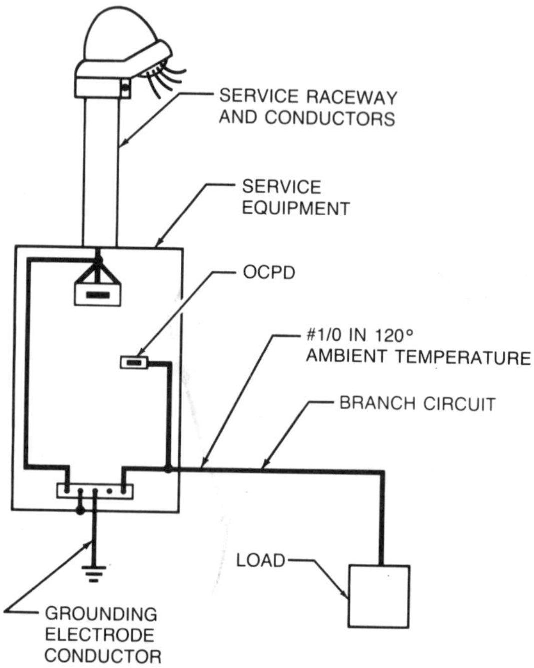

SELECTING AMPACITIES FOR HIGH AMBIENT TEMPERATURE

What is the ampacity of a #1/0 conductor passing through an ambient temperature of 120°F? Use TW, THWN, and THHN Cu.

Step 1. *Table 310-16, Ampacity Correction Factors:*
#1/0 TW Cu = 125 A × 58% = 72.5 A
#1/0 THWN Cu = 150 A × 75% = 112.5 A
#1/0 THHN Cu = 170 A × 82% = 139.4 A

Answer. **Ampacities:**
#1/0 TW Cu = 72.5 A
#1/0 THWN Cu = 112.5 A
#1/0 THHN Cu = 139.4 A

Figure 2-8. Insulation on a conductor determines the ampacity of the conductor after derating due to high ambient temperature surrounding the raceway.

Conduit—Tables 3A, 3B, 3C, 4, and 5, Ch. 9

Conduits are selected by the size of conductors and type of insulation. Table 3A, 3B, or 3C is used for conductors of the same size. The square inch area of conductors based on the size and type of insulation is used to size conduit for combination conductors per Table 5, Ch. 9. Table 4 is used to select the size of the conduit based on the total square inch area of the pulled conductors. For example, there are nine #10 TW Cu conductors routed through a ¾″ conduit per Table 3A. There are eleven #10 THWN Cu conductors per Table 3B. The overall outside diameter of a #10 THWN Cu conductor is smaller than a #10 TW, therefore allowing more conductors to be pulled in the ¾″ conduit.

What size conduit is required for six #1/0 THWN Cu conductors?

Step 1. *Table 3B, Ch. 9:*
Six #1/0 THWN Cu conductors = 2″ conduit

Answer. **Conduit = 2″**

Table 5 and Table 4, Ch. 9 are used to size conduit for conductors of different sizes or with other types of insulation. For example, the square inch area of four #6 THWN conductors and five #4 RHW conductors are selected from Table 5. Table 4 is then used to size the conduit.

The area of a #6 THWN Cu conductor is .0519 square inches per Table 5. The area of a #4 RHW Cu conductor is .1605 square inches. The square inch area of each conductor is multiplied by the number of conductors and added to obtain the total area in square inches. The size of the conduit is based on the total square inch area and selected from the over two conductor 40% fill column listed in Table 4.

What size conduit is required for four #6 THWN and five #4 RHW conductors?

Step 1. *Table 5, Ch. 9.*
#6 THWN = .0519 sq. in. × 4 = .2076 sq. in.
#4 RHW = .1605 sq. in. × 5 = .8025 sq. in.
1.0101 sq. in.

Step 2. *Table 4, Ch. 9:*
1.0101 sq. in. = 2″ conduit

Answer. **Conduit = 2″**

Care must be taken where conductors are the same size but the insulation is of different types. The square inch area of each conductor, based on the type of insulation, is calculated per Table 5, Ch. 9. The total square inch area of the conductors is compared to the square inch area of conduit for combination conductors listed in Table 4, Ch. 9.

What size conduit is required for eight #4 conductors? Two conductors each are THW, THWN, RHW, and XHHW.

Step 1. *Table 5, Ch. 9:*

#4 THW = .1087 × 2 =	.2174 sq. in.
#4 THWN = .0845 × 2 =	0.169 sq. in.
#4 RHW = .1605 × 2 =	0.321 sq. in.
#4 XHHW = .0845 × 2 =	0.169 sq. in.
Total =	.8764 sq. in.

Step 2. *Table 4, Ch. 9:*
.8764 sq. in. = 2″ conduit

Answer. **Conduit = 2″**

If eight #4 THWN Cu conductors with the same insulation are used, the size conduit is 1½″ per Table 3B, Ch. 9.

Higher temperature rated conductors, such as THHN 90°C insulation, can be used to replace RHW conductors routed in an existing conduit installed in an old building. In an existing building, five #12 RHW Cu conductors are pulled through a ¾″ conduit. Ten #12 conductors are needed to supply additional loads. Eighteen #12 THHN Cu conductors can be repulled through the existing ¾″ conduit. The number of #12 THHN conductors that can be pulled safely through the existing conduit depends upon the number of bends and how far the conduit is run from the service panel to the equipment. See Figure 2-9.

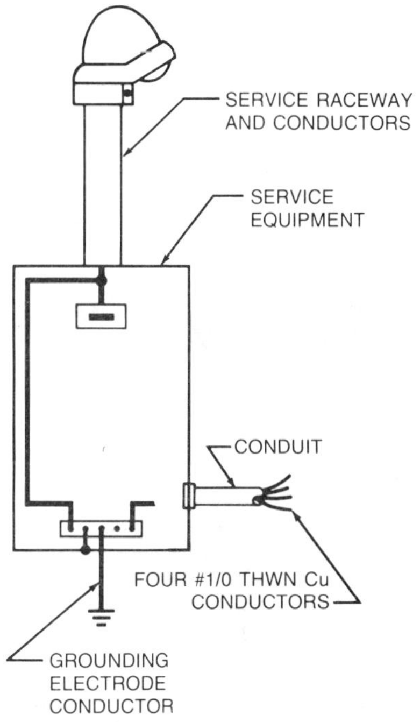

SIZING CONDUITS

What size conduit is required for four #1/0 THWN Cu conductors?

Step 1. *Table 3B, Ch. 9:*
Four #1/0 THWN Cu = 1½″ conduit

Answer. **Conduit = 1½″**

Figure 2-9. Tables 3A, 3B, and 3C of Ch. 9 are used to size conduit with same size conductors. Table 4 is used to size conductors of different sizes, and Table 5 is used to select cross-sectional areas of conductors.

OVERCURRENT PROTECTION—240, PART A

Conductors used for feeder circuits shall be protected from short circuits, ground faults, and overloads. Overcurrent protection devices are sized to protect feeder circuit conductors at their ampacity ratings unless the circuit complies with one of the exceptions to 240-3. Overcurrent protection devices are normally connected in the feeder circuit at the point where conductors receive their source of power. Circuit breakers and fuses are used to provide overcurrent protection for feeder circuit conductors and components.

Circuit Breakers—240, Part G

Circuit breakers used to protect feeder circuit conductors are sized for the calculated load. The overcurrent protection device is selected on the rating of the computed load or the ampacity of the feeder circuit conductors. The computed load, in VA or A, is determined by calculating the loads at noncontinuous duty (100%) or continuous duty (125%) and applying demand factors where applicable. See Figure 2-10.

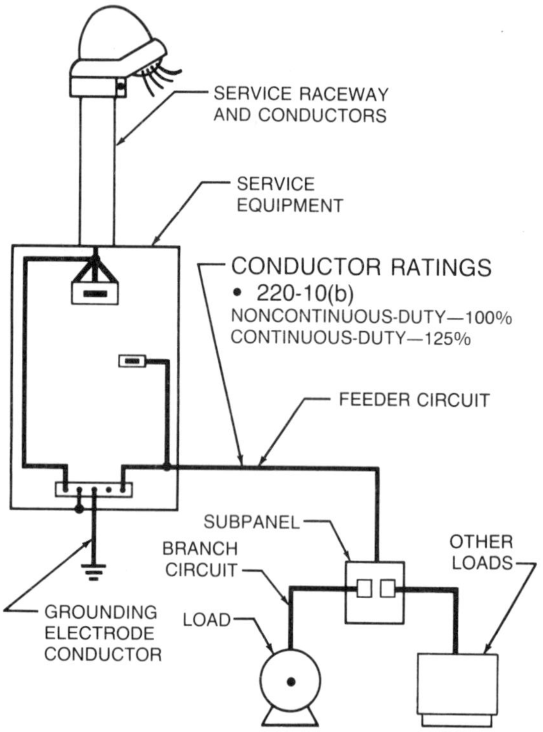

Figure 2-10. The ampacity ratings of feeder circuit conductors are determined by multiplying the VA rating of the load by 100% for noncontinuous-duty and 125% for continuous-duty operation. Demand factors can be applied where applicable.

Conductor size is selected from Table 310-16 based on the calculated load of the feeder. For example, a feeder circuit with a total connected load of 123 A requires #1 THWN Cu conductors to supply the feeder circuit load. The overcurrent protection device can be sized from the computed load of 123 A, which requires a 125 A circuit breaker per 240-6. The overcurrent protection device can also be selected from the ampacity of the conductors supplying the feeder circuit. A 150 A circuit breaker is required per 240-3, Ex. 4 based on 130 A #1 THWN Cu conductors per Table 310-16.

Circuit breakers hold approximately three times their rating for different periods of time based on the frame size of the unit. For example, an 80 A circuit breaker holds 240 A (80 A × 3 = 240 A) for about four to nine seconds, depending on the voltage. This time allows motors, air conditioners, and similar equipment with current values below 240 A to be started and accelerated to running speed without tripping the circuit breaker.

Fuses—240, Parts E and F

Fuses are selected and sized to protect feeder circuit conductors at their ampacity ratings. Conductor sizes are determined by the total calculated load, which are computed at noncontinuous duty, continuous duty, and by applying demand factors where applicable. Fuses shall not be selected above the ampacity of the feeder circuit conductors without applying one of the exceptions to 240-3. A 300 A overcurrent protection device will protect #300 kcmil THWN Cu conductors serving a feeder circuit with an ampacity of 285 A per conductor from Table 310-16 per 240-3, Ex. 4.

Fuses hold five times their rating for different periods of time based on the type of fuse used. A nontime delay fuse holds five times its rating for one-fourth to two seconds, depending on the type. A nontime delay fuse is not designed with time delay characteristics and will not allow a motor or similar load to start and run if acceleration time requires more than two seconds. A time delay fuse is equipped with time delay characteristics and will hold five times its rating for 10 seconds. For example, a motor or similar load has an inrush current of 415 A. When the power is connected to the unit, a 90 A nontime delay or time delay fuse holds 450 A (90 A × 5 = 450 A) and normally will start and run the load. For nontime delay fuses, the load is accelerated in one-fourth to two seconds. The load can take up to 10 seconds to accelerate when time delay fuses are used. See Figure 2-11.

■ *Master Electrician's Test Tip.* Overcurrent protection devices shall not exceed the ampacity of the conductors without applying one of the exceptions to 240-3.■

Eight exceptions in 240-3 permit overcurrent protection devices to be sized above the ampacity of the conductors they protect. The eight exceptions are

Ex. 1 .... Control conductors
Ex. 2.......................... Transformers
Ex. 3 .... Power loss hazard
Ex. 4 .... Next higher size
Ex. 5 .... Taps
Ex. 6............ Motors and air conditioners
Ex. 7 .... Capacitors
Ex. 8 .... Welders

Note that 240-3, Ex. 4 permits the use of listed overcurrent protection devices in 240-6. For example, a parallel feeder circuit of five #350 kcmil THWN Cu

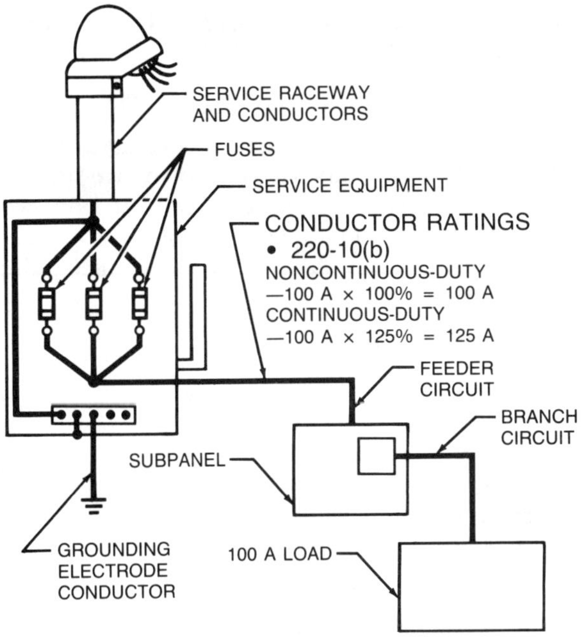

FUSES
NTDF holds 500% rating for 1/4–2 seconds
TDF holds 500% rating for 10 seconds

Figure 2-11. The size fuses required to protect feeder circuit conductors is determined by multiplying the load in VA by 100% for noncontinuous-duty and 125% for continuous-duty operation. Fuses hold five times their rating for different periods of time based on the type used.

conductors per phase has a total ampacity of 1550 A (310 A × 5 = 1550 A). No standard device from 1200 A to 1600 A is listed in 240-6; however, there are listed devices by UL that can be used. Section 240-6 now allows a listed 1500 A device per 110-3(b) to be used even though not listed in 240-6. By using a 1500 A device, 50 A is lost instead of 350 A (1550 A − 1200 A = 350 A). Without this change in 240-3, Ex. 4, most inspectors require a 1200 A device because there is no 1500 A standard device listed in 240-6.

A circuit breaker with adjustable or changeable trip ratings shall have load-circuit conductors of an ampacity at least equivalent to the highest trip rating for which the CB might be used. This requirement is in 240-6 and indicates that the CB rating shall not exceed the ampacity of the circuit conductors without complying with one of the exceptions of 240-3.

VOLTAGE DROP—215

Where feeder circuit conductors are routed a great distance from the service equipment to the load, voltage drop is considered to ensure that the proper voltage of the circuit is maintained. The greater distance conductors are run between overcurrent protection devices and equipment, the greater the resistance of each conductor will be. The amount of voltage drop is calculated to determine if larger conductors are required to compensate for the long run of conductors in the circuit.

Single-phase Circuits—215-2(b), FPN 2

Voltage drop in a 2-wire or 3-wire, 1ϕ feeder circuit is determined by applying the following formula and substituting the proper values.

$$VD = \frac{2 \times L \times R \times I}{1000}$$

where

L = length in feet
R = resistivity per Table 8, Ch. 9
I = current
1000 = 1000′ or less of wire

As longer conductors are routed, the resistance of each conductor increases as a result of opposition to the flow of current.

By using larger conductors, the diameter of each conductor is increased, providing more paths for the flow of current and less opposition to the movement of electrons. The resistance of a conductor is inverse to its diameter. Voltage drop for branch circuit or feeder circuit conductors is limited to 3% of supply voltage per 210-19(a), FPN 4 and 215-2(b), FPN 2. Overall voltage drop is limited to 5% of supply voltage by the same references. See Figure 2-12.

What is the voltage drop in a 120/240 V, 1ϕ, 3-wire feeder circuit with a load of 58 A and #6 THWN Cu conductors routed a distance of 125′?

Step 1. *215-2(b), FPN 2; Table 8, Ch. 9:*

$$VD = \frac{2 \times L \times R \times I}{1000}$$

$$VD = \frac{2 \times 125' \times .491 \times 58\text{ A}}{1000} = 7.1195\text{ V}$$

Step 2. *215-2(b), FPN 2:* 240 V × 3% = 7.2 V

Answer. **VD = 7.2 V**

Step 3. Checking percentage

$$\% = \frac{VD}{V} \times 100$$

$$\% = \frac{7.1195\text{ V}}{240\text{ V}} \times 100 = 2.97$$

Answer. **2.97 falls below 3% and #6 THWN Cu can be used**

Another method of determining voltage drop is to use the following formula:

$$VD = \frac{2 \times L \times K \times I}{CM}$$

where

L = length in feet
K = resistivity constant
for copper use 12
for aluminum use 18
I = current
CM = circular mils of wire

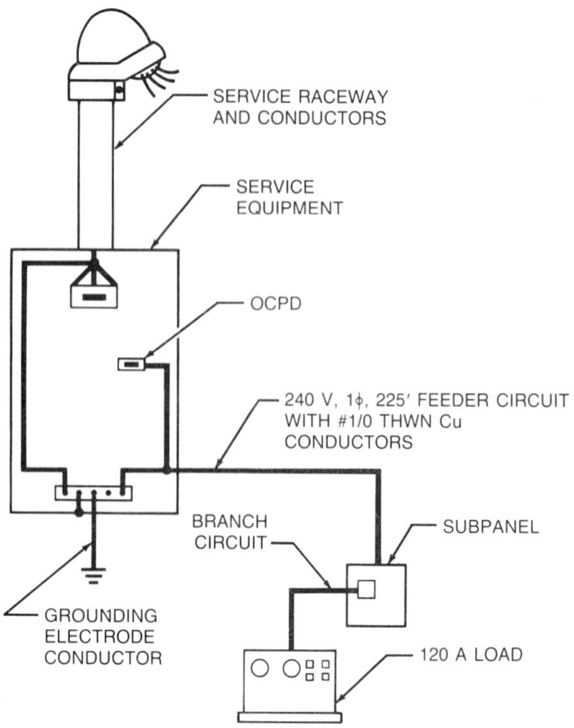

VD FOR 1ϕ CONDUCTORS

What is the voltage drop of a 240 V, 1ϕ, 225′ feeder circuit routed to a subpanel?

Step 1. *215-2(b), FPN 2:* 240 V × 3% = 7.2 V

Step 2. *215-2(b), FPN 2; Table 8, Ch. 9:*

$$VD = \frac{2 \times L \times R \times I}{1000}$$

$$VD = \frac{2 \times 225' \times .122\ \Omega \times 120\ A}{1000} = 6.588\ V$$

Answer. **VD = 6.588 V**

Figure 2-12. Voltage drop is limited to 3% of supply voltage for 1ϕ feeder circuit conductors.

What is the voltage drop for three #1/0 THWN Cu conductors routed 175′ to supply a 130 A load in a 120/240 V feeder circuit?

Step 1. $VD = \frac{2 \times L \times K \times I}{CM}$

$$VD = \frac{2 \times 175' \times 12 \times 130\ A}{105{,}600} = 5.17\ V$$

Step 2. *215-2(b), FPN 2:* 240 V × 3% = 7.2 V

Answer. **VD = 5.17 V**

Step 3: Checking percentage

$$\% = \frac{VD}{V} \times 100$$

$$\% = \frac{5.17\ V}{240\ V} \times 100 = 2.15$$

Answer. **2.15 falls below 3% and #1/0 THWN Cu can be used**

Three-phase Circuits—215-2(b), FPN 2

The same formula is used for finding voltage drop in 3ϕ circuits as 1ϕ circuits, except the voltage drop is multiplied by .866. Voltage drop between any two phase conductors is multiplied by .866, which is derived by dividing $\sqrt{3}$ by 2 (1.732/2 = .866). See Figure 2-13.

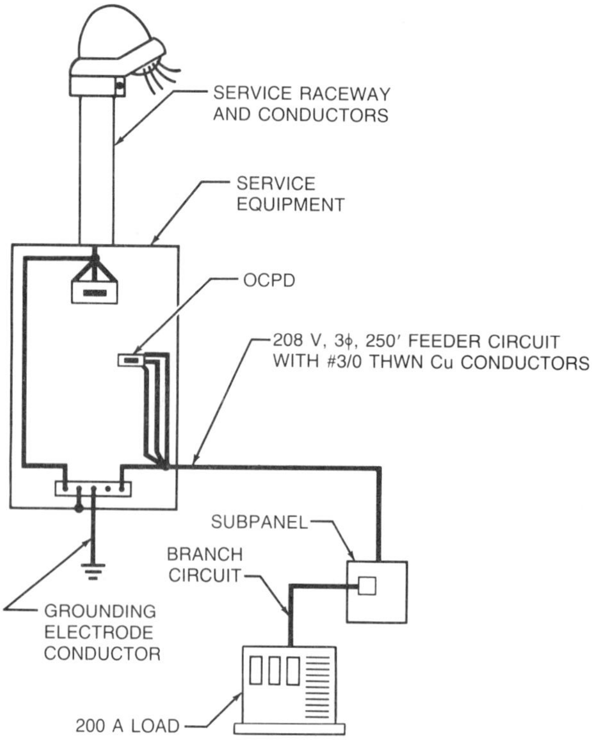

VD FOR 3ϕ CONDUCTORS

What is the voltage drop of a 208 V, 3ϕ, 200′ feeder circuit routed to a subpanel?

Step 1. *215-2(b), FPN 2:* 208 V × 3% = 6.24 V

Step 2. *215-2(b), FPN 2; Table 8, Ch. 9:*

$$VD = \frac{2 \times L \times R \times I}{1000}$$

$$VD = \frac{2 \times 200' \times .0766\ \Omega \times 200\ A}{1000} = 6.13\ V$$

Step 3. *215-2(b), FPN 2:* $\frac{\sqrt{3}}{2} = .866$

Step 4. 6.13 V × .866 = 5.3 V

Answer. **VD = 5.3 V**

Figure 2-13. Voltage drop is limited to 3% × .866 of supply voltage for 3ϕ feeder circuit conductors.

What are the voltage drop and size THWN Cu conductors required to supply a 168 A load in a 3ϕ, 120/208 V feeder circuit that is routed 185′?

Step 1. $VD = \frac{2 \times L \times R \times I}{1000}$

$$VD = \frac{2 \times 185' \times .0967 \times 168 \text{ A}}{1000} = 6.010872$$

$$VD = 6.010872 \times .866 = 5.2054151 \text{ V}$$

Step 2. *215-2(b), FPN 2:* 208 V × 3% = 6.24

Answer. **VD = 5.2054151 V**

Step 3. Checking percentage

$$\% = \frac{VD}{V} \times 100$$

$$\% = \frac{5.2054151 \text{ V}}{208 \text{ V}} \times 100 = 2.5026$$

Answer. **2.5026 falls below 3% and #2/0 THWN Cu can be used**

Note: The voltage drop for feeder circuits shall not exceed 3% of circuit voltage. Branch circuit voltage drop shall not exceed 3% of the system voltage. The voltage drop for the feeder and branch circuits shall not exceed 5% to the farthest outlet.

TAPS—240-21

The general rule of 240-21 requires overcurrent protection devices to be installed at the point at which the conductor receives its supply. Where taps from feeder circuit conductors are made, exceptions allow smaller conductors to be tapped from the larger conductors without installing overcurrent protection devices. Each exception has specific requirements that are applied.

10′ Tap Rule

The 10′ tap rule is applied for tapped conductors extending no more than 10′ from the point of the tap to the terminals of the equipment or overcurrent protection device. The 10′ tap rule is the most commonly used tap since it can be any size wire. A #12 THWN Cu wire can be tapped from a 600 A feeder circuit as long as it terminates to a 20 A overcurrent protection device. See Figure 2-14. The following requirements are applied when using the 10′ tap rule:

1. Conductors shall be 10′ or less in length.
2. Conductors shall be installed in conduit.
3. Conductors shall be equivalent to terminals or devices to which they terminate.
4. Conductors shall have sufficient ampacity to supply the load.
5. OCPDs on the line side of the tap conductors shall not exceed 1000% of the tap conductor's ampacity.

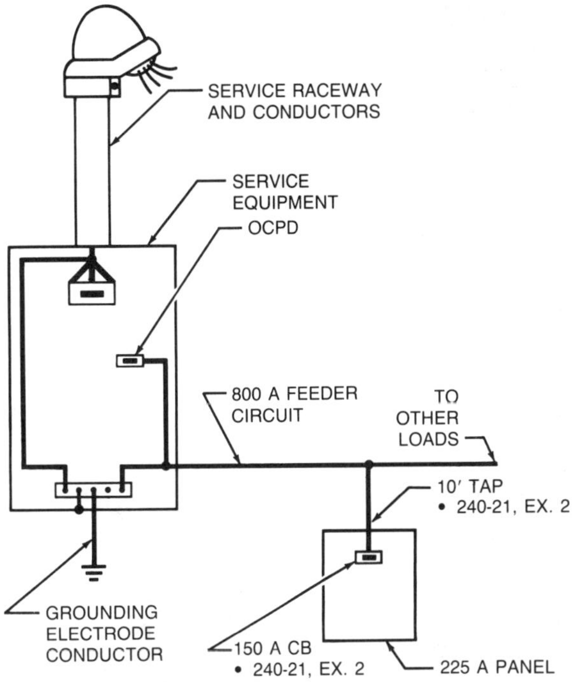

10′ TAP	
Tap terminates to 150 A CB	#1/0 THWN Cu
Tap terminates to lugs of panel	#4/0 THWN Cu

Figure 2-14. A 10′ tap can be made with any size conductor.

A 10′ tap using THWN Cu conductors is made from an 800 A feeder circuit to supply power to the lugs of a 225 A, 3ϕ panelboard. What size conductors is required for the tap?

Step 1. Lugs of panelboards = 225 A

Step 2. *240-21, Ex. 2, (1)(2a); Table 310-16:* Tap conductors shall be 225 A

Answer. **Conductors = #4/0 THWN Cu**

■ *Master Electrician's Test Tip.* Tap conductors shall be 10′ or less in length per 240-21, Ex. 2. Note that the length of the conductors is measured, not the length of the conduit. ■

25′ Tap Rule

The 25′ tap rule is applied where a length of conductors greater than 10′ is needed to reach the electrical equipment. The disadvantage of the 25′ tap rule is that the tapped conductors shall be at least ⅓ the ampacity of the feeder circuit conductors. This size tap requires larger conductors, conduits, and components to complete the electrical circuit. The tap conductors are terminated in a main circuit breaker or fusible disconnect switch. The 25′ tap rule is used where wall space prevents a fusible switch or panelboard to be mounted using the 10′ tap rule. See Figure 2-15.

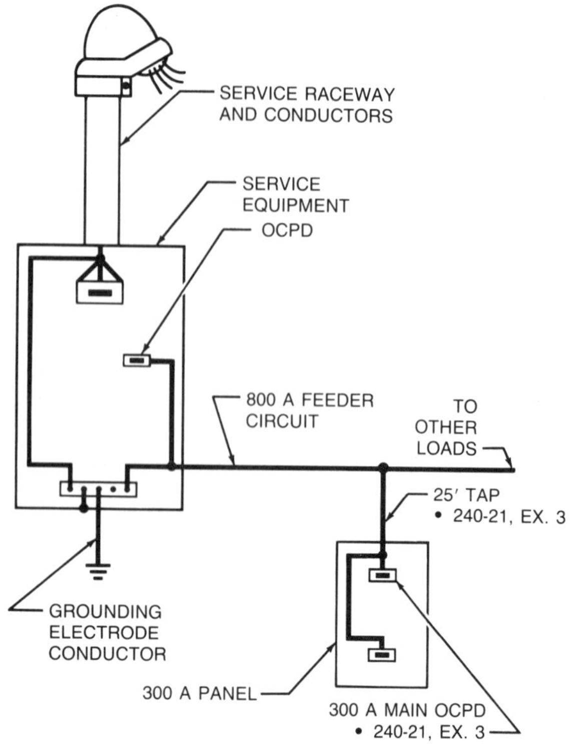

25′ TAP	
Tap ampacity 800 A × 1/3 = 266.7 A	#300 kcmil THWN Cu
Tap terminates to main OCPD	300 A main OCPD

Figure 2-15. A 25′ tap conductor shall be at least 1/3 the ampacity of feeder circuit conductors and terminate in a main overcurrent protection device.

What size THWN Cu conductors is required to make a tap from a 1200 A feeder circuit applying the 25′ tap rule?

Step 1. *240-21, Ex. 3b:* ⅓ × 1200 A = 400 A

Step 2. *240-21, Ex. 3; Table 310-16:*
#600 kcmil THWN Cu = 420 A

Answer. **Conductors** = **#600 kcmil THWN Cu**

■ *Master Electrician's Test Tip.* The ⅓ rating of the tap conductors can be selected from the ampacity of the feeder circuit conductors or the overcurrent protection devices.■

Where the ampacity and size of the feeder circuit conductors are increased to prevent voltage drop and the overcurrent protection device is sized from the calculated load, the ⅓ rating can be selected from the device or increased conductors. The method requiring the smaller conductors can be used to size the tap.

What size THWN Cu conductors and CB is required for a 25′ tap from a feeder circuit with #500 kcmil THWN Cu conductors protected with a 300 A overcurrent protection device? The conductors are increased to limit the voltage drop.

Step 1. *240-21, Ex. 3b:*
⅓ × 300 A = 100 A

Step 2. *240-21, Ex. 3c; Table 310-16:*
#3 THWN Cu = 100 A

Step 3. *240-21, Ex. 3c; 240-6:*
100 A = standard device

Answer. **Conductors** = **#3 THWN Cu**
CB = **100 A**

25′ Tap with Transformer—240-21, Ex. 8

A 25′ tap can be made without installing an overcurrent protection device at the point of the tap. The tapped conductors of the primary plus the transformer and conductors of the secondary shall not exceed 25′. The tapped primary conductors shall have an ampacity rating at least ⅓ the feeder conductors of the overcurrent protection device protecting the circuit. The tapped conductors from the secondary of the transformer are equal to ⅓ the feeder circuit conductors or overcurrent protection devices when multiplied by the ratio of the secondary to primary voltage. The tapped secondary conductor terminates in a single circuit breaker or set of fuses. See Figure 2-16.

A feeder circuit is supplied with #350 kcmil THWN Cu conductors and is protected with 350 A time delay fuses. The transformer is rated at 50 kVA with a primary of 480 V, 1ϕ and a secondary of 240 V, 1ϕ. What size THWN Cu conductors and circuit breaker is required for the primary and secondary, applying the 25′ tap rule of Ex. 8 to 240-21?

Primary

Step 1. *240-21, Ex. 8a:* ⅓ × 310 A = 103 A

Answer. **Conductors** = **#2 THWN Cu**

Secondary

Step 1. *240-21, Ex. 8b:*
480 V ÷ 240 V × ⅓ × 310 A = 207 A

Step 2. *240-21, Ex. 8b; Table 310-16:*
#4/0 THWN Cu = 230 A

Answer. **Conductors** = **#4/0 THWN Cu**

OCPD

Step 3. *240-21, Ex. 8e; 240-3, Ex. 4; 240-6:*
250 A is the next higher standard size

Answer. **OCPD** = **250 A CB**

100′ Tap Rule—240-21, Ex. 10

Taps from feeder circuits located in high bay areas of manufacturing buildings with walls over 35′ high can be made. The length of the tap is over 25′ but not over 100′. The tap conductors shall not be routed over 25′ horizontally and 75′ vertically. The tap shall be made at least 30′ from the floor, and conductors routed in a raceway that cannot penetrate walls, floors, or ceilings. The tapped conductors are not permitted to be spliced and shall be at least #6 copper or #4 aluminum. The ampacity of the conductors is required to be at

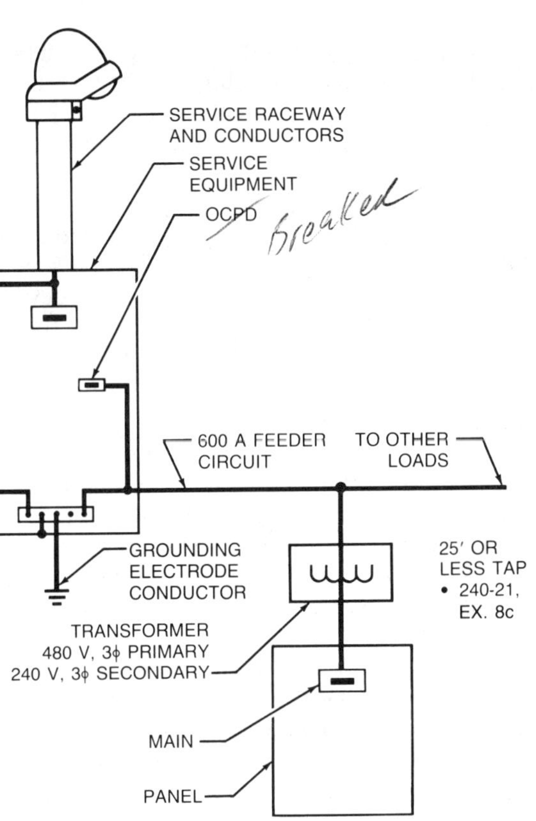

25′ TAP WITH TRANSFORMER	
PRIMARY CONDUCTORS	
Tap ampacity 600 A × $^1/_3$ = 200 A	#3/0 THWN Cu
SECONDARY CONDUCTORS	
Tap ampacity $\frac{480\text{ V}}{240\text{ V}}$ × 600 A × $^1/_3$ = 400 A	#600 kcmil THWN Cu

Figure 2-16. A 25′ tap with transformer requires $^1/_3$ ampacity based upon feeder conductors or overcurrent protection device for primary conductors. Secondary conductors are based on $^1/_3$ of the ratio between primary and secondary voltage and current and shall terminate in a main overcurrent protection device.

least ⅓ the overcurrent protection device protecting the feeder circuit. The tap conductors shall be terminated in a single circuit breaker or set of fuses.

What size THWN Cu conductors is required to tap #400 kcmil THWN Cu conductors supplying a feeder circuit protected with a 350 A overcurrent protection device?

Step 1. *240-21, Ex. 10a:* ⅓ × 350 A = 116.7 A

Step 2. *240-21, Ex. 10a; Table 310-16:*
#1 THWN Cu = 130 A

Answer. **Conductors = #1 THWN Cu**

Can the tapped conductors be routed 15′ horizontally and 50′ vertically and run 10′ horizontally to a single circuit breaker?

Step 1. *240-21, Ex. 10b:*
Horizontally = 25′
Vertically = 50′
Total = 75′

Answer. **Yes**

25′ Tap Rule from Secondary of Transformer—240-21, Ex. 11

Taps can be taken from the secondary side of a transformer whose primary side is supplied from an overcurrent protection device located in the service equipment. The tapped conductors are over 10′ but no more than 25′ in length. They shall be equivalent to the transformer's secondary output in full-load amperage. The tapped conductors from the secondary side shall terminate in a single circuit breaker, set of fuses, or a group of overcurrent protection devices that do not exceed the secondaries' full-load current rating. The 25′ tap rule for conductors tapped from the secondary side of transformers classified as separately derived systems is limited to industrial locations. See Figure 2-17.

Can six 100 A circuit breakers protect two #400 kcmil THWN Cu conductors parallel per phase that are tapped 25 ′ from a 120/208 V, 3ϕ, 4-wire secondary? The transformer is rated at 225 kVA.

Step 1. $I = \frac{kVA \times 1000}{V}$

$$I = \frac{225\text{ kVA} \times 1000}{208\text{ V} \times 1.732} = 625\text{ A}$$

Step 2. *240-21, Ex. 11b; Table 310-16:*
#400 kcmil THWN Cu = 335 A
335 A × 2 per phase = 670 A

Step 3. *240-21, Ex. 11b, 11c:*
100 A CB × 6 = 600 A

Answer. **Yes**

PANELBOARDS—384, PART B

Panelboards can be used as power panels or lighting and appliance branch circuit panels. Panelboards shall comply with the requirements governing installation, protection, and location of wiring methods and components. The number of mains is limited for lighting and appliance branch circuit panels. The number of mains for power panels is only limited at the service.

Lighting and Appliance—384-14; 384-16

A panelboard is considered a lighting and appliance panelboard by multiplying the total number of overcurrent protection devices rated at 30 A or less, that have neutral connections, by 10%. If the number of 30 A or less overcurrent protection devices with neutral connections exceeds 10%, the panel is classified as a lighting and appliance branch circuit panelboard.

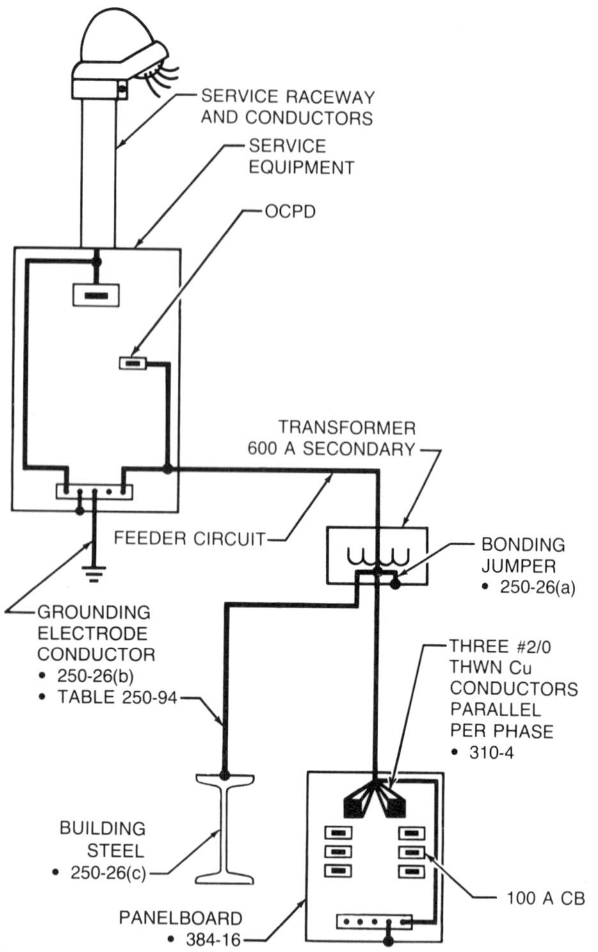

NUMBER OF CB'S

How many 100 A CBs are permitted?

Step 1. *240-21, Ex. 11:* $\frac{600 \text{ A output}}{100 \text{ A}} = 6$ CBs

Answer. **CBs = 6**

Figure 2-17. The secondary conductors are equal to the secondary output of the transformer. The overcurrent protection device cannot exceed the secondary output.

These panelboards are protected by an individual overcurrent protection device on the supply side. For example, a panelboard with a 225 A main that protects 225 A busbars in a panelboard prevents the busbars from overloads.

Lighting and appliance branch circuit panelboards can have two mains protecting the panel as long as their combined rating does not exceed the rating of the panelboard buses. The main overcurrent protection device protecting a lighting and appliance panelboard can be installed in the panelboard or be located remote and ahead of the circuit. A remotely located main shall not exceed the rating of the busbars in the lighting and appliance panelboard. The feeder circuit conductors shall be equivalent to the ampacity of the busbars in the lighting and appliance panelboard. For example, a 200 A main overcurrent protection device is located outside under the meter base and protects three #3/0 THWN Cu conductors routed to a 200 A lighting and appliance panelboard. The 200 A outside main can serve as the overcurrent protection device for the feeder circuit conductors and busbars in the lighting and appliance panelboard because it has an equivalent rating. See Figure 2-18.

Is a 42 circuit panelboard that has two 20 A, two 15 A, and two 30 A overcurrent protection devices with neutral connections classified as a lighting and appliance panelboard?

Step 1. *384-14:* 42 devices × 10% = 4.2

Step 2. *384-15:*
Two 15 A devices
Two 20 A devices
Two 30 A devices
Six devices with neutral connections

Step 3. *384-14; 384-15; 384-16:*
Six devices with neutral loads exceed 4.2

Answer. **Yes**

Power—384-14; 384-15

Power panels supply circuits to electrical equipment and subpanels. Ten percent of their overcurrent protection devices are rated at 30 A or less with neutral connections. They do not supply power to lighting fixtures or appliances with circuits rated 30 A or less with neutral connections.

How many circuits with neutral connections rated at 15, 20, 25, or 30 A can be installed in a power panelboard? The panelboard has a capacity of 30 single-pole circuit breakers.

Step 1. *384-14; 384-15:* 30 × 10% = 3

Answer. **Circuits = 3**

Power panelboards can be used as distribution centers to supply all types of loads. For example, a power panelboard can be installed and have six mains supplying power to other loads in a building. One 150 A overcurrent protection device (main) can be used to supply power to a heating unit and an 80 A device can be installed to supply power to an air conditioner. A 60 A device can be utilized for a circuit to cooking equipment, and a 175 A device can be installed to protect a feeder circuit to a subpanel. Two 100 A overcurrent protection devices can be used to supply circuits to other types of utilization equipment. This panelboard is classified as a power panelboard because there are no overcurrent protection devices rated 30 A or less with neutral connections.

■ *Master Electrician's Test Tip.* Power panels are not required to have a main installed where the 10′ tap rule is applied per 240-21, Ex. 2. Lighting and appli-

LIGHTING AND APPLIANCE PANELBOARD
How many circuits rated at 30 A or less are permitted in a lighting and appliance panelboard?
Step 1. *384-14; 384-15:* 30 CBs × 10% = 3
Answer. **Circuits rated at 30 A or less = 3**

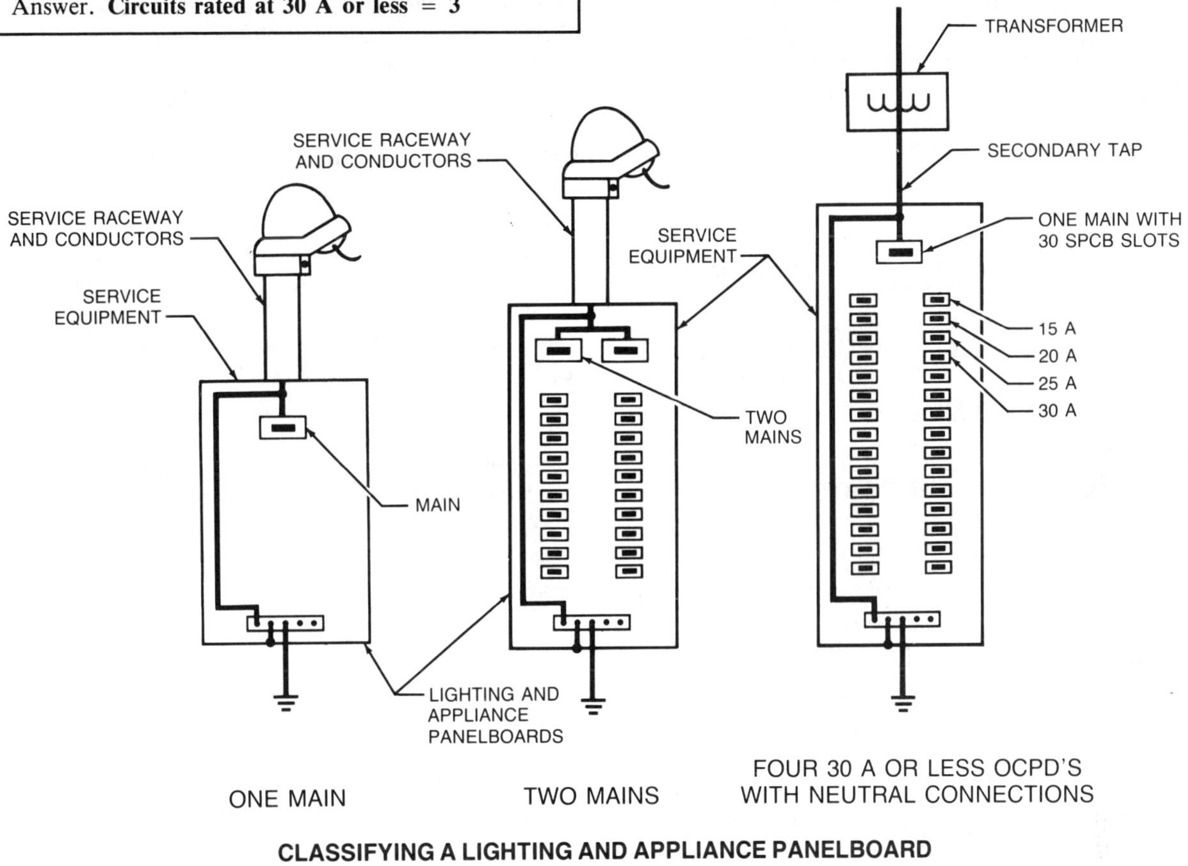

Figure 2-18. Lighting and appliance panelboards shall not have more than two main overcurrent protection devices installed on the supply side.

ance panelboards are required to have a main installed for a 10′ tap. Power panelboards are not limited to 42 circuits per 384-15 when not used as service equipment.■

SUBPANELS—384, PARTS A AND B

Subpanels are usually located downstream from the service equipment. It is too expensive to route branch circuits downstream to the location where lighting fixtures, appliances, and utilization equipment are installed. Feeder circuits are run from the service equipment to subpanels with overcurrent protection devices installed to protect the feeder conductors and busbars in the subpanels to various electrical loads.

Phases—384-3

Hot, ungrounded phase conductors can be any color but white, gray, green, or green with a yellow tracer per 310-12(c). The ungrounded conductors can be connected in any arrangement to the terminals in the subpanel where the supply voltage for the feeder circuit is rated for a 3-wire, 1ϕ or 3- or 4-wire wye system. Any ungrounded conductor can be connected to phase A, B, or C in the subpanel.

For a 3- or 4-wire delta system supply voltage, the high leg shall connect to the B phase and terminate in the subpanel. The high leg is identified by an orange color or other effective means per 215-8 and 384-3(e). The busbar phase arrangement is A, B, C, from front to back, left to right, or top to bottom. This arrangement and location of the high leg is required so the high leg is phase B and in the middle when viewed from the back or front. This safety precaution prevents electricians from accidentally connecting a 120 V piece of equipment to a high leg that is 208 V to ground. The ungrounded phase conductors are sized for the calculated load per 220-10(b) and selected from Table 310-16. A good rule of thumb is that the high leg terminates to phase C on the utility side and to phase B on the customer's side. See Figure 2-19.

What size THWN Cu conductors is required to supply a feeder circuit to a subpanel with a total connected load of 200 A for the hot ungrounded phase conductors? The load on the feeder circuit has been computed at 100% for noncontinuous duty and demand factors applied where permitted. Continuous-duty loads were calculated at 125%. All loads totaled 200 A.

Step 1. *210-10(b):* Calculated load = 200 A

Step 2. *Table 310-16:* #3/0 THWN Cu = 200 A

Answer. **Conductors = #3/0 THWN Cu**

Neutral—250-61(b)

The grounded neutral conductor can be white or gray per 200-7 and 310-12(a) when routed from the service equipment to the subpanel. The neutral conductor in sizes #4 and larger can be marked with white or gray tape or sprayed white or gray.

Grounded neutral conductors, #4 and larger per 200-6, in junction boxes or enclosures shall be white or gray per 200-7 and 310-12(a). *Note:* Conductors #6 and smaller are gray or white in their entire length when used as a grounded neutral conductor. The grounded neutral conductor is terminated to an isolated busbar in the subpanel enclosure per 250-61(b) and 384-20.

This type of connection is provided to complete a path for the flow of current from the subpanel to the common grounding busbar in the service equipment per 250-50(a), 250-53, and 250-54. The neutral is sized for the 120 V loads per 220-22 and used per Note 10 to Tables 310-16–310-19. The neutral conductor shall be capable of carrying the maximum unbalanced current from the phase conductors. See Figure 2-20.

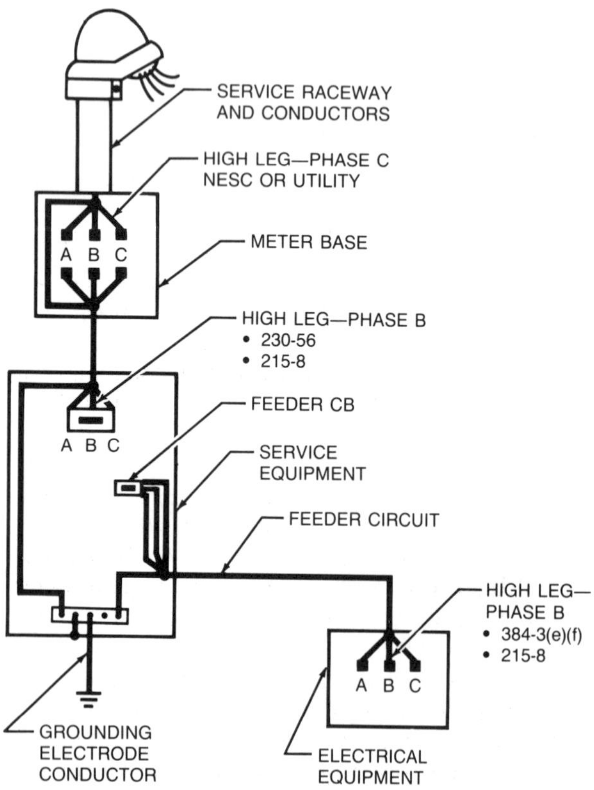

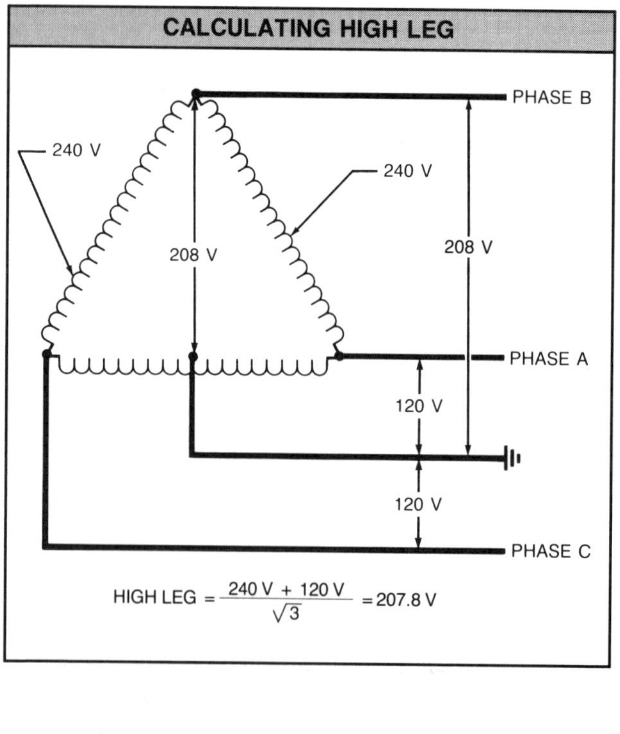

Figure 2-19. The high leg must be terminated to phase C for the metering of the utility company and to phase B on the load side of the service equipment.

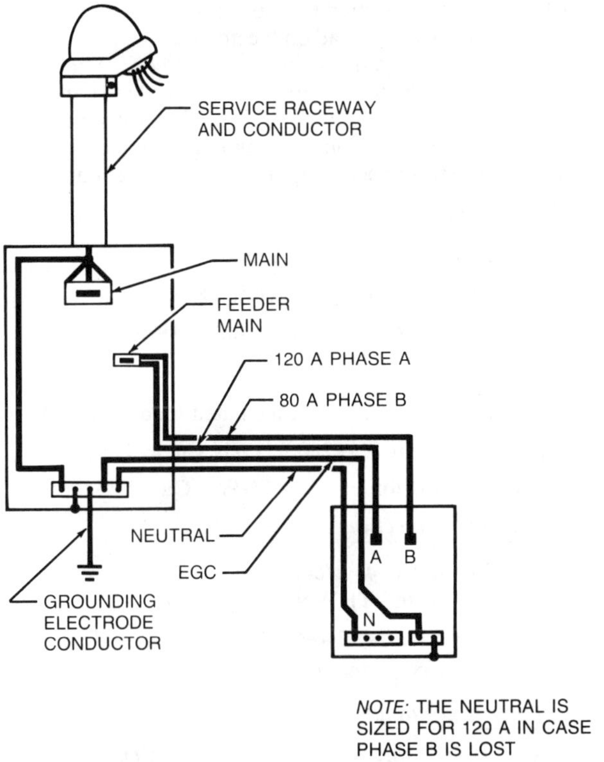

CALCULATING UNBALANCED NEUTRAL CURRENT

What is the maximum unbalanced neutral current for phases A and B?

Step 1. *220-22; Note 10 to Tables 310-16–310-19:*

Phase A =	120 A
Phase B =	80 A
	40 A

Answer. **Unbalanced neutral current = 40 A**

Figure 2-20. The neutral conductor shall be sized large enough to carry the unbalanced current flow plus the maximum current flow for a lost phase.

What size THWN Cu conductors is required to supply a neutral load of 150 A in a feeder circuit to a subpanel? The loads have been calculated per 220-10(b) and 220-22.

Step 1. *220-10(b); 220-22:* Calculated load = 150 A

Step 2. *Table 310-16:* #1/0 THWN Cu = 150 A

Answer. **Conductors = #1/0 THWN Cu**

If the current in the ungrounded phase conductor of a 120 V circuit is 80 A, the current in the neutral conductor is 80 A. For the current in the neutral conductor to be unbalanced, there must be two or three ungrounded hot phase conductors. For example, if the current in phase A is 100 A and the current in phase B is 78 A, the current in the neutral conductor is 22 A. The neutral is sized for the maximum unbalanced current of phase A or B. If phase B is lost, the neutral carries 100 A and the size of the conductor is selected from this ampacity rating.

What is the unbalanced current in the neutral conductor where current in phase A is 90 A and current in phase B is 128 A?

Step 1. *220-22:*

Phase B =	128 A
Phase A =	− 90 A
	38 A

Answer. **Unbalanced current = 38 A**

To find the unbalanced neutral current in a 3ϕ, 4-wire system, use the following formula:

$$A = \sqrt{I^2L_1 + I^2L_2 + I^2L_3 - (I_{L_1} \times I_{L_2}) - (I_{L_2} \times I_{L_3}) - (I_{L_3} \times I_{L_1})}$$

What is the unbalanced neutral current in a 4-wire wye system with phase A carrying 68 A, phase B carrying 88 A, and phase C carrying 96 A?

Step 1. *220-22:*

$$A = \sqrt{I^2L_1 + I^2L_2 + I^2L_3 - (I_{L_1} \times I_{L_2}) - (I_{L_2} \times I_{L_3}) - (I_{L_3} \times I_{L_1})}$$

$$A = \sqrt{68^2 + 88^2 + 96^2 - (68 \times 88) - (88 \times 96) - (96 \times 68)}$$

$$A = \sqrt{4624\text{ A} + 7744\text{ A} + 9216\text{ A} - 5984\text{ A} - 8448\text{ A} - 6528\text{ A}}$$

$$A = \sqrt{21{,}584\text{ A} - 20{,}960\text{ A}}$$

$$A = \sqrt{624\text{ A}} = 24.9\text{ A}$$

Answer. **24.9 A**

GROUNDING—384-20

Equipment grounding conductors from the service equipment to the subpanel can be green, green with yellow stripes, or bare per 250-57(b) and 310-12(b). Equipment grounding conductors #6 or smaller are required to be green, green with yellow stripes or bare per 250-57(b), Ex. 1 and 310-12(b), Ex. 1. Equipment grounding conductors #4 and larger can be taped or colored green for their entire length when used in junction boxes or terminated in enclosures per 250-57(b), Ex. 1b, 1c, and 310-12(b), Ex. 1b, 1c. The #4 or larger conductors can be stripped their entire length in junction boxes and enclosures per 250-57(b), Ex. 1a and 310-12(b), Ex. 1a.

Equipment grounding conductors are terminated to a grounding busbar bonded to the metal enclosure of the subpanel per 250-79(a) and 384-20. The equipment bonding jumper is sized per 250-79(e) which requires the jumper size to be selected from Table 250-95 based on the size of the overcurrent protection devices protecting the feeder circuit. The equipment grounding conductor routed from the service equipment to the subpanel is selected from the size overcurrent protection devices protecting the feeder circuit conductors per Table 250-95. It shall not carry any current except for a ground fault between an ungrounded phase conductor and metal raceways, enclosures, junction boxes, etc. See Article 100 for the definition of an equipment grounding conductor. See Figure 2-21.

What size copper equipment grounding conductor is required in a feeder circuit protected by a 200 A overcurrent protection device?

Step 1. *240-3, Ex. 4; 240-6:* OCPD = 200 A

Step 2. *Table 250-95:* 200 A device = #6 Cu

Answer. **ECG = #6 Cu**

Height—380-8(a)

The height of a subpanel is determined by the main or circuit breaker in the ON position. The handle of the main or circuit breaker shall not be higher than 6′-6″ in the ON position when measured from the finished grade or floor. The NEC® does not mention how close to the floor or finished grade that an overcurrent protection device can be installed. Overcurrent protection devices are required to be readily accessible to personnel per 240-24(a) where this requirement is maintained.

A feeder circuit is routed to a subpanel located in an industrial complex. The load on the ungrounded phase conductors is 278 A and the neutral load is 159 A. What size THWN Cu conductors is required for the phase conductors and neutral conductor? What size overcurrent protection device and equipment grounding conductor is required? Select a copper equipment grounding conductor.

Phase conductors

Step 1. *220-10(b):* Calculated load = 278 A

Step 2. *Table 310-16:* #300 kcmil THWN Cu = 285 A

Answer. **Conductors = #300 kcmil THWN Cu**

Neutral conductor

Step 3. *200-10(b); 220-22:* Calculated load = 159 A

Step 4. *Table 310-16:* #2/0 THWN Cu = 175 A

Answer. **Conductor = #2/0 THWN Cu**

Overcurrent protection device

Step 5. *240-3, Ex. 4; 240-6:*
#300 kcmil THWN Cu = 285 A

Step 6. *240-3, Ex. 4; 240-6:* 285 A = 300 A device

Answer. **OCPD = 300 A**

Equipment grounding conductor

Step 7. *Table 250-95:* 300 A device = #4 Cu

Answer. **EGC = #4 Cu**

SIZING EQUIPMENT GROUNDING CONDUCTORS

What size equipment grounding conductor is required for a feeder circuit protected by a 150 A CB?

Step 1. *Table 250-95:* 150 A CB = #6 Cu or #4 Al

Answer. **EGC = #6 Cu or #4 Al**

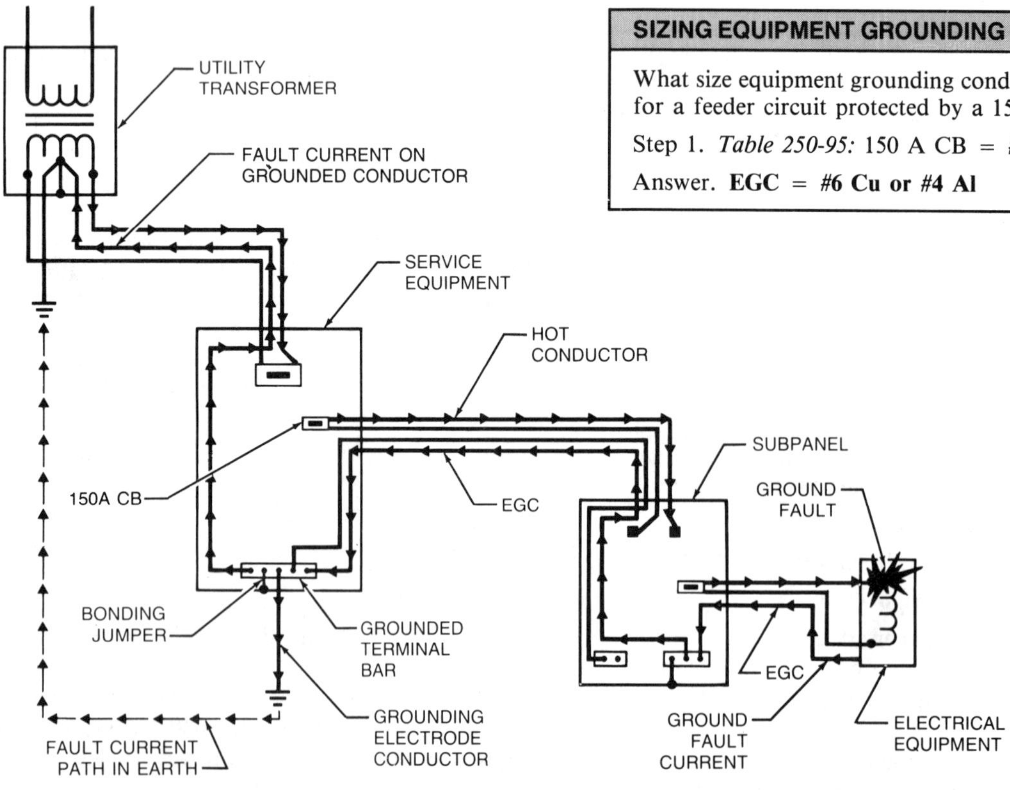

Figure 2-21. Equipment grounding conductors carry fault current from the load to the grounded terminal bar of the service equipment.

CALCULATIONS

Feeder loads are computed according to their method of use. Loads on feeder circuits are calculated at either continuous duty or noncontinuous duty. Demand factors may be applied to loads figured at noncontinuous duty under certain conditions of use. The total VA rating of the loads calculated at continuous duty, and noncontinuous duty with demand factors, is used to size the conductors, overcurrent protection devices, and other components needed to complete and protect feeder circuits from point to point. Loads on feeder circuits are determined by applying the standard or optional calculation per 220, Part B.

STANDARD CALCULATION: COMMERCIAL AND INDUSTRIAL—220, PART B

The loads on a commercial feeder circuit are determined by dividing the loads into six groups. These loads are

1. Lighting loads................220, Part B
2. Receptacle loads...............220-3(c)(5)
3. Special loads...................220-10(b)
4. Heating or A/C load.............220-21
5. Motor loads..... 430-22(a); 430-24; 430-25
6. Largest motor load................220-14

Lighting Loads

The total VA rating of the lighting loads is determined by dividing the loads into five groups. These loads are

1. General lighting loads.... Table 220-3(b); 220-10(b)
2. Show window loads...............220-12
3. Track lighting loads..............410-102
4. Outside lighting loads...........220-10(b)
5. Sign lighting loads.... 220-3 FPN; 600-6(c)

Lighting Loads—Table 220-3(b). The general lighting load is found by multiplying the total square footage of the building by the VA per square foot per Table 220-3(b). The VA per square foot is selected by each individual occupancy listed in Table 220-3(b). For example, a store building has a VA rating of 3 VA per square foot and an office building has a VA rating of 3.5 VA per square foot. If the number of receptacles is not known, another 1 VA per square foot is added for the receptacle outlet loads.

■ *Master Electrician's Test Tip.* Most examination boards will not allow the demand factors of Table 220-13 to be applied for receptacle loads determined at 1 VA per square foot.■

What is the VA rating for a 1600 sq. ft. barber shop with receptacles used at continuous duty?

Step 1. *Table 220-3(b):* 1600 sq. ft. × 3 = 4800 VA

Step 2. *220-10(b):* 4800 VA × 125% = 6000 VA

Answer. **Rating = 6000 VA**

■ *Master Electrician's Test Tip.* For occupancies not listed in Table 220-3(b), the VA rating for lighting loads is determined per 210-22(b), 220-3(c), or 220-3(c)(5).■

Show Window Loads—220-12. The show window load is found by multiplying the linear footage of the show window by 200 VA. Note that the 200 VA is not increased by 125%. *Note:* See page 9 for number of outlets method.

What is the VA rating for 65 linear feet of show window?

Step 1. *220-12:* 65′ × 200 VA = 13,000 VA

Answer. **Rating = 13,000 VA**

Lighting Track Loads—410-102. The lighting track load is found by multiplying the length of the track by 180 VA and dividing this total by 2. Note that this total VA rating is not increased by 125%.

What is the VA rating for 32′ of lighting track installed in an office?

Step 1. *410-102:*

$$VA = \frac{\text{track length} \times 180\ \text{VA}}{2'}$$

$$VA = \frac{32' \times 180\ \text{VA}}{2'} = 2880\ \text{VA}$$

Answer. **Rating = 2880 VA**

Outside Lighting Loads—210-22(b); 220-3(c)(3). The outside lighting load is found by multiplying the lighting fixtures per 210-22(b), 220-3(c)(3), or 220-3(c)(5). This VA rating is multiplied by 125% for lighting used for three hours or more and by 100% for lighting used for less than three hours per Article 100.

What is the VA rating for 42 electric discharge lighting fixtures used to illuminate a commercial parking lot six hours per night? (Each ballast is rated at 180 VA.)

Step 1. *210-22(b): VA = no. of ballast* × 180 VA
VA = 42 × 180 VA = 7560 VA

Step 2. *220-10(b):* 7560 VA × 125% = 9450 VA

Answer. **Rating = 9450 VA**

Sign Lighting Loads—600-6(c). The sign lighting load is computed at a minimum of 1200 VA or the nameplate rating, whichever is greater. For nameplate ratings less than 1200 VA, the VA rating is figured at 1200 VA. For nameplate ratings above 1200 VA, the VA rating is increased by 125%. The branch circuit load shall be rated at 16 A (20 A CB × 80% = 16 A) for signs used for three hours or more. The branch circuit load can be rated at 20 A (20 A CB × 100% = 20 A) for signs used less than three hours.

What is the VA rating for a 11,520 VA sign used for seven hours each night?

Step 1. *600-6(c); 220-10(b):*
11,520 VA × 125% = 14,400 VA

Answer. **Rating = 14,400 VA**

■ *Master Electrician's Test Tip.* Some examination boards require the sign circuit load of 1200 VA only where the problem states ground level access. Others require the sign load to be included where the problem has any type of commercial occupancy.■

Receptacle Loads—220-3(c)(5)

The receptacle load is found by dividing the receptacle outlets into three groups. The first group of receptacle outlets is computed at 125% for continuous-duty operation. The second group of receptacle outlets is calculated at 100% for noncontinuous-duty operation. If the VA rating of the receptacle outlets exceeds 10,000 VA, a demand factor of 50% is applied for VA exceeding 10,000 VA per Table 220-13. Each receptacle outlet is figured at 180 VA per 220-3(c)(5). The third group of receptacle outlets is identified as multioutlet assemblies. These are classified as either light- or heavy-duty operation. See 220-13 and Table 220-11.

Noncontinuous Duty—220-3(c)(5); 220-10(b). Receptacles used for three hours or less are considered to operate at noncontinuous duty. A 180 VA per outlet is computed at 100% times the number of receptacles. The first 10,000 VA are computed at 100% per Table 220-13. Each receptacle outlet is calculated at 180 VA and the total VA rating of the receptacles shall comply with Table 220-13. The Table requires any VA rating of 10,000 A or less to be computed at 100%.

What is the VA rating used for 55 receptacles connected to a feeder circuit?

Step 1. *220-3(c)(5):* 55 × 180 VA = 9900 VA

Step 2. *220-13:* 10,000 VA or less = 100%
9900 VA × 100% = 9900 VA

Answer. **Rating = 9900 VA**

Demand Factor—Table 220-13. Where each receptacle outlet is multiplied by 180 VA times 100% and the VA rating exceeds 10,000 VA, the VA rating above 10,000 is computed at 50%. This demand factor is permitted since all receptacles are not in use at one time.

What is the demand load in VA for 220 receptacles that are connected to a feeder circuit?

Step 1. *220-3(c)(5):*
220 × 180 VA = 39,600 VA

Step 2. *220-10(b):*
39,600 VA × 100% = 39,600 VA

Step 3. *Table 220-13:*

First 10,000 VA × 100% =	10,000 VA
Next 29,600 VA × 50% =	14,800 VA
	24,800 VA

Answer. **Demand load = 24,800 VA**

Continuous Duty—220-3(c)(5); 220-10(b). Receptacles used for three hours or more are considered to operate at continuous duty. Receptacles operating at continuous duty are calculated at 180 VA per outlet and this VA rating is increased by 125%. There is no demand factor permitted for receptacles operating at continuous duty because there is no rest period.

What is the VA rating of 90 receptacles used at continuous duty?

Step 1. *220-3(c)(5):* 90 × 180 VA = 16,200 VA

Step 2. *Table 220-13:* No demand permitted

Step 3. *220-10(b):*
16,200 VA × 125% = 20,250 VA

Answer. **Rating = 20,250 VA**

Multioutlet Assembly—220-3(c), Ex. 1. The multioutlet assembly is calculated at 180 VA per foot for heavy duty and at 180 VA each 5′ for light duty.

What is the VA rating for 30 receptacles operating at noncontinuous duty?

Step 1. *220-3(c)(5):*
$VA = \textit{no. of receptacles} \times 180\ VA$
$VA = 30 \times 180\ VA = 5400\ VA$

Step 2. *Table 220-13:* No demand permitted

Answer. **Rating = 5400 VA**

What is the VA rating for 60 receptacles used at continuous duty in an office building?

Step 1. *220-3(c)(5):*
$VA = \textit{no. of receptacles} \times 180\ VA$
$VA = 60 \times 180\ VA = 10{,}800\ VA$

Step 2. *220-3(a):* 10,800 VA × 125% = 13,500 VA

Answer. **Rating = 13,500 VA**

What is the VA rating for 80′ of multioutlet assembly used at heavy duty in a TV repair shop?

Step 1. *220-3(c), Ex. 1:*
$VA = \textit{length of track} \times 180\ VA$
$VA = 80' \times 180\ VA = 14{,}400\ VA$

Answer. **Rating = 14,400 VA**

What is the VA rating for 220′ of multioutlet assembly used at light duty in a commercial building?

Step 1. *220-3(c), Ex. 1:*

$$VA = \frac{\textit{length of assembly}}{5'} \times 180\ VA$$

$$VA = \frac{220'}{5'} \times 180\ VA = 7920\ VA$$

Answer. **Rating = 7920 VA**

Special Loads—220-10(b)

Special loads are calculated at continuous or noncontinuous duty. Under certain conditions of use, a number of special loads can have a demand factor applied. Special loads used at continuous duty are computed at 125%. Those used at noncontinuous duty are computed at 100% with a demand factor applied under specific conditions. The VA rating of the demand factor shall be equivalent to the VA rating of the two larger units for certain pieces of equipment.

What is the VA rating of a 7920 VA machine used at noncontinuous duty?

Step 1. *220-3(a); 220-10(b):*
7920 VA × 100% = 7920 VA

Answer. **Rating = 7920 VA**

What is the VA rating of a 7920 VA machine used at continuous duty?

Step 1. *220-3(a); 220-10(b):*
7920 VA × 125% = 9900 VA

Answer. **Rating = 9900 VA**

What is the VA demand load of ten 7560 VA cooking units used at noncontinuous duty in a commercial kitchen?

Step 1. *Table 220-20:* 10 units = 65% demand factor
7560 VA × 10 × 65% = 49,140 VA

Answer. **Demand load = 49,140 VA**

Heating or A/C Load—220-21

The total VA rating of the heating load is compared to the VA rating of the air conditioning load and the smaller of the two is dropped per 220-21. For example, a heating load of 15 kW is larger than an air conditioning load of 6.7 kW; therefore, the air conditioning load is dropped.

What is the VA rating for a feeder with a heating load of 20 kW and an air conditioning load of 8 kW?

Step 1. *220-21:*
Heating load = 20 kW
A/C unit load = 8 kW

Answer. **Rating = 20 kW**

The air conditioning load of 8 kW is dropped per 220-21, and the heating load of 20 KW is added to the other loads.

■ *Master Electrician's Test Tip.* Some testing authorities consider the air conditioning load eligible to be the largest load even if it has a lower rating than the heating load.■

Motor Loads—430-22(a); 430-24; 430-25

Each motor is computed at 100% of the full-load current rating obtained from Table 430-148 or 430-150. The full-load current ratings in each table are based on the horsepower and voltage ratings of each motor. For example, a 10 HP, 230 V, 3ϕ squirrel-cage induction motor has a full-load current rating of 28 A per Table 430-150. All full-load current ratings are added together to obtain the total motor load when there is more than one motor on a feeder circuit.

What is the VA rating for six 7½ HP, 230 V, 3ϕ motors connected to a 230 V, 3ϕ feeder circuit with other loads?

Step 1. *430-6(a); 430-25:* 7½ HP = 22 A

Step 2. *430-24; 430-25:*
$VA = V \times \sqrt{3} \times A \times \text{no. of motors}$
$VA = 230\text{ V} \times 1.732 \times 22\text{ A} \times 6 = 52{,}536\text{ VA}$

Answer. **Rating = 52,536 VA**

Note: The largest motor is calculated at 25% of the largest motor's FLC rating.

Largest Motor Load—220-14

The largest motor load is calculated at 25% of the largest motor's FLC rating per 220-14. If a hermetically sealed compressor is the largest motor, 25% of the compressor's FLC is computed per 440-34. This 25% total is added to the other loads to obtain the total VA rating for the feeder circuit.

Calculating Feeder Circuit

The following is a step-by-step procedure for determining the VA rating of a feeder circuit applying the standard calculation for a commercial building.

A 120/208 V, 3ϕ, 4-wire feeder circuit supplies a 3000 sq. ft. store area on the second floor of a multistory building. The office has the following loads which operate at continuous duty:

40′ of lighting track .... 1ϕ, 120 V
10 outside lights at 225 VA each ...... 1ϕ, 120 V
25 linear feet of show window .... 1ϕ, 120 V
18 duplex receptacles .... 1ϕ, 120 V

The store also has the following loads which operate at noncontinuous duty:

120 duplex receptacles .... 1ϕ, 120 V
60 A of special loads per phase
one 7½ HP motor .... 3ϕ, 208 V
one 5 HP motor .... 3ϕ, 208 V
one 10 HP motor .... 3ϕ, 208 V
25 kW heating unit.................. 3ϕ, 208 V
28 A A/C unit....................... 3ϕ, 208 V

What is the VA rating for the feeder circuit?

Lighting loads

Step 1. *Table 220-3(b):* 3000 sq. ft. × 3 VA = 9000 VA

Step 2. *Table 220-11:* No demand permitted for office

Step 3. *220-10(b):* 9000 VA × 125% = 11,250 VA

Step 4. *220-12:* 25′ × 200 VA = 5000 VA

Step 5. *410-102:* $\dfrac{40' \times 180\text{ VA}}{2} = 3600\text{ VA}$

Step 6. *220-10(b):* 225 VA × 10 × 125% = 2812.5 VA
Lighting loads = **22,662.5 VA**

Receptacle loads

Step 1. *220-3(c)(5):* 120 × 180 VA × 100% = 21,600 VA

Step 2. *Table 220-13:*
First 10,000 VA × 100% = 10,000 VA
Next 11,600 VA × 50% = 5800 VA

Step 3. *220-3(c)(5):* 18 × 180 VA × 125% = 4050 VA
Receptacle loads = **19,850 VA**

Special loads

Step 1. *220-10(b):*
VA = 60 A × 208 V × 1.732 = 21,600 VA
Special loads = **21,600 VA**

Largest heating or A/C load

Step 1. *220-15; 220-21:* Heating load = **25,000 VA**
A/C load = 28 A × 208 V × 1.732 = 10,080 VA

Motor loads

Step 1. *430-6(a); Table 430-150:*
5 HP = 15.2 A × 208 V × 1.732 × 110% = 6,019.2 VA
7½ HP = 22 A × 208 V × 1.732 × 110% = 8,712 VA
10 HP = 28 A × 208 V × 1.732 × 110% = 11,088 VA
Motor loads = **25,819.2 VA**

Largest motor load

Step 1. *220-14; 430-25:* 11,088 VA × 25% = **2772 VA**

Answer. **Feeder circuit VA rating = 117,703.7 VA**

The VA rating of the feeder circuit is 117,703.7 VA. The amperage for the feeder circuit is determined by dividing the VA by the voltage. The other elements that complete the feeder circuit are based on the amperage.

A step-by-step procedure for finding the VA and amperage ratings is shown in the form, STANDARD CALCULATION—COMMERCIAL AND INDUSTRIAL. See Figure 2-22.

What is the load in VA and amps for a 5200 sq. ft. store building with grade level access supplied by a 120/208 V, 3ϕ, 4-wire feeder circuit? The store building has the following loads.
80′ track lighting
20 outside lights at 175 VA each (continuous-duty)
35 linear feet show window
150 general-purpose duplex receptacles
42 duplex receptacles (continuous-duty)
40′-0″ light duty multioutlet assembly
120 A per phase of 208 V special loads (continuous-duty)
30 A of 120 V special loads (noncontinuous-duty)
40 kW, 3ϕ heating unit
38 A, 3ϕ A/C unit
One 5 HP, 3ϕ motor
One 10 HP, 3ϕ motor
One 15 HP, 3ϕ motor

STANDARD CALCULATION—ONE-FAMILY DWELLING IN MULTIFAMILY COMPLEXES

The loads on feeder circuits for dwelling units in multifamily complexes are determined by dividing the loads into three groups. These are

1. General lighting and receptacle loads . . . 220-3(b); Table 220-3(b)
2. Small appliance loads . . . . 220-16(a)(b)
3. Special appliance loads . . . . 220-10(b)

General lighting and receptacle loads are found by multiplying the square footage of the dwelling unit by 3 VA per square foot per 220-3(b) and Table 220-3(b). Small appliance loads are figured at 1500 VA each per 220-16(a)(b). Special appliance loads are figured at 100% for noncontinuous duty and at 125% for continuous duty per 220-10(b). Demand factors are applied for certain appliances per 220-17.

Four loads are selected, if present in the three groups of loads, to which demand factors can be applied. These are
General lighting and receptacle loads . . . . Table 220-11
Cooking equipment loads . . . . Table 220-19
Dryer load . . . . Table 220-18
Fixed appliance loads 220-17

Six loads are selected from the three groups of loads and are totaled to determine the load on feeder circuits to dwelling units. These loads are:
General lighting loads . . . . Table 220-3(b)
Cooking equipment loads . . . . Notes to Table 220-19
Dryer load . . . . 220-18; Table 220-18
Fixed appliance loads 220-17
Heating or A/C load 220-21
Largest motor load 220-14

These loads are grouped and the VA rating of each load is determined from the computations of the four demand factors, plus the heating or A/C load and the largest motor load. The load for the neutral is calculated for 120 V loads only, except for four pieces of 240 V, 1ϕ equipment, which are the range, cooktop, oven, and dryer per 220-22.

The following is a step-by-step procedure for finding the service load in VA for a one-family dwelling using the standard calculation.

A feeder circuit is routed from a fusible disconnect located beneath the meter base to a subpanel inside a 2850 sq. ft. one-family dwelling. The supply voltage on the feeder circuit is 120/240 V, 1ϕ. Find the VA rating of the feeder circuit for this dwelling with the following loads:
Two small appliance circuits
One laundry circuit
25 kW heating unit . . . . 240 V
26 A air conditioner . . . . 240 V
8 kW wall oven . . . . 240 V
6 kW cooktop . . . . 240 V
6.5 kW water heater . . . . 240 V
1.2 kW disposal . . . . 120 V
1.4 kW compactor . . . . 120 V
1.6 kW dishwasher . . . . 120 V
1.2 kW attic fan . . . . 120 V
1.5 kW water pump . . . . 120 V
5 kW dryer . . . . 240 V
3.6 kW blower motor . . . . 240 V

Separate the loads into three groups and apply demand factors to the four loads for phases A and B. The three groups are

1. General lighting and receptacle loads . . . . 220-3(b); Table 220-3(b)
   2850 sq. ft. × 3 VA = 8550 VA
2. Small appliance loads . . . . 220-16(a)(b)
   220-16(a): 1500 VA × 2 = 3000 VA
   220-16(b): 1500 VA × 1 = 1500 VA
3. Special appliance loads . . . . 220-10(b)
   25 kW heating unit . . . . 240 V
   26 A air conditioner . . . . 240 V
   8 kW wall oven . . . . 240 V

STANDARD CALCULATION—COMMERCIAL AND INDUSTRIAL

1. LIGHTING LOADS—220, Part B

General lighting loads—Table 220-3(b); 220-10(b)
5200 sq. ft. × 3 VA × 125 % = 19,500 VA

Show window loads—220-12
35 ′ × 200 VA = 7000 VA

Track lighting loads—410-102
80 ′ (2′) × 180 VA = 7200 VA

Outside lighting loads—220-10(b)
175 VA × 20 units × 125 % = 4375 VA

Sign lighting loads—220-3, FPN; 600-6(c)
1200 VA × 100 % = 1200 VA

(total) 39,275 VA | PHASES 39,275 VA | NEUTRAL 39,275 VA

2. RECEPTACLE LOADS—220-3(c)(5)

Noncontinuous duty—220-10(b); Table 220-13
150 × 180 VA = 27,000 VA = — VA

Applying demand factors—Table 220-13
First 10,000 VA × 100% = 10,000 VA = 10,000 VA
Remaining 17,000 VA × 50% = 8500 VA
Continuous duty—220-10(b) # 42 × 180 VA × 125% = 9450 VA
Multioutlet assembly—220-3(c)
Heavy duty = — ′ × 180 VA = — VA
Light duty = 40 ′ (5′) × 180 VA = 1440 VA

(total) 29,390 VA | 29,390 VA | 29,390 VA

3. SPECIAL LOADS—220-10(b)

Load = 43,200 VA × 125 % = 54,000 VA
Load = 3600 VA × 100 % = 3600 VA
— = — VA × — % = — VA
— = — VA × — % = — VA
— = — VA × — % = — VA

(total) 57,600 VA | 57,600 VA | 3600 VA

4. HEATING OR A/C LOAD—220-21 40,000 VA | 40,000 VA | — VA

5. MOTOR LOADS—430-22(a); 430-24; 430-25 (3ϕ only)

5 HP = 16.72 A × — % × 208 V × $\sqrt{3}$ = 6019.2 VA
10 HP = 30.8 A × — % × 208 V × $\sqrt{3}$ = 11,088 VA
15 HP = 46.2 A × — % × 208 V × $\sqrt{3}$ = 16,632 VA
— = — A × — % × — V × $\sqrt{3}$ = — VA

(total) 33,739.2 VA | 33,739.2 VA | — VA

6. LARGEST MOTOR LOAD—220-14 (3ϕ only)

ϕ 42 A × 110 % × 208 V × $\sqrt{3}$ × 25% = 4158 VA | 4158 VA | — VA
N — A × — % × — V × $\sqrt{3}$ × 25% = — VA | — VA | — VA

TOTAL = 204,162.2 VA | 72,265 VA

1ϕ service: I = $\frac{VA}{V}$
PHASES I = $\frac{— \text{ VA}}{\text{V}}$ = — A
NEUTRAL I = $\frac{— \text{ VA}}{\text{V}}$ = — A

3ϕ service: I = $\frac{VA}{V \times \sqrt{3}}$
I = $\frac{204{,}162.2 \text{ VA}}{208 \text{ V} \times \sqrt{3}}$ = 567 A
I = $\frac{72{,}265 \text{ VA}}{208 \text{ V} \times \sqrt{3}}$ = 200.7 A

If over 200 A—220-22; First 200 A × 100% = 200 A
Remaining .7 A × 70% = .49 A
200.49 A

Figure 2-22. The standard calculation is used to determine the VA and A for feeder circuits to individual occupancies located in a commercial or industrial complex.

6 kW cooktop....................240 V
6.5 kW water heater..............240 V
1.2 kW disposal..................120 V
1.4 kW compactor................120 V
1.6 kW dishwasher...............120 V
1.2 kW attic fan..................120 V
1.5 kW water pump..............120 V
5 kW dryer.......................240 V
3.6 kW blower motor.............240 V

General Lighting Loads—Table 220-3(b). The general lighting load is determined by multiplying the square footage of the dwelling unit by 3 VA per Table 220-3(b).

Step 1. *Table 220-3(b):*
2850 sq. ft. × 3 VA = 8550 VA

Step 2. *220-16(a):* 1500 VA × 2 = 3000 VA

Step 3. *220-16(b):* 1500 VA × 1 = 1500 VA
13,050 VA

Step 4. Applying demand factors. *Table 220-11:*
First 3000 VA × 100% = 3000 VA
Next 10,050 VA × 35% = 3517.5 VA
Total = 6517.5 VA

Answer. **General lighting loads = 6517.5 VA**

Cooking Equipment Loads—Table 220-19. The cooking equipment load is determined by the kW rating of the unit and the demand factors of Table 220-19. Column B has a demand factor for any range with a rating of less than 3½ kW. Column C has demand factors for ranges rated at 3½ kW to 8¾ kW. Column A has demand factors for ranges over 8¾ to 12 kW.

Step 1. *Table 220-19, Column C:*
Two units = 65% demand factor
8 kW + 6 kW × 1000 × 65% = 9100 VA

Answer. **Cooking equipment loads = 9100 VA**

Dryer Load—Table 220-18. The dryer load is determined by the kW rating of each dryer. Dryers with a kW rating of less than 5000 kW must have their rating increased to 5000 VA. Dryers with a kW rating greater than 5000 VA are computed with the greater rating. Table 220-18 requires four or fewer dryers to be computed at 100% of their kW rating.

Step 1. *Table 220-18:* Four or fewer dryers = 100%

Step 2. *220-18:* 5 kW = 5 kVA
5 kVA × 1000 = 5000 VA

Answer. **Dryer load = 5000 VA**

Note: A dryer with a rating less than 5000 VA must have the rating increased to 5000 VA per 220-18.

Fixed Appliance Loads—220-17. Fixed appliance loads are determined by multiplying the VA rating of four or more fixed appliances by 75%. Fixed appliances are appliances such as dishwashers, disposals, and compactors. The exception to 220-17 will not allow heating units, air conditioner units, dryer or cooking equipment loads to have this demand factor of 75% applied. Note that electrical equipment that do not connect to the general-purpose receptacles or small appliance circuits and are exempt by the exception to 220-17 are considered fixed appliance loads.

Step 1. Total all loads. *220-17:*
6.5 kW water heater = 6500 VA
1.2 kW disposal = 1200 VA
1.4 kW compactor = 1400 VA
1.6 kW dishwasher = 1600 VA
1.2 kW attic fan = 1200 VA
1.5 kW water pump = 1500 VA
3.6 kW blower motor = 3600 VA
Total loads = 17,000 VA

Step 2. Applying demand factors. *220-17:*
17,000 VA × 75% = 12,750 VA

Answer. **Fixed appliance loads = 12,750 VA**

Note: There are four or more fixed appliances.

Heating or A/C Load—220-21. The heating and air conditioning loads are compared to verify which has the greater VA rating. The smaller VA rating of the two is dropped per 220-21. Where the heating load has the greater rating, the air conditioning load is dropped and only used for sizing the branch circuit.

Step 1. *220-21:* Heating load = 25,000 VA
A/C load = 6240 VA

Answer. **Heating or A/C load = 25,000 VA**

Largest Motor Load—220-14. The largest motor must be selected from one of the motors in the dwelling unit. Note that the blower motor in a heating unit is considered a fixed appliance and can be lumped with the fixed appliances and reduced 75%. The reason that this is permitted is because the blower motor can be used during the heating or cooling cycle or as a circulating fan.

■ *Master Electrician's Test Tip.* Some examination boards consider the heating elements and blower motor as one load, and the blower motor is not eligible to be considered the largest motor load or fixed appliance. ■

The largest motor can be a disposal, a compactor, an attic fan, a water pump, etc. The largest motor load must be added to the other five loads at 25% of the motor's full-load current rating per 430-22(a), 430-24, and 430-25.

Step 1. *220-14:*
Blower motor load = 3600 VA × 25% = 900 VA

Answer. **Blower motor load = 900 VA**

The total VA for the service load is found by adding all loads.

General lighting loads =	6517.5 VA
Cooking equipment loads =	9100 VA
Dryer load =	5000 VA
Fixed appliance loads =	12,750 VA
Heating or A/C load =	25,000 VA
Largest motor load =	900 VA
Total loads =	59,267.5 VA

Phases A and B of the feeder circuit are sized based on the 59,267.5 VA load. The neutral conductor is sized based on adding all 120 V loads in the six loads used to calculate phases A and B.

Step 1. General lighting load. *Table 220-11:*
6517.5 VA

Step 2. Cooking equipment load. *220-22:*
9100 VA × 70% = **6370 VA**

Step 3. Dryer equipment load. *220-22:*
5000 VA × 70% = **3500 VA**

Step 4. Fixed appliance loads. *220-17:*

Disposal =	1200 VA
Compactor =	1400 VA
Dishwasher =	1600 VA
Attic fan =	1200 VA
Water pump =	1500 VA
	6900 VA

6900 VA × 75% = **5175 VA**

Step 5. Largest heat or A/C load. *220-21:*
120 V load = **0**

Step 6. Largest motor load. *220-14; 430-25:*
1500 VA × 25% = **375 VA**
Total load = **21,937.5 VA**

Find the following elements for the feeder circuit.

1. Amperage for phases A and B
2. Size THWN Cu conductors for feeder (use max. size)
3. Size PVC conduit
4. Size fuses for phases A and B
5. Size of fuses for disconnect
6. Size subpanel
7. Size of equipment grounding conductor using copper

Amperage for phases A and B

Step 1. $I = \frac{VA}{V}$

$$I = \frac{59{,}267.5 \text{ VA}}{240 \text{ V}} = 246.9 \text{ A}$$

Answer. **Amperage = 246.9 A**

Amperage for neutral

Step 1. $I = \frac{VA}{V}$

$$I = \frac{21{,}937.5 \text{ VA}}{240 \text{ V}} = 91 \text{ A}$$

Answer. **Amperage = 91 A**

Conductors

Step 1. *Tables 310-16–310-19:*
Phases A and B = #250 kcmil THWN Cu
Neutral = #2 THWN Cu
Note: 250-23(b) requires #2 neutral

Answer. **Conductors = two #250 kcmil and one #2 THWN Cu**

(Note 3 is not applied.)

Conduit (PVC)

Step 1. *Table 5, Ch. 9:*
#250 kcmil THWN =
.4026 sq. in. × 2 = .8052 sq. in.
#2 THWN =
.1182 sq. in. × 1 = .1182 sq. in.
.9234 sq. in.

Step 2. *Table 4, Ch. 9:*
2″ Conduit = 1.34 sq. in. area

Answer. **Conduit = 2″**

Fuses

Step 1. *250-3(a), Ex. 1:* #250 kcmil THWN Cu = 255 A
240-6: 255 A = 300 A fuses

Answer. **Fuses = 300 A**

Disconnect

Step 1. *Chart:* 300 A fuses = 400 A disconnect

Answer. **Disconnect = 400 A**

Subpanel

Step 1. *Chart:* 300 A panelboard

Answer. **Subpanel = 300 A**

EGC

Step 1. *Table 250-95:*

Answer. **EGC = #4 Cu**

A step-by step procedure for finding the VA and amperage ratings is shown in the form, STANDARD CALCULATION—ONE-FAMILY DWELLING. See Figure 2-23.

What is the VA and amp rating for a 2550 sq. ft. dwelling unit with a 240 V, 1ϕ service with the following loads?

8.5 kW cooktop .... 240 V, 1ϕ
6.5 kW oven .... 240 V, 1ϕ
5 kW dryer .... 240 V, 1ϕ
5 kW water heater .... 240 V, 1ϕ
15 kW heating unit .... 240 V, 1ϕ
4.7 kW A/C unit .... 240 V, 1ϕ
840 VA blower motor .... 240 V, 1ϕ
420 VA disposal .... 120 V, 1ϕ
708 VA compactor .... 120 V, 1ϕ
1450 VA dishwasher .... 120 V, 1ϕ
828 VA attic fan .... 120 V, 1ϕ
780 VA overhead garage door motor .. 120 V, 1ϕ

OPTIONAL CALCULATION—ONE-FAMILY DWELLING

The load in VA or amps for a one-family dwelling unit can by determined by applying the optional calculation. This calculation will produce a smaller VA than the standard calculation. Loads are separated into two groups: other loads, and heating or A/C load. Other loads include all loads except heating or A/C. The heating or A/C load chosen is the larger of the two.

STANDARD CALCULATION—ONE-FAMILY DWELLING

1. GENERAL LIGHTING LOADS—Table 220-3(b)

2550 sq. ft. × 3 VA = 7650 VA

Small appliance loads—220-16(a)

1500 VA × 2 circuits = 3000 VA

Laundry load—220-16(b)

1500 VA × 1 = 1500 VA

12,150 VA

Applying demand factors—Table 220-11

First 3000 VA × 100% = 3000 VA

Next 9150 VA × 35% = 3202.5 VA

Remaining — VA × 25% = — VA

		PHASES	NEUTRAL
(total)	6202.5 VA	6202.5 VA	6202.5 VA

2. COOKING EQUIPMENT LOADS—Table 220-19; Notes

Col. A — VA demand = — VA

Col. B — VA × — % = — VA

Col. C 15,000 VA × 65 % = 9750 VA

(total) 9750 VA | 9750 VA × 70% = 6825 VA

3. DRYER LOAD—220-18; Table 220-18

5000 VA × 100 % = 5000 VA | 5000 VA × 70% = 3500 VA

4. FIXED APPLIANCE LOADS—220-17

Dishwasher = 1450 VA

Disposal = 420 VA

Compactor = 708 VA

Water heater = 5000 VA

Attic fan = 828 VA

Garage motor = 780 VA

Blower motor = 840 VA

______ = ______ VA

(120 V Loads × 75%)

(total) 10,026 VA × 75% = 7519.5 VA | 7519.5 VA | 3139.5 VA

5. HEATING OR A/C LOAD—220-21

Heating unit = 15,000 VA × 100% = 15,000 VA

A/C unit = 4700 VA × 100% = 4700 VA

Heat pump = — VA × 100% = — VA

(largest load) 15,000 VA | 15,000 VA | — VA

6. LARGEST MOTOR LOAD—220-14

ϕ 840 VA × 25% = 210 VA | 210 VA

N 828 VA × 25% = 207 VA | 207 VA

TOTAL = 43,682 VA | 19,874 VA

1ϕ service: $I = \frac{VA}{V}$

PHASES $I = \frac{43{,}682\ VA}{240\ V} = 182\ A$

NEUTRAL $I = \frac{19{,}874\ VA}{240\ V} = 82.8\ A$

If over 200 A—220-22; First 200 A × 100% = 200 A

Remaining ______ A × 70% = ______ A

______ A

Figure 2-23. The standard calculation is used to determine the VA and A for feeder circuits to dwelling units located in an apartment complex.

Other Loads—Table 220-30

The VA rating of the other loads is found by taking the first 10,000 VA at 100% and all remaining loads at 40%. The total is the total VA rating of the other loads.

What is the VA rating for 32,800 VA, which is added together and grouped in other loads?

Step 1. *Table 220-30:*

First 10,000 VA × 100% =	10,000 VA
Next 22,800 VA × 40% =	9120 VA
	19,120 VA

Answer. **Rating = 19,120 VA**

Heating or A/C Load—Table 220-30

The heating load is determined by taking 65% of three or fewer heating units or by taking 40% of four or more heating units. The total VA rating of the heating load is then compared to 100% of the air conditioning load per Table 220-30(3)(4).

What is the VA load for three or fewer heating units with a total rating of 25 kW?

Step 1. *Table 220-30(3):*
25 kW × 1000 × 65% = 16,250 VA

Answer. **Load = 16,250 VA**

What is the VA rating of four or more heating units with a total load of 35 kW?

Step 1. *Table 220-30(4):*
35 kW × 1000 × 40% = 14,000 VA

Answer. **Rating = 14,000 VA**

The air conditioner load is figured at 100% of the nameplate rating, regardless of the number present.

What is the VA rating for a compressor in a central air conditioner with a full-load rating of 42 A (supply is 240 V, 1ϕ)?

Step 1. *Table 220-30(1):*
42 A × 240 V × 100% = 10,080 VA

Answer. **Rating = 10,080 VA**

Calculating Feeder Circuit Load—220-30; Table 220-30. The feeder circuit load may be calculated by the optional method per 220-30 and Table 220-30.

A 120/240 V, 1ϕ, 3-wire feeder circuit supplies a subpanel in a dwelling unit. Find the VA rating for the feeder circuit. The 2850 sq. ft. dwelling unit has the following loads:

Cooktop	8000 VA, 240 V
Oven	6000 VA, 240 V
Dryer	5000 VA, 240 V
Water heater	6500 VA, 240 V
Blower motor	360 VA, 240 V
Heating unit	25000 VA, 240 V
A/C	6240 VA, 240 V
Disposal	1200 VA, 120 V
Compactor	1400 VA, 120 V
Dishwasher	1600 VA, 120 V
Attic fan	1200 VA, 120 V
Water pump	1500 VA, 120 V

Other loads

General lighting loads = 2850 sq. ft. × 3 VA = 8550 VA, 120 V
Small appliance loads = 1500 VA = 3000 VA, 120 V
Laundry loads = 1500 V × 1 = 1500 VA, 120 V

Cooktop	8000 VA, 240 V
Oven	6000 VA, 240 V
Dryer	5000 VA, 240 V
Water heater	6500 VA, 240 V
Blower motor	360 VA, 240 V
Heating unit	0 VA
A/C	0 VA
Disposal	1200 VA, 120 V
Compactor	1400 VA, 120 V
Dishwasher	1600 VA, 120 V
Attic fan	1200 VA, 120 V
Water pump	1500 VA, 120 V
Total =	45,810 VA

Other loads

Step 1. *Table 220-30:*

First 10,000 VA × 100% =	10,000 VA
Next 35,810 VA × 40% =	14,324 VA
	24,324 VA

Answer. **Other loads = 24,324 VA**

Heating or A/C load

Step 1. *Table 220-30(2):*

25,000 VA × 65% =	16,250 VA
6240 VA × 100% =	6 240 VA
	16,250 VA

Answer: **Heating load = 16,250 VA**

Total load

Step 1. Adding loads

Other loads =	24,324 VA
Heating load =	16,250 VA
	40,574 VA

Answer. **Total load = 40,574 VA**

Find the following elements of the feeder circuit.

Conduit (Use THWN Cu)

Step 1. *220-2; Text:* $I = \frac{VA}{V}$

$$I = \frac{40{,}574 \text{ VA}}{240} = 169 \text{ A}$$

Step 2. *Table 310-16:* #2/0 THWN Cu = 175 A

Step 3. Neutral load is 91 A per Standard Calculation.

Step 4. *Table 310-16:* #3 THWN Cu = 100 A

Step 5. *Table 5, Ch. 9:*
#2/0 THWN Cu =
.2265 sq. in. × 2 = 0.453 sq. in.
#3 THWN Cu =
.0995 sq. in. × 1 = .0995 sq. in.
Total = .5525 sq. in.

Step 6. *Table 5, Ch. 9:*
$1\frac{1}{4}''$ C = 60 sq. in.

Answer. **Conduit = $1\frac{1}{4}''$**

Size fuses

Step 7. *240-3, Ex. 4:* #2/0 THWN Cu = 175 A
240-6: 175 A fuses

Answer. **Fuses = 175 A**

Note: Note 3 to Tables 310-16–310-19 allows #1/0 THWN Cu conductors for phases A and B with a #3 THWN Cu conductor for the neutral. The conduit is 1¼″ with a 150 A fuse. Some authorities allow a 175 A OCPD. The following is a comparison in the selection of feeder circuit components applying Note 3 to Tables 310-16–310-19.

Size disconnect

Step 8. *Chart:* 175 A = 200 A disconnect
150 A = 200 A disconnect

Answer. **Disconnect = 200 A**

Size subpanel

Step 9. *Chart:* 175 A CB = 200 A panelboard
150 A = 150 A panelboard

Answer. **Panelboard = 200 A or 150 A**

Size EGC

Step 10. *Table 250-95:* 175 A CB = #6 Cu
150 A CB = #6 Cu

Answer. **EGC = #6**

A step-by-step procedure for finding the VA and amp ratings is shown in the form, OPTIONAL CALCULATION—ONE-FAMILY DWELLING (NEW OR EXISTING). See Figure 2-24. Refer to Figure 2-23 for sizing the neutral.

What is the VA and amp rating for a 2550 sq. ft. dwelling unit with a 240 V, 1ϕ service with the following loads?

8.5 kW cooktop .... 240 V, 1ϕ
6.5 kW oven .... 240 V, 1ϕ
5000 VA dryer .... 240 V, 1ϕ
5000 VA water heater .... 240 V, 1ϕ
15 kW heating unit .... 240 V, 1ϕ
4.7 kW A/C unit .... 240 V, 1ϕ
840 VA blower motor .... 240 V, 1ϕ
420 VA disposal .... 120 V, 1ϕ
708 VA HP compactor .... 120 V, 1ϕ
1450 VA dishwasher .... 120 V, 1ϕ
828 VA attic fan .... 120 V, 1ϕ
780 VA HP overhead
garage door motor .... 120 V, 1ϕ

Mobile Homes—Parts A and B

The branch circuits inside mobile homes are normally computed and installed by the manufacturer. The loads for mobile homes are calculated by the requirements of 550-13. The ampacity of the feeder circuit is figured by dividing the VA rating of the mobile home by the voltage. The size feeder circuit to a switchgear for a number of mobile homes is determined per Table 550-22.

Individual Mobile Home—550-13. The feeder circuit load for a mobile home is calculated by the following steps:

1. The lighting load and general-purpose receptacles are computed by multiplying the square footage by 3 VA per 550-13(a).
2. Ranges are figured on a demand factor based on the kW rating per 550-13(b)(5).
3. Cooktops, ovens, disposals, dishwashers, water heaters, and dryers are computed at 75% of the nameplate rating where there are four or more per 550-13(b)(4).
4. Heating, air conditioning, and motors are calculated at nameplate ratings per 550-13(b)(2).
5. Omit the smaller between the heating or air conditioning load per 550-13(b)(2). (*Note:* Check operation of blower motor.)
6. Add 25% of the largest motor load per 550-13 (b)(3). Add total VA and divide by the voltage to derive amperage for feeder circuit from the service equipment to the mobile home.

Note: The neutral load can be figured by applying the standard calculation for a dwelling unit.

Applying Note 3 to 215-2. Feeder circuits supplying power from the service equipment for a mobile home to the panelboard located in the mobile home can have Note 3 to Tables 310-16–310-19 applied to size the conductors. The total VA rating of the mobile home shall be at least 100 A to 200 A in rating.

What size THWN Cu conductors is used for a feeder circuit supplying power to a mobile home having a total load of 156 A?

Step 1. *Note 3 to Tables 310-16–310-19:*
156 A = #1/0 THWN Cu

Answer. **Conductors = #1/0 THWN Cu**

■ *Master Electrician's Test Tip.* Some examination boards permit the size of the overcurrent protection device to be selected from the ampacity of the conductors based on Note 3 to Tables 310-16–310-19. ■

A step-by-step procedure for finding the VA and amp ratings is shown in the form, STANDARD CALCULATION—MOBILE HOME. See Figure 2-25.

An 11′-0″ × 60′-0″ mobile home has a 120/240 V, 1ϕ service and the following loads. Find the load in VA and amps for the feeder circuit.

980 VA disposal .... 120 V, 1ϕ
1020 VA compactor .... 120 V, 1ϕ
1650 VA dishwasher .... 120 V, 1ϕ
180 VA vent fan .... 120 V, 1ϕ
12,000 VA cooktop .... 240 V, 1ϕ
10,000 VA oven .... 240 V, 1ϕ
5000 VA water heater .... 240 V, 1ϕ
5500 VA dryer .... 240 V, 1ϕ
10,000 VA heating unit .... 240 V, 1ϕ
600 VA blower motor .... 240 V, 1ϕ
4600 VA A/C unit .... 240 V, 1ϕ

Mobile Homes Parks—550-22; Table 550-22. A feeder circuit that connects the main service to a subpanel supplying power to a number of mobile home sites is computed at 16,000 VA for each site, or the calculated load for each site per 550-13, whichever is greater. For example, if the calculated load for a

OPTIONAL CALCULATION—ONE-FAMILY DWELLING (NEW OR EXISTING)

1. OTHER LOADS—220-30(b)

General lighting loads—220-30(b)(2)

2550 sq. ft. × 3 VA = 7650 VA

Small appliance and laundry loads—220-30(b)(1)

1500 VA × 3 circuits = 4500 VA

Special loads—220-30(b)(3)(4)

Dishwasher = 1450 VA

Disposal = 420 VA

Compactor = 708 VA

Water heater = 5000 VA

Cooktop = 8500 VA

Oven = 6500 VA

Dryer = 5000 VA

Blower motor = 840 VA

Attic fan = 828 VA

O.H. door mtr. = 780 VA

(total) 42,176 VA

Applying demand factors—Table 220-30(4)

First 10,000 VA × 100% = 10,000 VA

Remaining 32,176 VA × 40% = 12,870.4 VA

(total) 22,870.4 VA

PHASES

22,870.4 VA

2. HEATING OR A/C LOAD—Table 220-30(1)(2)(3)

Heating units (3 or less) = 15,000 VA × 65% = 9750 VA

Heating units (4 or more) = — VA × 40% = — VA

A/C unit = 4700 VA × 100% = 4700 VA

Heat pump = — VA × 100% = — VA

(largest load) 9750 VA 9750 VA

TOTAL = 32,620.4 VA

NEUTRAL

(Loads from Standard Calculation)

1. General lighting = 6202.5 VA
2. Cooking equipment = 6825 VA
3. Dryer = 3500 VA
4. Fixed appliance = 3139.5 VA
5. Heating or A/C = — VA
6. Largest motor = 207 VA

TOTAL = 19,874 VA

82.8 A

1ϕ service: I = $\frac{VA}{V}$

PHASES I = $\frac{32{,}620.4 \text{ VA}}{240 \text{ V}}$ = 135.9 A

NEUTRAL I = $\frac{19{,}874 \text{ VA}}{240 \text{ V}}$ = 82.8 A

135.9 A

Figure 2-24. The optional calculation is used to determine the VA and A for feeder circuits to dwelling units located in an apartment complex. Refer to Figure 2-23 for determining the VA and A ratings of the neutral.

STANDARD CALCULATION—MOBILE HOME

1. GENERAL LIGHTING LOADS—550-13(a); 550-13(b)(1)

660 sq. ft. × 3 VA = 1980 VA

Small appliance loads—550-13(a)

1500 VA × 2 circuits = 3000 VA

Laundry load—550-13(a)

1500 VA × 1 = 1500 VA

6480 VA

Applying demand factors—550-13(a)

First 3000 VA × 100% = 3000 VA

Next 3480 VA × 35% = 1218 VA

Remaining — VA × 25% = — VA

(total) 4218 VA — PHASES 4218 VA; NEUTRAL 4218 VA

2. FREE-STANDING RANGE—550-13(b)(5)—(nameplate rating)

Range = — VA — PHASES — VA; NEUTRAL — VA

3. APPLIANCE LOADS—550-13(b)(4)

Dishwasher = 1650 VA

Disposal = 980 VA

Water heater = 5000 VA

Cooktop = 12,000 VA

Oven = 10,000 VA

Dryer = 5500 VA

— = — VA

— = — VA

— = — VA

(total) 35,130 VA × 75% = 26,347.5 VA — PHASES 26,347.5 VA; NEUTRAL 14,422.5 VA

4. HEATING OR A/C LOAD—550-13(b)(2)

Heating unit = 10,000 VA × 100% = 10,000 VA

A/C unit = 4600 VA × 100% = 4600 VA

Heat pump = — VA × 100% = — VA

(largest load) 10,000 VA — PHASES 10,000 VA; NEUTRAL — VA

5. MOTOR LOADS—550-13(b)(2)

Vent fan = 1.5 A × 120 V = 180 VA

Compact. = 8.5 A × 120 V = 1020 VA

Blower mtr. = 2.5 A × 240 V = 600 VA

— = — A × — V = — VA

(total) 1800 VA — PHASES 1800 VA; NEUTRAL — VA

6. LARGEST MOTOR LOAD—550-13(b)(3)

ϕ 1020 VA × 25% = 255 VA — PHASES 255 VA

N 1020 VA × 25% = 255 VA — NEUTRAL 255 VA

TOTAL = PHASES 42,620.5 VA; NEUTRAL 18,895.5 VA

1ϕ service: $I = \frac{VA}{V}$

PHASES $I = \frac{42{,}620.5\ VA}{240\ V}$ = 177.6 A

NEUTRAL $I = \frac{18{,}895.5\ VA}{240\ V}$ = 78.7 A

Figure 2-25. The standard calculation is used to determine the VA and A for feeder circuits to mobile homes.

mobile home is 15,300 VA, the VA rating for the site is computed at 16,000 VA. If the VA rating for a mobile home is 18,850 VA, the VA rating is taken at 18,850 for the site because 18,850 VA is greater than 16,000 VA.

Demand factors can be applied for mobile home sites based on the number of mobile homes on the site. For example, a demand factor of 29% times the total VA rating of all the mobile homes on six sites can be applied. This reduced rating can be used to size the feeder circuit and components.

What is the VA rating for a feeder circuit supplying power to a subpanel with the following loads for each mobile home site? OCPDs are sized from conductor size.

Site	Computed Load	Size Feeder Circuit to Each Mobile Home
Lot 1	26,000 VA	125 A OCPD
Lot 2	21,500 VA	100 A OCPD
Lot 3	42,000 VA	175 A OCPD
Lot 4	32,500 VA	150 A OCPD
Lot 5	16,600 VA	100 A OCPD
Lot 6	48,000 VA	200 A OCPD
Lot 7	27,800 VA	125 A OCPD
Lot 8	52,000 VA	225 A OCPD
Lot 9	31,200 VA	150 A OCPD
Lot 10	24,000 VA	100 A OCPD

Step 1. *550-22(a):*

Lot 1 =	26,000 VA
Lot 2 =	21,500 VA
Lot 3 =	42,000 VA
Lot 4 =	32,500 VA
Lot 5 =	16,600 VA
Lot 6 =	48,000 VA
Lot 7 =	27,800 VA
Lot 8 =	52,000 VA
Lot 9 =	31,200 VA
Lot 10 =	24,000 VA
	321,600 VA

Step 2. *Table 550-22:* 10 sites = 27% demand factor

Step 3. *Table 550-22:* 321,600 VA × 27% = 86,832 VA

Answer. **VA rating = 86,832 VA**

Note: The neutral is calculated on the 120 V load.

Find the following elements of the feeder circuit based on the VA rating. Supply is 120/240 V, 1ϕ.

1. Size conductors for phases A and B using THWN Cu
2. Size main CB
3. Size switchgear

Conductors

Phases A and B

Step 1. *550-22; Text:* $I = \frac{VA}{V}$

$$I = \frac{86{,}832 \text{ VA}}{240 \text{ V}} = 361.8 \text{ A}$$

Step 2. *Table 310-16:* #500 kcmil = 380 A

Answer. **Conductors = #500 kcmil**

CB

Step 1. *240-3, Ex. 4; 240-6:* 380 A = 400 A CB

Answer. **CB = 400 A**

Switchgear

Step 2. *Text:* 400 A

Answer. **Switchgear = 400 A**

Feeder Circuits

Date ____________________

Chapter 2—TEST

Name ____________________________________

1. What is the amperage of a 120 V, 1ϕ feeder circuit that supplies a load of 3120 VA?
2. What is the amperage of a 240 V, 1ϕ feeder circuit with a total connected load of 15,600 VA?
3. What is the amperage of a 240 V, 3ϕ feeder circuit supplying a total load of 128,000 VA?
4. What is the amperage of a 208 V, 3ϕ feeder circuit serving a load that pulls 31.8 kVA for six hours each day?
5. What is the VA rating for a feeder circuit supplying two noncontinuous loads with 12,000 VA for lighting and 8000 VA for appliances? The supply voltage is 120/240 V, 1ϕ.
6. What is the VA rating of a 120/208 V feeder circuit serving a continuous load of 86 A per phase?
7. What is the demand factor for 205 general-purpose receptacles serving noncontinuous-duty outlets in an office building?
8. What is the amperage of a neutral load in a feeder circuit to an apartment complex that has a connected neutral load of 350 A?
9. What is the neutral load for a feeder circuit that supplies 295 A for loads without harmonic currents and 250 A for loads with harmonic currents?
10. What is the neutral load for a range that has a demand load of 10.4 kW?
11. What is the neutral load for a 5-wire, 2ϕ feeder circuit with a connected load of 175 A per phase?
12. What is the unbalanced neutral load for a feeder circuit with 129 A on phase A and 119 A on phase B?
13. What is the unbalanced neutral load for a 120 V feeder circuit with a load of 18 A on phase A?
14. What is the maximum unbalanced neutral load for a feeder circuit with 175 A on phase A and 150 A on phase B?
15. What is the neutral load for two 150 A ungrounded conductors tapped from a 120/208 V, 3ϕ feeder circuit?
16. What is the load diversity for twenty-two #1/0 THWN current-carrying Cu conductors pulled in the same raceway?
17. What is the load for each conductor in the same conduit where eighteen #4 THWN Cu conductors are pulled to a junction box (no load diversity)?
18. What size conduit is required for eight #3 TW Cu conductors?
19. Eight #3 TW Cu conductors in 2″ conduit are to be pulled out and replaced with #3 THHN Cu conductors. How many #3 THHN Cu conductors can be repulled through the 2″ conduit?
20. What is the ampacity of a #2 THWN Cu conductor connected to a 60°C circuit breaker?

_____ **21.** What is the ampacity of a #1/0 THHN Cu conductor connected to a 75 °C rated terminal?

_____ **22.** What size THWN Cu conductor is required for a 60 °C, 100 A circuit breaker?

_____ **23.** What is the ampacity of seven #4 THHN Cu current-carrying conductors routed through the same conduit (no load diversity)?

_____ **24.** What is the ampacity of six #12 THHN Cu conductors routed through a conduit exposed to an ambient temperature of 121 °F? Three of the conductors are current-carrying.

_____ **25.** What is the ampacity of eight #10 THHN Cu current-carrying conductors pulled in conduit routed through a smelting plant with an ambient temperature of 110 °F?

_____ **26.** Conductors in conduit terminate to 60 °C terminals of a device. What size overcurrent protection device is required for #6 THHN Cu conductors?

_____ **27.** What size conduit is required for twenty-two #12 THHN Cu conductors routed to a processing machine?

_____ **28.** What size conduit is required for fourteen #10 and sixteen #12 THHN Cu conductors?

_____ **29.** What size overcurrent protection device is required for three #400 kcmil THWN Cu conductors supplying a lighting and appliance panelboard?

_____ **30.** How many amps will a 150 A TDF hold for 10 seconds before tripping open the circuit (rule of thumb)?

_____ **31.** How many amps will a 200 A circuit breaker hold before tripping open the circuit (rule of thumb)?

_____ **32.** What is the voltage drop for a 208 V, 1ϕ feeder circuit with #3/0 THWN Cu conductors routed 302′-0″ and supplying a load of 185 A?

_____ **33.** What is the voltage drop for a 240 V, 3ϕ feeder circuit with #4/0 THWN Cu conductors routed 200′-0″ and supplying a load of 202 A?

_____ **34.** To operate with reasonable efficiency when supplying voltage to a load, what voltage drop is permissible for a 240 V, 1ϕ feeder circuit?

_____ **35.** What size THWN Cu conductors is required to tap a feeder circuit with an ampacity of 1200 A? The load to be served by the tap is a 46 A motor (10′ tap) used at continuous duty.

_____ **36.** What size THWN Cu conductors is required to make a 25′ tap from an 800 A feeder circuit?

_____ **37.** What size THWN Cu conductors is required to make a 100′ tap from a feeder supplied by #500 kcmil THWN Cu conductors?

_____ **38.** What size main circuit breaker is required to protect a lighting and appliance panelboard with #250 kcmil THWN Cu conductors tapped from the secondary of a 125 kVA, 120/208 V, 3ϕ transformer? Use the 25′ tap rule.

_____ **39.** A 25′ tap with primary and secondary plus transformer is made from a 225 A feeder circuit. The transformer primary is 480 V, 1ϕ and the secondary is 208 V, 1ϕ. What size primary feeder tap and secondary tap is required?

_____ **40.** How many 20 A circuit breakers with neutral connections can be installed in a 36 circuit panelboard without adding an in-line main?

_____ **41.** How many 100 A 2-pole mains can be installed in a power panel used as service equipment?

_____ **42.** What size THWN Al conductors is required to supply a noncontinuous-duty neutral load of 192 A?

_____ **43.** What size copper EGC is required for three #2/0 THWN Cu conductors parallel per phase and routed in three conduits? The circuit is protected by a 500 A overcurrent protection device.

__________ **44.** What is the total VA rating for the lighting loads on a feeder circuit with the following loads?

Store area . . . . 1600 sq. ft.
Show window lighting . . . . 20 linear feet
Track lighting . . . . 40 linear feet
Outside lighting on feeder. . . . . six 180 VA ballasts used at continuous duty
Sign . . . . 2100 VA

__________ **45.** What is the total VA rating for the receptacle loads on a feeder circuit with the following loads?

32 duplex receptacles used at continuous duty
52 duplex receptacles used at noncontinuous duty
80 linear feet of multioutlet assembly used at light duty

__________ **46.** What is the total VA rating for the receptacle loads on a feeder circuit with the following loads?

142 duplex receptacles used at noncontinuous duty
25 duplex receptacles used at continuous duty
100 linear feet of multioutlet assembly used at heavy duty

__________ **47.** What is the total VA rating for the motor loads on a feeder circuit with the following loads?

10 HP, 230 V, 3ϕ
15 HP, 230 V, 3ϕ
20 HP, 230 V, 3ϕ
25 HP, 230 V, 3ϕ

__________ **48.** What is the total VA rating for a feeder circuit in a 3000 sq. ft. office with the following loads? Use the form, STANDARD CALCULATION—COMMERCIAL AND INDUSTRIAL.

Show window lighting . . . . 42 linear feet
Track lighting . . . . 35 linear feet
Outside lighting . . . . 10 units at 180 VA each used at continuous duty
Sign . . . . 3200 VA
117 duplex outlets used at noncontinuous duty
12 duplex outlets used at continuous duty
35 linear feet of multioutlet assembly used at heavy duty
1600 VA, 120 V, 1ϕ copier used at continuous duty
1200 VA, 120 V, 1ϕ coffee pot used at continuous duty
2400 VA, 120 V, 1ϕ personal computer used at continuous duty
1800 VA, 120 V, 1ϕ personal computer used at continuous duty
1500 VA, 120 V, 1ϕ printer used at continuous duty
¾ HP, 208 V, 3ϕ motor
1 HP, 208 V, 3ϕ motor
2 HP, 208 V, 3ϕ motor
25 kW, 3ϕ heating unit
18 A, 3ϕ A/C unit

_______________ **49.** What is the total VA rating for a feeder circuit in a 2800 sq. ft. one-family dwelling with the following loads? Use the form, STANDARD CALCULATION—ONE-FAMILY DWELLING.

Range . . . . 12,000 VA, 240 V, 1ϕ
Water heater . . . . 6,000 VA, 240 V, 1ϕ
Dryer . . . . 5500 VA, 240 V, 1ϕ
Disposal . . . . 1200 VA, 120 V, 1ϕ
Compactor . . . . 1250 VA, 120 V, 1ϕ
Dishwasher . . . . 1600 VA, 120 V, 1ϕ
Vent hood . . . . 240 VA, 120 V, 1ϕ
Attic fan . . . . 1440 VA, 120 V, 1ϕ
Heating unit . . . . 20,000 VA, 240 V, 1ϕ
A/C unit . . . . 5040 VA, 240 V, 1ϕ
Blower motor . . . . 720 VA, 240 V, 1ϕ

_______________ **50.** What is the total VA rating for a feeder circuit in a 2800 sq. ft. one-family dwelling with the following loads? Use the form, OPTIONAL CALCULATION—ONE-FAMILY DWELLING (NEW OR EXISTING).

Range . . . . 12,000 VA, 240 V, 1ϕ
Dryer . . . . 5500 VA, 240 V, 1ϕ
Disposal . . . . 1200 VA, 120 V, 1ϕ
Compactor . . . . 1250 VA, 120 V, 1ϕ
Dishwasher . . . . 1600 VA, 120 V, 1ϕ
Vent hood . . . . 240 VA, 120 V, 1ϕ
Attic fan . . . . 1440 VA, 120 V, 1ϕ
Heating unit . . . . 20,000 VA, 240 V, 1ϕ
Blower motor . . . . 720 VA, 240 V, 1ϕ
A/C unit . . . . 5040 VA, 240 V, 1ϕ

_______________ **51.** What is the total VA rating for the feeder circuit to an apartment complex of twenty 950 sq. ft. dwelling units with the following loads? Use the form, STANDARD CALCULATION—MULTIFAMILY DWELLING.

Range . . . . 12,000 VA, 240 V, 1ϕ
Dryer . . . . 5500 VA, 240 V, 1ϕ
Disposal . . . . 1200 VA, 120 V, 1ϕ
Compactor . . . . 1250 VA, 120 V, 1ϕ
Dishwasher . . . . 1600 VA, 120 V, 1ϕ
Vent hood . . . . 240 VA, 120 V, 1ϕ
Attic fan . . . . 1440 VA, 120 V, 1ϕ
Heating unit . . . . 20,000 VA, 240 V, 1ϕ
A/C unit . . . . 5040 VA, 240 V, 1ϕ
Blower motor . . . . 720 VA, 240 V, 1ϕ

_______________ **52.** What is the total VA rating for the feeder circuit to an apartment complex of twenty 950 sq. ft. dwelling units with the following loads? Use the form, OPTIONAL CALCULATION—MULTIFAMILY DWELLING. *Note:* See problem 51 for sizing the neutral.

Range . . . . 12,000 VA, 240 V, 1ϕ
Dryer . . . . 5500 VA, 240 V, 1ϕ
Disposal . . . . 1200 VA, 120 V, 1ϕ
Compactor . . . . 1250 VA, 120 V, 1ϕ
Dishwasher . . . . 1600 VA, 120 V, 1ϕ
Vent hood . . . . 240 VA, 120 V, 1ϕ
Attic fan . . . . 1440 VA, 120 V, 1ϕ
Heating unit . . . . 20,000 VA, 240 V, 1ϕ
A/C unit . . . . 5040 VA, 240 V, 1ϕ
Blower motor . . . . 720 VA, 240 V, 1ϕ

Use for Problem 48

STANDARD CALCULATION—COMMERCIAL AND INDUSTRIAL

1. LIGHTING LOADS—220, Part B

General lighting loads—Table 220-3(b); 220-10(b)

____ sq. ft. × ____ VA × ____% = ____ VA

Show window loads—220-12

____′ × 200 VA = ____ VA

Track lighting loads—410-102

$\frac{\_\_\_\_'}{2'}$ × 180 VA = ____ VA

Outside lighting loads—220-10(b)

____ VA × ____ units × ____% = ____ VA

Sign lighting loads—220-3, FPN; 600-6(c)

____ VA × ____% = ____ VA

		PHASES	NEUTRAL
(total)	____ VA	VA	VA

2. RECEPTACLE LOADS—220-3(c)(5)

Noncontinuous duty—220-10(b); Table 220-13

#____ × 180 VA = ____ VA = ____ VA

Applying demand factors—Table 220-13

First 10,000 VA × 100% = ____ VA = ____ VA

Remaining ____ VA × 50% = ____ VA

Continuous duty—220-10(b) #____ × 180 VA × 125% = ____ VA

Multioutlet assembly—220-3(c)

Heavy duty = ____′ × 180 VA = ____ VA

Light duty = $\frac{\_\_\_\_'}{5'}$ × 180 VA = ____ VA

(total) ____ VA | VA | VA

3. SPECIAL LOADS—220-10(b)

____ = ____ VA × ____% = ____ VA

____ = ____ VA × ____% = ____ VA

____ = ____ VA × ____% = ____ VA

____ = ____ VA × ____% = ____ VA

____ = ____ VA × ____% = ____ VA

(total) ____ VA | VA | VA

4. HEATING OR A/C LOAD—220-21 ____ VA | VA | VA

5. MOTOR LOADS—430-22(a); 430-24; 430-25

				3ϕ only		
____ =	____ A ×	____% ×	____ V	× $\sqrt{3}$	=	____ VA
____ =	____ A ×	____% ×	____ V	× $\sqrt{3}$	=	____ VA
____ =	____ A ×	____% ×	____ V	× $\sqrt{3}$	=	____ VA
____ =	____ A ×	____% ×	____ V	× $\sqrt{3}$	=	____ VA

(total) ____ VA | VA | VA

6. LARGEST MOTOR LOAD—220-14

				3ϕ only			PHASES	NEUTRAL
ϕ	____ A ×	____% ×	____ V	× $\sqrt{3}$	× 25% =	____ VA	VA	VA
N	____ A ×	____% ×	____ V	× $\sqrt{3}$	× 25% =	____ VA	VA	VA
						TOTAL =	VA	VA

1ϕ service: $I = \frac{VA}{V}$

PHASES $I = \frac{\_\_\_\_ VA}{V}$ = ____ A

NEUTRAL $I = \frac{\_\_\_\_ VA}{V}$ = ____ A

3ϕ service: $I = \frac{VA}{V \times \sqrt{3}}$

$I = \frac{\_\_\_\_ VA}{V \times \sqrt{3}}$ = ____ A

$I = \frac{\_\_\_\_ VA}{V \times \sqrt{3}}$ = ____ A

If over 200 A—220-22; First 200 A × 100% = 200 A

Remaining ____ A × 70% = ____ A

A

STANDARD CALCULATION—ONE-FAMILY DWELLING

Use for Problem 49

1. GENERAL LIGHTING LOADS—Table 220-3(b)

______ sq. ft. × 3 VA = ______ VA

Small appliance loads—220-16(a)

______ VA × ______ circuits = ______ VA

Laundry load—220-16(b)

______ VA × 1 = ______ VA

______ VA

Applying demand factors—Table 220-11

First 3000 VA × 100% = 3000 VA

Next ______ VA × 35% = ______ VA

Remaining ______ VA × 25% = ______ VA

	PHASES		NEUTRAL
(total) ______ VA	VA		VA

2. COOKING EQUIPMENT LOADS—Table 220-19; Notes

Col. A ______ VA demand = ______ VA

Col. B ______ VA × ______% = ______ VA

Col. C ______ VA × ______% = ______ VA

(total) ______ VA | VA | × 70% = | VA

3. DRYER LOAD—220-18; Table 220-18

______ VA × ______% = ______ VA | VA | × 70% = | VA

4. FIXED APPLIANCE LOADS—220-17

Dishwasher = ______ VA

Disposal = ______ VA

Compactor = ______ VA

Water heater = ______ VA

______ = ______ VA

______ = ______ VA

______ = ______ VA

______ = ______ VA

(120 V Loads × 75%)

(total) ______ VA × 75% = ______ VA | VA | | VA

5. HEATING OR A/C LOAD—220-21

Heating unit = ______ VA × 100% = ______ VA

A/C unit = ______ VA × 100% = ______ VA

Heat pump = ______ VA × 100% = ______ VA

(largest load) ______ VA | VA | | VA

6. LARGEST MOTOR LOAD—220-14

ϕ ______ VA × 25% = ______ VA | VA

N ______ VA × 25% = ______ VA | ______ | | VA

TOTAL = | VA | | VA

1ϕ service: I = $\frac{VA}{V}$

PHASES I = $\frac{\text{\_\_\_\_\_\_ VA}}{V}$ = ______ A

NEUTRAL I = $\frac{\text{\_\_\_\_\_\_ VA}}{V}$ = ______ A

If over 200 A—220-22; First 200 A × 100% = 200 A

Remaining ______ A × 70% = ______ A

A

Use for Problem 50

OPTIONAL CALCULATION—ONE-FAMILY DWELLING (NEW OR EXISTING)

1. OTHER LOADS—220-30(b)

General lighting loads—220-30(b)(2)

________ sq. ft. × 3 VA = ________ VA

Small appliance and laundry loads—220-30(b)(1)

________ VA × ________ circuits = ________ VA

Special loads—220-30(b)(3)(4)

Dishwasher = ________ VA

Disposal = ________ VA

Compactor = ________ VA

Water heater = ________ VA

________ = ________ VA

________ = ________ VA

________ = ________ VA

________ = ________ VA

________ = ________ VA

________ = ________ VA

(total) ________ VA

Applying demand factors—Table 220-30(4)

First 10,000 VA × 100% = 10,000 VA

Remaining ________ VA × 40% = ________ VA

(total) ________ VA

PHASES

VA

2. HEATING OR A/C LOAD—Table 220-30(1)(2)(3)

Heating units (3 or less) = ________ VA × 65% = ________ VA

Heating units (4 or more) = ________ VA × 40% = ________ VA

A/C unit = ________ VA × 100% = ________ VA

Heat pump = ________ VA × 100% = ________ VA

(largest load) ________ VA — VA

TOTAL = VA

A

NEUTRAL

(Loads from Standard Calculation)

1. General lighting = ________ VA
2. Cooking equipment = ________ VA
3. Dryer = ________ VA
4. Fixed appliance = ________ VA
5. Heating or A/C = ________ VA
6. Largest motor = ________ VA

TOTAL = VA

A

1ϕ service: $I = \frac{VA}{V}$

PHASES $I = \frac{\_\_\_\_\ VA}{V}$ = ________ A

NEUTRAL $I = \frac{\_\_\_\_\ VA}{V}$ = ________ A

STANDARD CALCULATION—MULTIFAMILY DWELLING

Use for Problem 51

1. GENERAL LIGHTING LOADS—Table 220-3(b)

_______ sq. ft. × 3 VA × _______ units = _______ VA

_______ sq. ft. × 3 VA × _______ units = _______ VA

_______ sq. ft. × 3 VA × _______ units = _______ VA

_______ VA

Small appliance loads—220-16(a)

_______ VA × _______ circuits × _______ units = _______ VA

Laundry load—220-16(b)

_______ VA × 1 × _______ units = _______ VA

_______ VA

Applying demand factors—Table 220-11

First 3000 VA × 100% = 3000 VA

Next _______ VA × 35% = _______ VA

Remaining _______ VA × 25% = _______ VA

(total) _______ VA | **PHASES** _______ VA | **NEUTRAL** _______ VA

2. COOKING EQUIPMENT LOADS—Table 220-19, Notes

Col. A _______ units = _______ VA demand = _______ VA

Col. B _______ VA × _______ units × _______% = _______ VA

Col. C _______ VA × _______ units × _______% = _______ VA

(total) _______ VA | _______ VA × 70% = _______ VA

3. DRYER LOAD—220-18; Table 220-18

_______ VA × _______ units × _______% = _______ VA | _______ VA × 70% = _______ VA

4. FIXED APPLIANCE LOADS—220-17

Dishwasher = _______ VA

Disposal = _______ VA

Compactor = _______ VA

Water heater = _______ VA

_______ = _______ VA

_______ = _______ VA

(total) _______ VA × _______ units × 75% = _______ VA | _______ VA | (120 V Loads × units × 75%) _______ VA

5. HEATING OR A/C LOAD—220-21

Heating unit = _______ VA × 100% × _______ units = _______ VA

A/C unit = _______ VA × 100% × _______ units = _______ VA

Heat pump = _______ VA × 100% × _______ units = _______ VA

(largest load) _______ VA | _______ VA | _______ VA

6. LARGEST MOTOR LOAD—220-14

ϕ _______ VA × 25% = _______ VA | _______ VA

N _______ VA × 25% = _______ VA | | _______ VA

TOTAL = _______ VA | _______ VA

1ϕ service: $I = \frac{VA}{V}$

PHASES $I = \frac{\_\_\_\_ VA}{V} =$ _______ A

NEUTRAL $I = \frac{\_\_\_\_ VA}{V} =$ _______ A

3ϕ service: $I = \frac{VA}{V \times \sqrt{3}}$

$I = \frac{\_\_\_\_ VA}{V \times \sqrt{3}} =$ _______ A

$I = \frac{\_\_\_\_ VA}{V \times \sqrt{3}} =$ _______ A

If over 200 A—220-22; First 200 A × 100% = 200 A

Remaining _______ A × 70% = _______ A

_______ A

Use for Problem 52

OPTIONAL CALCULATION—MULTIFAMILY DWELLING

1. OTHER LOADS—220-32

General lighting loads—220-32(c)(2) — **PHASES**

_________ sq. ft. × 3 VA × _____ units = _________ VA — VA

_________ sq. ft. × 3 VA × _____ units = _________ VA

_________ sq. ft. × 3 VA × _____ units = _________ VA

_________ sq. ft. × 3 VA × _____ units = _________ VA

Small appliance and laundry loads—220-32(c)(1)

_________ VA × _____ circuits × _____ units = _________ VA — VA

Special loads—220-32(c)(3)

Dishwasher = _________ VA

Disposal = _________ VA

Compactor = _________ VA

Water heater = _________ VA

_____________ = _________ VA

_____________ = _________ VA

_____________ = _________ VA

_____________ = _________ VA

_____________ = _________ VA

_____________ = _________ VA

(total) _________ VA × _____ units = _________ VA — VA

2. HEATING OR A/C LOAD—220-32(c)(5)

Heating unit = _________ VA × 100% × _____ units = _________ VA

A/C unit = _________ VA × 100% × _____ units = _________ VA

Heat pump = _________ VA × 100% × _____ units = _________ VA

(largest load) _________ VA — VA

TOTAL = VA

Applying demand factors—Table 220-32

_________ VA × _________% = _________ VA

NEUTRAL

(Loads from Standard Calculation)

1. General lighting = _________ VA
2. Cooking equipment = _________ VA
3. Dryer = _________ VA
4. Fixed appliance = _________ VA
5. Heating or A/C = _________ VA
6. Largest motor = _________ VA

TOTAL = VA

A

1ϕ service: $I = \frac{VA}{V}$

PHASES $I = \frac{\_\_\_\_\_\_ VA}{V}$ = _________ A

NEUTRAL $I = \frac{\_\_\_\_\_\_ VA}{V}$ = _________ A

3ϕ service: $I = \frac{VA}{V \times \sqrt{3}}$

$I = \frac{\_\_\_\_\_\_ VA}{V \times \sqrt{3}}$ = _________ A

$I = \frac{\_\_\_\_\_\_ VA}{V \times \sqrt{3}}$ = _________ A

If over 200 A—220-22; First 200 A × 100% = 200 A

Remaining _______ A × 70% = _________ A

A

Motor Circuits

Chapter 3

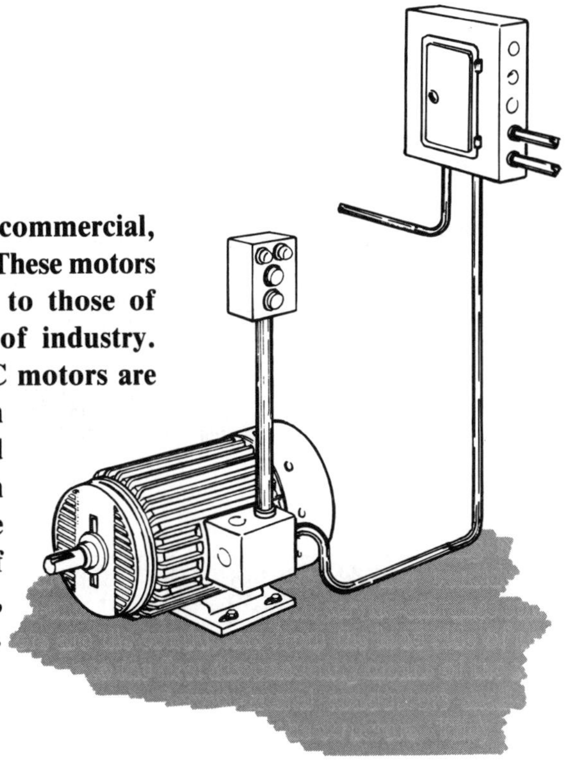

The number of motors used in residential, commercial, and industrial locations increases each year. These motors range in size from fractional-horsepower to those of sufficient horsepower to meet the needs of industry. The various types and sizes of AC and DC motors are protected by proper design and installation of motor circuits that include properly rated conductors, controls, starters, protection devices, and disconnecting means. Article 430 covers the design and installation of electrical systems containing motor circuits, controllers, and other important elements.

CONDUCTORS—430, PART B

Conductors supplying motors shall have the capacity rated in amps to supply the full-load current rating of the motor. Conductors supplying motors are selected at 60°C, 75°C, or 90°C per Table 310-16. Conductors serving motor loads are exposed to high inrush and continuous-duty operating currents, which require larger conductors to stand the heat and strain on starting and driving loads.

Single Motors—430-22(a)

Conductors supplying a single motor are sized by multiplying the full-load current rating of the motor by 125%. The full-load current rating of a motor is selected from Table 430-148 for 1ϕ motors and Table 430-150 for 3ϕ motors. The full-load current ratings are based on the horsepower, voltage, and phase of each motor.

■ *Master Electrician's Test Tip.* The full-load current ratings in Tables 430-147–430-150 are used to size conductors and not the running current listed on the nameplate of the motor. For example, the full-load current rating of a 40 HP, 230 V, 3ϕ motor is 104 A. The full-load current rating of a 60 HP, 460 V, 3ϕ motor is 77 A. The higher the voltage rating is, the lower the full-load current rating will be.■

Motors can have four to ten times their full-load current rating when they are started. For example, a motor with a 40 A full-load current rating can have a starting inrush current rating of 400 A (40 A × 10 = 400 A). When the motor comes up to its running speed, it will pull 40 A if fully loaded.

Note that the text above Table 430-148 and footnotes below Table 430-150 require 200 V motors to have the full-load current ratings listed in the 230 V columns of the Tables increased by 115%. Motors of 208 V are required to have the full-load current ratings in the 230 V column increased by 110%. For example, the full-load current rating of a 20 HP, 3ϕ motor rated at 200 V is 62.1 A (54 A × 115% = 62.1 A) and the same motor rated at 208 V has a full-load current rating of 59.4 A (54 A × 110% = 59.4 A). See Figure 3-1.

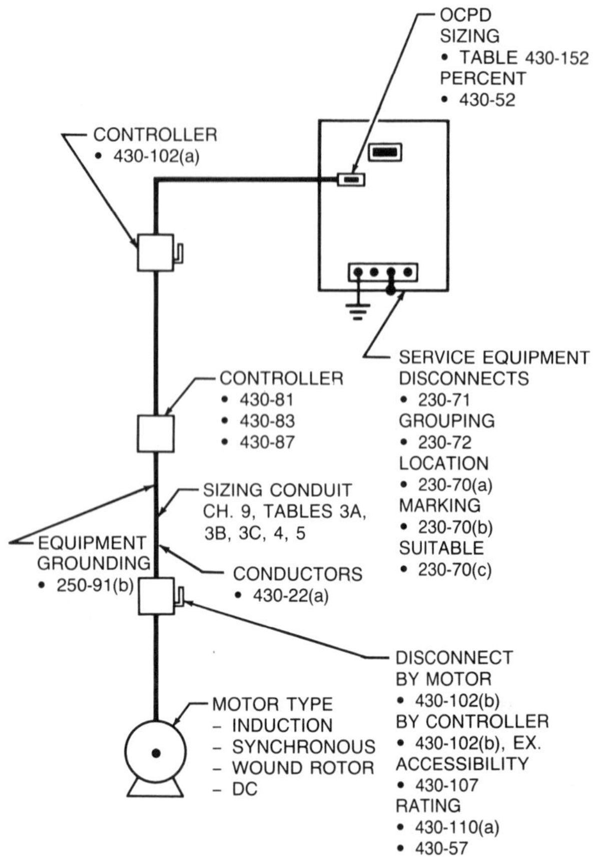

SIZING CONDUCTORS FOR ONE MOTOR—430-22(a)

What size THWN Cu conductors is required to supply power to a 40 HP, 230 V, 3ϕ motor with a 1.15 service factor and code letter G?

Step 1. *430-6(a); Table 430-150:* 40 HP = 104 A

Step 2. *430-22(a):* 104 A × 125% = 130 A

Step 3. *Table 310-16:* #1 THWN Cu = 130 A

Answer. **Conductors = #1 THWN Cu**

Figure 3-1. Conductors supplying power for one motor shall have an ampacity rating of at least 125% of the full-load current rating of the motor.

What size THWN Cu conductors is required to supply a 25 HP, 208 V, 3ϕ motor?

Step 1. *430-6(a); Table 430-150:*
25 HP = 68 A × 110% = 74.8 A

Step 2. *430-22(a):* 74.8 A × 125% = 93.5 A

Step 3. *Table 310-16, 75°C Column:*
#3 THWN Cu = 100 A

Answer. **Conductors = #3 THWN Cu**

The full-load current ratings listed in Table 430-150 for synchronous motors are based upon unity power factor. Synchronous motors can be operated at values of power factor below 100%. Synchronous motors operating with a 90% power factor shall have their full-load current ratings increased 110%, and motors with 80% power factor are required to be increased by 125%.

What is the full-load current rating and what size THWN Cu conductors is required for a 75 HP, 460 V, 3ϕ synchronous motor rated at 80% power factor?

FLC

Step 1. *Table 430-150:* 75 HP = 78 A

Step 2. *Table 430-150, Footnotes:* 80% PF = 125%
78 A × 125% = 97.5 A

Answer. **FLC = 98 A**

Conductors

Step 3. *430-22(a):* 97.5 A × 125% = 121.9 A

Step 4. *Table 310-16, 75°C Column:*
#1 THWN Cu = 130 A

Answer. **Conductors = #1 THWN Cu**

Two or More Motors—430-24

For two or more motors tapped from a feeder circuit, the largest motor is multiplied by 125% and the full-load current ratings of the remaining motors are added to this computation. The taps from the feeder circuit can be made in 10′ or 25′ lengths per 430-28. Any size 10′ tap can be made from larger feeder circuit conductors. A feeder circuit with #250 kcmil THWN Cu conductors can be #10 THWN Cu conductors tapped and routed 10′ or less to a 30 A fusible disconnect.

For tapped conductors routed more than 10′ in length but not more than 25′, the tapped conductors shall be at least ⅓ of the feeder conductors. For example, a 25′ tap from #250 kcmil THWN Cu conductors having an ampacity of 255 A is rated at least 85 A (⅓ × 255 A = 85 A). See Figure 3-2.

What size THWN Cu conductors is required to supply the lugs of a panelboard that provides power to a 10 HP, 15 HP, and 25 HP, 230 V, 3ϕ group of motors?

Step 1. *430-6(a); Table 430-150:*
10 HP = 28 A FLC
15 HP = 42 A FLC
25 HP = 68 A FLC

Step 2. *430-24:* 68 A × 125% = 85 A
Plus other motors 42 A
28 A
Total = 155 A

Step 3. *Table 310-16:* #2/0 THWN Cu = 175 A

Answer. **Conductors = #2/0 THWN Cu**

What size THWN Cu conductors is required to supply busbars in a gutter that provides power to a 30 HP synchronous motor with an 80% power factor and a 50 HP synchronous motor with a 90% power factor? Both motors are 230 V, 3ϕ.

Step 1. *430-6(a); Table 430-150:*
30 HP = 63 A
50 HP = 104 A

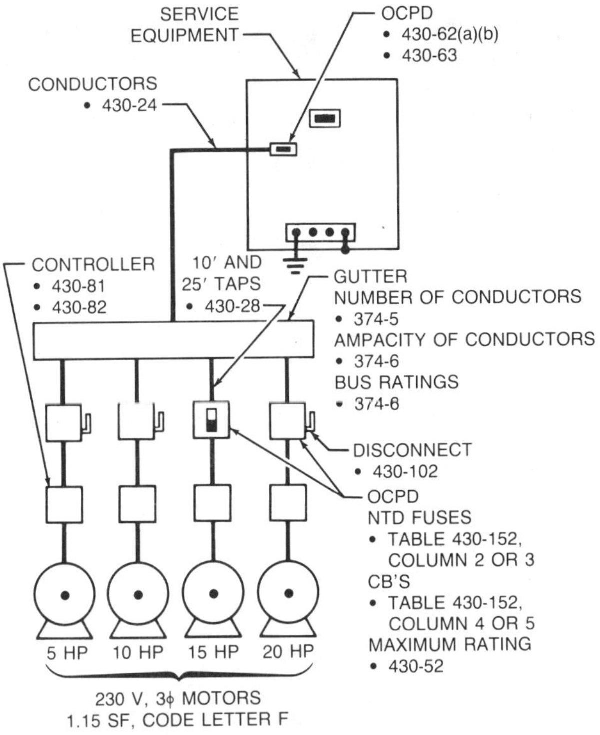

SIZING CONDUCTORS FOR TWO OR MORE MOTORS—430-24

What size THWN Cu conductors is required to supply power on a feeder circuit to a gutter serving a group of motors?

Step 1. *430-6(a); Table 430-150:*
5 HP = 15.2 A
10 HP = 28 A
15 HP = 42 A
20 HP = 54 A

Step 2. *430-24:*
54 A × 125% =
67.5 A + 15.2 A + 28 A + 42 A = 152.7 A

Step 3. *Table 310-16:* #2/0 THWN Cu = 175 A

Answer. **Conductors = #2/0 THWN Cu**

Figure 3-2. Conductors supplying power on a feeder circuit serving a group of motors shall have an ampacity of at least 125% of the largest motor plus the full-load current rating of the remaining motors.

Step 2. *Table 430-150, Footnotes:*
80% PF = 125%
90% PF = 110%
63 A × 125% = 78.75 A FLC
104 A × 110% = 114.4 A FLC

Step 3. *430-24:* 114.4 A × 125% = 143 A
Plus other motors = 78.75 A
Total = 221.75 A

Step 4. *Table 310-16, 75°C Column:*
#4/0 THWN Cu = 230 A

Answer. **Conductors = #4/0 THWN Cu**

Motors with Other Loads—430-25

A motor with other loads is computed by multiplying the largest motor by 125% plus the other loads calculated at continuous duty, noncontinuous duty with demand factors, or both, per 430-25, 430-24, and 430-22(a). See Figure 3-3.

A 10 HP, 230 V, 3ϕ motor has a full-load current rating of 28 A per Table 430-150. The load for this motor is 35 A (28 A × 125% = 35 A) per 430-22(a). If the other loads consist of 19 A of continuous duty and 18 A of noncontinuous duty, the total amperage of the continuous loads is 24 A (19 A × 125% = 23.75 A) per 220-10(b) and the noncontinuous load is 18 A (18 A × 100% = 18 A) per 220-10(b). The total load is 77 A.

Step 1. Motor load. *Table 430-150:* 10 HP = 28 A

Step 2. *430-25; 430-22(a):* 28 A × 125% = 35 A

Step 3. Other loads. *220-10(b):*
19 A × 125% = 23.75 A
18 A × 100% = 18 A
Total = 41.75 A

Step 4. Total loads. *430-22(a); 220-10(b):*
Motor load = 35 A
Other loads = 41.75 A
Total = 76.75 A

Answer. **Total load = 77 A**

What is the total load and what size THWN Cu conductors is required to supply a 50 HP, 550 V, 3ϕ motor with a 50 A continuous-duty load and a 45 A noncontinuous-duty load?

Loads

Step 1. Motor loads. *Table 430-150:* 50 HP = 52 A

Step 2. *430-25; 430-22(a):* 52 A × 125% = 65 A

Step 3. Other loads. *220-10(b):*
50 A × 125% = 62.5 A
45 A × 100% = 45 A
Total = 107.5 A

Step 4. Total loads. *430-22(a); 220-10(b):*
Motor load = 65 A
Other loads = 107.5 A
Total = 172.5 A

Answer. **Total load = 173 A**

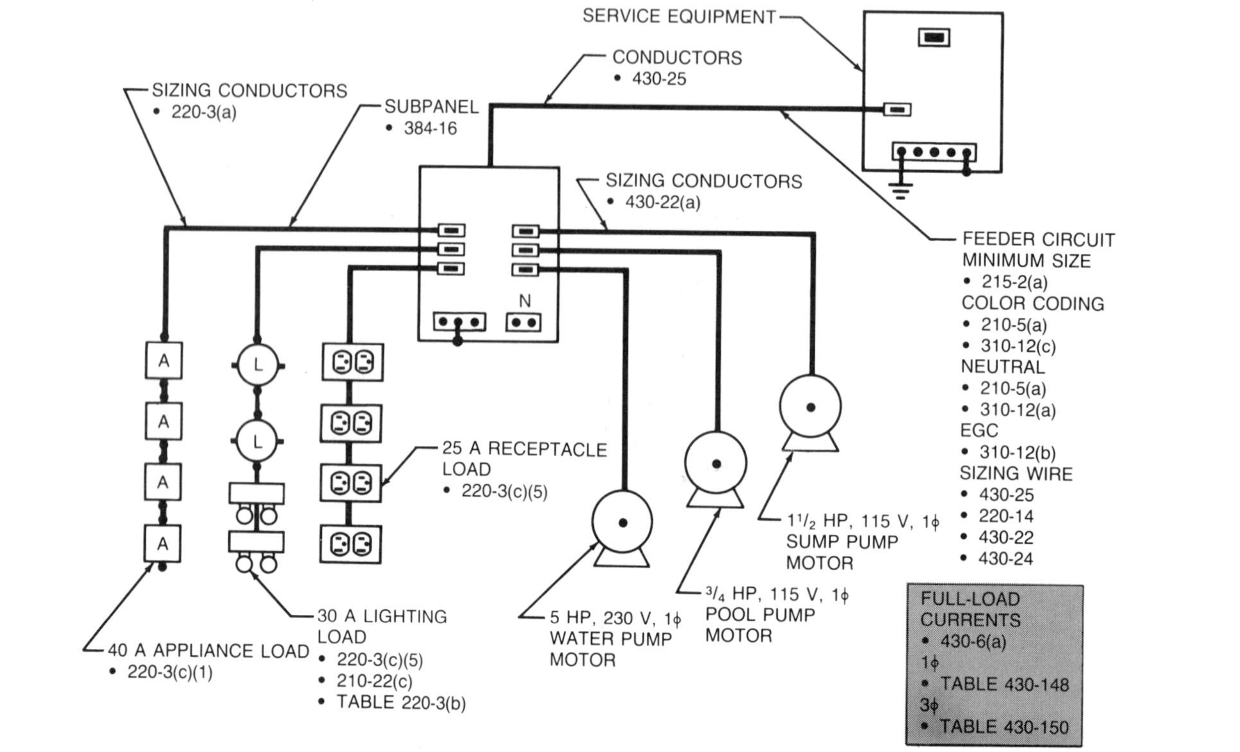

SIZING CONDUCTORS FOR MOTORS AND OTHER LOADS—430-25

What size THWN Cu conductors is required to supply the feeder circuit loads? All loads are continuous duty.

Step 1. Motor loads. *430-6(a); Table 430-148:*
5 HP = 28 A
1½ HP = 20 A
¾ HP = 13.8 A

Step 2. Other loads. *220-10(b):*
Lighting loads = 30 A × 125% = 37.5 A
Receptacle loads = 25 A × 125% = 31.25 A
Appliance loads = 40 A × 125% = 50 A

Step 3. Sizing feeder conductors. *430-25:*
28 A × 125% = 35 A + 20 A + 13.8 A + 37.5 A + 31.25 A + 50 A = 187.55 A

Step 4. *Table 310-16:* #3/0 THWN Cu = 200 A

Answer. **Conductors = #3/0 THWN Cu**

Figure 3-3. Conductors supplying power to motors and other loads shall have an ampacity of at least 125% of the largest motor plus the full-load current rating of the remaining motors in the group and other loads calculated at noncontinuous- or continuous-duty operation.

Conductors

Step 1. *Table 310-16:* #2/0 THWN Cu = 175 A

Answer. **Conductors = #2/0 THWN Cu**

Two or more motors with other loads are figured by multiplying the largest motor of the group by 125% and adding this total to the remaining motors in the group, which are computed at continuous duty, noncontinuous duty with demand factors, or both, per 430-25 and 430-24.

What is the total load for one 15 HP, one 20 HP, and one 25 HP, 230 V, 3ϕ motors with other loads of 23 A at continuous duty and 26 A at noncontinuous duty?

Step 1. Motor loads. *Table 430-150:*
15 HP = 42 A
20 HP = 54 A
25 HP = 68 A

Step 2. *430-25; 430-24:*
68 A × 125% = 85 A
54 A
42 A
Total = 181 A

Step 3. Other loads. *220-10(b):*
23 A × 125% = 28.75 A
26 A × 100% = 26 A
Total = 54.75 A

Step 4. Total loads. *430-24; 220-10(b):*
Motor loads = 181 A
Other loads = 54.75 A
Total = 235.75 A

Answer. **Total load = 236 A**

What is the total load and what size THWN Cu conductors is required to serve a 5 HP, 7½ HP, and 10 HP, 208

V, 3ϕ group of motors with a 31 A continuous-duty load and 39 A noncontinuous-duty load?

Loads

Step 1. Motor loads. *Table 430-150 and Footnotes:*
5 HP = 15.2 A × 110% = 16.72 A
$7\frac{1}{2}$ HP = 22 A × 110% = 24.2 A
10 HP = 28 A × 110% = 30.8 A

Step 2. *430-25; 430-24:*
30.8 A × 125% = 38.5 A
24.2 A
16.72 A
Total = 79.42 A

Step 3. Other loads. *220-10(b):*
31 A × 125% = 38.75 A
39 A × 100% = 39 A
Total = 77.75 A

Step 4. Total loads. *430-24; 220-10(b):*
Motor loads = 79.42 A
Other loads = 77.75 A
Total = 157.17 A

Answer. **Total load = 157 A**

Conductors

Step 1. *Table 310-16:* #2/0 THWN Cu = 175 A

Answer. **Conductors = #2/0 THWN Cu**

Motors with Duty Cycles—Table 430-22(a), Ex.

Table 430-22(a), Exception lists four applications in the Classification of Service column. These are (1) Short-time Duty, (2) Intermittent Duty, (3) Periodic Duty, and (4) Varying Duty. Duty cycle motors in this table operate for 5, 15, 30, or 60 minutes. See Figure 3-4.

A 15 minute rated motor operating for a full 15 minutes must rest for 45 minutes before it is run again (60 − 15 = 45 minutes). The same motor that does not operate for the full 15 minutes does not have to rest for a full 45 minutes before it can be run again. For example, a 5 minute rated 30 HP, 460 V, 3ϕ motor has a full-load current rating of 40 A. The motor is used to raise and lower a roll of paper on a printing press. The load for this motor is determined by multiplying 40 A by 110% (40 A × 110% = 44 A).

What is the load and what size THWN Cu conductors is required to supply a 100 HP, 440 V, 3ϕ short-time duty motor? The motor has a 30 minute rating.

Load

Step 1. *Table 430-150:* 100 HP = 124 A

Step 2. *Table 430-22(a), Ex.:* 124 A × 150% = 186 A

Answer. **Load = 186 A**

Conductors

Step 1. *Table 310-16:* #3/0 THWN Cu = 200 A

Answer. **Conductors = #3/0 THWN Cu**

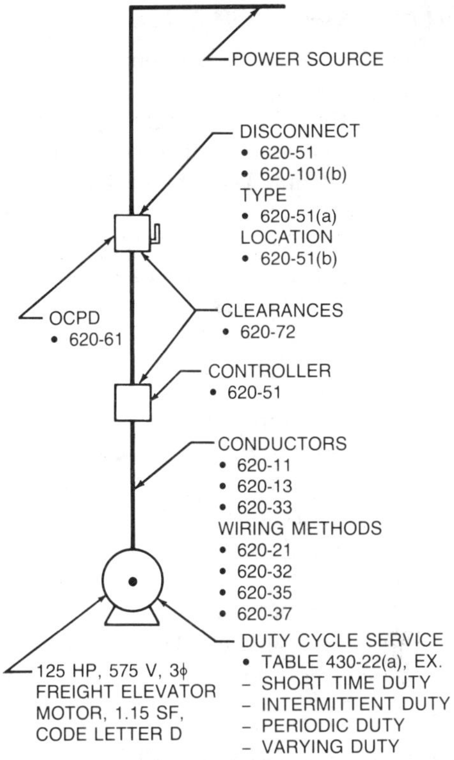

SIZING CONDUCTORS FOR A MOTOR WITH DUTY CYCLE SERVICE—TABLE 430-22(a), EX.

What size THWN Cu conductors is required to supply power to a freight elevator motor used for 30 minutes at intermittent duty?

Step 1. *430-6(a); Table 430-150:* 125 HP = 125 A

Step 2. *430-22(a); Table 430-22(a), Ex.:*
125 A × 90% = 112.5 A

Step 3. *Table 310-16:* #2 THWN Cu = 115 A

Answer. **Conductors = #2 THWN Cu**

Figure 3-4. Conductors supplying a motor with a duty cycle service shall have ratings equal to the percentages listed in Table 430-22(a), Ex.

OVERCURRENT PROTECTION DEVICE—430, PART D

Overcurrent protection devices are selected with a rating large enough to hold the inrush current of a motor and allow the motor to start and accelerate up to its running speed. The inrush current may be four to ten times the full-load current rating of the motor utilized. Overcurrent protection devices are sized larger than the conductors supplying such motors so they will not trip open due to the high inrush current.

Requirements—430-6(a); 430-17

Section 430-6(a) lists the requirements for general motor applications. This section refers to Tables 430-147–430-150 for the full-load current ratings of motors to select the overcurrent protection device to allow the motor to start and run. These tables are also used to select the full-load current ratings to size the conductors to individual motors or groups of motors. Section 430-6(a) refers to the nameplate of motors to select the full-load current ratings to size the overloads.

■ *Master Electrician's Test Tip.* When a motor has a full-load current rating listed on its nameplate, this rating is used to size the overloads. The full-load current ratings in Tables 430-147–430-150 are used to size the conductors and the overcurrent protection device to the motor.■

The full-load current ratings in Tables 430-147–430-150 are used to size switches and panels to hold the overcurrent protection device to permit motors to start and run. These full-load current ratings are also used to select and size disconnects to disconnect motor circuits.

Locked Rotor Current Values—Table 430-151

Table 430-151 lists the locked rotor current (LRC) values of motors based on horsepower and voltage. The locked rotor current values are based on six times the full-load current ratings from Table 430-150, which are also based on horsepower and voltage. For example, a 60 HP, 480 V, 3ϕ motor has a locked rotor current rating of 462 A per Table 430-151. This value of locked rotor current can be verified by multiplying the full-load current rating of 77 A (per Table 430-150) times 6 (77 A × 6 = 462 A).

What is the locked rotor current of a 40 HP, 230 V, 3ϕ motor per Table 430-151?

Step 1. *Table 430-150:* 40 HP = 104 A
104 A × 6 = 624 A

Step 2. *Table 430-151:* 40 HP = 624 A

Answer. **LRC = 624 A**

■ *Master Electrician's Test Tip.* Some testing authorities use the locked rotor current values of Table 430-151 for examinations instead of the motor's code letter to determine the locked rotor currents of motors unless otherwise stated.■

Locked Rotor Current Values—Table 430-7(b)

Table 430-7(b) lists kVA per horsepower, which is based on the code letter of the motor. See Figure 3-5.

A 60 HP, 480 V, 3ϕ motor with code letter G has a maximum of 6.29 kVA per horsepower. The minimum is 5.6 kVA per horsepower. What is the locked rotor current based on the motor's code letter? (Use maximum size.)

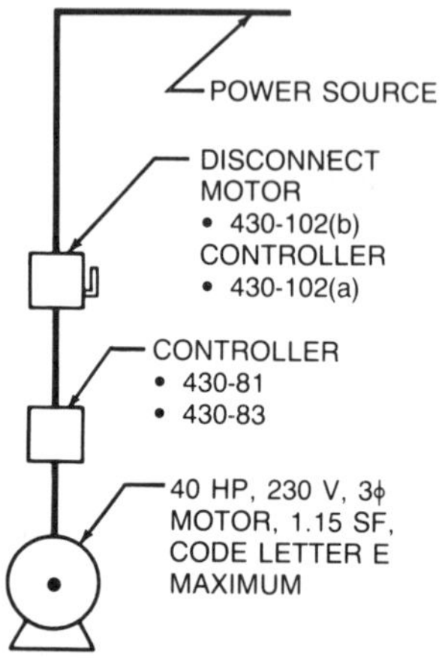

DETERMINING LOCKED ROTOR CURRENT OF A MOTOR—TABLE 430-151

What is the LRC of a 40 HP, 230 V, 3ϕ motor?

HP and V

Step 1. *Table 430-150:* 40 HP = 104 A

Step 2. *Table 430-151 and Footnotes:*
104 A × 6 = 624 A

Step 3. *Table 430-151, Column 5:* LRC = 624 A

Answer. **LRC = 624 A**

Code letter

Step 1. *Table 430-7(b):*
Code letter E = 4.99 kVA per HP

Step 2. $LRC = \frac{kVA \times 1000 \times HP}{V \times \sqrt{3}}$

$$LRC = \frac{4.99\text{ kVA} \times 1000 \times 40\text{ HP}}{230\text{ V} \times 1.732} = 501.5\text{ A}$$

Answer. **LRC = 501.5 A**

Figure 3-5. The locked rotor current of a motor can be determined by Table 430-151 based on HP and voltage, or by Table 430-7(b) based on the code letter.

Step 1. $LRC = \frac{kVA \times HP \times 1000}{V \times \sqrt{3}}$

$$LRC = \frac{6.29\text{ kVA} \times 60\text{ HP} \times 1000}{480\text{ V} \times 1.732} = 454\text{ A}$$

Answer. **LRC = 454 A**

What is the locked rotor current of a 40 HP, 230 V, 3ϕ motor with code letter S per Table 430-7(b)? Use minimum rating.

Step 1. *Table 430-7(b):* Code letter S = 16 kVA per HP

Step 2. $LRC = \frac{kVA \times HP \times 1000}{V \times \sqrt{3}}$

$$LRC = \frac{16\text{ kVA} \times 40\text{ HP} \times 1000}{230\text{ V} \times 1.732} = 1608\text{ A}$$

Answer. **LRC = 1608 A**

■ *Master Electrician's Test Tip.* Table 430-151 is based on six times the full-load current ratings listed in Table 430-150, which are based on the horsepower and voltage of each motor. Table 430-7(b) will usually produce lower locked rotor currents for code letters A through G and higher locked rotor currents for code letters H through V.■

Selecting Devices—Table 430-152; 430-52

Section 430-52 lists the minimum percentages that motors can be protected per Table 430-152. Exception 1 to 430-52 permits the next higher overcurrent protection device to be selected where the percentages of Table 430-152 do not correspond to a standard device listed in 240-6. For example, a motor with a full-load current rating of 54 A per Table 430-150 times 250% for a CB per Table 430-152 is equal to 135 A. By applying 430-52 and Table 430-152, the minimum size CB is 125 A. The size overcurrent protection device per 430-52, Ex. 1 is 150 A CB. This exception is only applied if the motor accelerating the driven load trips open the overcurrent protection device.

Exception 2 to 430-52 can be applied when motors fail to start and run due to the code letter or driven load. Motors with full-load current ratings of 100 A or less are allowed to have their ratings multiplied by 400% to find the size overcurrent protection device. If the full-load current rating is over 100 A, 430-52, Ex. 2c requires the full-load current rating of the motor to be multiplied by 300%. The full-load current rating of 54 A × 400% equals 216 A. The maximum size CB is 200 A per 240-6. Note that 300% or 400% cannot be exceeded. See Figure 3-6.

The five columns in Table 430-152 are used to select the size overcurrent protection device to allow motors to start and run. The first column, Types of Motors, includes

1. Squirrel-cage induction motors (1ϕ)
2. Synchronous
3. Wound-rotor
4. DC

The starting methods by which motors are started are listed in Column 1. These are

1. Full voltage
2. Resistor
3. Reactor
4. Autotransformer

The most important requirement in Column 1 of Table 430-152 is the code letter of the motor. When there is no code letter on the nameplate of the motor, the percentage in the Table based on the type of protection device selected is maximum as it is for code letters F to V. Code letters B to E have the next lower rating while code letter A has the minimum rating of all the code letters in the Table. Column 2 (Nontime Delay Fuses) lists the percentages by which nontime

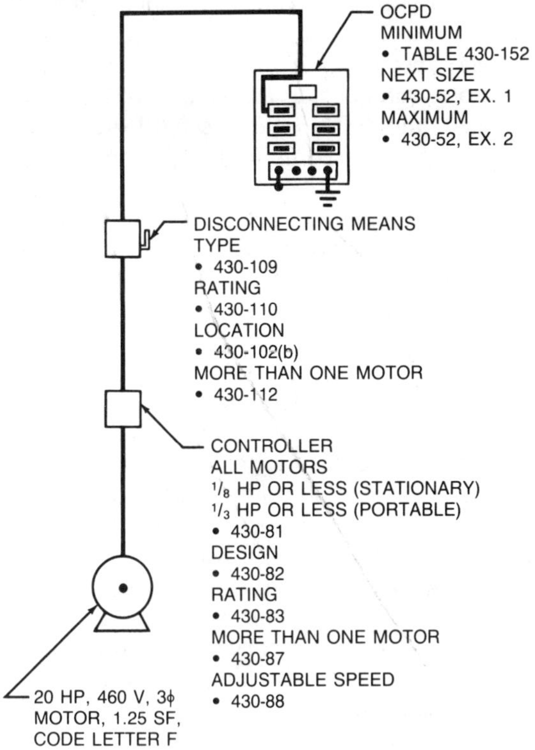

SELECTING MINIMUM, NEXT, AND MAXIMUM SIZE OCPD TO START AND RUN A MOTOR

What size CB is required to start and accelerate the motor up to its running speed using the minimum, next, and maximum size?

Minimum

Step 1. *430-6(a); Table 430-150:* 20 HP = 27 A

Step 2. *Table 430-152:* 27 A × 250% = 67.5 A

Step 3. *240-3, Ex. 6; 240-6:*
60 A CB is the next lower standard size

Answer. **CB = 60 A CB**

Next

Step 1. *430-52, Ex. 1; 240-6:*
70 A CB is the next higher standard size

Answer. **CB = 70 A CB**

Maximum

Step 1. *430-52, Ex. 2; 240-6:* 27 A × 400% = 108 A

Step 2. *240-6:*
100 A CB is the next lower standard size

Answer. **CB = 100 A CB**

Figure 3-6. The minimum size OCPD to start and run a motor is determined by multiplying the full-load current of the motor by the percentages in Table 430-152. The percentage selected is based on the OCPD chosen for the motor circuit.

delay fuses can be selected based on the requirements in Column 1. Nontime delay fuses hold five times their rating for ¼ to 2 seconds. For example, a 150 A nontime delay fuse holds 750 A (150 A × 5 = 750 A). If the motor will not start and run, 430-52, Ex. 2a can be applied, and 400% times the full-load current of the motor can be used up to a 600 A nontime delay fuse.

The percentages listed in Column 3 [Dual Element (Time Delay Fuse)] are based on the requirements of Column 1. Time delay fuses hold five times their ratings for 10 seconds. This time delay allows most motors to start and run. If the percentages in Column 3 will not start and accelerate the motor up to running speed, 430-52, Ex. 2b will allow the motor's full-load current rating to be increased to 225%.

The percentages in Column 4 (Instantaneous Trip Breaker) times the motor's full-load current is 700% except for DC motors. Instantaneous trip breakers do not have thermal overload protection. They only trip open due to inrush current, short circuit, or ground-fault currents that exceed their trip open settings. An instantaneous trip circuit breaker holds three times its rating in the low setting, five times its rating in the mid range setting, and ten times its rating in the high setting. If the settings in Table 430-152 will not allow a motor to start and run, 430-52 will permit up to 1300% times the full-load current of the motor.

Table 430-152, Column 5 lists the percentages for inverse time circuit breakers. These breakers hold approximately three times their rating for periods of time based on their frame size. Where the percentage in Table 430-152 will not hold the inrush current of the circuit, 430-52, Ex. 2c allows 400% times the full-load current of the motor to be used for motors with a 100 A or less full-load current rating. Only 300% times the full-load current of the motor can be used where the full-load current rating is over 100 A. See Figure 3-7 for the application of Table 430-152 in selecting overcurrent protection devices to allow motors to start and run.

■ *Master Electrician's Test Tip.* The percentage in 430-52, Ex. 2 shall not be exceeded. If the percentage times the full-load current of the motor, per Table 430-150, does not correspond with a standard size device per 240-6, the next lower rating shall be used.■

Sizing Nontime Delay Fuses—Table 430-152, Column 2. The type of motor, number of phases, starting method, and code letter must be known before the percentage in Column 2 is selected. See Figure 3-8.

What size nontime delay fuses is required to allow a 10 HP, 230 V, 3ϕ squirrel-cage induction motor with full voltage starting and code letter H to start and run?

Step 1. Type of motor (squirrel-cage induction motor). *Table 430-152, Column 1:*

Step 2. Number of phases (three). *Table 430-152, Column 1:*

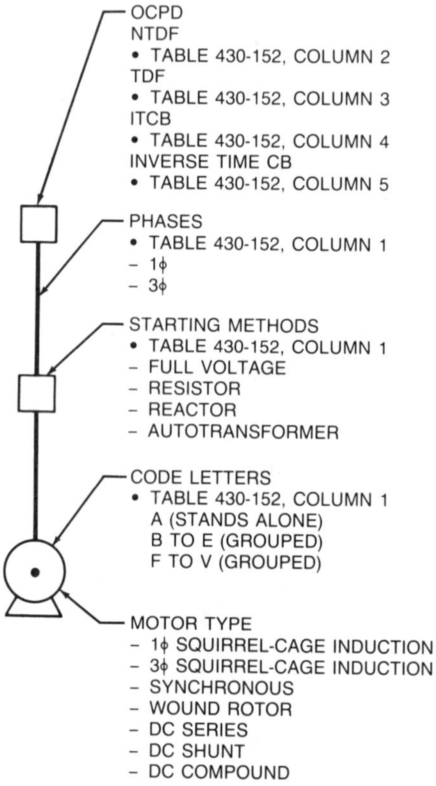

SELECTING OCPD'S TO START AND RUN A MOTOR

How many amps will a 200 A fuse or CB hold to allow a motor to start and run (rule of thumb)?

Fuse

Step 1. *Text:* Fuses = 200 A × 5 = 1000 A

Answer. **Fuse = 1000 A**

CB

Step 1. *Text:* CB = 200 A × 3 = 600 A

Answer. **CB = 600 A**

Note: A 200 A fuse will allow a motor with less than 1000 A LRC to start and run. A 200 A CB will allow a motor with less than 600 A LRC to start and run under normal starting conditions.

Figure 3-7. Fuses at five times their rating and circuit breakers at three times their rating can be used to start and run a motor with locked rotor current of the motor less than this value.

Step 3. Starting method (full voltage). *Table 430-152, Column 1:*

Step 4. Code letter (H). *Table 430-152, Column 1:*

Step 5. Percent (300%). *Table 430-152, Column 2:*

Step 6. Full-load current (28 A). *Table 430-150, Column 3:*

Step 7. *Table 430-150, Column 3; Table 430-152, Column 2:* 28 A × 300% = 84 A
Minimum size = 80 A

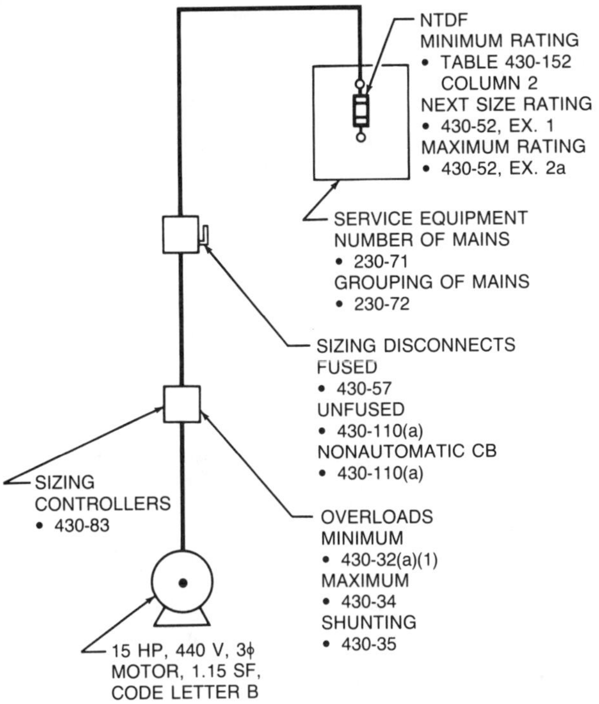

SIZING NTDF'S TO START AND RUN A MOTOR

What size nontime delay fuse is required to start and run the motor using the minimum, next, and maximum size?

Minimum

Step 1. *430-6(a); Table 430-150, Column 4:*
15 HP = 21 A

Step 2. *430-52; Table 430-152, Column 2:*
21 A × 250% = 52.5 A

Step 3. *240-3, Ex. 6; 240-6:*
50 A is the next lower standard size

Answer. **NTDF = 50 A**

Next

Step 1. *430-52, Ex. 1; 240-3, Ex. 6; 240-6:*
60 A is the next higher standard size

Answer. **NTDF = 60 A**

Maximum

Step 1. *430-52, Ex. 2a:* 21 A × 400% = 84 A

Step 2. *240-3, Ex. 6; 240-6:*
80 A is the next lower standard size

Answer. **NTDF = 80 A**

Figure 3-8. The minimum size nontime delay fuse to start and run a motor is sized from 430-52 and Table 430-152. If the ampacity of Table 430-152 does not correspond to a standard size device, the next size device is used per 430-52, Ex. 1. If the motor's starting current trips the overcurrent protection device, the higher percentages per 430-52, Ex. 2 can be applied.

Step 8. *430-52, Ex. 1; 240-3, Ex. 6; 240-6:*
Next size 90 A

Answer. **Minimum size NTDF = 80 A**
Next size NTDF = 90 A

Applying 430-52, Ex. 2a. If the percentage in Table 430-152, Column 2 will not allow the motor to start and run, 430-52, Ex. 2a can be applied. The standard device per 240-6 shall not exceed 600 A per 430-52, Ex. 2a after 400% times the full-load current has been applied. If 400% times the full-load current does not correspond to a standard size per 240-6, the next lower size is selected.

What size overcurrent protection device, using NTDFs, is required for a 10 HP, 230 V, 3ϕ induction motor to start and run? (Use maximum size.)

Step 1. *430-52, Ex. 2a:* 28 A FLC × 400% = 112 A

Step 2. *240-3, Ex. 6; 240-6:* Next lower rating = 110 A

Answer. **OCPD = 110 A NTDF**

Sizing Time Delay Fuses—Table 430-152, Column 3. To select the percentages in Table 430-152, Column 3, find the type of motor, number of phases, starting method, and code letter. See Figure 3-9.

What size overcurrent protection device, using TDFs, is required to permit a 15 HP, 440 V, 3ϕ wound-rotor motor used for reduced starting with no code letter to start and run?

Step 1. Type of motor (wound-rotor). *Table 430-152, Column 1:*

Step 2. Number of phases (three). *Table 430-152, Column 1:*

Step 3. Starting method (wound-rotor). *Table 430-152, Column 1:*

Step 4. Code letter (none). *Table 430-152, Column 1:*

Step 5. Percent (150%). *Table 430-152, Column 3:*

Step 6. Full-load current (21 A). *Table 430-150, Column 4:*

Step 7. *Table 430-150, Column 4; Table 430-152, Column 3:* 21 A × 150% = 31.5 A
Minimum size = 30 A

Step 8. *430-52, Ex. 1; 240-3, Ex. 6; 240-6:*
Next size = 35 A

Answer. **OCPD minimum size = 30 A TDF**
Next size = 35 A TDF

If time delay fuses will not hold the inrush current of the motor and allow the motor to start and run, 430-52, Ex. 2b can be applied. By applying 430-52, Ex.

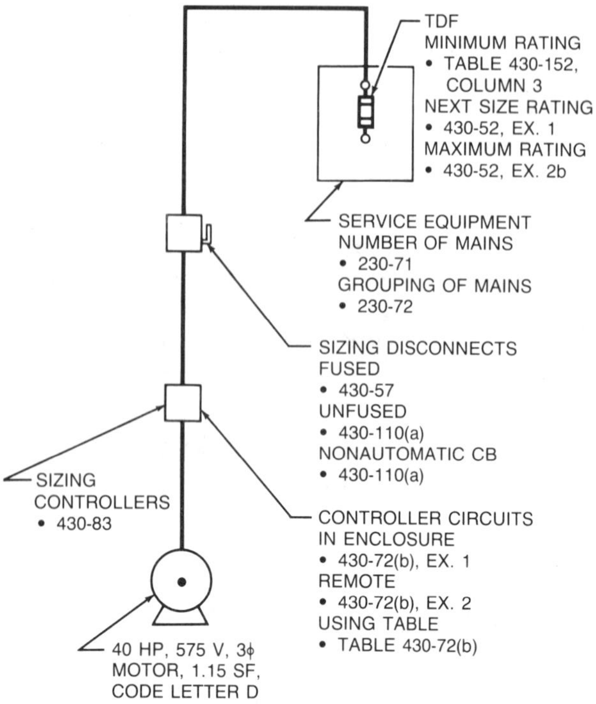

SIZING TDF'S TO START AND RUN A MOTOR

What size time delay fuse is required to start and run the motor using the minimum, next, and maximum size?

Minimum

Step 1. *430-6(a); Table 430-150, Column 5:*
40 HP = 41 A

Step 2. *430-52; Table 430-152, Column 3:*
41 A × 175% = 71.8 A

Step 3. *240-3, Ex. 6; 240-6:*
70 A is the next lower standard size

Answer. **TDF = 70 A**

Next

Step 1. *430-52, Ex. 1; 240-3, Ex. 6; 240-6:*
80 A is the next higher standard size

Answer. **TDF = 80 A**

Maximum

Step 1. *430-52, Ex. 2b:* 41 A × 225% = 92 A

Step 2. *240-3, Ex. 6; 240-6:*
90 A is the next lower standard size

Answer. **TDF = 90 A**

Figure 3-9. The minimum size time delay fuse to start and run a motor is sized from 430-52 and Table 430-152. If the ampacity of Table 430-152 does not correspond to a standard size device, the next size device is used per 430-52, Ex. 1. If the motor's starting current trips the overcurrent protection device, the higher percentages per 430-52, Ex. 2 can be applied.

2b, the full-load current of the motor can be increased to 225%. Note that 225% shall not be exceeded in selecting the time delay fuses. If 225% times the full-load current rating of the motor does not correspond to a standard size time delay fuse, the next lower size is selected. For example,

Step 1. *430-52, Ex. 2b:* 21 A × 225% = 47.25 A

Step 2. *240-3, Ex. 6; 240-6:*
Next lower rating = 45 A

Answer. **TDF = 45 A**

Instantaneous Trip Circuit Breaker—Table 430-152, Column 4. Instantaneous trip circuit breakers are designed to allow motors that have high inrush current due to the driven load or code letter to start and run. See Figure 3-10.

What setting of an instantaneous trip circuit breaker is required to allow a 50 HP, 460 V, 3ϕ synchronous motor with full voltage starting and code letter C to start and run?

Step 1. Type of motor (synchronous). *Table 430-152, Column 1:*

Step 2. Number of phases (three). *Table 430-152, Column 1:*

Step 3. Starting method (full voltage). *Table 430-152, Column 1:*

Step 4. Code letter (C). *Table 430-152, Column 1:*

Step 5. Percent (700%). *Table 430-152, Column 4:*

Step 6. Full-load current (65 A). *Table 430-150, Column 4:*

Step 7. *Table 430-150, Column 4; Table 430-152, Column 4:* 65 A × 700% = 455 A
430-52, Ex. 1; 240-3, Ex. 6; 240-6:
Permitted setting = 455 A

Answer. **Setting = 455 A ITCB**

The minimum setting in Table 430-152, Column 4 is 175%. The next setting is 250% and the maximum is 700%. If none of the settings in Table 430-152, Column 4 will permit the motor to start and run, 430-52, Ex. will permit a setting not to exceed 1300% of the motor's full-load current rating. The 1300% setting of 430-52, Ex. can be applied as follows:

Step 1. *430-52:* 65 A × 1300% = 845 A

Step 2. *240-3, Ex. 4; 430-52, Ex.:*
Maximum setting = 845 A

Answer. **Setting = 845 A**

Inverse Time Circuit Breaker—Table 430-152, Column 5. Inverse time circuit breakers are designed to provide overload, ground-fault, and short-circuit protection. These CBs will hold up to three times their ratings based upon their frame size and supply voltage.

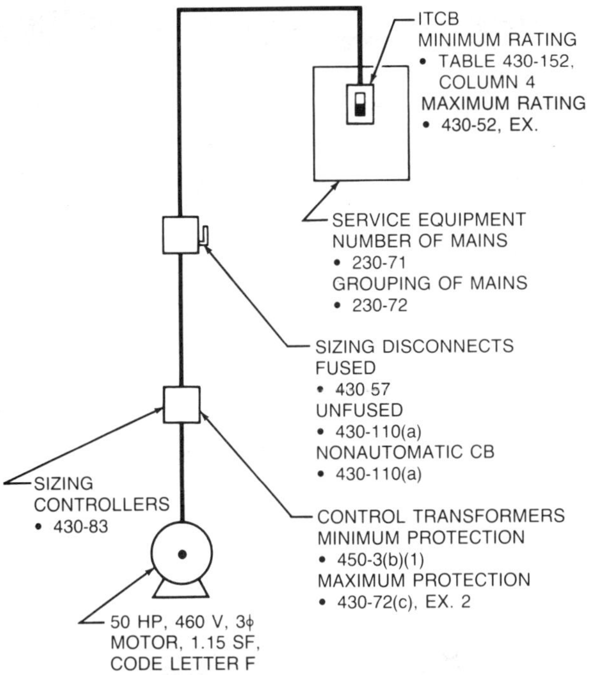

SIZING ITCB'S TO START AND RUN A MOTOR

What size ITCB is required to start and run the motor using the minimum and maximum size?

Minimum

Step 1. *430-6(a); Table 430-150, Column 4:*
50 HP = 65 A

Step 2. *430-52; Table 430-152, Column 4:*
65 A × 700% = 455 A

Answer. **ITCB = 455 A**

Maximum

Step 1. *240-3, Ex. 6; 430-52, Ex.:*
65 A × 1300% = 845 A

Answer. **ITCB = 845 A**

Figure 3-10. The minimum size instantaneous trip CB to start and run a motor is sized from 430-52, Ex. and Table 430-152. If the motor's starting current trips the overcurrent protection device, the higher percentages per 430-52, Ex. can be applied.

What size overcurrent protection device, using a circuit breaker, is required to allow a 60 HP, 550 V, 3ϕ induction motor with resistor starting and code letter F to start and run?

Step 1. Type of motor (induction). *Table 430-152, Column 1:*

Step 2. Number of phases (three). *Table 430-152, Column 1:*

Step 3. Starting method (resistor). *Table 430-152, Column 1:*

Step 4. Code letter (none). *Table 430-152, Column 1:*

Step 5. Percent (250%). *Table 430-152, Column 5:*

Step 6. Full-load current (62 A). *Table 430-150, Column 5:*

Step 7. *Table 430-150, Column 5; Table 430-152, Column 5:* 62 A × 250% = 155 A
Minimum size = 150 A

Step 8. *430-52, Ex. 1; 240-3, Ex. 6; 240-6:*
Next size = 175 A

Answer. **OCPD minimum size = 150 A CB**
Next size = 175 A CB

Where a motor with a full-load current rating of 100 A or less starts and trips open the circuit breaker due to high inrush starting current, apply 430-52, Ex. 2c to increase the CB rating to 400%. See Figure 3-11. For motors with a full-load current rating of over 100 A the CB rating can be increased 300%. For example,

Step 1. *430-52, Ex. 2c:* 62 A × 400% = 248 A

Step 2. *240-3, Ex. 6; 240-6:* Next lower rating = 225 A

Answer. **CB = 225 A**

Single Motor—430-52

The rating or setting of an overcurrent protection device for an individual motor circuit is determined by the percentage listed in 430-52. The general rule of 430-52 is the overcurrent protection device shall not exceed the percentages for each type of device listed in Table 430-152. For example,

A 20 HP, 230 V, 3ϕ motor has a full-load current rating of 54 A per Table 430-150, and the percentage used for a time delay fuse is 175% per Table 430-152. What size time delay fuse is required?

Step 1. *Table 430-150, Column 3:* 20 HP = 54 A

Step 2. *Table 430-152, Column 3:*
54 A × 175% = 94.5 A

Step 3. *240-3, Ex. 6; 240-6:* 90 A TDF

Answer. **TDF = 90 A**

Note: See 430-52, Ex. 1.

Where values in Table 430-152 do not correspond with a standard overcurrent protection device, 430-52, Ex. 1 is applied and the next size overcurrent protection device is selected. By applying this exception, a 100 A TDF is selected from 240-6.

■ *Master Electrician's Test Tip.* Some testing authorities automatically apply 430-52, Ex. 1 and use the next higher size overcurrent protection device. Other testing authorities apply 430-52 and use the next lower rating for selecting the minimum size overcurrent protection device.■

What size circuit breaker is required to allow a 25 HP, 208 V, 3ϕ motor with code letter F and full voltage starting to start and run? Apply 430-52, Ex. 1.

Step 1. *Table 430-150 and Footnotes:*
25 HP = 68 A × 110% = 74.8 A

Step 2. *Table 430-152:* 74.8 A × 250% = 187 A

Step 3. *430-52, Ex. 1; 240-3, Ex. 6; 240-6:* 200 A CB

Answer. **CB = 200 A**

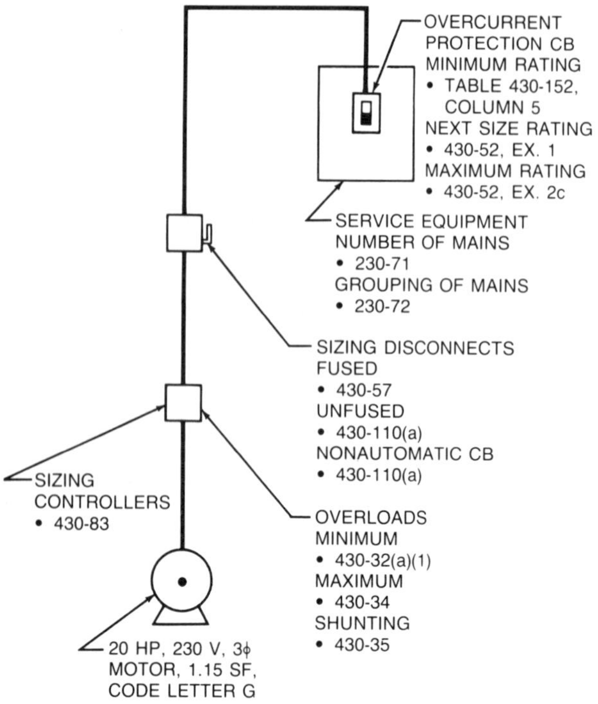

SIZING CB'S TO START AND RUN A MOTOR

What size CB is required to start and run the motor using the minimum, next, and maximum size?

Minimum

Step 1. *430-6(a); Table 430-150, Column 3:*
20 HP = 54 A

Step 2. *430-52; Table 430-152, Column 5:*
54 A × 250% = 135 A

Step 3. *240-3, Ex. 6; 240-6:*
125 A is the next lower standard size

Answer. **CB = 125 A**

Next

Step 1. *430-52, Ex. 1; 240-3, Ex. 6; 240-6:*
150 A is the next higher standard size

Answer. **CB = 150 A**

Maximum

Step 1. *430-52, Ex. 2c:* 54 A × 400% = 216 A

Step 2. *240-3, Ex. 6; 240-6:*
200 A is the next lower standard size

Answer. **CB = 200 A**

Figure 3-11. The minimum size circuit breaker to start and run a motor is sized from 430-52 and Table 430-152. If the ampacity of Table 430-152 does not correspond to a standard size device, the next size device is used per 430-52, Ex. 1. If the motor's starting current trips the overcurrent protection device, the higher percentages per 430-52, Ex. 2 can be applied.

If the percentages of Table 430-152 and 430-52, Ex. 1 will not allow the motor to start and run, 430-52, Ex. 2 is applied. Motors have high inrush current due to driven loads or code letters per Table 430-7(b).

What maximum size circuit breaker is required using 430-52, Ex. 2 for the same 25 HP motor previously discussed?

Step 1. *Table 430-150 and Footnotes:*
25 HP = 68 A × 110% = 74.8 A

Step 2. *430-52, Ex. 2c:* 74.8 A × 400% = 299.2 A

Step 3. *240-3, Ex. 6; 240-6:* 250 A CB

Answer. **CB = 250 A**

Note: By applying 430-52 and Table 430-152, the minimum size CB is 175 A. By applying 430-52, Ex. 1 the next size CB is 200 A, and by applying 430-52, Ex. 2 the maximum size CB is 250 A.

Two or More Motors—430-62(a)

Two or more motors can be connected to a feeder circuit instead of routing individual circuits to each motor. This type of installation is not as expensive as running individual circuits to each motor. See Figure 3-12. For two or more motors connected to a feeder circuit, the overcurrent protection device shall be sized large enough to allow the largest motor to start and run and permit the remaining motors of the group to start and run. For example, three induction motors are supplied by a 460 V feeder. The motors are 20 HP, 30 HP, and 40 HP. The full-load current, per Table 430-150, is 27 A for the 20 HP, 40 A for the 30 HP, and 52 A for the 40 HP motor. To determine the size overcurrent protection device for the feeder circuit, the overcurrent protection device for the largest motor is added to the full-load current rating of the remaining motors in the group. If this total ampacity does not correspond to a standard device per 240-6, the next lower rating is used.

■ *Master Electrician's Test Tip.* Section 430-62(a) does not list an exception that will allow the next size overcurrent protection device to be used where the total ampacity of the overcurrent protection device of the largest motor plus the remaining motors of the group does not correspond to a standard overcurrent protection device, per 240-6.■

If the 40 HP motor is marked with code letter H and started by applying full voltage, the overcurrent protection device is sized by multiplying 52 A × 175% per Table 430-152. The result of the calculation is 91 A (52 A × 175% = 91 A) and the size overcurrent protection device for the 40 HP motor is 100 A, per

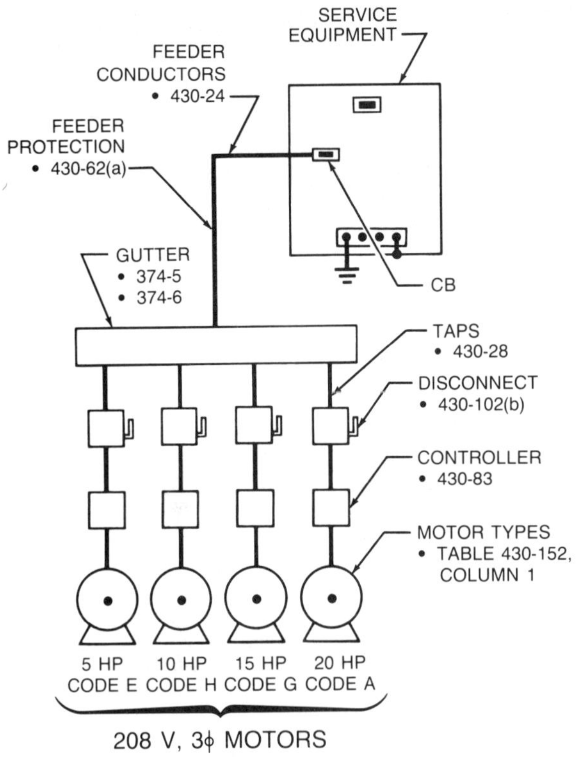

SIZING CB'S FOR FOUR MOTORS SUPPLIED BY A FEEDER CIRCUIT

What size CB is required for the feeder circuit supplying the four motors?

Step 1. *430-6(a); Table 430-150:*
5 HP = 15.2 A × 110% = 16.72 A
10 HP = 28 A × 110% = 30.8 A
15 HP = 42 A × 110% = 46.2 A
20 HP = 54 A × 110% = 59.4 A

Step 2. *430-52; Table 430-152, Column 5; 240-6; 240-3, Ex. 6:*
46.2 A × 250% = 115.5 A = 125 A CB
59.4 A × 150% = 89.1 A = 90 A CB
16.72 A × 200% = 33.44 A = 35 A CB
30.8 A × 250% = 77 A = 80 A CB

Step 3. *430-62(a):*
125 A + 59.4 A + 30.8 A + 16.72 A = 231.92 A

Step 4. *240-3, Ex. 6; 430-62(a); 240-6:*
225 A CB is the next lower standard size

Answer. **CB = 225 A**

Figure 3-12. The overcurrent protection device for a feeder circuit is determined by adding the largest overcurrent protection device of one of the motors in the group and the full-load current ratings of the remaining motors. The next lower size device is selected per 240-6 if the calculated rating does not correspond to a standard size device.

430-52, Ex. 1 and 240-6. The 100 A overcurrent protection device of the 40 HP motor is added to the 40 A of the 30 HP motor and the 27 A of the 30 HP motor for a total of 167 A (100 A + 40 A + 27 A = 167 A). The size overcurrent protection device using time delay fuses is the next standard device below 167 A. This is a 150 A device per 240-6. Therefore, the feeder circuit is protected by 150 A time delay fuses.

■ *Master Electrician's Test Tip.* Some examination boards require the largest overcurrent protection device for any one motor in a group to be selected at the next lower standard size below the percentages in Table 430-152. This overcurrent protection device rating is then added to the full-load current ratings of the other motors in the group, and the next lower standard size is selected.■

If two or more of the motors on the feeder are to be started at the same time, time delays in the control circuit are installed to start the motors at different periods of time. However, 430-52, Ex. 2 can be applied to increase the rating of the overcurrent protection device of the largest motor by multiplying the 52 A by 225% (52 A × 225% = 117 A). The next lower size overcurrent protection device is 110 A per 240-6. The size time delay fuses for the feeder circuit is determined by adding the 110 A overcurrent protection device to 40 A and to 27 A (110 A + 40 A + 27 A = 177 A). The next lower rated device per 240-6 is 175 A time delay fuses.

These fuses allow all three motors to start under normal starting conditions. This is based on time delay fuses holding five times their rating for 10 seconds. Time delay fuses rated at 175 A hold 875 A for 10 seconds and will allow the starting currents of the three motors at six times their full-load current ratings to start and run. The starting current of the three motors can be determined by the following procedure:

Step 1. *Table 430-150; Table 430-152:*

20 HP = 27 A × 6 =	162 A
30 HP = 40 A × 6 =	240 A
40 HP = 52 A × 6 =	312 A
Total =	714 A

Answer. **TDF = 714 A**

The 175 A time delay fuses hold 875 A (175 A TDF × 5 = 875 A) for 10 seconds, which is greater than the 714 A of starting current for the three motors.

The feeder circuit can be routed to a gutter (374-5) where taps, per 430-28, are made to individual disconnects with fuses that are sized to allow each motor to start and run. Taps from the feeder circuit can be connected to the lugs of a panelboard, and an individual circuit breaker can be sized to permit each motor to start and accelerate its driven load.

What size time delay fuses are required to allow a 10 HP, 5 HP, 3 HP, and 20 HP, 230 V, 3ϕ group of motors on a feeder circuit to start and run? The motors are equipped with code letter F and started by applying full voltage.

Step 1. Type of motor (induction). *Table 430-152, Column 1:*

Step 2. Number of phases (three). *Table 430-152, Column 1:*

Step 3. Starting method (full voltage). *Table 430-152, Column 1:*

Step 4. Code letter (F). *Table 430-152, Column 1:*

Step 5. Percent (175%). *Table 430-152, Column 3:*

Step 6. Full-load current of largest motor (54 A). *Table 430-150, Column 3:*

Step 7. *Table 430-150, Column 3; Table 430-152, Column 3:*
54 A × 175% = 94.5 A
430-52, Ex. 1; 240-3, Ex. 6; 240-6:

Step 8. Full-load currents of remaining motors. *Table 430-150:*
3 HP = 9.6 A
5 HP = 15.2 A
10 HP = 28 A

Step 9. Sizing feeder overcurrent protection device. *430-62(a):*
100 A TDF + 9.6 A + 15.2 A + 28 A = 152.8 A

Step 10. *240-3, Ex. 6; 430-62(a); 240-6:*
150 A TDF

Answer. **TDF = 150 A**

OVERLOAD PROTECTION—430, PART C

Running overload protection devices are designed to protect the windings of motors from excessive heating. This heating can result from the motor inlets and outlets becoming covered by a blanket of lint or dust or from a bearing beginning to lock on the motor or the driven load. Motor running overcurrent protection devices are set to open the motor's control circuit and disconnect the windings from the power source when excessive heating occurs in the motor windings due to overloads.

Minimum Rating—430-32(a)(1)

Running overload protection devices are selected at 115% or 125% of the full-load current rating that is listed on the motor's nameplate. The percentage selected is based on the temperature rise or service factor of the motor. See Figure 3-13.

A percentage of 125% times the full-load current of the motor is used for motors with a temperature rise of 40°C or a service factor of not less than 115%. A percentage of 115% is applied for motors with a temperature rise of over 40°C or a service factor of less than 115%. A percentage of 115% times the full-load current rating is used for motors without a temperature rise or a service factor listed on its nameplate.

SELECTING MINIMUM SIZE OVERLOADS	
SERVICE FACTOR	TEMPERATURE RISE
Less than 1.15, use 115%	40°C, use 125%
1.15 or greater, use 125%	Over 40°C, use 115%
No SF listed, use 115%	No temp. rise listed, use 115%

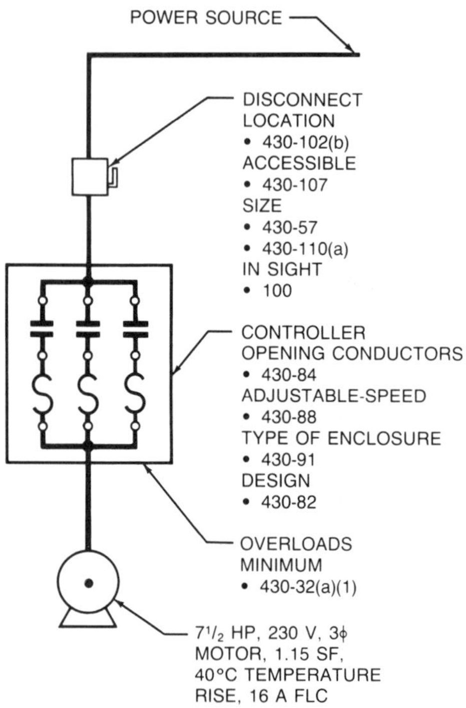

SIZING OVERLOADS TO PROTECT MOTOR WINDINGS

What size overloads, using the minimum rating, is required to protect the motor?

Step 1. *430-6(a); 430-32(a)(1):* FLC = 16 A
1.15 SF = *FLC* × 125%
40 °C temperature rise = *FLC* × 125%

Step 2. *430-32(a)(1):* 16 A × 125% = 20 A

Answer. **Overloads = 20 A**

Figure 3-13. The minimum size overloads based on service factors is found by multiplying the full-load current rating of the motor by 125% for a service factor not less than 1.15 and by 115% for a service factor less than 1.15.

For example, a motor with a 54 A full-load current rating and a temperature rise of 40°C or a service factor of 115% requires that the overloads be selected at 67.5 A (54 A × 125% = 67.5 A). Motors with a temperature rise of 50°C or a service factor of 110% have their overloads selected at 62.1 A (54 A × 115% = 62.1 A).

What size overloads is required to protect a 40 HP, 575 V, 3ϕ motor with a temperature rise of 50°C?

Step 1. *430-6(a); Table 430-150:* 40 HP = 41 A

Step 2. *430-32(a)(1):* 41 A × 115% = 47.15 A

Answer. **Overloads = 47.15 A**

What size overloads is required to protect a 60 HP, 440 V, 3ϕ motor with a service factor of 1.15?

Step 1. *430-6(a); Table 430-150:* 60 HP = 77 A

Step 2. *430-32(a)(1):* 77 A × 125% = 96.25 A

Answer. **Overloads = 96.25 A**

Temperature Rise. Temperature rise is based on motors located at elevations of 3300′ or less. Motors with 40°C are derated 10% for every 1000′ above 3300′. For example, a motor having a full-load current rating of 68 A and located at an elevation of 4300′ is loaded at not more than 61.2 A. This is determined by multiplying 68 A by 10% (68 A × 10% = 6.8 A) and subtracting from 68 A (68 A − 6.8 A = 61.2 A).

What is the full-load current rating of an 88 A motor located at an elevation of 9900′?

Step 1. 88 A × 66% = 58.08 A

Step 2. 88 A − 58.08 A = 29.92 A

Answer. **FLC = 29.92 A**

Service Factor. Service factor is the percentage of overload that a motor can endure without damage to its windings. For example, a motor with a full-load current rating of 42 A with a service factor of 1.15 will take an overload of 48.3 A (42 A × 115% = 48.3 A). The same motor with a 1.10 service factor will take an overload of 46.2 A (42 A × 110% = 46.2 A). The same motor with a service factor of 1.00 will not take a load above 42 A (42 × 100% = 42 A).

What is the overload rating for a 5 HP, 230 V, 3ϕ motor with a service factor of 1.10 and a full-load current rating of 9 A?

Step 1. 9 A × 115% = 10.35 A

Answer. **Overload = 10.35 A**

A fuse or circuit breaker can be used to provide overload protection for a motor. For a motor with a full-load current rating (nameplate rating) of 32 A and a service factor of 1.25, the overload protection, using fuses, shall not exceed 40 A (32 × 125% = 40 A). The fuse or circuit breaker rating shall not exceed the minimum of 115% or 125% times the full-load current of the motor based on the temperature rise or service factor. For example, a motor with a full-load current rating of 26 A and a temperature rise of 40°C is protected from overload by a 30 A overcurrent protection device (26 A × 125% = 32.5 A) per 430-32(a)(1). The overload protection device shall not exceed 32.5 A. Since there is no 32.5 A standard overcurrent protection device listed in 240-6, the next lower overcurrent protection device (30 A) is used.

What size overcurrent protection device is required to protect a motor with a temperature rise of 40°C and a service factor of 1.15? The motor has a full-load current rating of 46 A.

Step 1. *430-32(a)(1):* 46 A × 125% = 57.5 A

Step 2. *240-6:* 50 A OCPD

Answer. **OCPD = 50 A**

Maximum Rating—430-34

Should a motor fail to start and accelerate the driven load, a rating above the percentage listed in 430-32(a)(1) can be applied. See Figure 3-14. For motors marked with a temperature rise of 40°C or a service factor not less than 1.15, 140% times the full-load current listed on the nameplate of the motor is applied. Motors with a temperature rise of more than 40°C or with a service factor less than 1.15 can have the overloads increased to 130% times the full-load current rating of the motor per 430-34. Motors without the temperature rise or service factor listed on the nameplate shall have the overloads computed at not less than 130%.

What maximum size overloads is required for a 40 HP, 208 V, 3ϕ motor with a temperature rise of 30°C and full-load current rating of 85 A?

Step 1. *430-34:* 85 A × 130% = 110.5 A

Answer. **Maximum size overloads = 110.5 A**

Shunting Overloads—430-35

Overloads can be *shunted* (bypassed) for large motors driving inertia loads which accelerate the driven load slowly. The shunting may be nonautomatic or automatic. In nonautomatic shunting, a double-throw switch bypasses the fuses while the motor accelerates the driven load. After the motor comes up to running speed, the double-throw switch is switched to the running position. Shunt-start switches are designed so the switch cannot be left in the starting position. Fuses or other overcurrent protection devices shall not be set over 400% of the motor's nameplate full-load current rating. For example, a 25 HP, 230 V, 3ϕ motor has a full-load current rating of 58 A and the overload protection is to be manually shunted during the starting period. The overload protection shall not exceed 232 A (58 A × 400% = 232 A). If fuses or a circuit breaker is used, the rating of the overcurrent protection device shall be 225 A or less per standard ratings listed in 240-6.

Overloads may be automatically shunted during the starting period of the motor, per 430-35(b), Ex., where specific design techniques are observed. See Figure 3-15. The following requirements shall be applied before the overloads can be shunted:

1. The overloads cannot be shunted where overload devices are available with sufficient time delay to allow the motor to start and run.

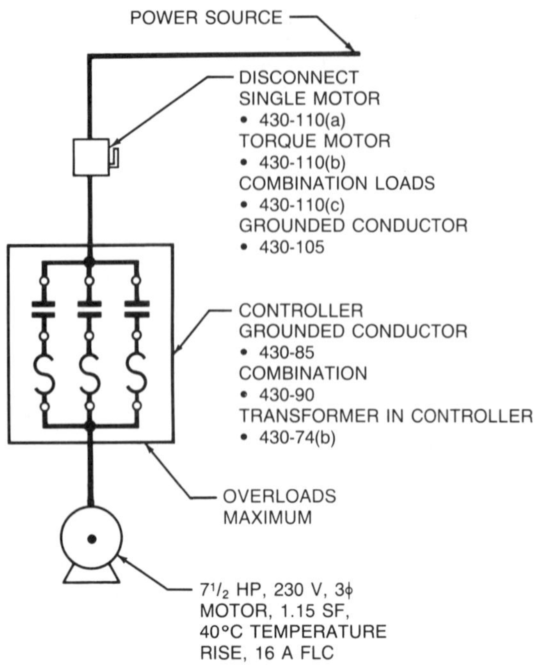

SELECTING MAXIMUM SIZE OVERLOADS	
SERVICE FACTOR	TEMPERATURE RISE
Less than 1.15 use 130%	40°C use 140%
1.15 or greater use 140%	Over 40°C use 130%
No SF listed use 130%	No temp. rise listed use 130%

SIZING OVERLOADS TO PROTECT MOTOR WINDINGS

What size overloads, using the maximum rating, is required to protect the motor?

Step 1. *430-6(a); 430-34:* FLC = 16 A
1.15 SF = *FLC* × 140%
40 °C temperature rise = *FLC* × 140%

Step 2. *430-34:* 16 A × 140% = 22.4 A

Answer. **Overloads = 22.4 A**

Figure 3-14. The maximum size overloads based on service factors is found by multiplying the full-load current rating of the motor by 140% for a service factor not less than 1.15 and by 130% for a service factor less than 1.15.

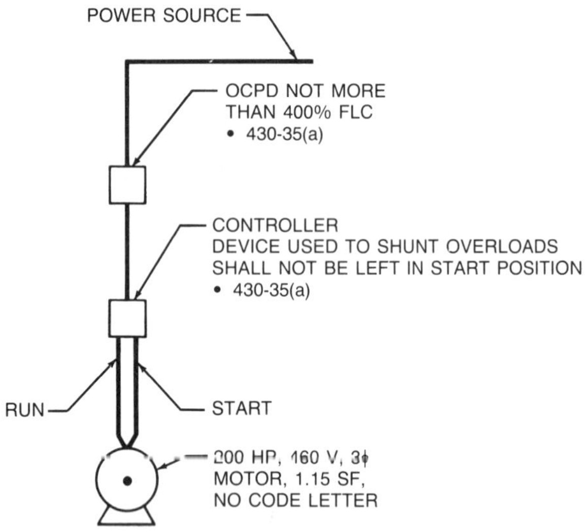

SHUNTING OVERLOADS—NONAUTOMATIC METHOD

What size OCPD is required to allow the motor to start during the shunting period?

Step 1. *430-6(a); Table 430-150:* 200 HP = 240 A

Step 2. *430-35(a):* 240 A × 400% = 960 A

Step 3. *240-3, Ex. 6; 240-6:*
800 A is the next lower standard rating

Answer. **OCPD = 800 A**

Figure 3-15. The overcurrent protection device supplying the motor circuit shall not exceed 400% of the motor's full-load current rating if left in the circuit during the shunting of the overloads.

2. Where the overloads are shunted, a rotation sensor shall be provided to reconnect the overload protection in the motor circuit should the motor fail to start and run. The listed system shall limit the time delay that the overload is shunted from the motor circuit to prevent damaging the windings.

3. Where the motor does not start and accelerate the driven load, a rotation detection system shall disconnect the automatic starting feature.

CONTROLLER—430, PART G

Motor controllers start a motor by either applying full voltage or utilizing a reduced starting method. Motor controllers can be any switch or device used to start and run a motor by making and breaking the contacts that connect the motor circuit. The general rule is that motor controllers are rated in horsepower equal to the horsepower of the motor. The control circuits can be wired to start or stop the motor manually or automatically. The overload protection can be incorporated in the controller to provide protection for the windings of the motor.

Rating—430-83

Controllers used to bridge the circuit to the motor windings can be an AC snap switch, a circuit breaker, disconnect switch, or magnetic starter. The rating of an AC snap switch shall be at least 125% of the motor's full-load current rating. For example, if a snap switch is used to start and stop a motor with a full-load current rating of 14 A, the snap switch is rated no less than 17.5 A (14 A × 125% = 17.5 A) per

430-83, Ex. 1. The 125% is derived by dividing 1 by 80% (1 ÷ 80% = 125%).

What size snap switch is required to control the circuit to a motor with a full-load current rating of 12.5 A?

Step 1. *430-83, Ex. 1:* 12.5 A × 125% = 15.6 A

Answer. **Switch = 20 A**

A disconnect switch used as a controller to start and stop a motor is rated in amperes or horsepower. A general-use disconnect switch is rated in amperes. For a 2 HP or less and 300 V or less motor, a general-use disconnect switch rated at least twice the full-load current rating of the motor can be used to start and stop the motor. A circuit breaker can be used to start and stop a motor. The circuit breaker shall be rated in amperes only. See 430-109 and 430-110 for requirements for sizing and selecting the disconnecting means for motors. See Figure 3-16.

What size general-use disconnect switch is required to start and stop a 2 HP, 208 V, 3ϕ motor with a full-load current rating of 7.48 A?

Step 1. *430-83, Ex. 1:* 7.48 A × 2 = 14.96 A

Answer. **Switch = 30 A**

What size circuit breaker is required to start and stop a 1 HP, 115 V, 1ϕ motor?

Step 1. *Table 430-148:* 1 HP = 16 A

Step 2. *430-83, Ex. 2:* 20 A CB

Answer. **CB = 20 A**

SIZING DEVICES TO CONTROL MOTORS

What size general-use switch is required to serve as a controller for the motor?

Step 1. *430-83, Ex. 1:* 13.8 A × 2 = 27.6 A

Step 2. *Chart:* 27.6 A = 30 A switch

Answer. **General-use switch = 30 A**

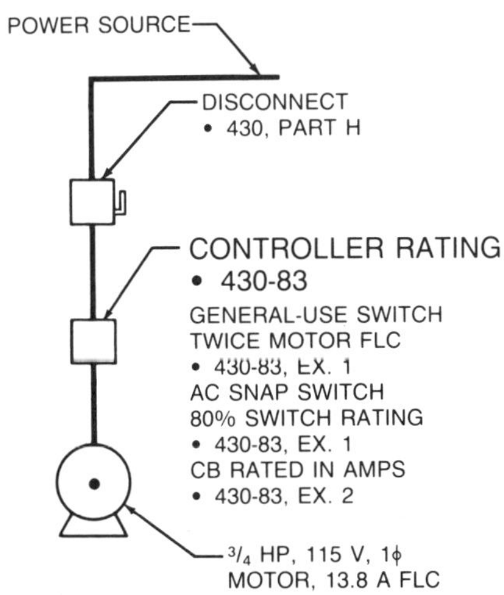

Figure 3-16. A general-use switch rated at least twice the full-load current rating of the motor can be used as a controller. A snap switch rated at least 125% of the motor's full-load current rating can be used as a controller or circuit breaker rated in amps.

Number of Motors—430-87

The general rule is that each motor shall have an individual controller. Specific conditions allow one controller to serve more than one motor. For a single machine operated by more than one motor, one controller can be used to control the motors of the entire machine.

A single controller can be used to start and stop several grouped motors located in the same room. All motors shall be located *in sight from* the controller. In sight from, within sight from, within sight, etc. means that one of the pieces of equipment is visible and no more than 50′-0″ from the other piece. A single controller can be used to start and stop several fractional horsepower motors on the same branch circuit per 430-53(a). Note that where a single controller starts and stops several motors, the controller shall be rated large enough to carry the total horsepower or current of all the motors. A switch shall have a horsepower rating equal to the sum of the horsepower ratings of all motors in the group. See 430-112 for requirements to size and select a disconnecting means to start and stop several motors. See Figure 3-17.

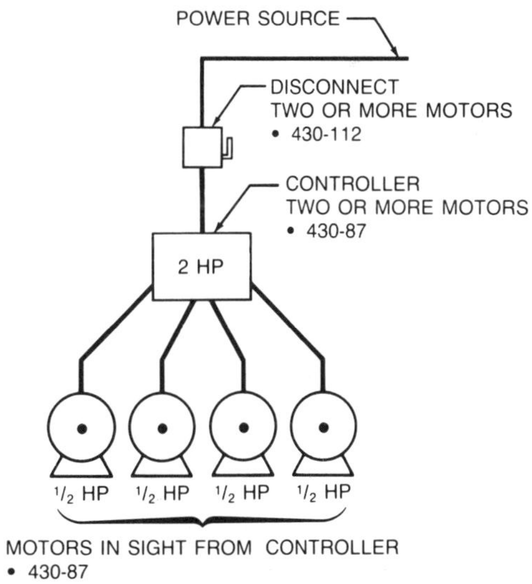

CONTROLLER SERVING SEVERAL MOTORS RATED AT 600 V OR LESS

What size controller rated in HP is required for the motors?

Step 1. *430-87:*

½ HP + ½ HP + ½ HP + ½ HP = 2 HP

Answer. **Controller = 2 HP**

Figure 3-17. A controller equivalent to the horsepower rating of all motors can be used to start and stop each motor. All motors shall be within sight of the controller and the disconnecting means.

DISCONNECTING MEANS—430, PART H

The general rule requires a disconnecting means for each motor. This requirement is to help assure safety for personnel working on the motor. The disconnecting means shall be reliable and easy to reach. The disconnect switch disconnects the controller and motor from the power source.

Location—430-102

A disconnect shall be provided in sight from the controller that starts and stops a motor. The motor shall be installed in sight from the controller, driven load, and the disconnecting means, *or* an additional disconnect shall be provided in sight from the motor and driven load.

Controller—430-102(a). For motors rated 600 V or less, a disconnect is required to be in sight from the controller. This means such equipment shall be visible and no more than 50′ from the other equipment per Article 100. Two exceptions to 430-102(a) allow the disconnecting means to be out of sight and more than 50′-0″ from the controller.

Exception 1 is for motors rated over 600 V, nominal. The disconnecting means can be located out of sight from the controller under the following conditions:

1. The controller is marked with a warning sign giving the location of the disconnect.
2. The disconnecting means can be locked in the open position.

Exception 2 allows a multimotor processing machine with a group of adjacent controllers to be served by a single disconnect. See Figure 3-18.

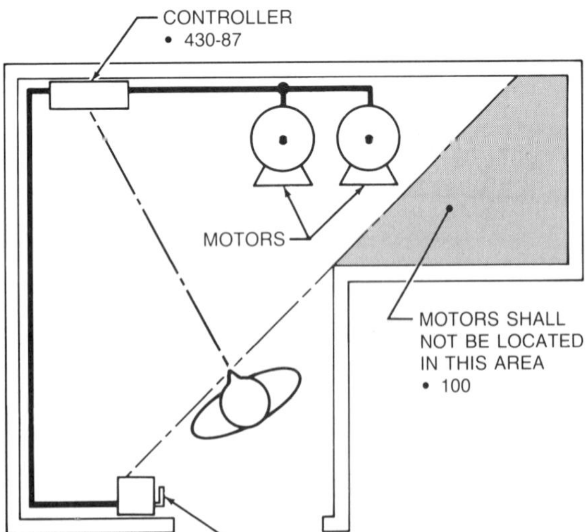

Figure 3-18. A disconnecting means shall be within sight of the controller to start and stop a motor. See 430-87 for a controller serving several motors.

Motor—430-102(b). A disconnecting means shall be provided in sight from the motor and the driven equipment. The disconnecting means in sight from the controller can serve as the disconnecting means for a motor not out of sight and located not more than 50′-0″ from the controller. If the motor is out of sight from the controller, an additional disconnect is required. The exception to 430-102(b) allows the motor to be located out of sight from the controller and no additional disconnecting means is required if the disconnect by the controller can be locked in the open position. See Figure 3-19.

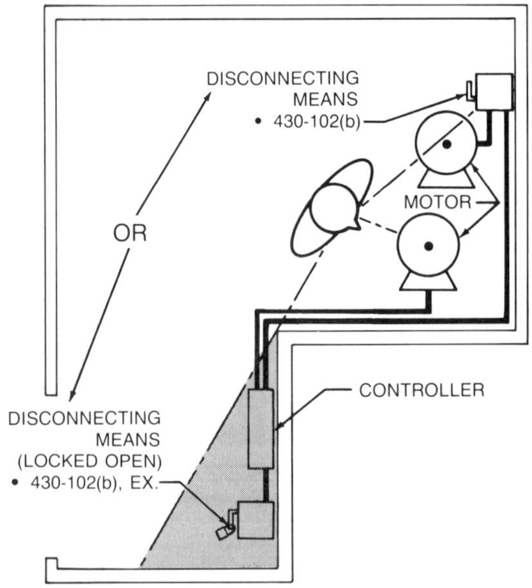

Figure 3-19. A motor shall be within sight of the disconnecting means for the controller or the disconnect for the controller shall be capable of being locked in the open position. If the disconnect by the controller cannot be locked in the open position and the motor is out of sight, a disconnect shall be placed in sight of the motor.

Accessibility—430-107

One of the disconnects that interrupts power to the controller and motor or motor only shall be *readily accessible*. Readily accessible indicates that the disconnect shall be reached easily and quickly without moving objects that prevent access. If the motor is located 6′-6″ or less from finished grade, the handle of the disconnect in the ON position shall not be mounted over 6′-6″ from finished grade per 380-8(a). If the motor is located on the wall above 6′- 6″ or in an attic or suspended ceiling, the disconnect shall be located adjacent to the motor. This prevents the electrician from climbing out of the attic space or suspended ceiling to reach the disconnect and shut off power to the motor. For this reason, the disconnect can be located by the motor for easy access to disconnect the motor and provide safety while serving the motor.

Rating—430-110

The disconnecting means for a motor can be a circuit breaker or disconnect switch. The disconnecting means is sized with an interrupting capacity great enough to disconnect the motor. Circuit breaker used as disconnecting means can be automatic or nonautomatic. The disconnecting means can also be a fused or nonfused disconnect. Different percentages times the full-load current of the motor are used to size the disconnect.

Fuses—430-57. A fused disconnect is sized by the rating of the fuses used to start and run the motor. See Figure 3-20. For example, if 45 A fuses are required to start and run a motor, the disconnecting means shall be 60 A to hold the fuses, per 430-57. A 100 A disconnect is required to hold 80 A fuses to permit a motor to start and run.

Larger disconnects are required to hold NTDFs than TDFs because of the percentages used for each device, per Table 430-152. For example, a 10 HP, 230 V, 3ϕ motor has a full-load current of 28 A per Table 430-150 and requires a multiplier of 175% to determine the size TDFs (28 A × 175% = 49 A). A 60 A disconnect is required to hold the 45 A minimum or 50 A next standard size fuses selected from 240-6, per 430-52. Standard size disconnects are

30 A   200 A   800 A   1600 A
60 A   400 A   1200 A   1800 A
100 A   600 A   1400 A

Where NTDFs are used to start and run the 28 A motor and also serve as the disconnecting means, a 100 A disconnect is required. This size disconnect is determined by multiplying 28 A by 300%, per Table 430-152 (28 A x 300% = 84 A). NTDFs are selected at 80 A or 90 A, per 430-52 and 240-6.

What size disconnect is required to hold time delay fuses to start and run a 15 HP, 208 V, 3ϕ motor with full voltage starting and equipped with code letter G? The disconnecting switch also serves as a disconnecting means for the motor.

Step 1. *430-6(a); Table 430-150 and Footnotes:*
15 HP = 46.2 A (42 × 110% = 46.2 A)

Step 2. *Table 430-152; 430-52, Ex. 1:*
46.2 A × 175% = 80.85 A

Step 3. *240-3, Ex. 6; 240-6:*
90 A is the next size standard fuse

Step 4. *430-57; Chart:* 100 A to hold 90 A fuses

Answer. **Disconnect = 100 A**

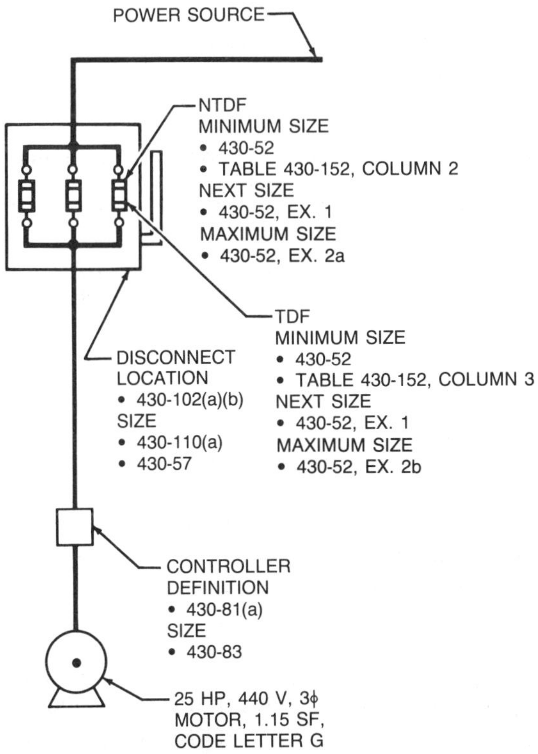

SELECTING DISCONNECTS BASED ON FUSE SIZE

What size disconnect is required for the motor using nontime delay or time delay fuses? (Use lower rating.)

NTDF

Step 1. *430-6(a); Table 430-150:* 25 HP = 34 A

Step 2. *430-57; Table 430-152, Column 2:*
34 A × 300% = 102 A

Step 3. *240-3, Ex. 6; 240-6; 430-57:*
100 A is the next lower standard size

Answer. **Disconnect = 100 A**

TDF

Step 1. *430-57; Table 430-152, Column 3:*
34 A × 175% = 59.5 A

Step 5. *240-3, Ex. 6; 240-6; 430-57:*
60 A is a standard size

Answer. **Disconnect = 60 A**

Note: Disconnect is sized by the rating of the fuses.

Figure 3-20. The disconnect is sized by the rating of the fuseholders required to hold the fuses in a disconnect.

What size disconnect, using nontime delay fuses, is required for the previous problem?

Step 1. *430-6(a); Table 430-150 and Footnotes:* 15 HP = 46.2 A (42 × 110% = 46.2 A)

Step 2. *Table 430-152; 430-52, Ex. 1:* 46.2 A × 300% = 138.6 A

Step 3. *240-3, Ex. 6; 240-6:* 150 A is the next standard size

Step 4. *430-57; Chart:* 200 A to hold 150 A fuses

Answer. **Disconnect = 200 A**

For a nonfused disconnect, the disconnect switch shall be equivalent to at least 115% of the full-load current rating of the motor. A motor with a full-load current rating of 65 A requires a 100 A disconnect (65 A × 115% = 74.75 A) per 430-110(a).

What size nonfused disconnect is required to disconnect a 40 HP, 460 V, 3ϕ motor?

Step 1. *430-6(a); Table 430-150:* 40 HP = 52 A

Step 2. *430-110(a):* 52 A × 115% = 59.8 A

Step 3. *430-110(a); Chart:* 59.8 A requires 60 A disconnect

Answer. **Disconnect = 60 A**

Circuit Breakers—430-110(a). The size of a circuit breaker used to disconnect a motor from its source of power is determined from the percentages listed in Table 430-152. The percentage selected is based on the information listed on the nameplate of the motor. For example, a motor with a 59 A full-load current rating requires a 150 A automatic circuit breaker to start and run the motor and also serve as the disconnecting means. The 150 A CB is determined by multiplying 59 A times 250%, per Table 430-152 (59 A × 250% = 147.5 A). Note that 430-52, Ex. 1 is applied.

If the disconnecting means to the motor is a nonautomatic circuit breaker, the circuit breaker shall be at least 115% of the motor's full-load current rating. See Figure 3-21. For example, a motor with a full-load current rating of 59 A requires a 70 A nonautomatic CB (59 A × 115% = 67.85 A) per 430-110(a). The next standard size CB above 67.85 A is 70 A per 240-6.

What size circuit breaker, also used as a disconnecting means, is required to allow a 15 HP, 440 V, 3ϕ motor to start and run? Apply 430-52, Ex. 1.

Step 1. *430-6(a); Table 430-150:* 15 HP = 21 A

Step 2. *240-3, Ex. 6; Table 430-152:* 21 A × 250% = 52.5 A

Step 3. *430-52, Ex. 1; 240-6:* 60 A is the next higher standard size

Answer. **CB = 60 A**

If the CB is nonautomatic, the size CB used as a disconnect is selected as follows:

Step 1. *430-110(a):* 21 A × 115% = 24.15 A

Step 2. *240-6:* 25 A is the next standard size

Answer. **CB = 25 A**

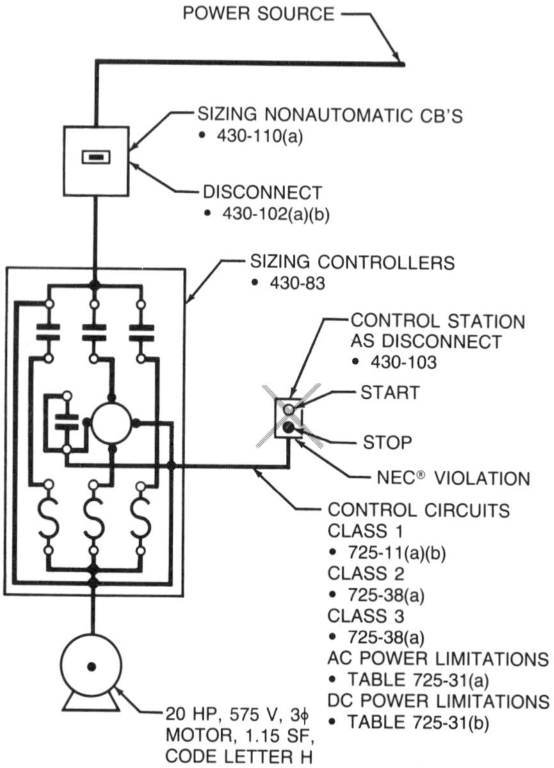

SIZING NONAUTOMATIC CB'S TO DISCONNECT A MOTOR'S POWER SOURCE

What size nonautomatic CB is required to disconnect the motor from the power source?

Step 1. *430-6(a); Table 430-150:* 20 HP = 22 A

Step 2. *430-110(a):* 22 A × 115% = 25.3 A

Step 3. *240-6:* 25 A CB

Answer. **CB = 25 A**

Note: A nonfused disconnect is sized by applying this procedure.

Figure 3-21. A nonautomatic CB shall be at least 115% of the motor's full-load current rating to be used as a disconnecting means for a motor.

■ *Master Electrician's Test Tip.* Where 115% times the full-load current rating of the motor does not correspond to a standard size CB, the next standard size per 240-6 is used.■

Number of Motors—430-112

The basic rule requires each motor to have a disconnecting means located within sight for personnel to service the motor. A disconnecting means by the controller and in sight from the motor can serve the controller and motor. If this disconnecting means can be locked open, it can serve as the disconnecting means for a motor located out of sight or more than 50′-0″ away. See Figure 3-22. Individual motors do not require a separate disconnecting means under the following conditions:

1. One disconnect can serve all the motors grouped with one single machine and operating together.
2. One disconnect can serve several motors grouped in the same room and in sight from the disconnect.
3. One disconnect can serve several fractional-horsepower motors supplied by a single branch circuit per 430-53(a).
4. The overcurrent protection device in the service equipment or subpanel can serve as the disconnecting means for motors rated ⅛ HP or less per 430-109, Ex. 1.
5. A single disconnect can serve the motors on a crane per 610-31 provided it is located in sight from the main collector wires, is readily operable from floor or ground level, and can be locked in the open position.

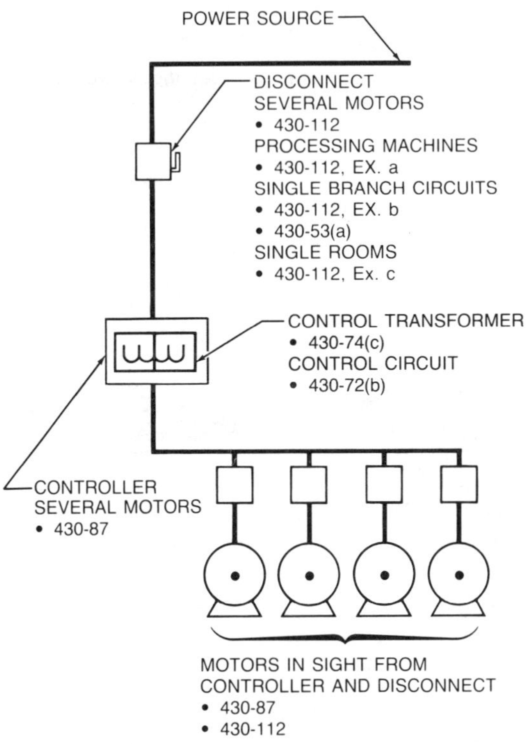

Figure 3-22. A single disconnect can be used as a disconnecting means for several motors located in a room and within sight.

CONTROLLER CIRCUITS—430, PART F

Control circuit conductors to operate the coils of magnetic starters or motor control centers can be routed in the enclosure of the equipment. They can also be routed through conduits to control devices to start and stop motors by energizing the coil to starters. The start and stop buttons can be mounted on the cover of magnetic starters or control centers or be mounted remotely for the convenience of starting or stopping a motor or motors at different locations.

Overcurrent Protection—430-72; 725-12

Motor control circuit conductors can be protected by the overcurrent protection device ahead of the motor circuit supplying the magnetic starter or the starter in a motor control center. For motor control circuit conductors not protected by the overcurrent protection device ahead of the starters, individual overcurrent protection shall be provided for the control circuit conductors. When using control voltage at a value less than that supplying the starter and motor, a low-voltage transformer can be installed to step the voltage down. If space prohibits the installation of a low-voltage transformer, a low voltage branch circuit can be run from another panel to the start and stop buttons. See Figure 3-23.

Controls Mounted on Enclosures—430-72(b), Ex. 1. Under certain conditions, control circuit conductors that control devices such as start and stop buttons installed in the cover of an enclosure housing a magnetic starter or motor control center are protected by the overcurrent protection device supplying power to the motor. The overcurrent protection device shall not exceed 400% of the ampacity of the control circuit conductors. Section 430-72(b), Ex. 1 only applies to control circuit conductors tapped from the line side of the magnetic starter and connected to the terminals of the start and stop stations. Table 310-17 is used when selecting the ampacity of conductors. The conductors are considered in free air due to available space in the open enclosure. See Figure 3-24.

For example, #14 THWN Cu conductors have an ampacity of 25 A per Table 310-17. The 60°C Column to Table 310-17 is used per Note 2 to Table 430-72(b). The size of the overcurrent protection device protecting the circuit conductors to the motor is 100 A (25 A × 400% = 100 A). The CB or fuse shall not exceed 400% per Table 430-72(b).

> What size overcurrent protection device (CB or fuses) is required to protect #12 THWN Cu control circuit conductors installed inside the enclosure of the magnetic starter? The control circuit conductors are tapped from the line side of the magnetic starter and are to be protected by the overcurrent protection device supplying the motor.
>
> Step 1. *430-72(b), Ex. 1; Note 2 to Table 430-72(b); Table 310-17, 60°C Column:*
> 30 A × 400% = 120 A
>
> Step 2. *240-3, Ex. 6; 240-6:*
> 110 A is the next lower standard size
>
> Answer. **OCPD = 110 A**

CONTROL TRANSFORMER PROTECTION	
FLC of primary = 9 A or more 125% × FLC	450-3(b)(1), Ex. 1
FLC of primary = less than 9 A but is 2 A or more 167% × FLC	450-3(b)(1), Ex. 1
FLC of primary = less than 2 A 300% or 500% × FLC	450-3(b)(1), Ex. 1 430-72(c), Ex. 2

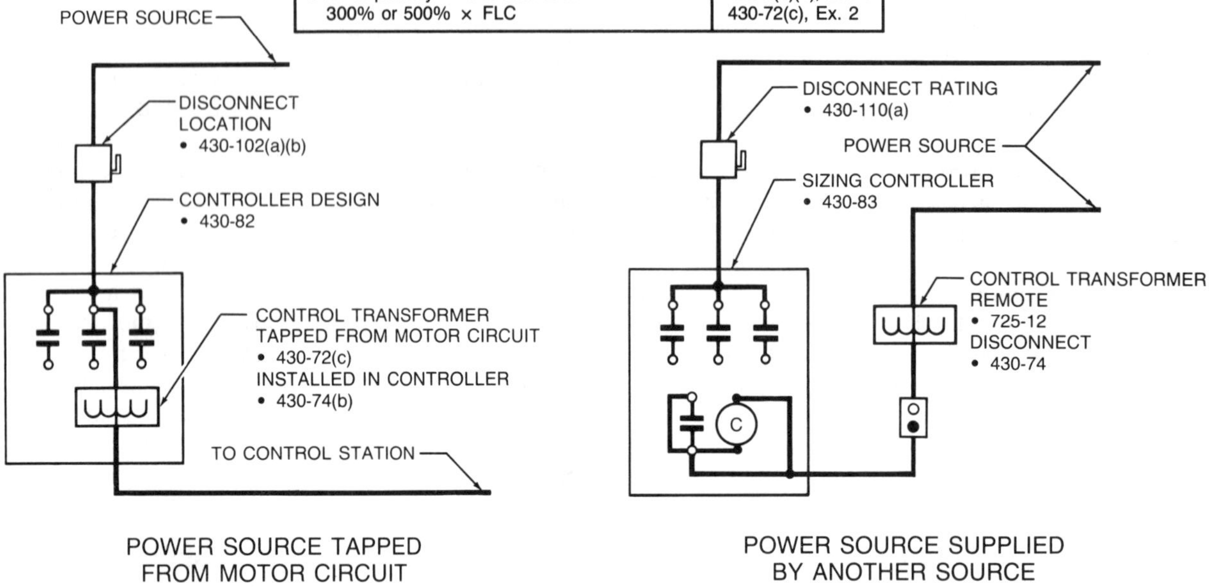

Figure 3-23. The power source to a control transformer can be tapped from the motor circuit or supplied by another source of power.

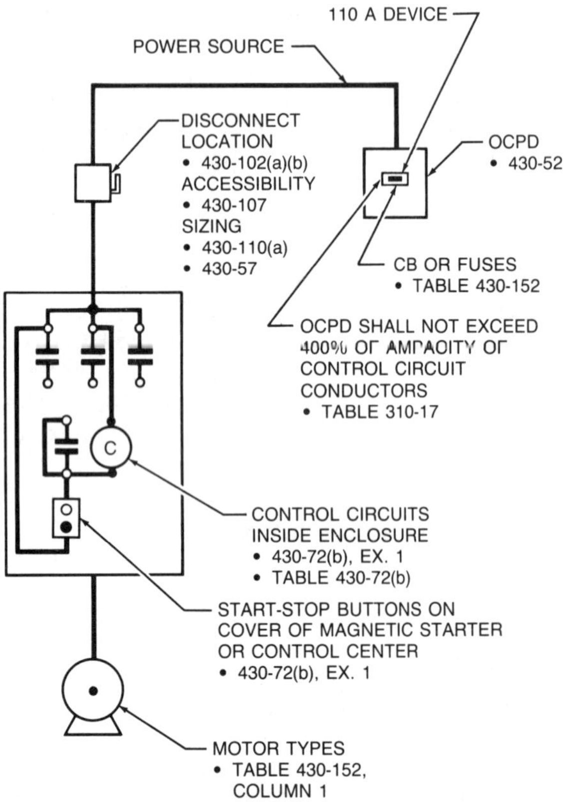

SIZING CONTROL CIRCUIT CONDUCTORS NOT EXTENDING BEYOND STARTER ENCLOSURE

What size control circuit conductors, using THHN Cu, is required in the enclosure?

Step 1. *430-72(b), Ex. 1; Table 430-72(b), Note 2: Ampacity of conductors* × 400%

Step 2. *Table 310-17, 60°C Column:*
#12 THHN Cu = 30 A

Step 3. *Table 430-72(b), Note 2:* 30 A × 400% = 120 A

Step 4. *240-3, Ex. 6; 240-6:*
110 A is the next lower standard size

Answer. **Conductors = #12 THHN Cu**

Figure 3-24. If the control conductors do not extend beyond the enclosure, the overcurrent protection device shall not exceed 400% of the ampacity of the control circuit conductors per Table 310-17.

Controls Remotely Mounted—430-72(b), Ex. 2. Where control devices such as start and stop buttons are mounted remotely from the enclosure of magnetic starters and motor control centers to permit different locations of control, the CBs or fuses supplying the power to the motor are permitted to protect the control circuit conductors. The CBs or fuses shall not exceed 300% of the ampacity of the control circuit conductors per Table 430-72(b), Note 3 and Table 310-16, 60°C Column.

The ampacity of control conductors tapped from the line side of a magnetic starter and routed remotely through conduit to start and stop buttons is selected from the 60°C Column in Table 310-16. This ampacity rating is based on 15 A for #14, 20 A for #12, and 30 A for #10 conductors per Table 310-16, Footnotes. For example, #14 THHN Cu conductors are considered protected by 45 A CBs or fuses per 430-72(b), Ex. 2 (15 A × 300% = 45 A). This is verified by Table 430-72(b), Column C.

What size THHN Cu control circuit conductors is required to be tapped ahead of a magnetic starter if the motor circuit conductors are served by a 50 A, 3-pole circuit breaker? (The control circuit conductors are routed remotely to start and stop stations.)

Step 1. *Table 310-16, Footnotes:* #12 = 20 A

Step 2. *240-3, Ex. 6; 430-72(b), Ex. 2; Table 430-72(b), Column C:* 20 A × 300% = 60 A

Answer. **Conductors = #12 THHN Cu**

Note: The 50 A CB serving the motor circuit is less than the 60 A overcurrent device permitted per Table 430-72(b), Column C. If the overcurrent protection device is sized at a greater value than 300% times the ampacity of the control circuit conductors to allow the motor to start and run, an overcurrent protection device block shall be installed with separate CBs or fuses or larger conductors shall be used. See Figure 3-25.

Control Transformers—430-72(c). A control transformer can be installed to use lower voltage than the voltage on the motor circuit. This is often done if the motor circuit voltage is over 600 V. The overcurrent protection device is placed in the primary or secondary side of the control transformer. If it is in the primary side per 430-72(c), Ex. 2, and the full-load current in the primary is less than 2 A, the overcurrent protection device can be set at not more than 500%. See Figure 3-26.

What size overcurrent protection device is permitted to protect the primary of a transformer having a full-load current rating of 1 A? (Use maximum size.)

Step 1. *240-3, Ex. 2; 450-3(b)(1), Ex. 1; 430-72(c), Ex. 2:* 1 A × 500% = 5 A

Step 2. *240-6, Ex.:* 3 A is the next lower size

Answer. **OCPD = 3 A fuses**

Note: 5 A is not a standard size fuse listed in 240-6, Ex. The next standard size below 5 A is selected. This is a 3 A fuse per 240-6, Ex.

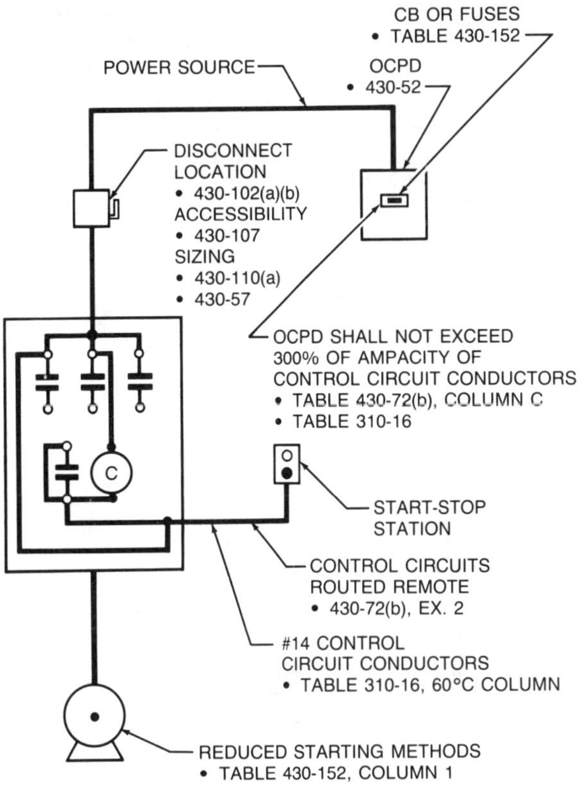

SIZING CONTROL CIRCUIT CONDUCTORS EXTENDING BEYOND CONTROLLER ENCLOSURE

What maximum size OCPD is required for the motor control circuit using THHN Cu conductors?

Step 1. *430-72(b), Ex. 2; Table 430-72(b), Note 3: Ampacity of conductors* × 300%

Step 2. *Table 310-16, 60°C Column, Footnotes:* #14 THHN Cu = 15 A

Step 3. *Table 430-72(b), Note 3:* 15 A × 300% = 45 A

Step 4. *240-3, Ex. 6; 240-6:* 45 A is a standard size

Answer. **OCPD = 45 A**

Figure 3-25. If the control conductors extend beyond the enclosure, the overcurrent protection device shall not exceed 300% of the ampacity of the control circuit conductors per Table 310-16.

The overcurrent protection device can also be installed on the secondary side of the transformer. The control circuit run from the secondary side of the transformer shall be a Class 1 circuit, per 725-11(a). Section 725-11(a)(1) references 450-3(b) for the protection of transformers utilizing a Class 1 circuit supplied from the secondary side.

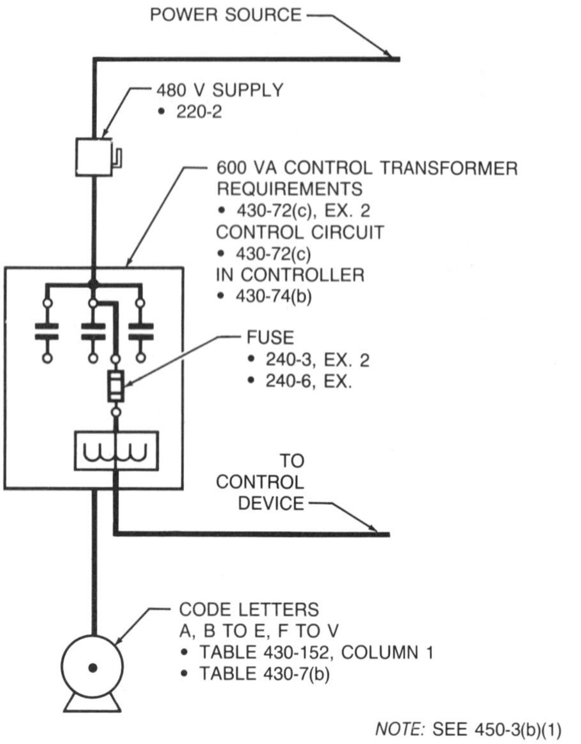

SELECTING OCPD OF PRIMARY TO PROTECT CONTROL CONDUCTORS OF SECONDARY	
PRIMARY FLC	PERCENT × FLC
9 A or more	125%
2 A and less than 9 A	167%
Less than 2 A	300% (See 430-72(c), Ex. 2)

SIZING OCPD'S IN PRIMARY SIDE OF CONTROL TRANSFORMER

What minimum and maximum size fuse is required in the primary of the control transformer?

Minimum

Step 1. *Text:* $I = \frac{600 \text{ VA}}{480 \text{ V}} = 1.25 \text{ A}$

Step 2. *430-72(b), Ex. 3; 450-3(b)(1), Ex. 1:* 1.25 A × 300% = 3.75 A

Step 3. *240-3, Ex. 2; 240-6, Ex.:* 3 A is the next lower standard size

Answer. **Fuse = 3 A**

Maximum

Step 1. *430-72(b), Ex. 3; 430-72(c), Ex. 2:* 1.25 A × 500% = 6.25 A

Step 2. *240-3, Ex. 2; 240-6, Ex.:* 6 A is the next lower standard size

Answer. **Fuse = 6 A**

Figure 3-26. If the transformer is 2-wire to 2-wire per 240-3, Ex. 2, the overcurrent protection device can be placed in the primary side of a control transformer per 450-3(b)1, Ex. 1 or 430-72(c), Ex. 2.

What size overcurrent protection device is required in the secondary side of a 30 V transformer having an output of 210 VA and supplying a Class 1 circuit for control circuits?

Step 1. Secondary output. $I = \frac{VA}{A}$

$$I = \frac{210 \text{ VA}}{30 \text{ V}} = 7 \text{ A}$$

Step 2. *430-72(c), Ex. 3; 725-11(a); 725-11(a)(1); 450-3(b)(1), Ex. 1:* 7 A × 167% = 11.69 A

Step 3. *240-6, Ex.:* 10 A is the next standard size below 11.69 A

Answer. **OCPD = 10 A fuses**

The 10 A fuses are selected because 167% times the secondary full-load current shall not be exceeded. The next lower standard size fuse below 11.69 A, per 240-6, Ex., is 10 A.

CAPACITORS—460, PART A

Current in induction motors is out of phase with line current, which causes the current to lag behind the line voltage. Capacitors are used to correct power factor for induction motors creating a *leading current.* Capacitors draw a leading current from the line; therefore, the current leads the voltage. The leading current of the capacitors counteracts the lagging current drawn by induction motors. This counteraction raises and improves the power factor. Capacitors are designed to draw current. They do not consume any power. Conductors supplying capacitors are sized to carry currents to correct poor power factor. They are protected with overcurrent protection, and a disconnect is provided for the safety of personnel servicing the capacitors. See Figure 3-27.

Ampacity—460-8(a)

Two rules apply to capacitors used to correct power factor to a motor:

1. Conductors supplying a capacitor shall have a current-carrying capacity of not less than 135% of the full-load current rating of the capacitor. For example, the rating used for sizing the conductors to a capacitor

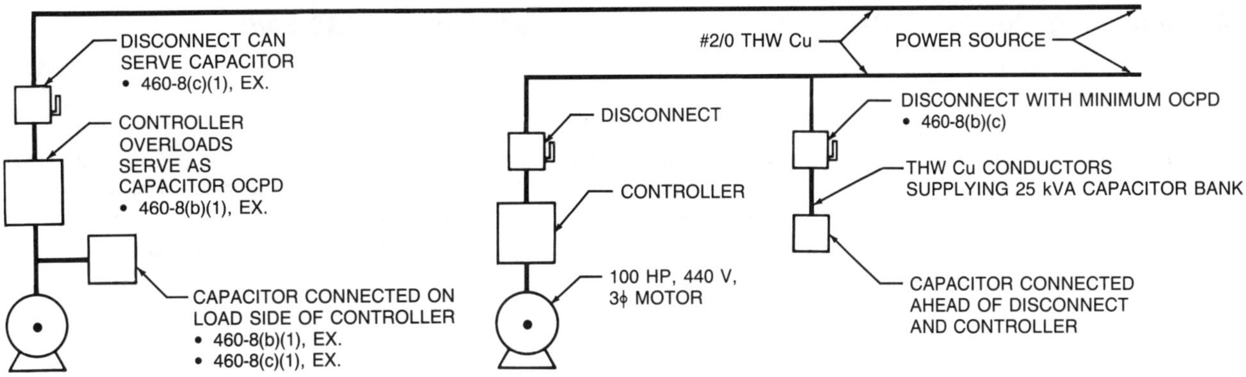

SIZING CONDUCTORS, OCPD'S, AND DISCONNECT FOR CAPACITOR CONNECTED AHEAD OF DISCONNECT AND CONTROLLER

What size THW Cu conductors, OCPD, and disconnect is required for the capacitor connected ahead of the disconnect and controller to the motor?

Conductors

Step 1. *Text:* $FLC = \dfrac{25\text{ kVA} \times 1000}{440\text{ V} \times 1.732} = 32.8\text{ A}$

Step 2. *460-8(a):* 32.8 A × 135% = 44 A

Step 3. *Table 310-16:* #8 THW Cu = 50 A

Step 4. *460-8(a); Table 430-150:*
#2/0 THW Cu = 175 A
175 A × ⅓ = 58 A

Step 5. *Table 310-16:* #6 THW Cu = 65 A

Answer. **Conductors = #6 THW Cu**

OCPD

Step 1. *460-8(b): A = FLC of capacitor* × 150%
32.8 A × 150% = 49.2 A

Step 2. *240-3, Ex. 7; 240-6:* 45 A is below 49.2 A

Answer. **OCPD = 45 A**

Note: OCPD can be set from 150% to 250% of FLC.

Disconnect

Step 1. *460-8(c)(4): FLC* × 135%
32.8 A × 135% = 44.28 A

Step 2. *Chart:* 60 A is above 44.28 A

Answer. **Disconnect = 60 A**

Figure 3-27. Capacitor size is determined by multiplying the FLC rating of the capacitor by 135% or taking ⅓ of the conductor's ampacity supplying the motor, and selecting whichever is greater at 150% to 250% of the capacitor's FLC.

with a full-load current of 24 A is 32.4 A (24 A × 135% = 32.4 A) per 460-8(a).

2. Multiply the ampacity of conductors supplying power to the motor by ⅓. For example, the rating used to size the conductors to a capacitor with #1/0 THWN Cu conductors supplying the motor circuit is 50 A (⅓ × 150 A = 50 A) per 460-8(a). See Table 310-16 for ampacities of conductors.

What size THWN Cu conductors is required to supply a 15 kVA capacitor bank that serves a 230 V, 3ϕ induction motor supplied by #3/0 THWN Cu conductors?

Rule 1

Step 1. FLC Rating. $I = \dfrac{kVA \times 1000}{V \times \sqrt{3}}$

$$I = \frac{15\text{ kVA} \times 1000}{230\text{ V} \times 1.732} = 37.7\text{ A}$$

Step 2. *460-8(a):* 37.7 A × 135% = 50.9 A

Step 3. *Table 310-16:* #6 THWN Cu = 65 A

Answer. **Conductors = #6 THWN Cu**

Rule 2

Step 1. *Table 310-16:* #3/0 THWN Cu = 200 A

Step 2. *460-8(a):* ⅓ × 200 A = 66.7 A

Step 3. *Table 310-16:* #4 THWN Cu = 85 A

Answer. **Conductors = #4 THWN Cu**

Overcurrent Protection—460-8(b)

The running overload protection device can serve as overcurrent protection for a capacitor connected on the load side of the magnetic starter. If the capacitor is connected on the line side of the magnetic starter, it shall be protected by individual fuses or CBs in the capacitor circuit conductors. Capacitors pull a high momentary inrush current when energized by the line. The overcurrent protection device set at 150% to 250% of the capacitor's full-load current rating will usually hold this inrush current under normal operating conditions. The setting of the overcurrent protection device should be as low as practical. The device is selected at a rating just high enough to prevent unnecessary tripping of the capacitor circuit.

Disconnecting Means—460-8(c)

Capacitors shall be provided with a disconnecting means. Where the capacitor is located on the load side of the disconnecting means used for the motor, the disconnect for the motor controller can serve as the disconnect for the capacitor. If the capacitor is not connected in this manner, an individual disconnecting means shall be provided. The disconnecting means shall have a rating of at least 135% of the capacitor's full-load current rating. For example, a capacitor with a full-load current rating of 37.7 A requires a 60 A disconnect (37.7 A × 135% = 50.9 A). The standard disconnect switch or CB is 60 A. See 240-6 for standard size CBs and the chart on the inside back cover of this book for selecting standard disconnects.

Branch Circuits—430-22(b)

Single-phase conductors supplying 1ϕ power to phase converters are sized at not less than 2.16 times the full-load current of the motor or the load. The input and output voltages shall be the same. When the voltages are not the same, the current is determined by multiplying the ratio of the output voltage to input voltage. Overcurrent protection devices are sized based on the load calculation. See Figure 3-28.

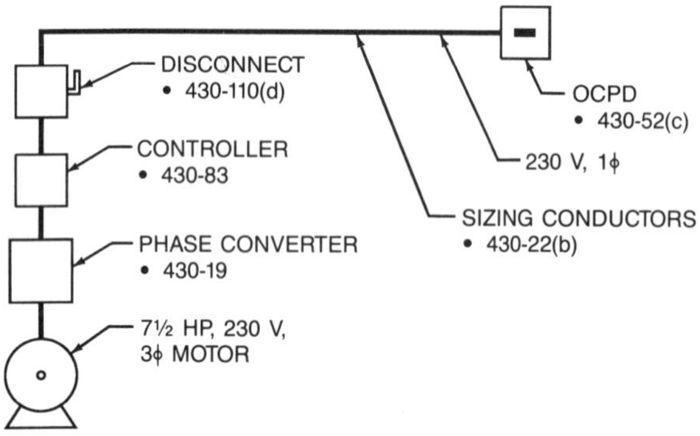

SIZING CONDUCTORS AND OCPD FOR PHASE CONVERTERS

What size THWN Cu conductors and OCPD are required to supply the phase converter with 230 V, 1ϕ power?

Step 1. *Table 430-150:* 7½ HP = 22 A

Step 2. *430-22(b):* 22 A × 2.16 = 47.52 A

Step 3. *Table 310-16:* #8 THWN Cu = 50 A

Answer. **Conductors = #8 THWN Cu**

Step 4. *430-52(c):* Load of motor = 47.52 A

Step 5. *430-52(c):* Next size = 50 A

Answer. **OCPD = 50 A**

Figure 3-28. Single-phase conductors supplying 1ϕ power to phase conductors are sized at not less than 2.16 times the full-load current of the motor or load.

Feeder Circuits—430-24(b)

Single-phase conductors supplying 1ϕ power to phase converters shall have an ampacity not less than 1.73 times the sum of the full-load current of all motors plus 25% of the largest motor in the group. The input and output voltage shall be the same. If they are not the same, current is determined by multiplying the ratio of output voltage to the input voltage. See Figure 3-29.

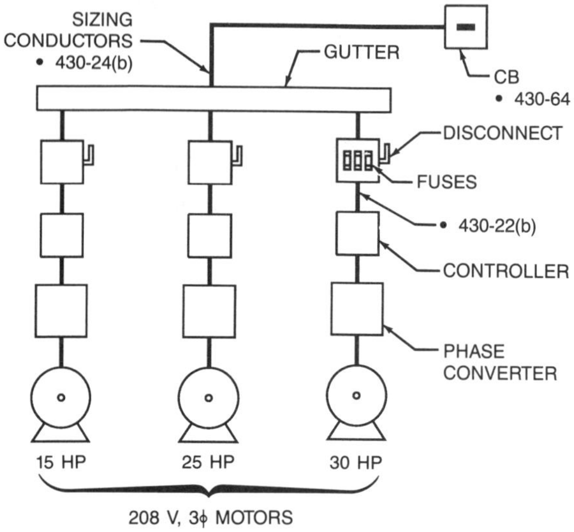

SIZING CONDUCTORS AND OCPD FOR PHASE CONVERTERS

What size THWN Cu conductors and OCPD are required to supply 1ϕ power to phase converters serving more than one motor in a group?

Step 1. *Table 430-150:*
15 HP = 46.2 A
25 HP = 74.8 A
30 HP = 88 A

Step 2. *430-24(b):* 1.73 of load amps of motors
88 A + 74.8 A + 46.2 A = 209 A
209 A × 1.73 = 361.57 A

Step 3. *430-24(b):*
25% of the largest motor plus sum of group
88 A × 1.732 × 25% = 38 A
361.57 A
399.57 A

Step 4. *Table 310-16:* Selecting conductors
#600 kcmil THWN Cu = 420 A

Answer. **Conductors = 600 kcmil THWN Cu**

Step 5. *430-165; 430-64:* Sizing load for OCPD
Load of motors = 399.57 A

Step 6. *430-165; 430-64; 240-6:* Selecting OCPD
Next standard size = 400 A

Answer. **OCPD = 400 A**

Figure 3-29. Single-phase conductors supplying 1ϕ power to phase converters are sized at not less than 1.73 times the sum of the full-load current of all motors plus 25% of the largest motor.

Motor Circuits

Date ____________________

Chapter 3–TEST

Name ____________________________________

Section	Answer

1. What are the four types of motors listed in Table 430-152?

2. What are the four starting methods listed in Table 430-152?

3. What are the three code letter groups used to size overcurrent protection devices for motors in Table 430-152?

4. What are the four types of overcurrent protection devices listed in Table 430-152?

5. What size THWN Cu conductors is required to supply power to a motor with the following nameplate?

HP	10	Temperature rise	50 °C
Voltage	230 V, 3ϕ	Code letter	D
FLC	19.5 A	RPM	—
Service factor	1.32	Power factor	—

6. What size THWN Cu conductors is required to supply power to a motor with the following nameplate?

HP	25	Temperature rise	40 °C
Voltage	208 V, 3ϕ	Code letter	C
FLC	62 A	RPM	—
Service factor	1.10	Power factor	—

7. What size THWN Cu conductors is required to supply power to a motor with the following nameplate?

HP	125	Temperature rise	40 °C
Voltage	440 V, 3ϕ	Code letter	E
FLC	98.5 A	RPM	—
Service factor	1.00	Power factor	—

________ **8.** What size THWN Cu conductors is required to supply power to a motor with the following nameplate?

HP	30	Temperature rise	50 °C
Voltage	200 V, 3ϕ	Code letter	B
FLC	78.5 A	RPM	—
Service factor	1.15	Power factor	—

________ **9.** What size THWN Cu conductors is required to supply power to a synchronous motor with the following nameplate?

HP	200	Temperature rise	40 °C
Voltage	460 V, 3ϕ	Code letter	—
FLC	201 A	RPM	—
Service factor	1.25	Power factor	80%

________ **10.** What size THWN Cu conductors is required to supply power on a feeder circuit having four motors with the following nameplates?

Motor 1		***Motor 3***	
HP	3	HP	10
Voltage	230 V, 3ϕ	Voltage	230 V, 3ϕ
FLC	8 A	FLC	19 A
Service factor	1.15	Service factor	1.15
Temperature rise	40 °C	Temperature rise	40 °C
Code letter	C	Code letter	A
RPM	—	RPM	—
Power factor	—	Power factor	—
Motor 2		***Motor 4***	
HP	5	HP	25
Voltage	230 V, 3ϕ	Voltage	230 V, 3ϕ
FLC	12 A	FLC	52 A
Service factor	1.15	Service factor	1.15
Temperature rise	40 °C	Temperature rise	40 °C
Code letter	B	Code letter	E
RPM	—	RPM	—
Power factor	—	Power factor	—

________ **11.** What size THWN Cu conductors is required to supply power on a feeder circuit having four motors with the following nameplates?

Motor 1		***Motor 3***	
HP	2	HP	15
Voltage	440 V, 3ϕ	Voltage	440 V, 3ϕ
FLC	3 A	FLC	21 A
Service factor	1.15	Service factor	1.15
Temperature rise	40 °C	Temperature rise	40 °C
Code letter	G	Code letter	H
RPM	—	RPM	—
Power factor	—	Power factor	—
Motor 2		***Motor 4***	
HP	7.5	HP	25
Voltage	440 V, 3ϕ	Voltage	440 V, 3ϕ
FLC	8 A	FLC	27 A
Service factor	1.15	Service factor	1.15
Temperature rise	40 °C	Temperature rise	40 °C
Code letter	D	Code letter	E
RPM	—	RPM	—
Power factor	—	Power factor	—

______ **12.** What size THWN Cu conductors is required to supply power on a feeder circuit having four motors with the following nameplates?

Motor 1		***Motor 3***	
HP	10	HP	20
Voltage	208 V, 3ϕ	Voltage	208 V, 3ϕ
FLC	20.5 A	FLC	40.5 A
Service factor	1.15	Service factor	1.15
Temperature rise	40 °C	Temperature rise	40 °C
Code letter	E	Code letter	G
RPM	—	RPM	—
Power factor	—	Power factor	—

Motor 2		***Motor 4***	
HP	15	HP	25
Voltage	208 V, 3ϕ	Voltage	208 V, 3ϕ
FLC	35.8 A	FLC	52.5 A
Service factor	1.15	Service factor	1.15
Temperature rise	40 °C	Temperature rise	40 °C
Code letter	F	Code letter	H
RPM	—	RPM	—
Power factor	—	Power factor	—

______ **13.** What size THWN Cu conductors is required to supply power on a feeder circuit having four motors with the following nameplates?

Motor 1		***Motor 3***	
HP	15	HP	25
Voltage	200 V, 3ϕ	Voltage	200 V, 3ϕ
FLC	42 A	FLC	68 A
Service factor	1.15	Service factor	1.15
Temperature rise	40 °C	Temperature rise	75 °C
Code letter	B	Code letter	D
RPM	—	RPM	—
Power factor	—	Power factor	—

Motor 2		***Motor 4***	
HP	20	HP	30
Voltage	200 V, 3ϕ	Voltage	200 V, 3ϕ
FLC	54 A	FLC	80 A
Service factor	1.25	Service factor	1.15
Temperature rise	50 °C	Temperature rise	40 °C
Code letter	C	Code letter	E
RPM	—	RPM	—
Power factor	—	Power factor	—

______ **14.** What size THWN Cu conductors is required to supply power on a feeder circuit having four motors with the following nameplates?

Motor 1
- HP 15
- Voltage . . . . 230 V, 3ϕ
- FLC . . . . —
- Service factor 1.15
- Temperature rise . . . . 40 °C
- Code letter . . . . B
- RPM . . . . —
- Power factor . . . . —

Motor 2
- HP 20
- Voltage . . . . 230 V, 3ϕ
- FLC . . . . —
- Service factor 1.15
- Temperature rise . . . . 40 °C
- Code letter C
- RPM . . . . —
- Power factor . . . . —

Motor 3
- HP 20
- Voltage . . . . 230 V, 3ϕ
- FLC . . . . —
- Service factor 1.10
- Temperature rise . . . . 50 °C
- Code letter . . . . D
- RPM . . . . —
- Power factor 90%

Motor 4
- HP 25
- Voltage . . . . 230 V, 3ϕ
- FLC . . . . —
- Service factor 1.10
- Temperature rise . . . . 75 °C
- Code letter . . . . E
- RPM . . . . —
- Power factor 80%

______ **15.** What size THWN Cu conductors is required to supply power on a feeder circuit having four motors with the following nameplates?

Motor 1
- HP 1
- Voltage . . . . 230 V, 1ϕ
- FLC . . . . 8 A
- Service factor 1.15
- Temperature rise . . . . 40 °C
- Code letter . . . . A
- RPM . . . . —
- Power factor . . . . —

Motor 2
- HP 1.5
- Voltage . . . . 230 V, 1ϕ
- FLC . . . . 10 A
- Service factor 1.15
- Temperature rise . . . . 40 °C
- Code letter . . . . B
- RPM . . . . —
- Power factor . . . . —

Motor 3
- HP 2
- Voltage . . . . 230 V, 1ϕ
- FLC . . . . 12 A
- Service factor 1.15
- Temperature rise . . . . 40 °C
- Code letter C
- RPM . . . . —
- Power factor . . . . —

Motor 4
- HP 10
- Voltage . . . . 230 V, 1ϕ
- FLC . . . . 50 A
- Service factor 1.15
- Temperature rise . . . . 40 °C
- Code letter . . . . D
- RPM . . . . —
- Power factor 80%

16. What size THWN Cu conductors is required to supply power on a feeder circuit with other loads of 50 A (noncontinuous duty) and 35 A (continuous duty) and four motors with the following nameplates?

Motor 1

HP 5
Voltage . . . . 208 V, 3ϕ
FLC . . . . 15.2 A
Service factor 1.15
Temperature rise . . . . 40 °C
Code letter C
RPM . . . . —
Power factor . . . . —

Motor 2

HP 7.5
Voltage . . . . 208 V, 3ϕ
FLC . . . . 22 A
Service factor 1.15
Temperature rise . . . . 40 °C
Code letter C
RPM . . . . —
Power factor . . . . —

Motor 3

HP 10
Voltage . . . . 208 V, 3ϕ
FLC . . . . 28 A
Service factor 1.15
Temperature rise . . . . 40 °C
Code letter C
RPM . . . . —
Power factor . . . . —

Motor 4

HP 15
Voltage . . . . 208 V, 3ϕ
FLC . . . . 42 A
Service factor 1.15
Temperature rise . . . . 40 °C
Code letter C
RPM . . . . —
Power factor . . . . —

17. What size THWN Cu conductors is required to supply power on a feeder circuit with other loads of 26 A (noncontinuous duty) and 19 A (continuous duty) and four motors with the following nameplates?

Motor 1

HP . . . . ¾
Voltage . . . . 230 V, 1ϕ
FLC . . . . 6 A
Service factor 1.15
Temperature rise . . . . 40 °C
Code letter . . . . B
RPM . . . . —
Power factor . . . . —

Motor 2

HP 1
Voltage . . . . 230 V, 1ϕ
FLC . . . . 7 A
Service factor 1.15
Temperature rise . . . . 40 °C
Code letter . . . . B
RPM . . . . —
Power factor . . . . —

Motor 3

HP 1.5
Voltage . . . . 230 V, 1ϕ
FLC . . . . 8 A
Service factor 1.15
Temperature rise . . . . 40 °C
Code letter . . . . B
RPM . . . . —
Power factor . . . . —

Motor 4

HP 2
Voltage . . . . 230 V, 1ϕ
FLC . . . . 10 A
Service factor 1.15
Temperature rise . . . . 40 °C
Code letter . . . . B
RPM . . . . —
Power factor . . . . —

18. What size THWN Cu conductors is required to supply power to a motor with the following nameplate? The motor operates on short-time duty with a 15-minute rating.

HP 100
Voltage . . . . 550 V, 3ϕ
FLC . . . . 80 A
Service factor 1.15
Temperature rise . . . . 40 °C
Code letter . . . . G
RPM . . . . —
Power factor 100%

_____ **19.** What size THWN Cu conductors is required to supply power to a motor with the following nameplate? The motor operates on periodic duty with a 30-minute rating.

HP	40	Temperature rise	40 °C
Voltage	440 V, 3ϕ	Code letter	H
FLC	52 A	RPM	—
Service factor	1.15	Power factor	—

_____ **20.** What size THWN Cu conductors is required to supply power to a motor with the following nameplate? The motor operates on intermittent duty with a 30-minute rating.

HP	50	Temperature rise	50 °C
Voltage	220 V, 3ϕ	Code letter	S
FLC	118 A	RPM	—
Service factor	1.32	Power factor	—

_____ **21.** What size THWN Cu conductors is required to supply power to a motor with the following nameplate? The motor operates on short-time duty with a five-minute rating.

HP	3	Temperature rise	40 °C
Voltage	230 V, 3ϕ	Code letter	E
FLC	8 A	RPM	—
Service factor	1.15	Power factor	—

_____ **22.** What is the LRC of a motor with the following nameplate?

HP	75	Temperature rise	40 °C
Voltage	230 V, 3ϕ	Code letter	D
FLC	178 A	RPM	—
Service factor	1.15	Power factor	—

_____ **23.** What is the LRC of a motor with the following nameplate? Use the code letter.

HP	50	Temperature rise	40 °C
Voltage	460 V, 3ϕ	Code letter	K (maximum)
FLC	55 A	RPM	—
Service factor	1.15	Power factor	—

_____ **24.** What size overloads is required to protect a motor with the following nameplate?

HP	25	Temperature rise	40 °C
Voltage	208 V, 3ϕ	Code letter	C
FLC	52 A	RPM	—
Service factor	1.15	Power factor	—

_____ **25.** What size overloads is required to protect a motor with the following nameplate?

HP	30	Temperature rise	—
Voltage	460 V, 3ϕ	Code letter	F
FLC	32 A	RPM	—
Service factor	—	Power factor	—

Transformer Circuits

Chapter **4**

Transformers are used in the electrical industry to provide power to dwelling units, apartment complexes, and commercial and industrial locations. They *transform* (change) high voltage distribution line power to lower voltage utilization power. This power supplies circuits that feed motors, lighting, appliances, and other types of loads. Conductors and overcurrent protection devices are sized and selected to protect the windings of transformers supplying power to various loads.

The primary overcurrent protection devices are used to trip open a ground fault or short circuit. The secondary overcurrent protection devices are sized to protect the windings of the transformer from overloads.

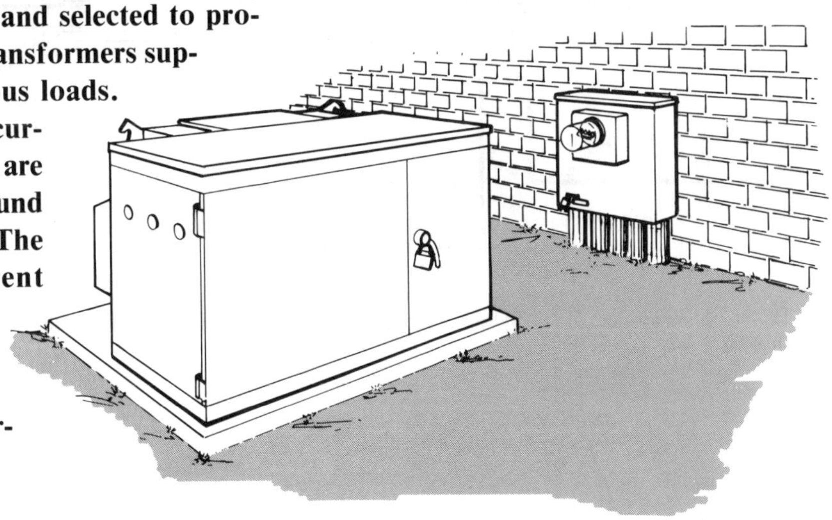

LOCATION—450-13

Transformers are located so that they are readily accessible to qualified personnel for inspection and maintenance. Without applying exceptions, 450-13 does not permit ladders to be used to gain access to transformers. Per 450-9, transformers are located in areas with sufficient cool air to prevent overheating of windings. Transformers that are exposed to physical damage shall be guarded per 450-8. Section 450-13 requires transformers to be installed on the floor, hung from the ceiling, or mounted on the wall at readily accessible locations.

450-13, Ex. 1

Dry-type transformers rated at 600 V or less can be located on *open walls*. Open walls are where the transformers are hung from the ceiling or mounted on the wall, and can be seen. Anyone entering the room or area can see the transformers hung or mounted on the open walls. With this type of installation, transformers do not have to be readily accessible. See Figure 4-1.

450-13, Ex. 2

Dry-type transformers rated at 600 V or less and not over 50 kVA can be installed in fire-resistant hollow spaces of buildings not permanently enclosed by the construction. Access by a scuttle hole, pull-down stairway, or similar device shall be available. The transformers are required to be accessible, but not readily accessible, per 450-13. Exception 2 to 450-13 and 300-22(c) allow transformers, under specific conditions, to be installed in suspended ceilings used for return air systems. See Figure 4-2.

TRANSFORMER RATINGS

The different ratings of transformers are used to calculate values when sizing conductors and overcurrent protection devices. Volts, amperage, and volt-amps based on loads in buildings are used to determine the rating of the transformer in kVA. The ratio of transformers due to the primary and secondary voltage determines the amps that can be taken from each side to supply loads. Known values are used with formulas

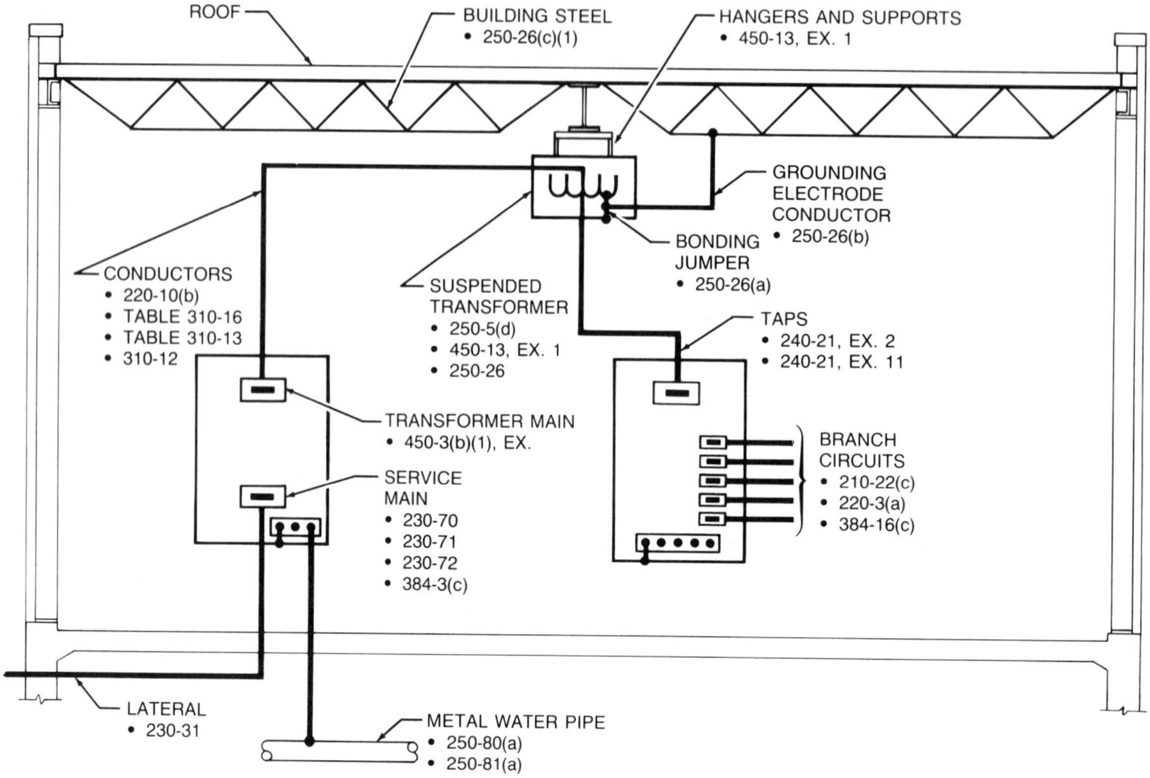

Figure 4-1. Transformers are hung from the ceiling or mounted to the wall per 450-13, Ex. 1 when there is not enough space available on the floor.

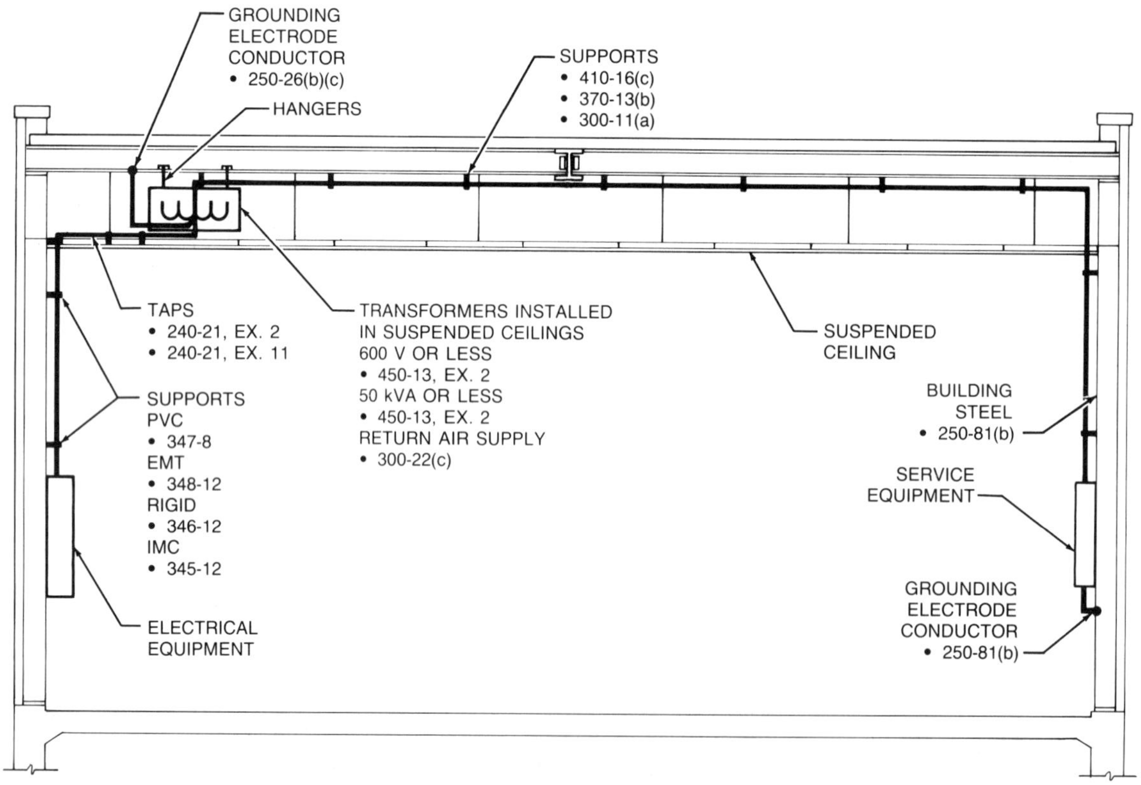

Figure 4-2. Transformers rated 50 kVA or less and 600 V or less are installed in fire-resistant hollow spaces of buildings per 450-13, Ex. 2.

to obtain unknown values needed to size transformers and transformer elements and circuits.

Full-load Current

The full-load current ratings of transformers can be obtained by two methods. One method is to select the full-load current rating of the transformer based on the voltage and kVA from the manufacturer's chart or catalog. See Figure 4-3. In the second method, kVA is divided by voltage to find the full-load current rating in amperes. See Figure 4-4.

■ *Master Electrician's Test Tip.* The second method is commonly used for finding full-load current ratings of transformers when taking the master's examination.■

TRANSFORMERS

SINGLE-PHASE

kVA RATING	AMPERES 120 V	240V	480 V	600 V
1	8.33	4.17	2.08	1.67
1½	12.5	6.25	3.13	2.50
2	16.7	8.33	4.17	3.33
3	25.0	12.5	6.25	5.00
5	41.7	20.8	10.4	8.33
7½	62.5	31.3	15.6	12.5
10	83.3	41.7	20.8	16.7
15	125	62.5	31.3	25.0
20	167	83.3	41.7	33.3
25	208	104	52.1	41.7
30	250	125	62.5	50
37½	313	156	78.0	62.5
50	417	208	104	83.3
75	625	313	156	125
100	833	417	208	167
150	1250	625	313	250
167	1392	696	348	278
200	1667	833	417	333
250	2083	1042	521	417
333	2775	1388	694	555
500	4167	2083	1042	833

THREE-PHASE

kVA RATING	AMPERES 208 V	240 V	480 V	600 V
3	8.3	7.2	3.6	2.9
6	16.6	14.4	7.2	5.8
9	25.0	21.6	10.8	8.7
15	41.6	36	18	14.4
20	55.6	48.2	24.1	19.3
25	69.5	60.2	30.1	24.1
30	83.0	72	36	28.8
37½	104	90.3	45.2	36.1
45	125	108	54	43
50	139	120	60.2	48.2
60	167	145	72.3	57.8
75	208	180	90	72
100	278	241	120	96.3
112.5	312	271	135	108
150	415	360	180	144
200	554	480	240	192
225	625	540	270	216
300	830	720	360	288
400	1110	960	480	384
500	1380	1200	600	480
750	2080	1800	900	720
1000	2780	2400	1200	960
1500	4150	3600	1800	1440
2000	5540	4800	2400	1920

15 kVA TRANSFORMER WITH 480 V, 3ϕ PRIMARY HAS AN FLC OF 18 A

Figure 4-3. Manufacturers' charts can be used to find full-load current ratings of transformers.

Chart or Catalog. If a chart or catalog is used to find full-load current ratings of transformers in amperes, the kVA and voltage must be known. The size of the transformer is determined by locating the column with the correct voltage and kVA rating, and then selecting the corresponding amperes. If voltage and amps are known, the kVA can be found by applying the same procedure. For example, the handy chart in Figure 4-3 shows that the amperage for a 30 kVA, 208 V, 3ϕ transformer is 83 A.

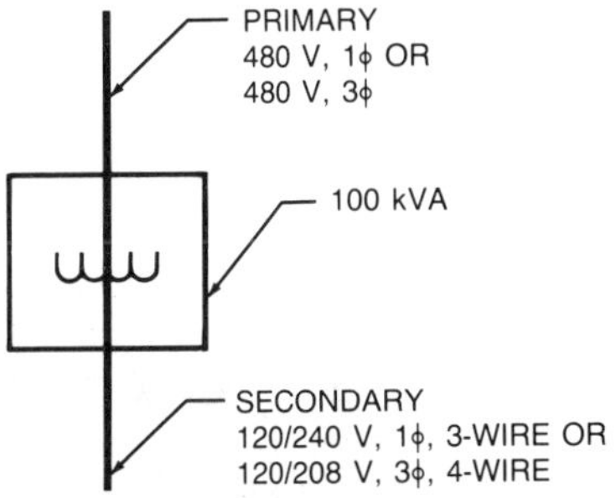

FINDING PRIMARY FLC

What is the FLC rating of the primary using the listed voltages?

Single phase

Step 1. $FLC = \frac{kVA \times 1000}{V}$

$$FLC = \frac{100\ \text{kVA} \times 1000}{480\ \text{V}} = 208\ \text{A}$$

Answer. **FLC = 208 A**

Three phase

Step 1. $FLC = \frac{kVA \times 1000}{V \times \sqrt{3}}$

$$FLC = \frac{100\ \text{kVA} \times 1000}{480\ \text{V} \times 1.732} = 120\ \text{A}$$

Answer. **FLC = 120 A**

FINDING SECONDARY FLC

What is the FLC rating of the secondary using the listed voltages?

Single phase

Step 1. $FLC = \frac{kVA \times 1000}{V}$

$$FLC = \frac{100\ \text{kVA} \times 1000}{240\ \text{V}} = 416.7\ \text{A}$$

Answer. **FLC = 416.7 A**

Three phase

Step 1. $FLC = \frac{kVA \times 1000}{V \times \sqrt{3}}$

$$FLC = \frac{100\ \text{kVA} \times 1000}{208\ \text{V} \times 1.732} = 277.8\ \text{A}$$

Answer. **FLC = 277.8 A**

Figure 4-4. Single-phase FLC = kVA × 1000/V. Three-phase FLC = kVA × 1000/V × $\sqrt{3}$.

Single Phase. Single phase voltage is either 110, 115, 120, or 220, 230, 240 V. Voltages of 110, 115, and 120 consist of two conductors: one hot and one neutral. Only one hot conductor can be pulled with the neutral conductor using this system. An additional hot conductor requires an additional neutral conductor for each circuit. Voltages of 220, 230, and 240 consist of two hot conductors or two hot conductors with one neutral. When two hot conductors are routed with one neutral, the neutral will carry the unbalanced neutral current of the two hot conductors. The full-load current rating of 1ϕ transformers is found by dividing kVA by voltage.

What is the full-load current rating of a 25 kVA, 240 V, 1ϕ transformer?

Step 1. *Text:* $I = \dfrac{kVA \times 1000}{V}$

$$I = \frac{25\text{ kVA} \times 1000}{240\text{ V}} = 104.2\text{ A}$$

Answer. **FLC = 104 A**

Note: By dividing by 240 V, amperage is evenly divided on phases A and B for sizing and selecting conductors and overcurrent protection devices.

What is the full-load current rating of a 5 kVA, 120 V, 1ϕ, 2-wire transformer?

Step 1. *Text:* $I = \dfrac{kVA \times 1000}{V}$

$$I = \frac{5\text{ kVA} \times 1000}{120\text{ V}} = 41.7\text{ A}$$

Answer. **FLC = 42 A**

Three Phase. Three phase voltages are usually 208, 240, 480, or 600 V. The 240 V range consists of voltages of 220, 230, or 240 V. The 480 V range consists of voltages of 440, 460, or 480 V. The 600 V range consists of voltages of 550, 575, or 600 V.

■ *Master Electrician's Test Tip.* Substitute the following values for voltage times 1.732 when solving 3ϕ problems:

for 208 V × 1.732, use 360
for 230 V × 1.732, use 398
for 240 V × 1.732, use 416
for 440 V × 1.732, use 762
for 460 V × 1.732, use 797
for 480 V × 1.732, use 831

Note: Some examining boards require that 1.732 be used when solving 3ϕ problems.■

What is the full-load current rating of a 250 kVA, 480 V, 3ϕ, transformer?

Step 1. *Text:* $I = \dfrac{kVA \times 1000}{V \times \sqrt{3}}$

$$I = \frac{250\text{ kVA} \times 1000}{480\text{ V} \times 1.732} = 300.8\text{ A}$$

Answer. **FLC = 301 A**

kVA

Transformers are rated in VA or kVA. They are rated in VA when less than 1000 VA and (generally) in kVA when above 1000 VA. For example, a 900 VA transformer is rated at 900 VA. A 20,000 VA transformer is rated at 20 kVA.

Single Phase. The kVA rating of 1ϕ transformers is found by dividing VA by 1000. The VA rating is found by multiplying kVA by 1000. Volts times amps equal volt amps ($V \times I = VA$). For example, the VA of a circuit with a full-load current rating of 130 A supplied by a 240 V, 1ϕ system is 31,200 VA (240 V × 130 A = 31,200 VA). The kVA of this circuit is 31.2 kVA (31,200 VA/1000 = 31.2 VA).

What is the kVA of a 120/208 V feeder circuit supplying a connected load of 220 A?

Step 1. *Text:* $VA = V \times I$
$VA = 220\text{ A} \times 208\text{ V} = 45{,}760\text{ VA}$

Step 2. *Text:* $kVA = \dfrac{VA}{1000}$

$$kVA = \frac{45{,}760\text{ VA}}{1000} = 45.76\text{ kVA}$$

Answer. **kVA = 45.76 kVA**

Three Phase. The kVA rating of 3ϕ transformers is found by dividing VA by 1000. The VA rating is found by multiplying the load amperage by voltage times the square root of three. For example, the amperage rating of a service load is 365 A and the supply voltage is 480 V, 3ϕ. The VA rating for this service load is 303,315 VA (365 A × 480 V × 1.732 = 303,315 VA). The kVA of the service load is 303.315 (303,315 VA/1000 = 303.315 kVA). The size of the transformer is selected from this kVA rating of the loads.

Ratio. The voltage ratio of transformers is the ratio of terminal voltage to secondary terminal voltage due to specified conditions of use. The current ratio in transformers is the ratio of primary current to secondary current, which is based on the voltage ratio.

What is the voltage ratio of a transformer with a primary voltage of 480 V and a secondary voltage of 240 V?

Step 1. *Text:* $Ratio = \dfrac{primary\ V}{secondary\ V}$

$$Ratio = \frac{480\text{ V}}{240\text{ V}} = 2{:}1$$

Answer. **Voltage ratio = 2:1**

The full-load current rating of transformers is found by dividing kVA by voltage. The ratio times the smaller current rating, based on the higher voltage, equals higher current based on the lower voltage. For example, a 15 kVA transformer with a primary voltage of 480 V, 1ϕ has a full-load current rating of 31 A (15 kVA × 1000/480 V = 31.25 A). The secondary voltage is 240 V with a full-load current rating of 63 A (15 kVA × 1000/240 V = 62.5 A). By using the ratio to determine the secondary current, the primary voltage is divided by the secondary voltage, which is a 2 to 1 ratio (480 V/240 V = 2). The primary current of 31.25 A is multiplied by 2 to arrive at 62.5 A. This verifies the ratio of current values in relation to the ratio of voltage values. See Figure 4-5.

What is the voltage ratio of a transformer with a primary of 480 V and a secondary of 240 V?

Step 1. *Text:* $Ratio = \dfrac{primary\ V}{secondary\ V}$

$$Ratio = \frac{480\text{ V}}{240\text{ V}} = 2{:}1$$

Answer. **Voltage ratio = 2:1**

FINDING TRANSFORMER RATIO AND kVA

What is the ratio of the transformer?

Step 1. $Ratio = \dfrac{primary}{secondary}$

$$Ratio = \frac{480\text{ V}}{208\text{ V}} = 2.3$$

Answer. **Ratio = 2.3:1**

What is the kVA rating of the transformer?

Step 1. $kVA = \dfrac{VA}{1000}$

$$kVA = \frac{70{,}000\text{ VA}}{1000} = 70\text{ kVA}$$

Answer. **kVA = 70 kVA**

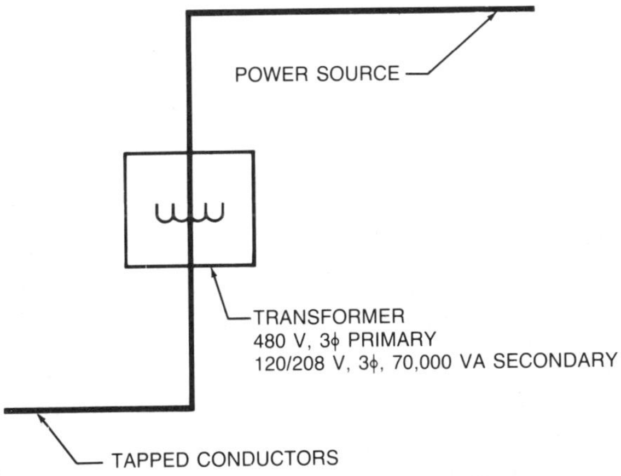

Figure 4-5. The kVA of transformers = VA/1000. The ratio = primary voltage/secondary voltage.

What is the secondary full-load current rating of a 20 kVA transformer with a 208 V, 1ϕ secondary, and a 480 V primary? Apply the ratio method for determining full-load current.

Step 1. $Primary\ FLC = \dfrac{kVA \times 1000}{V}$

$$Primary\ FLC = \frac{20\text{ kVA} \times 1000}{480\text{ V}} = 41.7\text{ A}$$

Step 2. $Secondary\ ratio = \dfrac{primary\ V}{secondary\ V}$

$$Secondary\ ratio = \frac{480\text{ V}}{208\text{ V}} = 2.3{:}1$$

Step 3. $Secondary\ FLC = primary\ FLC \times secondary\ ratio$

$$Secondary\ FLC = 41.7\text{ A} \times 2.3 = 95.9\text{ A}$$

Answer. **Secondary FLC = 96 A**

Note: By dividing kVA by secondary voltage, the secondary full-load current rating is 96 A (20 kVA × 1,000/208 V = 96 A).

CLASSIFYING PANELBOARDS—PARTS A AND B

Panelboards are either *power panelboards* or *lighting and appliance branch-circuit panelboards.* Power panels supply subpanels or equipment with neutral-connected loads rated above 30 A, or with 10% or more of their overcurrent protection devices rated at 30 A or less. Lighting and appliance panels are limited to two mains. The types of loads, overcurrent protection, and supply voltage determine the classification of a panelboard.

Power Panelboard—384-14; 384-15

Power panels commonly have double-pole or 3-pole mains to supply subpanels or equipment. No more than 10% of their overcurrent protection devices rated at 30 A or less shall supply neutral-connected loads. Power panels with 10′ taps from the secondary side of transformers have the advantage of tapped conductors being permitted to terminate to the lugs of the panel instead of the terminals of a main. For example, a panelboard with 42 single-pole slots has three branch circuits with 30 A or less overcurrent protection devices supplying neutral-connected loads. This is a power panelboard since there can be four (42 × 10% = 4.2) 30 A or less overcurrent protection devices with neutral-connected loads. See Figure 4-6.

What is the classification of a panelboard with 42 circuits and four combination overcurrent protection devices rated at 30 A or less for neutral-connected loads?

Step 1. *384-14:*

$Number\ of\ slots = single\text{-}pole\ slots \times 10\%$

$Number\ of\ slots = 42 \times 10\% = 4.2$

Step 2. *384-15:* Four 30 A or less single-pole slots with neutral connections

Answer. **Classification = power panelboard**

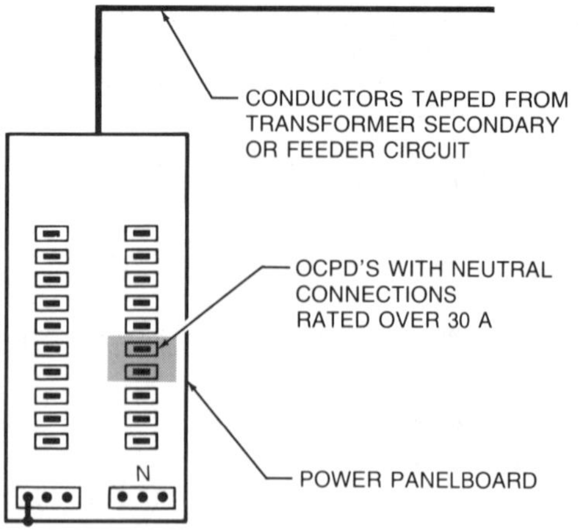

CLASSIFYING POWER PANELBOARDS

How many OCPDs rated at 30 A or less with neutral connections can be installed in the power panelboard?

Step 1. *384-14; 384-15:* 20 OCPDs × 10% = 2

Answer. **OCPDs = 2**

Figure 4-6. In a power panelboard, the total number of 30 A or less overcurrent protection devices with neutral connections shall not exceed 10% of the slots.

Lighting and Appliance Panelboard—384-14; 384-15

If over 10% of the single-pole slots of a panelboard rated at 30 A or less is used for branch circuits to lighting and appliance loads with neutral connections, the panelboard is classified as a lighting and appliance panelboard per 384-14 and 384-15. See Figure 4-7. For example, a 42 circuit panelboard supplies loads including eight 20 A single-pole circuits with neutral connections. Panelboard classification is determined by multiplying the 42 single-pole slots by 10% (42 × 10% = 4.2). There can be only four single-pole circuits rated at 30 A or less for use with neutral connections without classifying the panel as a lighting and appliance panelboard.

A 36 circuit panelboard has three 15 A, three 20 A, and three 30 A single-pole overcurrent protection devices supplying neutral-connected loads. How many single-pole slots rated at 30 A or less are permitted and what is the classification of the panelboard?

Slots

Step 1. *384-14:*

Number of slots = *single-pole slots* × 10%
Number of slots = 36 × 10% = 3.6

Answer. **Single-pole slots = 3**

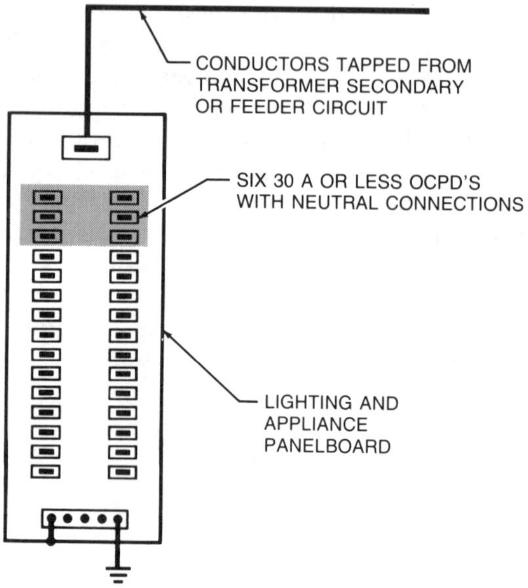

CLASSIFYING LIGHTING AND APPLIANCE PANELBOARDS

What is the classification of the panelboard?

Step 1. *384-14; 384-15:* 30 OCPDs × 10% = 3

Answer. **Panelboard = lighting and appliance**

Note: Six OCPDs are rated at 30 A or less with neutral connections, but only three are allowed to be a power panel.

Figure 4-7. In a lighting and appliance panelboard, the total number of 30 A or less overcurrent protection devices with neutral connections shall exceed 10% of the slots.

Classification

Step 2. *384-15:*

Three 15 A single-pole slots with neutral connections
Three 20 A single-pole slots with neutral connections
Three 30 A single-pole slots with neutral connections
Nine single-pole slots with neutral connections

Answer. **Single-pole slots = 9**
Classification = Lighting and appliance panelboard

■ *Master Electrician's Test Tip.* When determining the number of single-pole slots with 30 A or less overcurrent protection devices with neutral-connected loads, a single-pole overcurrent protection device counts as one, a double-pole counts as two, and a 3-pole counts as three.■

TAPS—240-21

Taps from the secondary side of transformers can supply power to panelboards, motor control centers, or other types of equipment. Taps supplying power to panelboards can be 10′ or 25′ in length from the secondary side of transformers. The 25′ tap rule is used where the location of the panelboards requires more than 10′ tap lengths from the secondary side of the transformer to the panelboard.

10′ Tap—240-21, Ex. 2

Taps from the secondary side of transformers can be made in 10′ or 25′ lengths. Where 10′ taps are used, the size of the tapped conductor is not limited. For example, #6 THWN Cu conductors can be tapped from a 600 kVA transformer, provided they do not exceed 10′ in length. The tapped conductors shall be installed in conduit and be equal to the size bus or the terminals of the overcurrent protection device to which they terminate.

Where there is no limitation to the number of mains permitted, 10′ taps can be taken from the secondary side of transformers to power panels. For example, a 300 A 10′ tap can be taken from the secondary side of a transformer and terminated to the lugs of a 300 A 1ϕ or 3ϕ power panel. There is no limit to the number of overcurrent protection devices permitted in the power panel. However, the loads on each overcurrent protection device shall not exceed the 300 A bus rating of the power panel. Note that a 10′ tap does not have to terminate to a main but can terminate to the lugs of a power panel. See 240-21, Ex. 2e. See Figure 4-8.

What size THWN Cu conductors, using the 10′ tap rule, is required from the secondary side of a transformer to supply a 225 A power panel where the tapped conductors terminate to the lugs of the bus?

Step 1. *240-21, Ex. 2b(2b):* Tap shall equal 225 A

Step 2. *Table 310-16:* #4/0 = 230 A

Answer. **Conductors = #4/0 THWN Cu**

What size THWN Cu conductors, using the 10′ tap rule, is required to supply a 400 A, 3ϕ, 4-wire power panel with a 350 A main?

Step 1. *240-21, Ex. 2b(2b):* Tap shall equal 350 A

Step 2. *Table 310-16:* #500 kcmil = 380 A

Answer. **Conductors = #500 kcmil THWN Cu**

What size THWN Cu conductors paralleled two times per phase is required to supply a 10′ tap from the secondary side of a transformer to the lugs of a 400 A power panel?

Step 1. *240-21, Ex. 2b(2a):* 400 A lugs

Step 2. *Table 310-16; 310-4:*
#3/0 THWN = 200 A × 2 = 400 A

Answer. **Conductors = two #3/0 THWN Cu**

25′ Tap—240-21, Ex. 11

When the 25′ tap is applied, the tapped conductors shall terminate in a main circuit breaker or a set of fuses in a switch. The tapped conductors can also be connected to the lugs of a panelboard. The rating of the overcurrent protection devices shall not exceed the

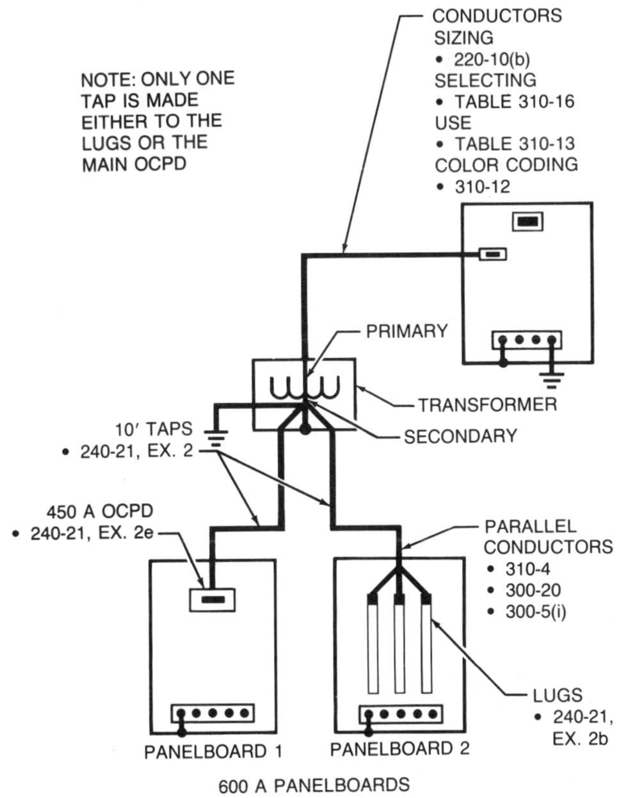

APPLYING 10′ TAP RULE

What size THWN Cu conductors paralleled three times per phase is required to supply power to panelboards 1 and 2?

Panelboard 1

Step 1. *240-21, Ex. 2:* Tap shall be equivalent to 450 A

Step 2. *310-4:* $\textit{Conductor ampacity} = \dfrac{\textit{total A}}{\textit{no. paralleled}}$

$$\textit{Conductor ampacity} = \frac{450\ \text{A}}{3} = 150\ \text{A}$$

Step 3. *Table 310-16:* #1/0 THWN Cu = 150 A

Answer. **Conductors = #1/0 THWN Cu**

Panelboard 2

Step 1. *240-21, Ex. 2:* Tap shall be equivalent to 600 A

Step 2. *310-4:* $\textit{Conductor ampacity} = \dfrac{\textit{total A}}{\textit{no. paralleled}}$

$$\textit{Conductor ampacity} = \frac{600\ \text{A}}{3} = 200\ \text{A}$$

Step 3. *Table 310-16:* #3/0 THWN Cu = 200 A

Answer. **Conductors = #3/0 THWN Cu**

Figure 4-8. The 10′ tap rule can be used to select taps supplying panelboards with lugs or mains.

secondary output of the transformer. The 25′ or less tapped conductors shall not exceed the secondary full-load current rating of the transformer. See Figure 4-9. For example, the 25′ tap rule is applied from the secondary side of a transformer with a 600 A output. The ampacity of the conductors shall be sized and based on the full 600 A output. The overcurrent protection device shall not exceed the 600 A output. Two #250 kcmil THWN Cu conductors paralleled per phase are used for the 25′ tap. Six 100 A overcurrent protection devices are installed in the power panel (6 × 100 A = 600 A) since they do not exceed 600 A, which is the secondary output of the transformer.

What size THWN Cu conductors and how many 50 A overcurrent protection devices are required for a 25′ tap from the secondary of a transformer with a 416 A output supplying a power panel (noncontinuous load)?

Conductors

Step 1. *240-21, Ex. 11b; Table 310-16:*
#600 kcmil = 420 A

Answer. **Conductors = #600 kcmil THWN Cu**

OCPDs

Step 2. *240-21, Ex. 11b:*

$$No. = \frac{output\ in\ A}{overcurrent\ protection\ device}$$

$$No. = \frac{416\ A}{50\ A} = 8.3$$

Answer. **OCPDs = 8**

25′ Taps Plus Transformer—240-21, Ex. 8

A 25′ tap can be made from a feeder circuit that includes the primary tap, transformer, and secondary tap conductors. The primary tap is sized by taking ⅓

APPLYING 25′ TAP RULE

What size tapped conductors and OCPDs is required for panelboards 1 and 2? Use THWN Cu.

Panelboard 1

Conductors

Step 1. *240-21, Ex. 11b:*
Tap shall be equivalent to 600 A

Step 2. *310-4: Conductor ampacity* $= \frac{total\ A}{no.\ paralleled}$

$$Conductor\ ampacity = \frac{600\ A}{2} = 300\ A$$

Step 3. *Table 310-16:* #350 kcmil THWN Cu = 310 A

Answer. ***Conductors* = #350 kcmil THWN Cu**

OCPDs

Step 1. *240-21, Ex. 11b, c:* Output = 600 A

Answer. ***240-6:* OCPDs = 600 A**

Panelboard 2

Conductors

Step 1. *240-21, Ex. 11b:*
#350 kcmil THWN Cu paralleled twice per phase

Answer. ***Table 310-16:***
Conductors = #350 kcmil THWN Cu

OCPDs

Step 1. *240-21, Ex. 11b, c:*

$$No.\ of\ OCPDs = \frac{secondary\ output}{size\ of\ OCPDs}$$

$$No.\ of\ OCPDs = \frac{600\ A}{100\ A} = 6$$

Answer. ***240-6:* OCPDs = six 100 A**

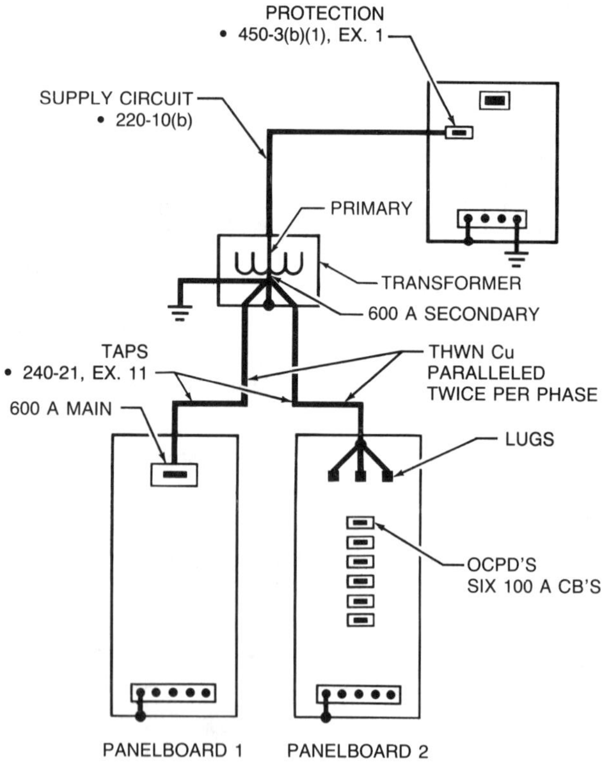

NOTE: ONLY ONE TAP IS MADE EITHER TO THE LUGS OR THE MAIN OCPD.

Figure 4-9. The 25′ tap rule can be used to supply power to lighting and appliance or power panels. Secondary windings are protected at 125% of the transformer's secondary FLC.

of the ampacity of the feeder circuit conductors or the overcurrent protection device of the feeder, whichever is the lowest in rating. For example, the conductors of a feeder circuit are increased as a result of voltage drop.

The overcurrent protection device is sized on the load and has a smaller rating than the ampacity of the conductors. The primary tap from the feeder circuit to the transformer is sized on the ampacity of the overcurrent protection device. The tap from the secondary side of the transformer to the panelboard is sized by dividing the primary voltage by the secondary voltage and multiplying by ⅓. Multiply the ampacity of the feeder circuit conductors or the overcurrent protection device (whichever is smaller) by this total. For example, the conductors of a feeder circuit are increased from #3/0 to #4/0 THWN Cu to allow for voltage drop. The overcurrent protection device is selected at 200 A, which is based on the ampacity of the #3/0 THWN Cu conductors that were selected on the load. The larger #4/0 THWN Cu conductors were selected to reduce the resistance in the wire due to the length between the service panel and equipment. The primary tap is determined by taking ⅓ of the 200 A device, which is 67 A (200 A × ⅓ = 66.7 A). The secondary tap is determined by the following procedure.

Primary = 480 V

Secondary = 240 V

$$\frac{480\ \text{V}}{240\ \text{V}} \times \tfrac{1}{3} \times 200\ \text{A} = 133\ \text{A}$$

The conductors of the primary tap shall be equivalent to 67 A and the secondary tap shall be equivalent to 133 A. These values are based on the ampacity of the overcurrent protection device. See Figure 4-10.

SIZING CONDUCTORS—220-3(a); 220-10(b)

Conductors supplying the primary side of transformers and taps from the secondary side of transformers are computed at noncontinuous or continuous duty. Continuous-duty loads are calculated at 125% of the total VA rating or amperage rating. Noncontinuous-duty loads are computed at 100%.

Primary Side—220-10(b)

Conductors supplying power between service equipment and the primary side of a transformer are computed on the kVA rating of the transformer or the load served. For example, the primary of a 480 V transformer has a full-load current rating of 108 A. Conductors supplying power to the primary are sized by multiplying 108 A by 125%, which equals 135 A (108 A × 125% = 135 A). The size conductor selected from Table 310-16 is #1/0 THWN Cu, which has an ampacity of 150 A.

If the primary full-load current rating of a 112 kVA, 240 V, 3ϕ transformer supplies a load of 200 A, the conductors are sized from the 269 A primary full-load current rating (112 kVA × 1000/240 V × 1.732 = 269.2 A) or the 200 A load.

Calculations of the two amperage ratings show that smaller conductors can be used when sized by the load. This is verified by multiplying 269 A by 125%, which is 336 A (269 A × 125% = 336.25 A). Based on this amperage, #500 kcmil THWN Cu conductors are required per Table 310-16. The 200 A load is multiplied by 125% to obtain 250 A (200 A × 125% = 250 A), which requires #250 kcmil THWN Cu conductors.

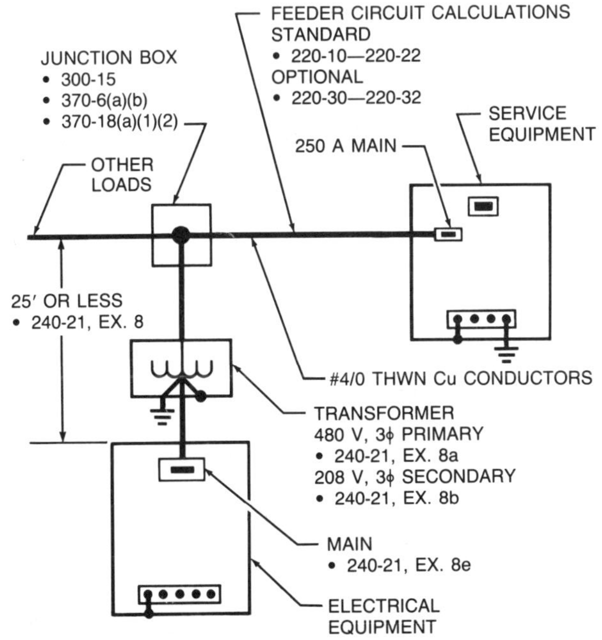

25′ TAP PLUS TRANSFORMER

What size THWN Cu conductors is required to supply power for the primary and secondary taps?

Primary

Step 1. *240-21, Ex. 8a; Table 310-16:*
230 A × ⅓ = 76.7 A
250 A × ⅓ = 83 A

Step 2. *Table 310-16:* #4 THWN Cu = 85 A

Answer. **Conductors = #4 THWN Cu**

Secondary

Step 1. *240-21, Ex. 8b:*

$$\frac{480\ \text{V}}{208\ \text{V}} \times \tfrac{1}{3} \times 230\ \text{A} = 176.9\ \text{A}$$

Step 2. *Table 310-16:* #3/0 THWN Cu = 200 A

Answer. **Conductors = #3/0 THWN Cu**

Figure 4-10. Primary and secondary tapped conductors are sized from the rating of the overcurrent protection device or conductors.

Where the maximum kVA of the transformer is not needed, the smaller size conductors based on the load can be used. See Figure 4-11.

What size THWN Cu conductors is required to supply the primary side of a 150 kVA, 480 V, 3ϕ transformer supplying continuous loads?

Step 1. *Text:* $I = \dfrac{kVA \times 1000}{V \times \sqrt{3}}$

$$I = \frac{150 \text{ kVA} \times 1000}{480 \text{ V} \times 1.732} = 180.5 \text{ A}$$

Step 2. *220-10(b):* 180.5 A × 125% = 225.6 A

Step 3. *Table 310-16:* #4/0 THWN = 230 A

Answer. **Conductors = #4/0 THWN Cu**

SIZING PRIMARY CONDUCTORS

What size THWN Cu conductors is required to supply power to the primary side of the transformer? (The load is continuous duty.)

Step 1. *Text:* $I = \dfrac{kVA \times 1000}{V \times \sqrt{3}}$

$$I = \frac{47.52 \text{ kVA} \times 1000}{480 \text{ V} \times 1.732} = 57 \text{ A}$$

Step 2. *220-10(b):* 57 A × 125% = 71 A

Step 3. *Table 310-16:* #4 THWN Cu = 85 A

Answer. **Conductors = #4 THWN Cu**

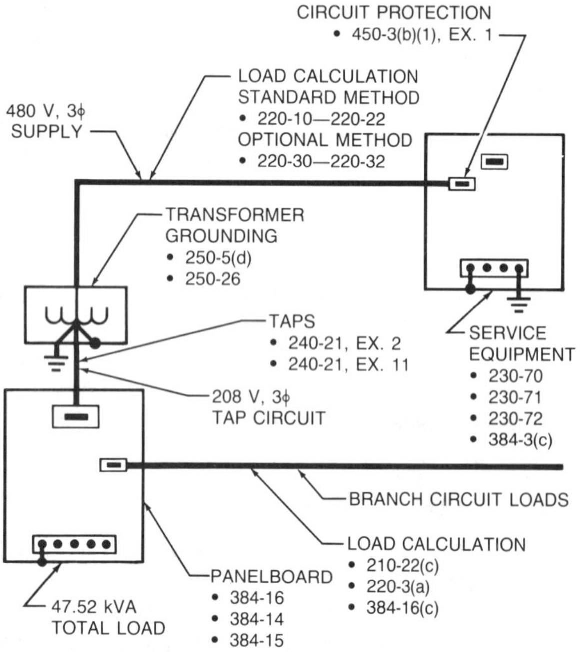

Figure 4-11. The size of conductors supplying power between the service equipment and primary side of transformers is determined by multiplying the load by 100% or 125%, based on noncontinuous or continuous duty.

Secondary Side—220-10(b)

Tapped conductors from the secondary side of transformers are usually sized on the load they supply. However, they can be sized on the secondary output of the transformer. Where the 10′ tap rule is applied, the tapped conductors can be sized on the load served or the output of the transformers. The tapped conductors for a 25′ tap shall have an ampacity equivalent to the secondary output of the transformer. The tapped conductors are sized on the period of time that they are used. Conductors supplying loads that operate for a period of less than three hours are selected at 100% of the total VA or amperage rating. For example, the conductors tapped for the secondary of a transformer supply power to a load of 62 A used at noncontinuous duty. The ampacity of the tapped conductors shall be at least 62 A (62 A × 100% = 62 A).

Tapped conductors from the secondary side of transformers that supply loads used for a period of three hours or more are computed at 125% of the total VA or amperage rating. See Figure 4-12. For example, conductors tapped from the secondary side of a transformer supplying a connected load of 102 A used for a period of eight hours each day shall have an ampacity of 128 A (102 A × 125% = 127.5 A). If the secondary side of a transformer has an output of 83 A, the ampacity of the tapped conductors is 104 A (83 A × 125% = 103.75 A) for continuous loads.

What size tapped conductors, using THWN Cu, is required to supply a continuous-duty load between a power panelboard and a 60 kVA, 208 V, 3ϕ transformer?

Step 1. *Text:* $I = \dfrac{kVA \times 1000}{V \times \sqrt{3}}$

$$I = \frac{60 \text{ kVA} \times 1000}{208 \text{ V} \times 1.732} = 166.7 \text{ A}$$

Step 2. *220-10(b):* 166.7 A × 125% = 208 A

Step 3. *Table 310-16:* #4/0 THWN = 230 A

Answer. **Conductors = #4/0 THWN Cu**

What size tapped conductors, using THWN Al, is required to serve a continuous-duty load between a lighting and appliance panelboard and a 100 kVA, 240 V, 3ϕ transformer?

Step 1. *Text:* $I = \dfrac{kVA \times 1000}{V \times \sqrt{3}}$

$$I = \frac{100 \text{ kVA} \times 1000}{240 \text{ V} \times 1.732} = 240.4 \text{ A}$$

Step 2. *220-10(b):* 240 A × 125% = 300 A

Step 3. *Table 310-16:* #350 kcmil THWN = 310 A

Answer. **Conductors = #350 kcmil Al**

■ *Master Electrician's Test Tip.* Taps taken from the secondary side of transformers are sized per 240-21, Ex. 2 or 11 and selected per Tables 310-16–310-19.■

PROTECTION—450-3(b)(1)	
FLC	PERCENT
9 A or more	125%
Less than 9 A but 2 A or more	167%
Less than 2 A	300%

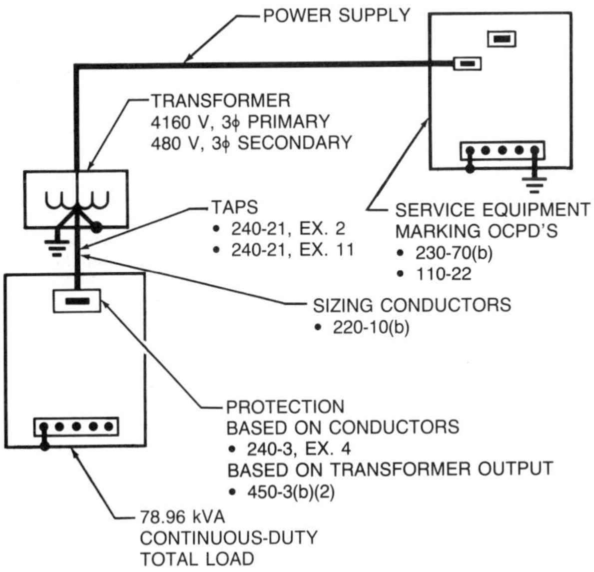

SIZING SECONDARY CONDUCTORS

What size THWN Cu conductors is required to supply power to the secondary side of the transformer?

Step 1. *Text:* $I = \dfrac{kVA \times 1000}{V \times \sqrt{3}}$

$$I = \frac{78.96 \text{ kVA} \times 1000}{480 \text{ V} \times 1.732} = 95 \text{ A}$$

Step 2. *220-10(b):* 95 A × 125% = 118.8 A

Step 3. *Table 310-16:* #1 THWN Cu = 130 A

Answer. **Conductors = #1 THWN Cu**

Figure 4-12. The size of conductors supplying power between the service equipment and secondary side of transformers is determined by multiplying the load by 100% or 125%, based on noncontinuous or continuous duty.

Branch Circuits—220-3(a). Branch circuits from a power panel or a lighting and appliance panelboard are computed at 100% of the load for noncontinuous duty and 125% of the load for continuous duty. Each branch circuit is calculated based on the period of time it is used. A branch circuit operated for less than three hours is classified as noncontinuous duty. A branch circuit operated for three hours or more is classified as continuous duty. For example, all loads on a branch circuit are 18 A, noncontinuous operation. The value of the load is selected at 18 A (18 A × 100% = 18 A). If the 18 A loads on the branch circuit operate for a period of three hours or more, the load is selected at 23 A (18 A × 125% = 22.5 A). See Figure 4-13.

What size tap, using THWN Cu conductors, is required from the secondary side of a transformer to a panelboard supplying the following loads?

1. 10 A of continuous-duty loads
2. 8 A of noncontinuous-duty loads
3. 16 A load that operates for five hours
4. 12 A load that operates for two hours
5. 14 A load that operates for four hours
6. 10 A motor load
7. 12 A motor load
8. 40 A continuous-duty lighting load
9. 25 A noncontinuous-duty receptacle load

Noncontinuous duty

Step 1. *220-3(a); 220-10(b):*

8 A × 100% =	8 A
12 A × 100% =	12 A
25 A × 100% =	25 A
Total load =	45 A

Continuous duty

Step 2. *220-10(b):*

10 A × 125% =	12.5 A
16 A × 125% =	20 A
14 A × 125% =	17.5 A
40 A × 125% =	50 A
Total load =	100 A

Motor load

Step 3. *430-24:*

10 A × 100% =	10 A
12 A × 100% =	12 A
Total load =	22 A

Largest motor load

Step 4. *220-14:* 12 A × 25% = 3 A

Total load = 3 A

Total loads

Step 5. *430-25; 220-10(b):*

Noncontinuous-duty load =	45 A
Continuous-duty load =	100 A
Motor load =	22 A
Largest motor load =	3 A
Total loads =	170 A

Conductors

Step 6. *Table 310-16:* #2/0 = 175 A

Answer. **Tap = #2/0 THWN Cu**

Feeder circuit loads supplying power for periods of three hours or more without a period of rest are calculated at 125% of each branch circuit load. The ampacity is selected on the loads of feeders from the branch circuits. See Figure 4-14.

What size THWN Cu conductors is required to tap the secondary side of a transformer to a panelboard that supplies the following loads?

SIZING TAP FROM TRANSFORMER TO PANELBOARD

What size THWN Cu conductors is required for the tap between the transformer and panelboard?

Step 1. Lighting loads. *220-10(b):*

10 A × 125% =	12.5 A
14 A × 125% =	17.5 A
22 A × 125% =	27.5 A
Total =	57.5 A

Step 2. Receptacle loads. *220-3(a); 220-10(b):*

9 A × 100% =	9 A
8 A × 100% =	8 A
10 A × 100% =	10 A
Total =	27 A

Step 3. Special loads. *220-3(a); 220-10(b):*

25 A × 125% =	31.25 A
30 A × 125% =	37.5 A
28 A × 125% =	35 A
Total =	103.8 A

Step 4. Motor loads. *430-25:*

16.7 A × 100% =	16.7 A
24 A × 100% =	24 A
Total =	40.7 A

Step 5. Largest motor load. *220-14:* 24 A × 25% = 6 A

Step 6. Total loads.

Lighting loads =	57.5 A
Receptacle loads =	27 A
Special loads =	103.8 A
Motor loads =	40.7 A
Largest motor load =	6 A
Total =	235 A

Step 7. *Table 310-16:* #250 kcmil THWN Cu = 255 A

Answer. **Conductors = #250 kcmil THWN Cu**

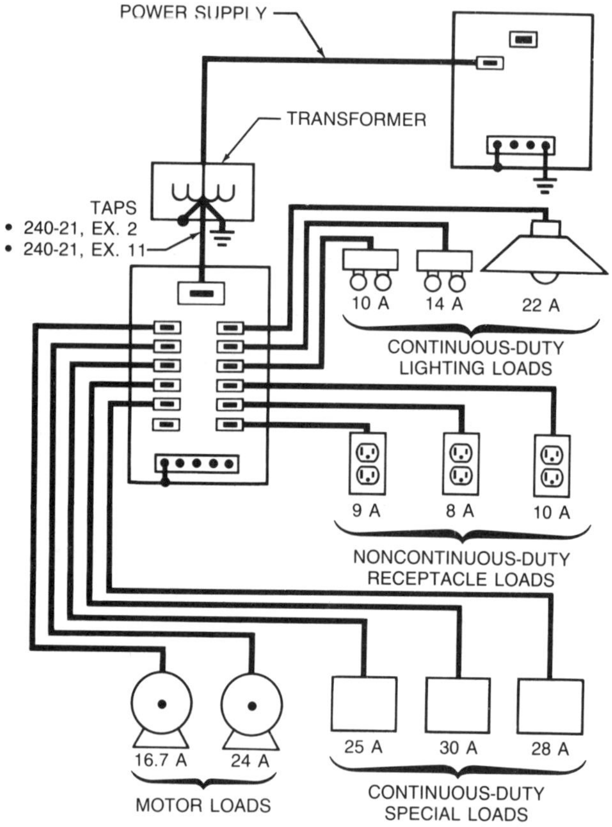

Figure 4-13. Tapped conductors between the secondary of the transformer and the electrical equipment supplied are sized at 100% of the load for noncontinuous duty and 125% for continuous duty per 220-10(b).

1. 70 A continuous-duty feeder circuit
2. 50 A noncontinuous-duty feeder circuit
3. 35 A noncontinuous-duty feeder circuit
4. 60 A continuous-duty feeder circuit

Noncontinuous duty

Step 1. *220-10:*

50 A × 100% = 50 A
35 A × 100% = 35 A

Continuous duty

Step 2. *220-10(b):*

70 A × 125% = 87.5 A
60 A × 125% = 75 A

Step 3. *Total loads:*

	50 A
	35 A
	87.5 A
	75 A
Total loads =	247.5 A

Step 4. *Table 310-16:* #250 kcmil THWN = 255 A

Answer. **Conductors = #250 kcmil THWN Cu**

OVERCURRENT PROTECTION—450-3

Overcurrent protection devices shall be provided in the primary and secondary sides of transformers. Overcurrent protection is installed in the primary side to protect against short circuits and ground faults. It is installed in the secondary side to protect the windings of the transformer from overloads. Overcurrent protection devices installed in the primary of a transformer can protect the secondary side for a 2-wire to 2-wire system under certain conditions. Overcurrent protection devices are sized and selected at ratings and settings that will protect conductors and windings of transformers from short circuits, ground faults, and overloads.

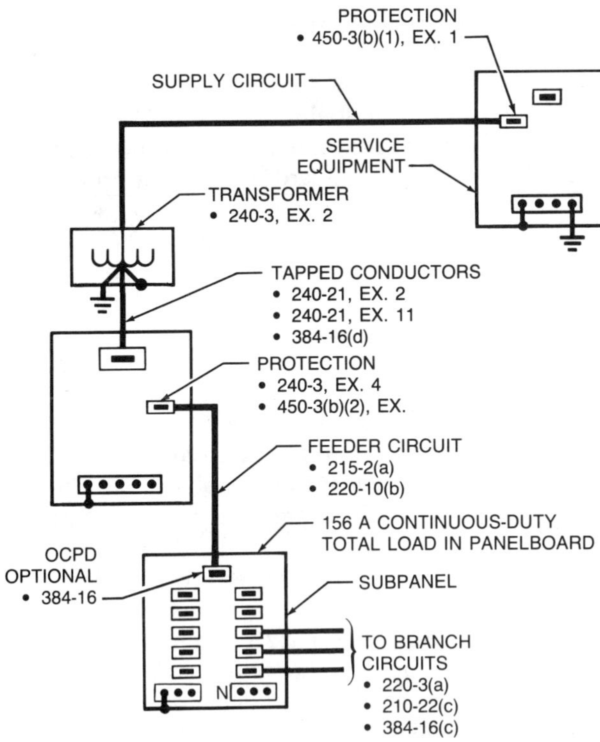

SIZING SUBPANEL CONDUCTORS

What size THWN Cu conductors is required to supply power to a subpanel fed from a panelboard served from a separately derived system?

Step 1. *220-10(b):* 156 A × 125% = 195 A

Step 2. *Table 310-16:* #3/0 THWN Cu = 200 A

Answer. **Conductors = #3/0 THWN Cu**

Figure 4-14. Feeder circuit conductors to a subpanel supplied from a panelboard served from a separately derived system are calculated at 100% or 125%, based on noncontinuous or continuous duty.

600 V or Less—450-3(b)

Transformers rated at 600 V or less in the primary and secondary side shall have overcurrent protection devices with ratings and settings selected on the percentages listed in 450-3(b)(1) or (2). Dangerous low magnitude ground faults can exist in systems of 600 V or less. Overcurrent protection devices in the primary and secondary side of transformers are sized to protect the primary and secondary windings from such currents that can create an overload.

Primary Side—450-3(b)(1). The percentage times the full-load current rating of the transformer in the primary side is based on three full-load current ratings. The first full-load current rating is for primary circuits of 9 A or more per 450-3(b)(1). The overcurrent protection devices for transformers with full-load current ratings of 9 A or more are increased by 125%. For example, a transformer with a primary full-load current rating of 32 A has a setting of 40 A (32 A × 125% = 40 A). This rating is equal to a 40 A standard size overcurrent protection device per 240-6.

The second full-load current rating is 2 A or more, but less than 9 A, per 450-3(b)(1), Ex. 1. The overcurrent protection devices for transformers with these full-load current ratings are increased by 167%. For example, a transformer with a full-load current rating of 7 A has a setting of 12 A (7 A × 167% = 11.69 A). This rating is equal to 10 A fuses per 240-6, Ex.

The third full-load current rating is less than 2 A per 450-3(b)(1), Ex. 1. The overcurrent protection devices for transformers with full-load current ratings of less than 2 A are increased by 300%. For example, a transformer with a full-load current rating of 1 A has a rating of 3 A (1 A × 300% = 3 A). This rating is equal to 3 A fuses per 240-6, Ex. See 430-72(c), Ex. 2, which allows 500% times the full-load current rating to be applied for a transformer supplying control circuits. See Figure 4-15.

■ *Master Electrician's Test Tip.* If 125% times the full-load current rating of a transformer does not correspond to a standard size overcurrent protection device, the next standard size above the rating can be used. The next lower standard overcurrent protection device is selected when the percentages of 167% and 300% do not correspond to a standard size per 240-6, Ex.■

What size overcurrent protection device is required in the primary side of a 50 kVA, 480 V, 3ϕ transformer?

Step 1. *Text:* $I = \dfrac{kVA \times 1000}{V \times \sqrt{3}}$

$$I = \frac{50 \text{ kVA} \times 1000}{480 \text{ V} \times 1.732} = 60.2 \text{ A}$$

Step 2. *450-3(b)(1), Ex. 1:* 60 A × 125% = 75 A

Step 3. *240-6:* 80 A is the next higher standard size

Answer. **OCPD = 80 A**

What size overcurrent protection device is required in the primary side of a 2 kVA, 240 V, 1ϕ transformer?

Step 1. *Text:* $I = \dfrac{kVA \times 1000}{V}$

$$I = \frac{2 \text{ kVA} \times 1000}{240 \text{ V}} = 8.3 \text{ A}$$

Step 2. *450-3(b)(1), Ex. 1:* 8.3 A × 167% = 13.86 A

Step 3. *240-6, Ex.:*
10 A is the next lower standard size device

Answer. **OCPD = 10 A**

What size overcurrent protection device is required in the primary side of a 1.2 kVA, 480 V, 3ϕ transformer?

Step 1. *Text:* $I = \dfrac{kVA \times 1000}{480 \text{ V} \times \sqrt{3}}$

$$I = \frac{1.2 \text{ kVA} \times 1000}{480 \text{ V} \times 1.732} = 1.4 \text{ A}$$

SELECTING OCPD OF PRIMARY TO PROTECT PRIMARY SIDE OF SEPARATELY DERIVED SYSTEMS	
PRIMARY FLC	PERCENT × FLC
9 A or more	125%
2 A and less than 9 A	167%
Less than 2 A	300%
For Control-Circuit Transformer, see 430-72(c), Ex. 2	

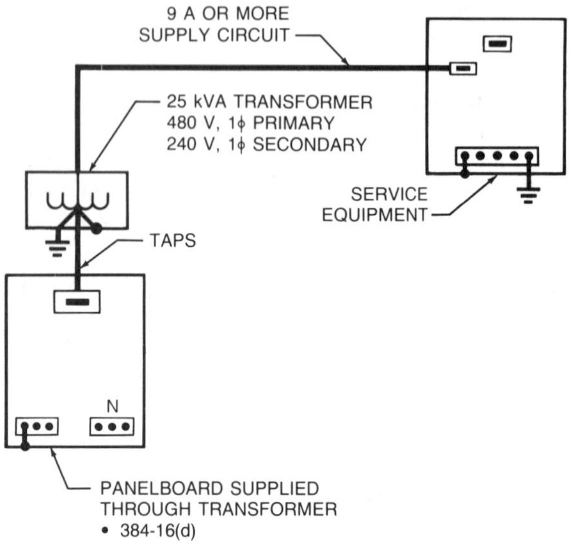

SELECTING OCPD'S FOR PRIMARY

What size overcurrent protection device is required for the primary side of the transformer?

Step 1. *Text:* $A = \dfrac{kVA \times 1000}{V}$

$$A = \frac{25 \text{ kVA} \times 1000}{480 \text{ V}} = 52 \text{ A}$$

Step 2. *450-3(b)(1), Ex. 1:* 52 A × 125% = 65 A

Step 3. *240-3, Ex. 2; 240-6:*
70 A is the next higher standard size

Answer. **OCPD = 70 A**

Figure 4-15. Overcurrent protection devices in the primary side of transformers are sized on the output current level in amperage times the selected percentage.

Step 2. *450-3(b)(1), Ex. 1:* 1.4 A × 300% = 4.2 A

Step 3. *240-6, Ex.:* 3 A is the next lower standard size

Answer. **OCPD = 3 A**

Note: If the secondary side of the transformer supplies a control circuit, see 430-72(c), Ex. 2, which permits 500% times the full-load current rating.

Secondary Side—450-3(b)(2). Overcurrent protection devices placed in the primary side of a 2-wire system can provide overcurrent protection in both the primary and secondary sides of the transformer per 240-3, Ex. 2. However, the percentages times the full-load current rating of the transformer shall not exceed 125% for transformers with full-load current ratings of 9 A or more. The exception to 450-3(b)(2) permits the next higher size standard overcurrent protection device where 125% does not correspond to a standard size overcurrent protection device per 240-6. Should the overcurrent protection devices in the primary circuit trip because of high inrush current, the full-load current rating is increased to 250%. See Figure 4-16.

■ *Master Electrician's Test Tip.* If 250% times the full-load current rating of the transformer does not correspond to a standard size overcurrent protection device, the next lower rating is selected. Some testing agencies require the conductors supplying the primary to be equivalent to 250% times the full-load current rating of the primary.■

Overcurrent protection devices are placed in the secondary side of a 2-wire to 2-wire system when a percentage exceeding 125% to 250% times the full-load current rating is applied. Overcurrent protection devices installed in the secondary shall not exceed 125% times the secondary full-load current rating of the transformer where protection in the primary and secondary side shall be provided. For example, a transformer has a full-load current rating of 83 A and is supplied by a 2-wire system. The secondary side has a 2-wire supply to 240 V, 1ϕ equipment. The size overcurrent protection device required in the primary side is 110 A (83 A × 125% = 103.8 A).

If the loads in the secondary side of the transformer have high inrush current that trips open the primary overcurrent protection devices, the overcurrent protection devices in the primary are increased to 200 A (83 A × 250% = 207.5 A). ***Warning:*** Overcurrent protection devices shall be placed in the secondary side at 125% times the full-load current rating of the transformer based on the secondary voltage divided into the kVA rating.

If the transformer in the previous example were 40 kVA, the secondary overcurrent protection devices would be 225 A (40 kVA × 1000/240 V × 125% = 208.3 A). *Note:* The next higher standard size overcurrent protection device is selected per 450-3(b)(2), Ex.

■ *Master Electrician's Test Tip.* Examining boards require 125% to be applied in the primary side or the secondary side of a 2-wire to 2-wire transformer.■

What size overcurrent protection device is required to protect the primary and secondary sides of a 2-wire to 2-wire, 60 kVA, 480 V and 240 V, 1ϕ transformer?

Step 1. *Text:* $I = \dfrac{kVA \times 1000}{V}$

$$I = \frac{60 \text{ kVA} \times 1000}{480 \text{ V}} = 125 \text{ A}$$

Step 2. *450-3(b)(2):* 125 A × 125% = 156.25 A

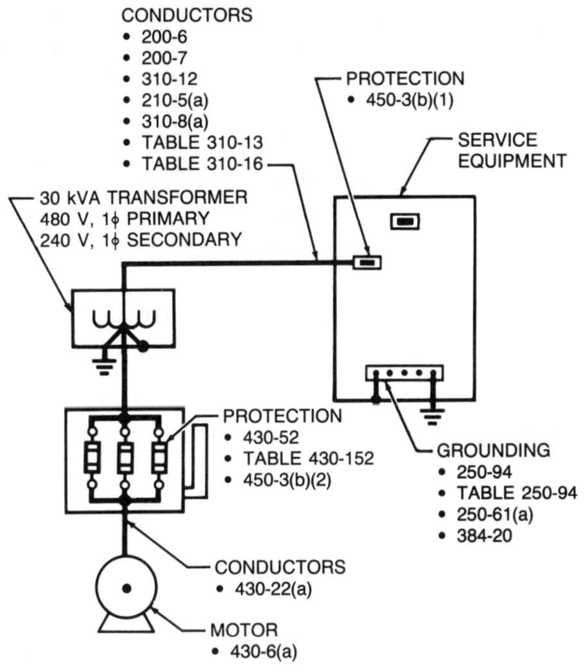

SELECTING MAXIMUM OCPD'S FOR PRIMARY

What size overcurrent protection devices is required in the primary and secondary with the primary using the maximum value?

<u>*Primary*</u>

Step 1. $I = \frac{kVA \times 1000}{V}$

$I = \frac{30 \text{ kVA} \times 1000}{480 \text{ V}} = 62.5 \text{ A}$

Step 2. *450-3(b)(2), Ex.:* 62.5 A × 250% = 156 A

Answer. ***240-6; 240-3, Ex. 2:*** **OCPD = 150 A**

<u>*Secondary*</u>

Step 1. $I = \frac{kVA \times 1000}{V}$

$I = \frac{30 \text{ kVA} \times 1000}{240 \text{ V}} = 125 \text{ A}$

Step 2. *450-3(b)(2), Ex.:* 125 A × 125% = 156 A

Answer. ***240-6; 240-3, Ex. 2:*** **OCPD = 175 A**

Figure 4-16. The overcurrent protection device in the primary side of 2-wire to 2-wire systems can be increased to 250% with the overcurrent protection device placed in the secondary side at 125% or less.

Step 3. *240-6:* 175 A is the next higher standard size

Answer. **OCPD = 175 A**

What is the maximum size overcurrent protection device permitted in the primary and the secondary of the transformer in the previous problem?

<u>*Primary FLC*</u>

Step 1. *Text:* $I = \frac{kVA \times 1000}{V}$

$I = \frac{60 \text{ kVA} \times 1000}{480 \text{ V}} = 125 \text{ A}$

Step 2. *450-3(b)(2):* 125 A × 250% = 312.5 A

Step 3. *240-3, Ex. 2; 240-6:*
300 A is the next lower standard size

Answer. **OCPD = 300 A**

<u>*Secondary*</u>

Step 1. *Text:* $I = \frac{kVA \times 1000}{240 \text{ V}}$

$I = \frac{60 \text{ kVA} \times 1000}{240 \text{ V}} = 250 \text{ A}$

Step 2. *450-3(b)(2), Ex.:* 250 A × 125% = 312.5 A

Step 3. *240-6:* 350 A is the next higher standard size

Answer. **OCPD = 350 A**

Over 600 V—450-3(a)

Two tables are used to size overcurrent protection devices for transformers with over 600 V on the primary side and over 600 V or 600 V or less on the secondary side. Percentages selected to size overcurrent protection devices are based on location of the transformer in an unsupervised or a supervised area.

Unsupervised Location—450-3(a)(1); Table 450-3(a)(1). An unsupervised location is a location where a maintenance electrician or an electrical engineer is not present or employed. Section 450-3(a)(1) requires overcurrent protection devices to be installed in the primary and secondary side per Table 450-3(a)(1). The percentages in Table 450-3(a)(1) are selected on the voltage and impedance of the transformer. For example, the percentage used to size the overcurrent protection device in the primary of a 12,500 V transformer with an impedance of 1.5 is 600% for a CB and 300% for fuses per Table 450-3(a)(1). Where the percentage times the full-load current does not correspond to a standard device per 240-6, the next size overcurrent protection device is selected per 450-3(a)(1), Ex. 1.

The greatest change is in the secondary side for voltage ratings of 600 V or less. The maximum percentage times the full-load current rating is 125% for CBs and fuses. This prevents the secondary windings of the transformer from being overloaded. Where no personnel is present, added loads are not usually properly calculated and the conductors to the transformers windings are overloaded.

The size of the overcurrent protection devices in the secondary side of a transformer with over 600 V primary is limited to 125% of the full-load current rating of the secondary whose voltage is 600 V or less. For example, 125% of the secondary full-load current rating is used to select the overcurrent protection devices for transformers with a primary of 13,500 V and secondary of 208 V. See Figure 4-17.

■ *Master Electrician's Test Tip.* Table 450-3(a)(1) requires the overcurrent protection devices in the secondary side to be selected at 125% of the full-load current rating per 450-3(a)(1). Where the percentages of Table 450-3(a)(1) do not correspond to a standard size device, the next higher size standard device is selected from 240-6. ■

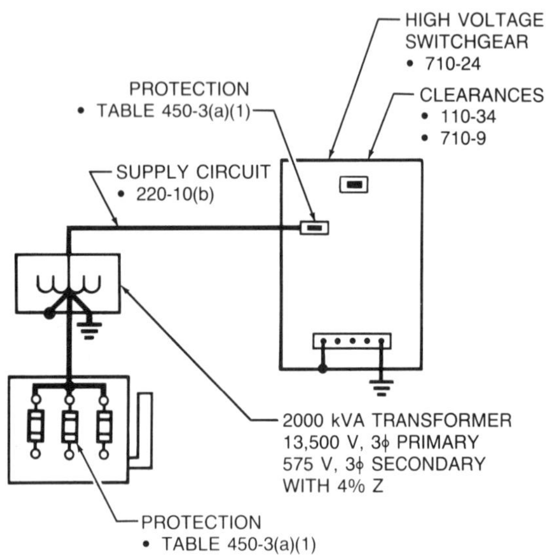

UNSUPERVISED LOCATION

SELECTING OCPD'S FOR PRIMARY AND SECONDARY (UNSUPERVISED LOCATION)

What size OCPDs is required in the primary and secondary sides of the transformer? Use CB and fuses.

Primary

Step 1. $I = \dfrac{kVA \times 1000}{V \times \sqrt{3}}$

$I = \dfrac{2000 \text{ kVA} \times 1000}{13{,}500 \text{ V} \times 1.732} = 85.5 \text{ A}$

Step 2. *Table 450-3(a)(1):* 85.5 A × 600% = 513 A

Answer. ***450-3(a)(1), Ex. 1; 240-6:* CB = 600 A**

Secondary

Step 1. $I = \dfrac{kVA \times 1000}{V \times \sqrt{3}}$

$I = \dfrac{2000 \text{ kVA} \times 1000}{575 \text{ V} \times 1.732} = 2008 \text{ A}$

Step 2. *Table 450-3(a)(1):* 2008 A × 125% = 2510 A

Answer. ***Table 450-3(a)(1); 240-6:* Fuses = 2500 A**

Figure 4-17. Combination overcurrent protection devices installed in the primary and secondary sides of high voltage transformers in unsupervised locations are selected by the values listed in Table 450-3(a)(1).

What size fuses is required for a 500 kVA, 3ϕ transformer with a 12,500 V primary, a 4160 V secondary, and an impedance of 2%? The transformer is in an unsupervised location.

Primary

Step 1. *Text:* $I = \dfrac{kVA \times 1000}{V \times \sqrt{3}}$

$I = \dfrac{500 \text{ kVA} \times 1000}{12{,}500 \text{ V} \times 1.732} = 23 \text{ A}$

Step 2. *Table 450-3(a)(1):* 23 A × 300% = 69 A

Step 3. *240-6:* 70 A is the next higher standard size

Answer. **Fuses = 70 A**

Secondary

Step 1. *Text:* $I = \dfrac{kVA \times 1000}{V \times \sqrt{3}}$

$I = \dfrac{500 \text{ kVA} \times 1000}{4160 \text{ V} \times 1.732} = 69.4 \text{ A}$

Step 2. *Table 450-3(a)(1):* 69 A × 250% = 172.5 A

Step 3. *240-6:* 175 A is the next higher standard size

Answer. **Fuses = 175 A**

What size CBs are required for a 650 kVA, 3ϕ transformer with a 4160 V primary, a 240 V secondary, and an impedance of 3%? The transformer is in an unsupervised location.

Primary

Step 1. *Text:* $I = \dfrac{kVA \times 1000}{V \times \sqrt{3}}$

$I = \dfrac{650 \text{ kVA} \times 1000}{4160 \text{ V} \times 1.732} = 90.2 \text{ A}$

Step 2. *Table 450-3(a)(1):* 90 A × 600% = 540 A

Step 3. *240-6:* 600 A is the next higher standard size

Answer. **CB = 600 A**

Secondary

Step 1. *Text:* $I = \dfrac{kVA \times 1000}{V \times \sqrt{3}}$

$I = \dfrac{650 \text{ kVA} \times 1000}{240 \text{ V} \times 1.732} = 1562.5 \text{ A}$

Step 2. *Table 450-3(a)(1); 240-6:*
1562.5 A × 125% = 1953.1 A
2000 A is the next higher standard size

Answer. **CB = 2000 A**

Supervised Locations—450-3(a)(2)b; Table 450-3(a)(2)b. A supervised location is a location with a plant electrician or plant engineer employed. The difference in Table 450-3(a)(1) and Table 450-3(a)(2)b is that voltage on the secondary side allows a higher percentage times the full-load current rating to size the overcurrent protection devices in the secondary side.

■ *Master Electrician's Test Tip.* When Table 450-3(a)(2)b is used to size overcurrent protection devices in the primary and secondary, and the percentages times the full-load current ratings do not correspond to a standard size device, the next lower rating per 240-6 is used. ■

Individual overcurrent protection devices can be placed in the primary side of transformers rated over 600 V. For fuses, the settings shall not exceed 250% of the primary's full-load current rating. For example, the overcurrent protection device in the primary of a transformer with a full-load current of 55 A requires fuses to be selected at 125 A (55 A × 250% = 137.5 A) or less per 450-3(a)(2). However, 450-3(a)(2)a, Ex. 1 permits the next higher standard rating above 137.5 A, which is 150 A fuses per 240-6.

For CBs, the settings shall not exceed 300% of the primary's full-load current rating. For example, a primary with a full-load current rating of 110 A requires the CB to be sized no greater than 300 A (110 A × 300% = 330 A). Exception 1 to 450-3(a)(2)a permits the next higher standard size above 330 A, which is 350 A per 240-6. See Figure 4-18.

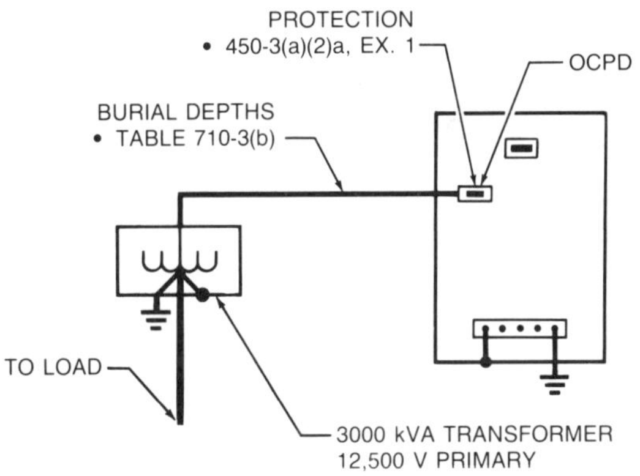

SELECTING OCPD'S FOR PRIMARY

What size overcurrent protection device is required for the transformer using individual protection in the primary side? Use CB and fuses.

<u>*CB*</u>

Step 1. $I = \dfrac{kVA \times 1000}{V \times \sqrt{3}}$

$I = \dfrac{3000 \text{ kVA} \times 1000}{12{,}500 \text{ V} \times 1.732} = 138.6 \text{ A}$

Step 2. *450-3(a)(2)a:* 138.6 A × 300% = 415.8 A

Answer. ***450-3(a)(2)a, Ex. 1:* CB = 450 A**

<u>*Fuses*</u>

Step 1. *450-3(a)(2)a:* 138.6 A × 250% = 346.5 A

Answer. ***450-3(a)(2)a, Ex. 1; 240-6:* Fuses = 350 A**

Figure 4-18. Circuit breakers are selected at 300% and fuses are selected at 250% of primary FLC per 450-3(a)(2)b for individual overcurrent protection devices installed in the primary side of high-voltage transformers.

Overcurrent protection devices, installed in the primary and secondary side of transformers with primaries rated over 600 V and secondaries rated over 600 V or 600 V or less, shall not exceed the percentages listed in Table 450-3(a)(2)b. The percentage selected is based on the voltage and impedance of the transformer. For example, an 800 kVA, 3ϕ, 12,500 V transformer with an impedance of 3% is required to have fuses selected at no more than 300% per Table 450-3(a)(2)b. See Figure 4-19.

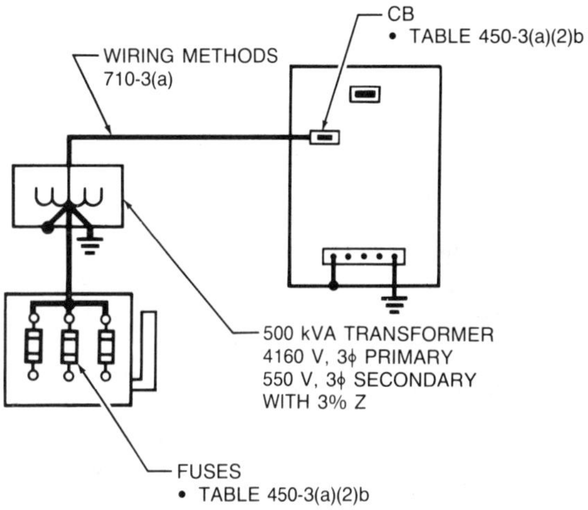

SUPERVISED LOCATION

SELECTING OCPD'S FOR PRIMARY AND SECONDARY (SUPERVISED LOCATION)

What size overcurrent protection devices is required in the primary and secondary side of the transformer?

<u>*Primary*</u>

Step 1. $I = \dfrac{kVA \times 1000}{V \times \sqrt{3}}$

$I = \dfrac{500 \text{ kVA} \times 1000}{4160 \text{ V} \times 1.732} = 69 \text{ A}$

Step 2. *Table 450-3(a)(2)(b):* 69 A × 600% = 414 A

Answer. ***450-3(a)(2)b; 240-6:* CB = 400 A**

<u>*Secondary*</u>

Step 1. $I = \dfrac{kVA \times 1000}{V \times \sqrt{3}}$

$I = \dfrac{500 \text{ kVA} \times 1000}{550 \text{ V} \times 1.732} = 524.9 \text{ A}$

Step 2. *Table 450-3(a)(2)b:*
524.9 A × 250% = 1312.3 A

Answer. ***450-3(a)(2)b; 240-6:* Fuses = 1200 A**

Figure 4-19. Combination overcurrent protection devices installed in the primary and secondary sides of high-voltage transformers in supervised locations are selected by the values listed in Table 450-3(a)(2)b.

What size fuses is required for a 750 kVA, 3ϕ transformer with a 4160 V primary, a 208 V secondary, and an impedance of 2%? The transformer is in a supervised location.

Primary

Step 1. *Text:* $I = \dfrac{kVA \times 1000}{V \times \sqrt{3}}$

$$I = \frac{750\text{ kVA} \times 1000}{4160\text{ V} \times 1.732} = 104\text{ A}$$

Step 2. *Table 450-3(a)(2)b:* 104 A × 300% = 312 A

Step 3. *240-6:* 300 A is the next lower standard size

Answer. **Fuses = 300 A**

Secondary

Step 1. *Text:* $I = \dfrac{kVA \times 1000}{V \times \sqrt{3}}$

$$I = \frac{750\text{ kVA} \times 1000}{208\text{ V} \times 1.732} = 2083.3\text{ A}$$

Step 2. *Table 450-3(a)(2)b; 240-6:*
2083 A × 250% = 5207.5 A
5000 A is the next lower standard size

Answer. **Fuses = 5000 A**

Control Transformers—430-72(c)

Control transformers are used to step higher voltage down to a lower voltage for control circuits energizing coils to make or break contacts that connect power to motors. Protection is provided with fuses or circuit breakers installed in the primary or secondary side of the transformer.

Primary Protection—430-72(c), Ex. 1 and 2. Control transformers rated less than 50 VA are considered protected by the overcurrent protection device that allows the motor to start and run. Where overcurrent protection is installed on the primary side of the control transformer, the overcurrent protection device shall not exceed 500% of the transformer's full-load current rating. See 430-72(c), Ex. 2. See Figure 4-20. For example, a motor control transformer with a primary full-load current rating of 1.5 A requires an overcurrent protection device installed on the primary side rated at 6 A (1.5 A × 500% = 7.5 A) per 240-6, Ex.

What size overcurrent protection device is required for the primary side of a control transformer with a primary full-load current rating of 1 A?

Step 1. *430-72(c), Ex. 2:* 1 A × 500% = 5 A

Step 2. *240-6, Ex.:* 3 A is the next lower standard size fuse

Answer. **OCPD = 3 A**

Secondary Protection—430-72(c), Ex. 3 and Ex. 4. The size overcurrent protection devices placed on the secondary side of a control transformer shall comply with 430-72(c), Ex. 3 or Ex. 4. If the control transformer is rated at 100 VA or less and 30 V or less, the control transformer and its secondary circuit is considered a Class 2 circuit and shall meet all requirements of 725-38. The full-load current rating of such a transformer is determined by dividing the total VA rating of the control transformer by the secondary voltage that supplies power to the stop and start stations, which energize the coil to a magnetic starter or other type of electrical apparatus.

Control circuits taken from the secondary side of a control transformer are *Class 1, Class 2,* or *Class 3 circuits.* A Class 1 circuit operates at 1000 VA or less with a voltage of 30 V or less per 725-11. A Class 2 circuit operates at 100 VA or less with a voltage of 30 V or less. A Class 3 circuit operates at 30 V to 150 V. Class 2 and Class 3 circuits are determined by their power limitations, which are based upon VA and voltage listed in Table 725-31(a) for AC circuits and Table 725-31(b) for DC circuits. See Figure 4-21. A control

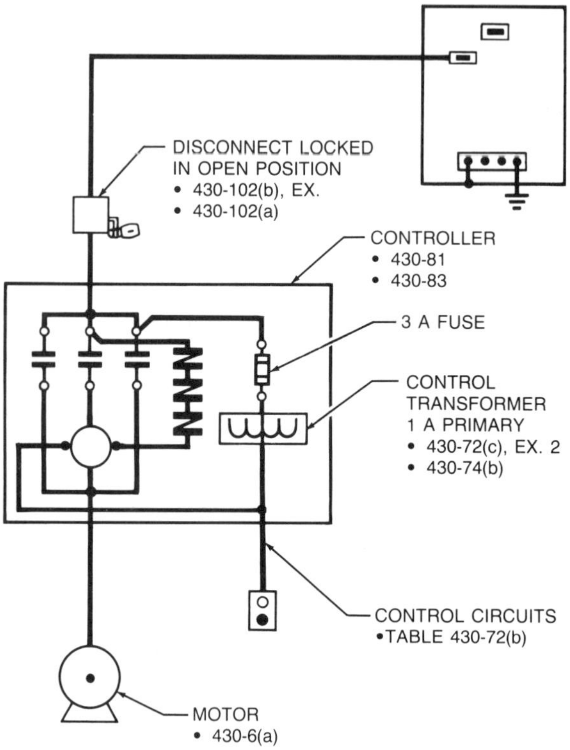

SELECTING OCPD'S FOR PRIMARY OF CONTROL TRANSFORMER

What maximum size overcurrent protection device is required in the primary side of the control transformer?

Step 1. *430-72(c), Ex. 2:* 1 A × 500% = 5 A

Step 2. *240-3, Ex. 2; 240-6, Ex.:*
3 A is the next lower standard size

Answer. **Fuses = 3 A**

Figure 4-20. Overcurrent protection devices installed in the primary side of control transformers with an FLC of less than 2 A shall not exceed 500% of the FLC.

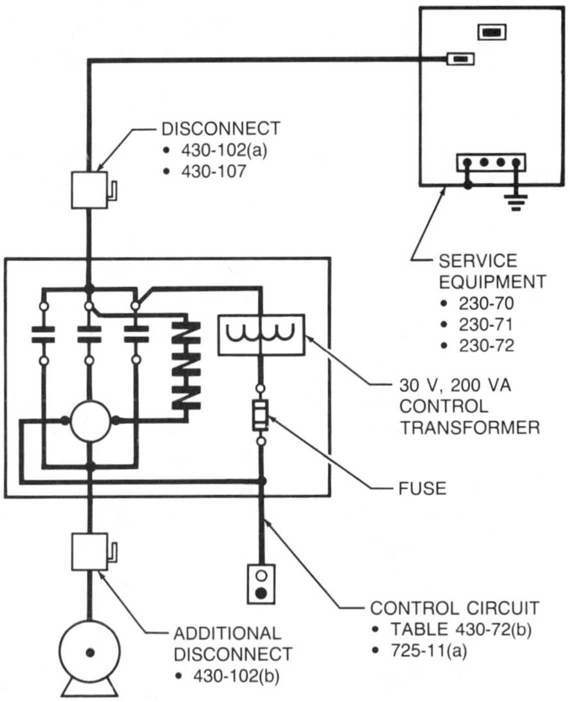

SELECTING OCPD'S FOR SECONDARY OF CONTROL TRANSFORMER

What size overcurrent protection device is required in the secondary side of the control transformer?

Step 1. *Text:* $I = \frac{VA}{V}$

$$I = \frac{200\ \text{VA}}{30\ \text{V}} = 6.7\ \text{A}$$

Step 2. *430-72(c), Ex. 3; 725-11(a):*
6.7 A × 167% = 11 A

Step 3. *240-3, Ex. 2; 430-72(c), Ex. 3; 725-11(a):*
10 A is the next lower standard size

Answer. **Fuses = 10 A**

Figure 4-21. Overcurrent protection devices installed in the secondary side of control transformers shall not exceed 167% of the secondary FLC for Class I circuits.

transformer with a secondary full-load current rating of 4 A requires a 6 A fuse per 240-6 (4 A × 1.67 = 6.68 A).

What size overcurrent protection device is required for the secondary side of a control transformer with a rating of 100 VA and an operating voltage of 30 V?

Step 1. *Text; 430-72(c), Ex. 3:* $I = \frac{VA}{V}$

$$I = \frac{100\ \text{VA}}{30\ \text{V}} = 3.33\ \text{A}$$

Step 2. *725-38:* 3.33 A × 167% = 5.6 A

Step 3. *240-6, Ex.:* 3 A is the next lower standard size fuse

Answer. **OCPD = 3 A**

GROUNDING—250-26

Where a separately derived system is used to step down a higher voltage to supply utilization type equipment, the secondary side of the transformer is connected to a grounding electrode. The grounding electrode conductor is sized properly to ensure a safe ground for the transformer during its operating period. Separately derived systems are installed at different locations in commercial and industrial plants and buildings to provide the necessary voltage to supply equipment, lighting circuits, receptacle circuits, etc.

The secondary side of a separately derived system is grounded properly to ensure that the secondary side of the transformer will provide system grounding and circuit grounding. The secondary windings of a separately derived system can be delta-connected or wye-connected.

Where the secondary windings of a separately derived system are connected for a closed- or open-delta, one of the windings provides for a midpoint tap with a 240 V coil so 120 V loads can be pulled from either side of the coil. Where the secondary windings of a separately derived system are connected in a closed-delta system and a midpoint tap transformer is not provided, any corner of the transformer can be connected to ground and provide the grounded transformer that is required by 250-5. One 240 V coil of the proper size can be provided with a midpoint tap so that neutral loads can be taken from the midpoint tap to each outside phase conductor of the coil. Two 120 V coils connected in series to form the secondary side can be used to provide two power sources. The two outer legs of the two transformers can be used to obtain 240 V. Either one of the two outer legs can be used to ground to obtain 120 V to supply 120 V neutral-connected loads.

For a wye system, three windings on the secondary side can be connected together and tapped to ground to form the wye. Any one of the outer legs of the wye can be measured to ground and 120 V will be obtained. By measuring the voltage phase to phase, 208 V can be obtained. This is verified by multiplying 120 V by $\sqrt{3}$ to obtain 208 V. If 277 V coils are used, 277 V is multiplied by $\sqrt{3}$ and the phase-to-phase voltage is 480 V.

Bonding—250-26(a)

Enclosures housing the windings of separately derived systems are bonded to the grounded circuit conductor of the system by a *bonding jumper.* A bonding jumper is a conductor used to connect the metal of enclosures. The enclosure of the separately derived system is bonded to the grounding electrode conductor, which is bonded to the grounded circuit conductor that bonds the enclosure of the separately derived system to the enclosure of the equipment that is supplied. The proper size bonding jumper ensures system grounding and circuit grounding in the system.

Sizing. Bonding jumper size is determined by the phase conductors that supply power from the windings of the separately derived system to the electrical equipment served. Section 250-79(d) requires the bonding jumper to be sized and selected from Table 250-94 and based upon the largest phase conductor between the transformer and the electrical equipment. For example, if the phase conductors are #1/0, the size of the bonding jumper is #6 Cu, which is selected from Table 250-94. Section 250-79(d) requires the bonding jumper to be 6′-0″ in length or less. Note that the size of the bonding jumper shall be equivalent to the grounding electrode conductor. All grounding electrode conductors are sized from the CM or kcmil ratings of phase conductors whether they are single conductors per phase or connected in parallel. See Figure 4-22.

If #250 kcmil, THW Cu conductors are used from the secondary side of a transformer to supply power to the electrical equipment, a #2 Cu bonding jumper is required from Table 250-94 per 250-79(d). Note that the grounding electrode conductor will also be #2 up to #1100 kcmil.

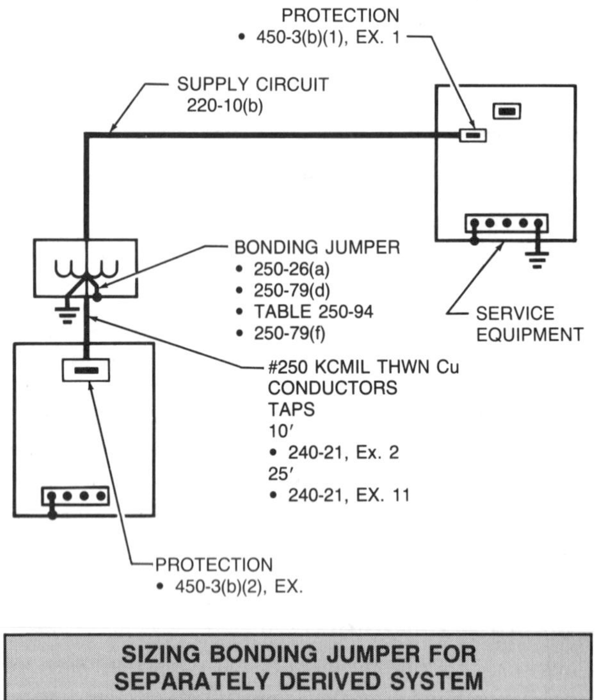

SIZING BONDING JUMPER FOR SEPARATELY DERIVED SYSTEM

What size copper bonding jumper is required to ground the transformer enclosure?

Step 1. *250-26(a); 250-79(d); Table 250-94:*
#250 kcmil THWN Cu requires #2 THWN Cu

Answer. **Bonding jumper = #2 THWN Cu**

Figure 4-22. The bonding jumper for a separately derived system is sized from the rating of the secondary conductors supplying the electrical equipment per 250-26(a).

■ *Master Electrician's Test Tip.* The bonding jumper and the grounding electrode conductor are the same size when the phase conductors and any parallel hookup or single run do not exceed #1100 kcmil. The bonding jumper is equal to 12½% of all kcmil ratings exceeding #1100 kcmil per phase.■

For phase conductors exceeding #1100 kcmil, the bonding jumper shall be equivalent to 12½% of the largest phase arrangement of conductors. For example, if the largest phase arrangement of conductors of any one phase is #2000 kcmil, the bonding jumper shall be at least #250 kcmil (2000 × .125 = 250). Note that Table 8, Ch. 9 is used when selecting bonding jumpers. The table lists kcmil ratings for conductors that are #250 kcmil and larger and circular mil ratings for conductors that are #4/0 or less. Kcmil values are determined by dividing CM values by 1000. CM values are determined by multiplying kcmil values by 1000.

What size copper bonding jumper is required for #4/0 conductors per phase supplying power to a motor control center supplied by a separately derived system?

Step 1. *250-79(d); Table 250-94:*
Phase conductors = #4/0

Step 2. *Table 250-94:* #4/0 requires #2 Cu

Answer. **Bonding jumper = #2 Cu**

Phase conductors are made up of 2400 kcmil Cu per leg between the secondary side of a separately derived system and the electrical equipment served. What size copper bonding jumper is required?

Step 1. *250-79(d):* 2400 kcmil × .125 = #300 kcmil

Step 2. *Table 8, Ch. 9:*
#300 kcmil is a standard size conductor

Answer. **Bonding jumper = #300 kcmil Cu**

Grounding Electrode Conductor—250-26(b)

For a separately derived system installed to step up or step down the voltage to supply utilization equipment, the secondary side of the transformer is connected to ground by a properly sized grounded electrode conductor. The grounding electrode conductor is selected from 250-94 or Table 250-94 based upon copper or aluminum conductors supplying the electrical equipment from the secondary side of the separately derived system. The grounding electrode conductor can be installed at the separately derived system and connected to ground, or it can be connected in the electrical equipment served and connected to ground.

If the transformer is hung from the ceiling or mounted on the wall and the building steel is closer to the transformer, the grounding electrode conductor is usually connected to the building steel from this point to limit the length of the grounding electrode conductor. If the electrical equipment is mounted on the wall and is closer to the building steel than the separately derived transformer, the grounding electrode conductor is connected to the building steel from the electrical equipment. See Figure 4-23.

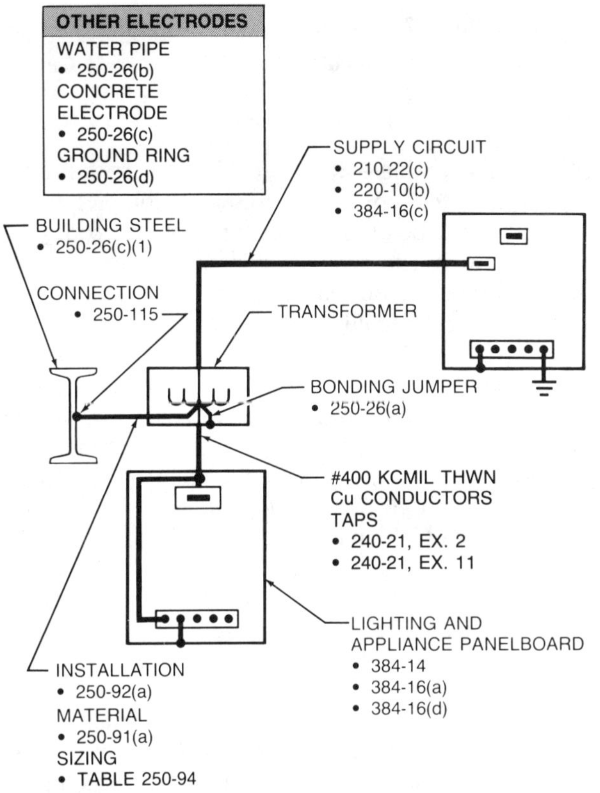

SIZING GROUNDING ELECTRODE CONDUCTOR FOR SEPARATELY DERIVED SYSTEM

What size copper grounding electrode conductor is required to ground the transformer secondary to the building steel?

Step 1. *250-26(b); Table 250-94:*
#400 kcmil requires #1/0 grounding electrode conductor

Answer. **Conductor** = ***#1/0 Cu***

Figure 4-23. The grounding electrode conductor is sized from Table 250-94 when connected to water pipe or building steel. It is sized from 250-81(c) when terminated to a concrete-encased electrode and 250-81(d) when terminated to a ground ring.

Sizing. If the grounding electrode conductor of the separately derived system is connected to building steel to ground the transformer, the grounding electrode conductor is selected from Table 250-94 based on the tap size. The tap is all conductors from the secondary side of the transformer to the load served. For example, the size grounding electrode conductor required for four #4/0 THWN Cu conductors tapped from the secondary side of a transformer is #2 Cu per Table 250-94. The type of grounding electrode selected to ground the secondary side of the transformer determines the size of the grounding electrode conductor. The types of grounding electrodes are

1. Building steel
2. Metal water pipe
3. Concrete-encased electrode
4. Ground ring
5. Driven rod

What size copper grounding electrode conductor, using each grounding electrode type, is required to ground the secondary side of a transformer with #250 kcmil Cu conductors supplying a panelboard?

Building steel

Step 1. *250-26(c)(1); Table 250-94:*
#250 kcmil requires #2 Cu

Answer. **Conductor** = ***#2 Cu***

Metal water pipe

Step 1. *250-26(c)(2); Table 250-94:*
#250 kcmil requires #2 Cu

Answer. **Conductor** = ***#2 Cu***

Concrete-encased electrode

Step 1. *250-26(c)(3); 250-81(c):*
Any size conductor requires #4 Cu

Answer. **Conductor** = ***#4 Cu***

Ground ring

Step 1. *250-26(c)(3):* Any size conductor requires #2 Cu

Answer. **Conductor** = ***#2 Cu***

Driven rod

Step 1. Driven rod. *250-26(c)(3); 250-94, Ex. l:*
Any size conductor requires #6 Cu

Answer. **Conductor** = ***#6 Cu***

Grounding Electrode—250-26(c)

The grounding electrode used to ground the secondary side of a separately derived system shall be located as close as possible to the area and the electrical equipment the system is to serve. The length of the grounding electrode conductor shall be as short as possible to ensure a low impedance path to the grounding electrode. This low impedance path helps open the overcurrent protection device in case of short circuits or ground faults. The low impedance path of the grounding electrode conductor and the low resistance of the grounding electrode stabilize the voltage to ground.

Building Steel—250-26(c)(1). The preferred ground at the service equipment is the water piping system. However, on a separately derived system, building steel is the number one choice. When a separately derived system is connected to building steel, all metal parts and objects within the area of the grounding electrode system and the electrical equipment served have a closer potential to ground.

The size of the grounding electrode conductor is selected from Table 250-94, which is based upon the tap conductors from the secondary side of the transformer to the equipment. This tap can be either 10′

or 25′. For example, if the tap conductors between the separately derived system and the equipment served are #3/0, the size of the grounding electrode conductor shall be at least #4 Cu. Note that #2 Al could be used where not exposed to dampness, wetness, moisture, or corrosion. Sections 250-92(a) and 250-112 require that the connection of the grounding electrode conductor to the building steel be accessible for inspection and removal to check out grounds. The bonding jumper and grounding electrode conductor are the same size when the phase conductor does not exceed 1100 kcmil.

What size copper grounding electrode conductor is required where the tap conductors between a separately derived system and the service equipment is three #400 kcmil THWN Cu conductors parallel per phase?

Step 1. *Table 250-94:* three #400 kcmil requires #3/0 Cu

Answer. **Grounding electrode conductor = #3/0 Cu**

Metal Water Pipe—250-26(c)(2). Where there is no building steel, metal water piping is the second choice. The size of the grounding electrode conductor is selected from Table 250-94 and based upon the phase conductors (tap conductors). For example, if the tapped secondary phase conductors are #1 THWN Cu conductors, the size of the grounding electrode conductor is #6 Cu per Table 250-94.

Warning: Care must be taken to ensure that the metal water piping or copper lines are connected to a metal conduit in the earth per 250-80(a) and 250-81(a).

What size copper grounding electrode conductor is required to ground the secondary side of a separately derived system to a water pipe with #3/0 Cu tap conductors from the transformer to the service equipment?

Step 1. *Table 250-94:* #3/0 conductors require #4 Cu

Answer. **Grounding electrode conductor = #4 Cu**

Other Electrodes—250-26(c)(3). Where building steel or a water pipe is not present, another type of electrode shall be used. Other types of electrodes include concrete-encased electrodes. For example, a #4 Cu conductor buried in the foundation and extended and connected to the secondary side of the separately derived system is acceptable. A rebar, ½″ in diameter and at least 20′-0″ in length, can also be used as a grounding electrode. A driven rod can be used per 250-83 and 250-84. Note that the size of the grounding electrode conductor is not required to be any larger than #6 per 250-94, Ex. 1.

Where a #4 Cu conductor or ½″ rebar is placed in the foundation and used as a grounding electrode, #4 Cu is the largest size grounding electrode conductor required to connect the concrete-encased electrode to the secondary side of the transformer. A #4 Cu grounding electrode conductor is also required to connect the secondary side of a transformer to the foundation steel used as the grounding electrode. A #4 conductor, at least 20′-0″ in length, is placed in the foundation and stubbed-out and left to be connected to the grounding electrode conductor.

What size grounding electrode conductor, using copper, is required to ground the secondary side of a separately derived system supplying power to electrical equipment in a machine shop?

Step 1. *250-26(c)(3); 250-81(c):*
#4 concrete encased electrode requires #4 Cu

Answer. **Grounding electrode conductor = #4 Cu**

Transformer Circuits

Date ____________________

Chapter 4—TEST

Name ____

1. What is the FLC rating of a 75 kVA, 240 V, 1ϕ transformer?
2. What is the FLC rating of a 112 kVA, 240 V, 3ϕ transformer?
3. What is the FLC rating of a 225 kVA, 480 V, 3ϕ transformer supplying power to a group of motors?
4. What is the FLC rating of a 300 kVA, 208 V, 3ϕ transformer supplying power to an office area?
5. What is the FLC rating of a 5 kVA, 120 V, 1ϕ transformer supplying power to a pumping system?
6. What is the FLC rating of a 500 kVA, 575 V, 3ϕ transformer supplying power to a workshop?
7. What is the kVA rating of a 240 V, 1ϕ feeder circuit supplying power to a load of 112 A?
8. What is the kVA rating of a 208 V, 3ϕ feeder circuit supplying power to a load of 225 A?
9. What is the ratio of a transformer having a 4160 V primary and 480 V secondary?
10. What is the ratio of a transformer having a 208 V primary and 480 V secondary?
11. What is the FLC rating of a 240 V secondary with a 2 to 1 ratio? The primary is 480 V with a 50 A rating.
12. What is the classification of a 42 circuit panelboard supplying three 20 A, single-pole loads with neutral connections?
13. A 36 circuit panelboard has three 30 A, two 20 A, and two 15 A overcurrent protection devices supplying neutral-connected loads. What is the classification of the panelboard?
14. What size tapped conductors is required to supply power to the lugs of a 300 A panelboard? (Use the 10′ tap rule.)
15. What size tapped conductors is required to supply power to the terminals of a 250 A CB installed in a 400 A panelboard?
16. How many 60 A CBs can be installed in a panelboard supplied by a 25′ tap from the secondary side of a transformer with a 400 A output?
17. What size THWN Cu conductors is required for a 25′ tap supplying a panelboard with a load of 450 A? The conductors are paralleled three times per phase.
18. What size primary tap is required from a 350 A feeder circuit where the primary tap, secondary tap, and transformer do not extend over 25′-0″ from the feeder?
19. What size secondary tap is required from the secondary side of a 240 V transformer with the primary tapped from a 480 V feeder circuit supplied by a 250 A overcurrent protection device? The secondary tap includes the primary tap and transformer.
20. What is the load for a tap from the secondary side of a transformer supplying a continuous-duty load of 87 A?

______ **21.** What is the load for a tap from the secondary side of a transformer supplying a noncontinuous-duty load of 110 A?

______ **22.** What is the load for a tap from the secondary side of a transformer supplying a noncontinuous-duty load of 138 A and continuous-duty load of 92 A?

______ **23.** What is the load for a tap from the secondary side of a 208 V, 3ϕ transformer supplying a continuous-duty load of 142 A and a 10 HP, 20 HP, and 30 HP motor?

______ **24.** What size THWN Cu conductors is required to supply power to the primary side of a 100 kVA, 480 V, 3ϕ transformer? The load is continuous duty.

______ **25.** What size THWN Cu conductors is required to supply twelve 1.5 A ballasts in fluorescent lighting fixtures operating at noncontinuous duty?

______ **26.** What size THWN Cu conductors is required for a tap from the secondary side of a transformer with the following loads?

52 A continuous-duty lighting
20 A noncontinuous-duty lighting
36 A continuous-duty sign
12 A continuous-duty receptacles
7 A motor load
10 A motor load
12 A motor load

______ **27.** What size overcurrent protection device is required for the supply circuit to an 80 kVA, 480 V, 3ϕ transformer?

______ **28.** What size overcurrent protection device is required for the supply circuit to a 1.92 kVA, 240 V, 1ϕ transformer?

______ **29.** What size overcurrent protection device is required for the supply circuit to a .768 kVA, 480 V, 1ϕ transformer?

______ **30.** What is the maximum size overcurrent protection device permitted in the primary side of a control transformer with a 1.5 A FLC?

______ **31.** What is the maximum size overcurrent protection device required in the primary side of a 2-wire to 2-wire transformer with a 22 A FLC?

______ **32.** What is the maximum size overcurrent protection device required in the primary and secondary sides of a 2-wire to 2 wire, 15 kVA, 480 to 240 V transformer?

______ **33.** What size circuit breaker is required to provide individual protection in the primary side of an 800 kVA, 2400 V, 3ϕ transformer in a supervised location?

______ **34.** What size fuses is required to provide individual protection in the primary side of a 600 kVA, 12,500 V, 3ϕ transformer in a supervised location?

______ **35.** What size circuit breaker is required in the secondary side of a 600 kVA, 480 V, 3ϕ transformer in an unsupervised location? The primary side is over 600 V.

______ **36.** What size fuses is required in the secondary side of a 500 kVA, 480 V, 3ϕ transformer in a supervised location? The primary is over 600 V with 6% impedance.

______ **37.** What size fuses is required in the secondary side of a 2500 kVA, 13.5 kV, 3ϕ transformer with a rated impedance of 8.9%? The location is supervised.

______ **38.** What size overcurrent protection devices is required for the primary (using circuit breakers) and the secondary (using fuses) sides of an 850 kVA, 3ϕ transformer with a rated impedance of 1.15%? The primary is 13,500 V and the secondary is 460 V. The location is unsupervised.

______ **39.** What size overcurrent protection devices is required for the primary (using circuit breakers) and secondary (using fuses) sides of a 950 kVA, 3ϕ transformer with a rated impedance of 10%? The primary is 12,500 V and the secondary is 550 V. The location is supervised.

______ **40.** What size overcurrent protection devices is required for the primary (using circuit breakers) and the secondary (using fuses) sides of an 1100 kVA, 3ϕ transformer with a rated impedance of 1.5%? The primary is 14,400 V and the secondary is 480 V. The location is unsupervised.

______ **41.** What size overcurrent protection devices is required for the primary (using fuses) and the secondary (using circuit breakers) sides of a 750 kVA, 3ϕ transformer with a rated impedance of 4%? The primary is 7200 V and the secondary is 480 V. The location is supervised.

______ **42.** What size overcurrent protection device is required in the primary side of a 1000 VA control transformer tapped from a 550 V supply?

______ **43.** What size bonding jumper is required to bond the enclosure of a separately derived system to a water pipe? The secondary is wired with #250 kcmil THWN Cu conductors.

______ **44.** What size bonding jumper is required for a separately derived system with tapped secondary conductors rated at #350 kcmil THWN Cu paralleled five times per phase?

______ **45.** What size copper grounding electrode conductor is required to ground a separately derived system with #4/0 taps to building steel?

______ **46.** What size copper grounding electrode conductor is required to ground a separately derived system with #3/0 taps to a nearby water pipe?

______ **47.** What size copper grounding electrode conductor is required to ground a separately derived system with #600 kcmil THWN Cu conductors to a driven rod?

______ **48.** What size grounding electrode conductor is required to ground a separately derived system with #400 kcmil taps to a concrete-encased electrode?

______ **49.** What size copper bonding and grounding electrode conductor is required to bond and ground a separately derived system with #300 kcmil THWN Cu taps to building steel?

______ **50.** What size copper bonding and grounding electrode conductor is required to bond and ground a separately derived system with #2100 kcmil Cu taps to building steel?

Service Entrance and Grounding

Chapter

5

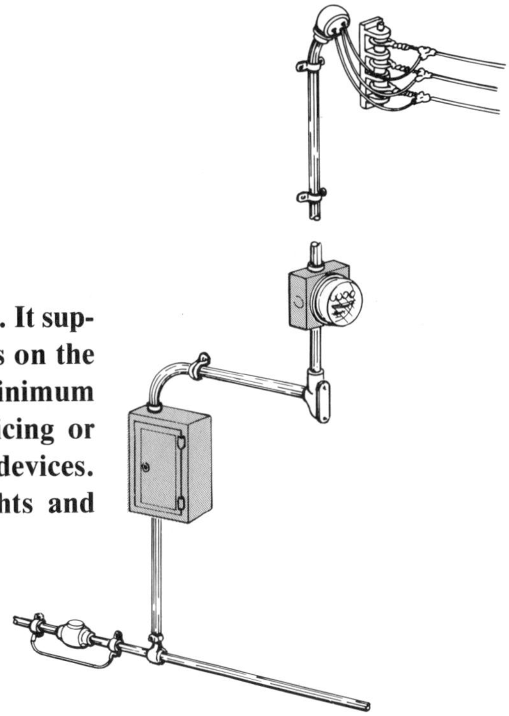

The service is the heart of the electrical system. It supplies power to all feeders and branch circuits on the premises. Service equipment shall have minimum clearances for the safety of personnel servicing or requiring access to overcurrent protection devices. Overhead drops shall have minimum heights and underground laterals shall have minimum depths for protection of service conductors. Service disconnects shall be grouped in one acceptable location for accessibility. The service shall be grounded.

NUMBER OF SERVICES—230-2

The number of service drops or laterals to a building is limited to one without meeting one of the exceptions in 230-2. There are seven exceptions to the general rules of 230-2 that allow more than one service to a building.

Exception 1: Fire pumps
Exception 2: Emergency systems
Exception 3: Multiple-occupancy buildings
Exception 4: Capacity requirements
Exception 5: Buildings of large area
Exception 6: Different characteristics
Exception 7: Two or more laterals

Fire Pumps—230-2, Ex. 1

An additional service drop or lateral is permitted to supply a fire pump. The additional service drop or lateral may be at a different location from the ordinary service drop that supplies power to lighting loads, receptacle loads, and other loads in the building.

The service equipment for the fire pump can be in a different location from the main service with six disconnecting means to supply power to fire pump equipment. However, 230-2 requires a plaque at the main service identifying the location of the service to supply the fire pump. The fire pump service shall have a disconnecting means with a plaque to identify the location of the main service. This is to prevent an electrician from de-energizing power from the main service and accidentally receiving an electrical shock from the service supplying the fire pumps.

In addition to 230-2, Ex. 1, *NFPA-20* indicates where fire pumps are supplied by a separate service, a service drop is run to the main service and an underground lateral is run to the service equipment that supplies the fire pump. With an overhead service drop supplying the main service and an underground lateral servicing the fire pump, a crane, dumpster, or truck could take down the service-drop conductors and leave the building without a service to the fire pump. Also, a digging machine digging a ditch for a water line would not take out both service laterals supplying power to the main service and the service supplying power to the fire pump.

NFPA-20 requires conductors supplying power to a fire pump to be encased by at least 2″ of concrete where entering and passing through a building. However, 2″ of concrete is not required where the conductors in conduit enter the fire pump room and extend to the fire pump controller, and from the controller to the fire pump itself.

Section 230-82, Ex. 5 allows a tap ahead of the main to supply power to a fire pump. However, 700-12(e), which deals with connections ahead of the service disconnecting means, does not allow the tap conductors to be installed in the same vertical section of a switchgear where the disconnecting means is located. Where the tap is permitted, the tap conductors shall be located in a vertical section that does not contain any disconnecting means for the service equipment. See *UL Standard 869 A*.

Section 230-90(a) requires the conductors to a fire pump to be protected with an overcurrent protection device having a rating or setting not higher than the allowable ampacity of the conductors. Exception 4 to 230-90(a) requires the overcurrent protection device for fire pumps to be selected or set to carry the locked rotor current of the motors. For example, a 100 HP, 460 V, 3ϕ motor used as a fire pump has a locked rotor current of 744 A per Table 430-151. Section 230-90(a), Ex. 4 requires a circuit breaker or fuse to be capable of holding above 744 A indefinitely.

NFPA-20 requires the rotor of the motor to turn and deliver water to extinguish fires as long as possible. As heat from a fire begins to heat the conductors and windings of a motor, the rotor will begin to slow down. The overcurrent protection device in the circuit that supplies power to a fire pump shall be capable of holding this increased current until the rotor is in the stalled position. When a ground fault occurs between the windings of the motor or from the windings to the metal frame of the motor enclosure, the overcurrent protection device opens the circuit that supplies power to the fire pump and prevents damage to the source of power delivering energy.

Emergency Systems—230-2, Ex. 2

Emergency systems are required by building codes to supply power in the event normal service is lost. Emergency systems provide power so that people can evacuate buildings without creating panic and chaos. These systems are covered under Sections 700 (emergency systems), 701 (legally required standby systems), and 702 (optional standby systems). Requirements for transferring service-entrance conductors from the emergency power service to the emergency system loads are found in 230-83 and 700-6.

Transfer equipment assures the successful shifting of loads from normal power to emergency power without back-feeding into the utility lines. All junction boxes and enclosures for emergency circuits are marked to show that they house components of an emergency circuit. Section 700-9(b) requires that all wiring for emergency systems be kept independently of the other wiring used to supply power to various loads in a building. For example, emergency system conductors shall not enter the same raceway, cable, box, or cabinet with other wiring per 700-9(b).

Five exceptions to the general rule of 700-9(b) allow conductors to be installed in transfer equipment enclosures, exit or emergency lighting fixtures supplied from two sources, and any common junction box attached to exit or emergency lighting fixtures that are supplied from two sources. Wiring of two or more emergency circuits supplied from the same source is permitted in the same raceway or in a common junction box attached to unit equipment that contains only the branch circuit supplying the unit equipment and the emergency circuit served by the unit equipment.

Section 700-12 lists the general requirements for sources of power for emergency systems. Section 700-12(a) lists the requirements for a storage battery. A storage battery shall supply the total load for a minimum period of 1½ hours without load voltage falling below 87% of the normal power.

Section 700-12(b) contains information about generator sets and contains five parts with specific rules. Part 1 requires a time delay feature that permits a 15-minute setting to be provided to avoid retransfer in case of short time reestablishment of the power source. Normally, 15 cycles are required to reestablish normal service after a failure. Part 2 requires the on-site fuel supply to be provided with not less than two hours of full demand operation of the system. Part 3 does not allow the prime mover to depend upon the public utility gas system to supply fuel or for the municipal water supply to provide water for cooling the system. Where dual-fed supplies are permitted, automatic transferring from one fuel supply to another shall be provided. The exception to Part 3 allows the authority having jurisdiction to vary these rules where the on-site fuel cannot be disturbed. See Figure 5-1.

Part 4 requires that storage batteries be equipped with automatic charging means independent of generator sets. In Part 5, a generator requiring more than 10 seconds to develop full power can be used, provided auxiliary power to energize the emergency system is provided until the generator can pick up the load and provide full power.

Part C deals with uninterrupted power supplies and refers to Section 700-12(a)(b) for requirements. Part D to 700-12 permits a separate service where the authority having jurisdiction feels that the service providing power cannot be interrupted easily. This section correlates with 230-2, Ex. 2. See Figure 5-2. Section 230-82, Ex. 5 permits a tap ahead of the main if the authority having jurisdiction will allow such a tap. The location of the tap at the service equipment shall comply with 700-12(e). Section 700-12(f) allows unit equipment to be placed in a room with a tap ahead of the switch that controls the lighting circuits to the lighting.

The exception to 700-12(f) allows three normal lighting circuits from the same panelboard with 120/208 V

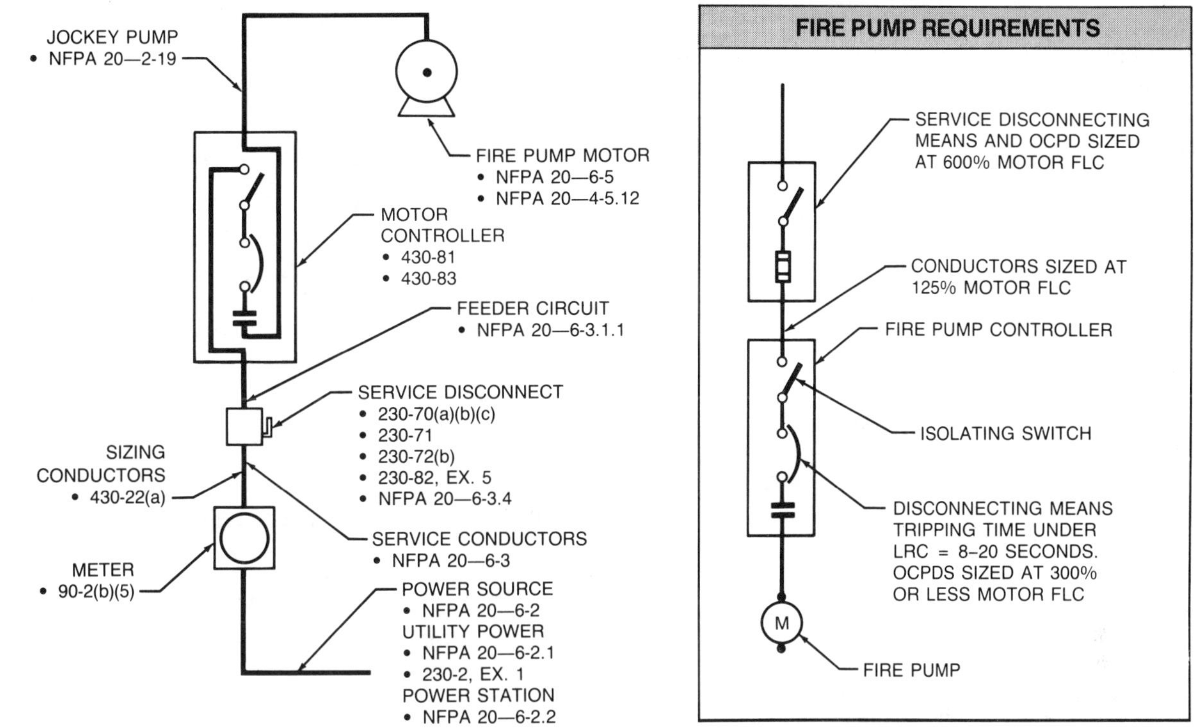

Figure 5-1. Exception 1 to 230-2 allows a separate service drop or lateral to a building to supply power to a fire pump.

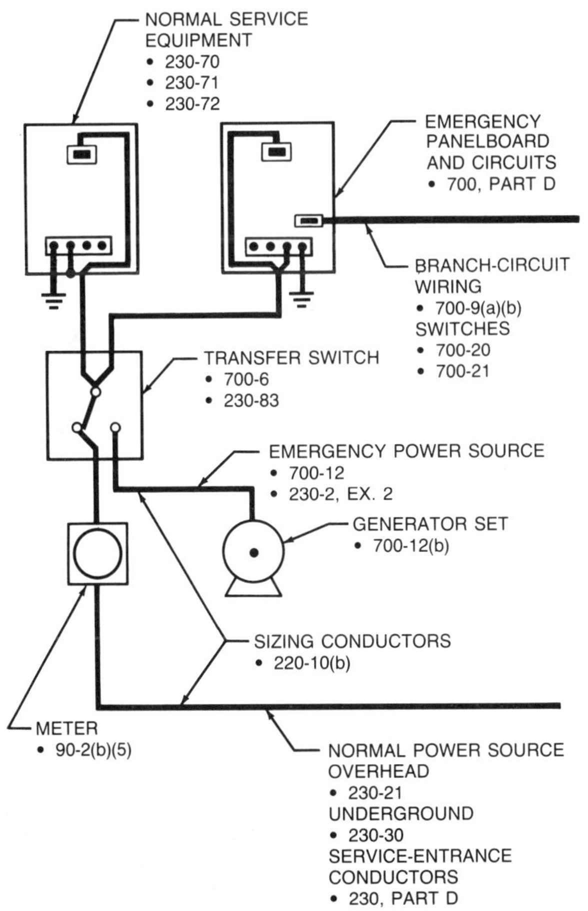

SIZING GENERATOR SETS USING RULE-OF-THUMB METHOD

What size generator is required to supply a 50 HP, 40 HP, 30 HP, 208 V, 3ϕ, and a 78 A lighting and receptacle load?

Step 1. *Table 430-150:*

50 HP = 130 A × 110% = 143 A
40 HP = 104 A × 110% = 114.4 A
30 HP = 80 A × 110% = 88 A

Step 2. Rule-of-thumb method.

First motor = 143 A × 125% =	178.75	A
Second motor = 114.4 A × 125% =	143	A
Third motor = 88 A × 125% =	110	A
Lighting and receptacle load = 78 A × 100% =	78	A
Total =	509.75	A

Step 3. Sizing generator.

$$\textit{Size generator} = \frac{A \times V \times \sqrt{3}}{1000}$$

$$\textit{Size generator} = \frac{509.75\text{ A} \times 208\text{ V} \times 1.732}{1000} = 183.51\text{ kW}$$

Answer. **Generator = 183.51 kW**

Figure 5-2. Exception 2 to 230-2 allows a second service drop or lateral to supply power for emergency circuits.

or 277/480 V to supply three rows of lights in one room without a switch being provided for each circuit. The additional circuit supplied by the battery packs shall be provided with a lock-on feature on the circuit breaker. Legally required standby systems (701) are intended to provide electrical power to aid in fire fighting, rescue operations, and other types of loads. Legally required standby systems have requirements different from emergency systems found in Article 700. Legally required standby systems are only required to provide power for 60 seconds or less instead of the 10 seconds or less required for emergency systems.

Section 701-10 allows wiring circuits to be pulled in the same conduit or junction boxes as the normal power. Article 702 deals with optional standby systems designed and used by an owner. These systems only supply power to the various loads that a user wants in case the normal power should fail. Section 702-9 allows the wiring of optional standby systems to occupy the same raceway, cable, boxes, and cabinets as the general wiring.

Multiple-Occupancy Buildings—230-2, Ex. 3

Multiple-occupancy buildings are commonly high-rise buildings and have more than one service permitted. Some high-rise buildings or multiple-occupancy buildings have as many as 12 individual services with six disconnecting means at each service location. However, 230-2 requires that a plaque be located at each service to identify the location of the other services. Tenants in a multiple-occupancy building shall have access to their overcurrent protection devices per 240-24(b), Ex. Section 230-72(c), Ex. requires that tenants have access to their overcurrent protection devices.

Section 230-91(c) requires that tenants in a multiple-occupancy building have access to overcurrent protection devices. For overcurrent protection devices installed in one mechanical room or one panel room, a maintenance person shall be available 24 hours a day or each tenant shall have a key and access to overcurrent protection devices at all times. See Figure 5-3.

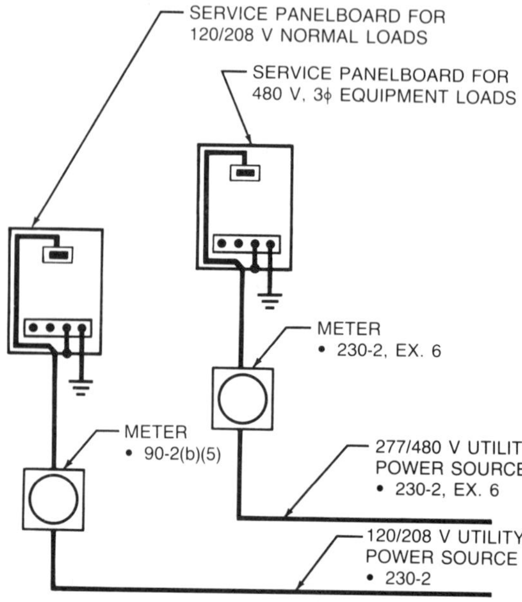

Figure 5-3. Exception 3 to 230-2 allows a second drop or lateral to supply power to multiple-occupancy buildings. Each service location is permitted to have six disconnecting means.

Capacity Requirements—230-2, Ex. 4

Two or more services shall be permitted to serve the building for additional capacity. The capacity requirements shall be in excess of 2000 A with a supply voltage of 600 V or less. For example, a 120/208 V wye system can supply a building that requires a load greater than 2000 A. A second service drop or lateral can supply the building to provide the additional amperage for the various loads in the building. Note that these two services can have six or fewer disconnecting means at each location. Section 230-71(a) allows no more than six disconnecting means per service grouped in any one location. Section 230-72 allows two to six disconnecting means at any one location either in one enclosure or five other disconnecting means in separate enclosures, making up a total of six.

Buildings of Large Area—230-2, Ex. 5

Buildings of large area require excessively long runs of feeder circuits from the service location to the subpanel. The voltage drop in such a long run would limit the power supplied to the subpanel. In this case a second drop, or *lateral,* can be run to the location where the subpanel is installed and power can be supplied with branch circuits to individual loads.

Different Characteristics—230-2, Ex. 6

Different characteristics include such items as different voltages, frequencies, or phases for different uses, and various rate schedules required by the utility for demand. For example, a building may have a service drop with a 120/208 V supply for lighting loads, small 208 V motors, and various other types of either 1ϕ or 3ϕ loads. The owner wishes to install some 480 V equipment. A second voltage coming in at 480/277 V can supply the building and be considered a lateral with six disconnecting means permitted for the service at that location per 230-71 and 230-72. Utility companies across the United States supply power with a frequency of 60 hertz. If a second service is needed because of higher frequency, then a second service is permitted.

A business has one service supplying the normal load and wants a second service for equipment so that a demand meter can be set and the owner would know what to charge to the heavy equipment loads. A second service is allowed under this condition of use. However, before a second service is permitted, the authority having jurisdiction and the local utility should be consulted to determine if the kVA rating is large enough for a second service.

Two or More Laterals—230-2, Ex. 2

A lateral is not considered service-entrance conductors per Article 100 and definitions in the NEC®. Therefore, two or more laterals rated #1/0 or larger can be run to a central location with six or fewer disconnecting means and the six or fewer laterals are considered one service.

The main problem with two or more laterals for one service is the method of metering. Most installations with two or more laterals for one service are metered by a master meter at the transformer's location of the utility. It is not feasible to provide six individual meters to six individual laterals that constitute one service lateral to a building.

TAPS AHEAD OF MAIN—230-82

Eight exceptions to 230-82 permit a tap to be connected ahead of the main service disconnecting means. Exception 1 permits cable limiters or other current-limiting devices. Exception 2 permits fuses and disconnecting means or circuit breakers to be used as service equipment and metering pedestals. Meters rated at 600 V or less and grounded per Article 250 can be tapped ahead of the main per Exception 3. Exception 4 permits current or voltage instrument transformers, high impedance, shunts, surge-protective devices, and other such equipment to be tapped ahead of the main.

Exception 5 permits fire pumps, emergency systems, exit lights, standby power, sprinkler head alarms, and similar systems to be tapped ahead of the main. Exception 6 permits interconnecting electrical power production sources to be tapped ahead of the main. See Articles 690 and 705.

Exception 7, in accordance with 230-76 and 230-78, permits a service disconnecting means to be tapped ahead of the main to provide power to a control circuit for energizing or de-energizing the disconnecting means of the service equipment. Exception 8 permits a tap ahead of the main to supply power to ground-fault protection systems if overcurrent protection and a disconnecting means are provided. All taps ahead of the main shall comply with 700-12(e) when used with emergency systems. See Figure 5-4.

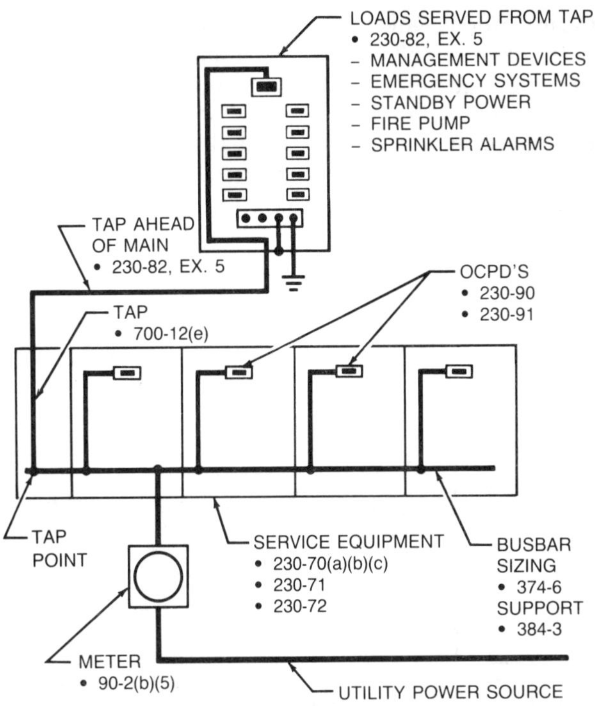

Figure 5-4. Section 230-82 allows taps to be made ahead of the main to supply power to certain types of loads. Exception 5 to 230-82 permits a tap ahead of the main to supply loads such as emergency, standby, and fire pumps.

SERVICE DROP—230, PART B

A service drop is all conductors extending from the utility service pole and transformer to the point of attachment on a building. The service-drop conductors tie to the service-entrance conductors at the weatherhead in the riser pipe connected to the meter base. Service-entrance conductors extend from the meter base to the service equipment. Service-drop conductors shall be of sufficient size and rating. Their point of attachment to the structure is specified from the finished grade. Minimum height requirements over property such as alleys, streets, railroad tracks, or water are specified. The NEC® and the National Electrical Safety Code list requirements for the proper installation of service-drop conductors.

Size and Rating—230-23

Service-drop conductors cannot be smaller than #8 Cu or #6 Al except for installations supplying limited loads of a single branch circuit. A #12 hard drawn copper or equivalent conductor is used. For example, a #12 hard drawn copper conductor could be used as a service drop to supply power to a telephone booth. This is only permitted when a single branch circuit is required and the supplying utility will go along with the local inspector in such an approval. The grounded or neutral conductor shall be sized by the requirements of Section 250-23(b), which requires the grounded neutral conductor to be as large as the grounding electrode conductor sized from Table 250-94 based on the service-entrance conductors.

Section 230-23(a) requires service-drop conductors to be large enough to carry the entire load without the temperature rising to a value that would deteriorate the insulation of the conductors. It is the responsibility of the designer and the electrical inspector to make sure that the loads in a building will not exceed service-entrance conductors and overload the service drop to a building. It is the responsibility of the utility company to provide a service drop that will provide the proper power due to diversity of loads to supply the building. See Figure 5-5.

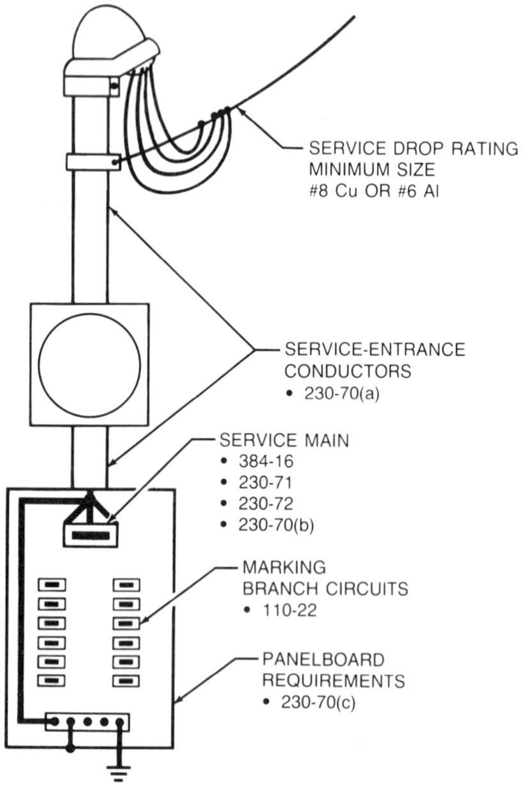

Figure 5-5. The minimum size service drop to a building is #8 Cu or #6 Al. Conductors for the service drop are sized according to the load of the building.

Point of Attachment—230-26

Section 230-26 requires the service-drop conductors to a building to have a minimum clearance as specified in Section 230-24(b) of the NEC®. In no case shall this point of attachment be less than 10′-0″ above finished grade.

Table 232-1 of the *National Electrical Safety Code* also deals with the minimum point of attachment of 10′-0″. Where a person can walk 15′-0″ directly toward the pole under the drop and measure up to a height of 12′-0″ or greater, the point of attachment to the building can be a minimum of 10′-0″. This requirement can be applied only where people have access.

Service Mast—230-28. A service mast used for the support of service-drop conductors shall be of sufficient strength to support the service-entrance conductors. This is often accomplished by using braces or guide wires to withstand the strain imposed on the service drop during bad weather. Most local codes and utility codes will not permit IMC to be used as a service mast. Service masts extend above the roof or overhang where the roof joists or side of the building are too low for the point of attachment to provide sufficient clearance.

Means of Attachment—230-27. Three basic methods are used to attach service-drop conductors to a building. An *I-bolt* can be bolted to a wall at any point above the 10′-0″ or 12′-0″ minimum required by most utilities to support the service drop. A *rafter support* attached to the end of a rafter can be used to provide the proper clearance. For installations requiring greater clearances, a *service mast* is used to attach the service-drop conductors. Most utilities and local inspection departments require the point of attachment to be located at least 6″ below the service-drop weatherhead and service-entrance conductors. Section 230-54 allows the drop to be located in any radius extending not more than 2′ from the weatherhead. See Figure 5-6.

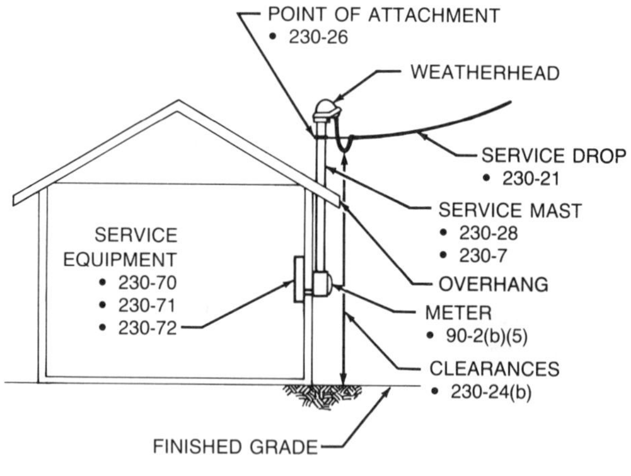

Figure 5-6. The service mast shall be strong enough to hold the point of attachment securing the service drop.

Clearances—230-24(b)

Service-entrance conductors shall provide clearances that ensure protection of conductors and safety of persons around or under the drop. Part A to 230-24 requires a minimum clearance of 8′-0″ for service-drop conductors passing over a flat roof. Exception 1 to 230-24(a) permits a minimum clearance of 3′-0″ when the roof has a slope of at least 4″ in 12″. The voltage between conductors shall not exceed 300 V for this reduced clearance.

Where the voltage between conductors is above 300 V, the drop shall be at least 8′-0″. Exception 2 to 230-24(a) permits the service drop to be attached to the service conduit extending through the roof. This is a service mast per 230-28. The drop shall not pass over more than 4′-0″ of the roof. The point of attachment to the drop shall be at least 18″ from the conduit and roof line. The clearances of service-drop conductors are based upon the voltage level to ground of the conductors. These voltage levels range from 0 V to 600 V. See Figure 5-7.

VOLTS TO GROUND			
VOLTAGE LEVELS	SERVICE ENTRANCE	RESIDENTIAL PROPERTY	PUBLIC STREETS
To 150 V	10′	12′	18′
151 V to 300 V	12′	12′	18′
301 V to 600 V	15′	15′	18′

Figure 5-7. Clearance from finished grade to the point of attachment of the service-drop conductors is based on voltage levels to ground.

150 V or Less to Ground—230-24(b). The voltage levels to ground for these service-drop conductors are voltages that are supplied from a wye or delta system with voltages of 120 V to ground or less. The coil voltage of the transformer windings is 120 V in a wye system. By multiplying 120 V by $\sqrt{3}$, 208 V phase-to-phase is derived (120 V × 1.732 = 207.8 V). Therefore, the voltage to ground will not exceed 120 V in most cases. In no case will the voltage to ground in this type of system exceed 150 V. Where a 240 V coil is tapped in the center to obtain 120 V from each ungrounded phase conductor to the grounded neutral point, only 120 V to ground can be obtained. The voltage to ground will never exceed 150 V for this type of system.

Where two 120 V single windings are in series to obtain 240 V, the center tap between the two windings provides a neutral connection to ground and 120 V can be obtained from either hot ungrounded conductor to the grounded point of the transformer. The voltage to ground in this type of system will never exceed 150 V to ground. The clearance of the service-drop conductors shall be at least 10′-0″ minimum from finished grade for voltages of 150 V or less to ground. Note that only people can have access underneath the service-drop conductors for the drop to be attached to the building at the 10′-0″ minimum.

151 V to 300 V to Ground—230-24(b). Voltage to ground exceeds 150 V for three transformers connected in a delta system or two transformers connected in an open-delta system. Voltage to ground exceeds 150 V because the high leg of an open- or closed-delta system travels through one 240 V winding and through half of the other 240 V winding (120 V). The voltage of the high leg is found by adding 240 V and 120 V and dividing by $\sqrt{3}$, which is 208 V (240 V + 120 V = 360 V/1.732 = 207.8 V). The minimum clearance from finished grade to the point of attachment of the service-drop conductors shall be at least 12′-0″ for this type of system.

301 V to 600 V to Ground—230-24(b). This type of system consists of three 480 V windings connected in a delta configuration with one phase conductor connected to ground. A voltage to ground for two of the conductors is 480 V. There is no voltage to ground for one conductor. As 480 V to ground can be obtained from two conductors, the minimum point of attachment is 15′-0″ above finished grade.

SERVICE LATERAL—230, PART C

Service-lateral conductors are conductors that extend from the pad-mounted transformer of the utility to the meter base, or tap cans mounted on the building. Note that service-lateral conductors are not considered service-entrance conductors. The service-entrance conductors are the conductors from the load side of the meter base to the first disconnecting means of the service equipment. Service-lateral conductors shall be rated to handle the load of the structure. They shall also meet burial depths to assure protection from physical damage. See Table 300-5 for burial depths.

Size and Rating—230-31(b)

Section 230-31(a) requires that service-lateral conductors be large enough to carry the current with a load that is computed in accordance with 220-10(b) and 220-22. Part B requires service-lateral conductors to be no smaller than #8 Cu or #6 Al. The exception to 230-31(b) permits a single branch circuit to serve small loads such as a telephone booth with a rating of not less than #12 Cu or #10 Al.

The grounded conductor in a service lateral is sized from Section 250-23(b), which requires the grounded neutral conductor to be as large as the grounded electrode conductor sized from Table 250-94 based upon the largest phase conductor. Section 230-30 requires service-lateral conductors to have an insulation that will not deteriorate due to moisture per 310-8(a) of the NEC®. The exception to 230-30 allows the grounded conductor to be uninsulated under conditions such as bare copper used in a raceway or in direct burial where the bare copper is suitable for the soil conditions.

A bare copper conductor can be used for direct burial without regard to soil conditions when part of an approved assembly for underground use. Aluminum conductors shall be part of a cable assembly identified for underground use in a raceway or for direct burial. It is the responsibility of the designer who calculates the load for the building to provide the proper service-entrance conductors to the meter base or tap can. (Check local codes.) It is the sole responsibility of the utility to provide an underground lateral large enough to provide power to the meter base or tap can.

Burial Depths—230-32

Underground service-lateral conductors shall be protected against damage per 300-5. Table 300-5 gives the minimum requirements for burying a service lateral. Direct burial cables shall be at least 24″ deep. Rigid metal conduit shall be at least 6″ deep. Intermediate metal conduit (IMC) shall be at least 6″ deep. Rigid nonmetallic conduit (usually Schedule 40 or 80), approved for direct burial, shall be at least 18″ deep.

Other approved raceways shall be at least 18″ deep. The minimum cover requirements for all raceway systems except rigid metal conduit and IMC are permitted to have the burial depth reduced by 6″ when a 2″ thick concrete pad (or equivalent) is placed in the trench over the underground conduit. Table 300-5(a) permits conduits to have a minimum covering of 4″ of concrete when placed beneath and not within 6″ of the edge of the slab. Table 300-5 requires a minimum covering of 24″ for raceway systems extending beneath highways, roads, alleys, driveways, and parking lots.

Lesser depths are permitted for cables and conductors of a service lateral rising for terminations or splices. For example, a service lateral rising and terminating into a tap can or extending from a tap can on the building and terminating into a meter base may be buried at lesser depths. Table 300-5 requires service laterals beneath airport runways to be at least 18″ below finished grade. See Figure 5-8.

■ *Master Electrician's Test Tip.* Most local codes and utility companies do not recognize requirements of 300-5(a) and Table 300-5. They require their own burial depths for service-entrance laterals from a transformer to a building structure.■

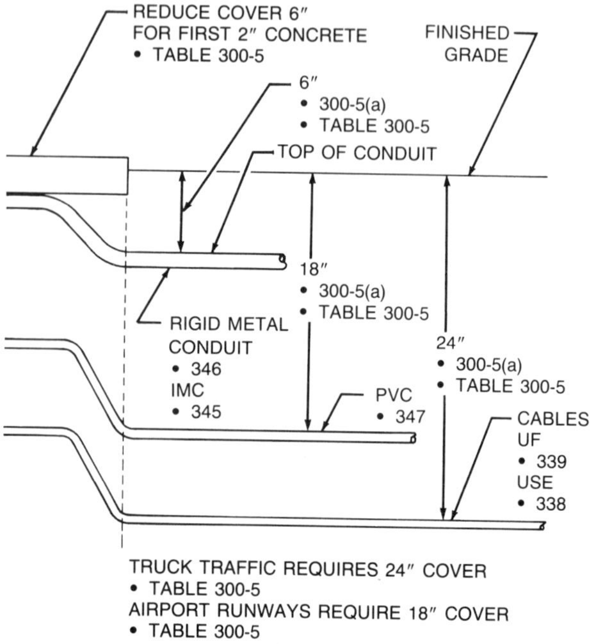

TABLE 300-5
One branch circuit rated 30 A and 300 V or less requires 12″ covering
30 V or less circuits require a 6″ covering

Figure 5-8. The minimum size lateral is #8 Cu or #6 Al. The covering of the wiring methods shall comply with Table 300-5.

OVERCURRENT PROTECTION—230, PART G

Overcurrent protection is provided to protect the service-entrance conductors from overloads or ground faults. Overloads occur where additional loads are added until the ampacity of the loads exceed the ampacity of the service-entrance conductors. *Short circuits* occur between phase conductors and *ground faults* occur between the phase conductors and a metal object. Overcurrent protection devices shall be sized from 240-3 and selected from 240-6.

Ungrounded Conductors—230-90(a)

Overcurrent protection devices shall be placed in each ungrounded hot phase conductor. This overcurrent protection device is sized and selected on the ampacity of the conductors, which is determined by the load served in the building. The rating of the overcurrent protection device shall not exceed the ampacity of the service-entrance conductors. However, 240-3, Ex. 4 allows the overcurrent protection device to be selected and sized on the ampacity of the service-entrance conductors.

For example, three #4/0 THWN Cu conductors have an ampacity of 230 A per Table 310-16. Section 240-3, Ex. 4 allows a 250 A overcurrent protection device to be selected from 240-6 based upon the 230 A of the #4/0 THWN Cu conductor. Section 430-63 allows the overcurrent protection device to be sized greater than the service-entrance conductors when supplying a motor load and a lighting or lighting and appliance load. The overcurrent protection device is selected and sized per 430-62.

What size overcurrent protection device is required for service-entrance conductors supplying a 100 HP, 75 HP, and 60 HP, 460 V, 3ϕ group of motors, 90 A lighting load, and 185 A receptacle and appliance load? (All loads are continuous duty. Use TDFs.)

Step 1. Motor loads. *Table 430-150:* 100 HP = 124 A
75 HP = 96 A
60 HP = 77 A

Step 2. Lighting, receptacle, and appliance loads. *220-10(b):*
Lighting load = 90 A × 125% = 112.5 A
Receptacle and appliance load = 185 A × 125% = 231.25 A

Step 3. Selecting overcurrent protection device. *430-62(a); 430-63; Table 430-152:*
124 A × 175% = 217 A = 225 A OCPD
225 A + 96 A + 77 A + 112.5 A + 231.25 A = 741.75 A

Step 4. *240-3, Ex. 6; 240-6:*
700 A is the next lower standard size

Answer. **OCPD = 700 A**

The ungrounded service conductors extend from the meter base or CT can to the first overcurrent protection device in the switchgear, service panel, or lighting and appliance panelboard. See Figure 5-9.

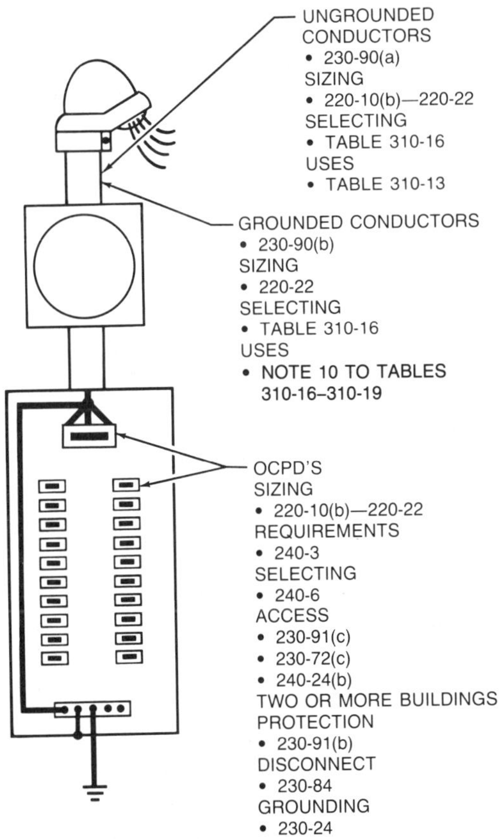

Figure 5-9. Each ungrounded phase conductor requires an overcurrent protection device. The devices shall be grouped in a way that will allow access by occupants.

Grounded Conductor—230-90(b)

The size of the neutral grounded conductor is determined by calculating all the 120 V loads of the system and adding them together to get the total kVA or ampacity. The grounded neutral conductor is then selected from Table 310-16 based upon the type of insulation used. The grounded neutral conductor shall be capable of carrying all the unbalanced currents from the phase conductors per 220-22. For example, if phase A carries 100 A and phase B carries 70 A, the unbalanced current the neutral would carry is 30 A.

The neutral is rated at 100 A based on phase A. After all 120 V loads are calculated and the ampacity of the neutral conductor is determined, the size of the neutral conductor shall be equivalent to the grounding electrode conductor per 250-23(b). In addition, it shall be at least 12½% of the largest phase conductor of phase A, B, or C for 3ϕ systems and phase A or B for 1ϕ systems where any arrangement of phase conductors exceeds #1100 kcmil for copper or #1750 kcmil for aluminum.

Section 230-90(b) does not allow an overcurrent protection device to be placed in the grounded service neutral conductor unless it simultaneously opens all conductors of the circuit. A 2-pole or 3-pole circuit breaker could be used because if one phase should short out or a ground fault should occur from any phase to ground, the overcurrent protection device would trip simultaneously, disconnecting all conductors of the circuit. Section 230-90(b) correlates with 240-22, Ex. 1, which lists the same requirements. If the service equipment supplying a motor is a 3-wire, corner-grounded delta system, the grounded conductor could be protected by a circuit breaker or fuse per 430-36. Section 240-22 permits the grounded conductor to be protected by a fuse or circuit breaker per 430-36.

Two or More Buildings—230-91(b)

Where two or more buildings are supplied with a feeder circuit from the main building, the overcurrent protection in the main building can serve as the protection for the feeder to the other buildings per 230-91(b).

What size overcurrent protection device is required in the service equipment of the main building for a feeder circuit with #2/0 THWN Cu conductors routed underground to a panelboard located in the second building approximately 100′-0″ away?

Step 1. *Table 310-16:* #2/0 conductors = 175 A

Step 2. *230-91(b); 240-6:* 175 A OCPD is required

Answer. **OCPD = 175 A**

Disconnect—230-84

A disconnecting means is required at the second building to disconnect the power circuit conductors routed from the service equipment located in the main building. This disconnecting means is required even though the overcurrent protection in the service equipment located in the main building is protecting the circuit conductors per 230-91(b). This disconnecting means shall be located on the outside of the building or immediately after entering the building per 230-70(a). The disconnecting means required by 230-84 can be a nonautomatic circuit breaker, a nonfusible disconnect, or an overcurrent protection device that provides additional overcurrent protection at the building. The panel in the building provides overcurrent protection for the branch circuits.

Grounding—250-24

Section 250-24 deals with two types of circuits going to an auxiliary building from a main building. These circuits are a single branch circuit and a feeder circuit. A single branch circuit can have an equipment ground pulled with it and a grounding electrode conductor does not have to be established to a grounding electrode at the second building. However, a feeder circuit going out to an auxiliary building shall have a grounding electrode system installed if one is not present on the premises. For example, if there is no water pipe, no building steel or concrete-encased electrode present, a driven rod may be used per 250-83 and 250-84.

If the grounding electrode conductor is connected to a concrete-encased electrode, the maximum size of the grounding electrode conductor is #4 per 250-81(c). If a grounding electrode conductor is connected to a ground ring circling the building, the maximum size grounding electrode conductor is #2 per 250-81(d). The size of the grounding electrode conductor is determined from Table 250-94 based upon the largest phase conductor in the power circuit from the service equipment in the main building to the panelboard located in the second building. The grounding electrode conductor is connected to a metal water pipe or building steel per 250-81(a) and (b). See Figure 5-10.

GROUND-FAULT PROTECTION—230-95

Ground-fault protection shall be provided for service equipment where the supply service from the utility is a solidly grounded wye system with a voltage of more than 150 V to ground but not exceeding 600 V phase-to-phase. This requirement pertains to a 480 V, 277 V, 4-wire wye-connected system. This requirement does not pertain to a 120/240 V, delta-connected system or a 120/208 V wye-connected system.

Ground-fault protection of services will not protect the conductors on the supply side of the service disconnecting means but will protect the circuit conductors on the load side of the service disconnecting means from line-to-ground faults. If the ground faults are low magnitude with a small amount of current flowing to ground, a large overcurrent protection device will not trip open without adding ground-fault protection.

Two basic types of ground-fault protection equipment are used in the electrical industry. The *window* type requires all phase conductors and grounded conductors to be run through the sensor relay. See Figure 5-11. The *donut* type is inserted around a bonding jumper from the equipment grounding bar, which is bonded to the service enclosure and extended to an isolated grounded terminal bar. All neutral conductors are terminated to this bar. Any fault current (partial short or bolted ground fault) that leaks from a phase conductor to ground will be sensed. The current will travel over the metal raceways or equipment grounds to the grounded terminal and through the donut current relay to the grounded neutral bar. The current relay will pick up this excess current and trip the overcurrent protection device open, clearing the circuits. See Figure 5-12.

Setting of 1200 A or Less—230-95(a)

The maximum setting is 1200 A. There is no minimum requirement, but it should be noted that if a set-

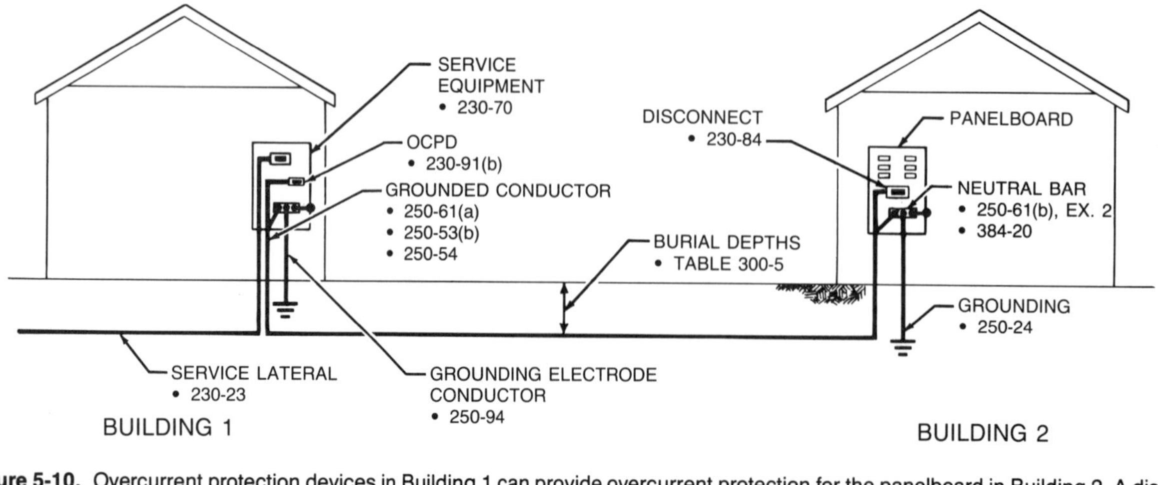

Figure 5-10. Overcurrent protection devices in Building 1 can provide overcurrent protection for the panelboard in Building 2. A disconnecting means shall be provided at Building 2.

RELAY REQUIREMENTS	
SETTING	CLEARING TIME
0 A to 1200 A	Clear a fault of 3000 A or more in 1 second

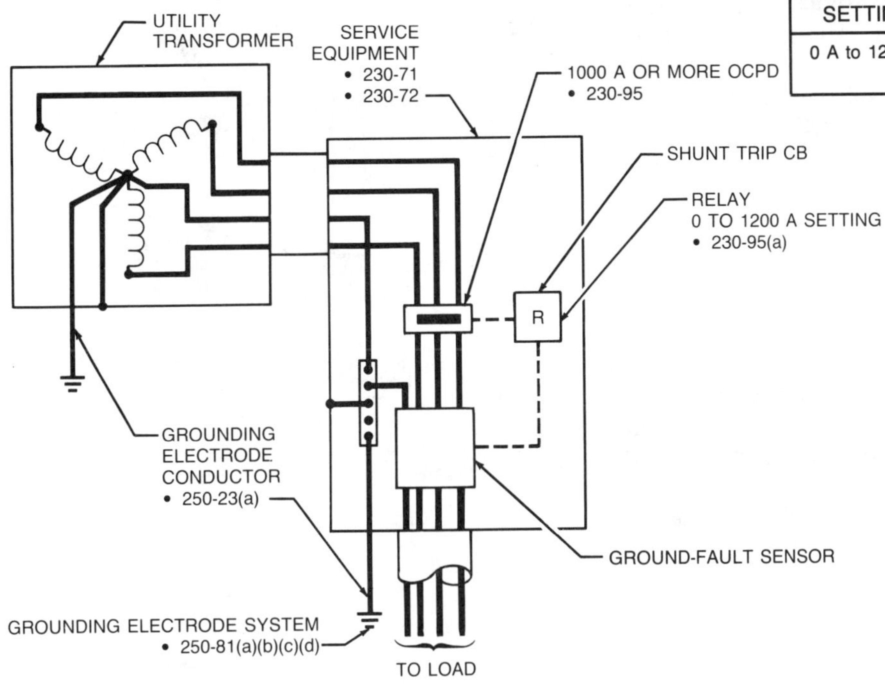

Figure 5-11. Ground-fault protection shall be provided with 277/480 V, 1ϕ, 4-wire services with overcurrent protection devices of 1000 A or more. A ground-fault sensor (window) can be used to encircle all service conductors, including the neutral.

RELAY REQUIREMENTS	
SETTING	CLEARING TIME
0 A to 1200 A	Clear a fault of 3000 A or more in 1 second

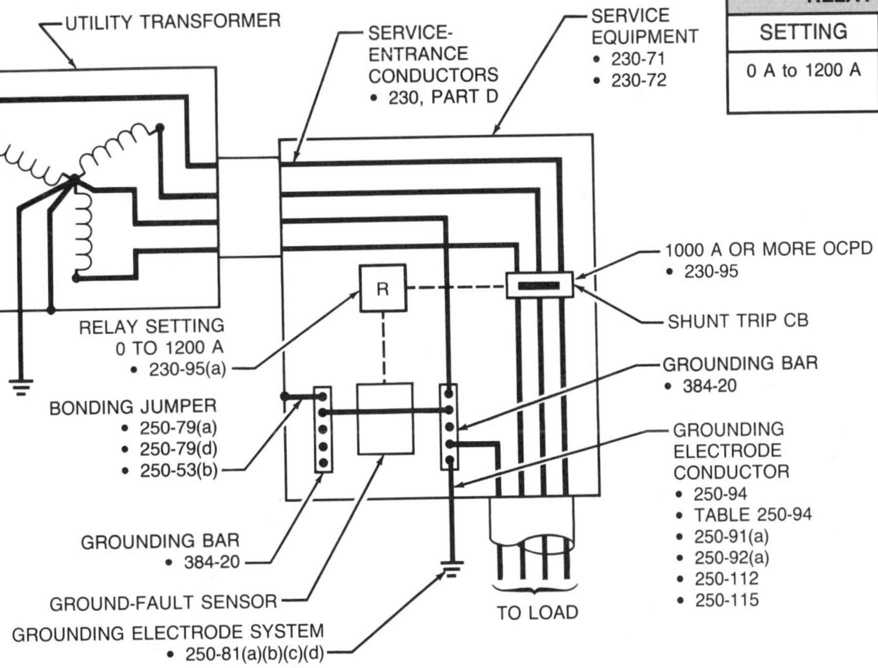

Figure 5-12. A ground-fault sensor encircling a bonding jumper conductor can be used to provide ground-fault protection for services of 1000 A or more supplied by 277/480 V systems.

ting is lower than 1200 A the likelihood of unwanted shutdowns might occur. The ground-fault protection equipment shall be capable of clearing a ground fault that is equal to or greater than 3000 A. This minimizes the amount of damage that could be done by an arcing ground fault.

Clearing Time of One Second—230-95(a)

The ground-fault system shall clear a fault of 3000 A or greater within one second to prevent damage to overcurrent protection devices, busbars, and service equipment. A study shall be made to determine the circuit impedance in short-circuit currents that would be available at the supply terminals so that the equipment and overcurrent protection of the proper interrupting rating is installed per 110-9 and 110-10.

Ground-fault protection is not required for disconnects rated less than 1000 A. For example, six 800 A disconnects may be used without installing ground-fault protection. Disconnecting means rated 1000 A or more are required to have ground-fault protection if supplied by a service having more than 150 V to ground with phase conductors rated less than 600 V phase-to-phase. Note that if a 900 A fuse is installed in a 1200 A disconnect, ground-fault protection is required based upon the 1200 A disconnect per 230-95.

METER BASE—90-2(b)(5)

Meter base installation requirements are regulated by the utility company. Meter bases are usually set at a maximum height of not more than 6′-6″ and a minimum height of not more than 4′-0″. However, this height can vary according to the local utility and local electrical ordinances. Meter bases should match the service-entrance conductors and the load of the service equipment. They are usually required to be located outside without special permission from the utility and authority having jurisdiction. The high leg of an open- or closed-delta system is required to be terminated to the right terminal in the meter base per 230-56, 384-3(e)(f), and 215-8.

The corner-grounded conductor of a corner-grounded delta system is usually terminated in the center phase in the meter base. The length of the conductors between a meter base and CT can be limited by the local utility. CTs located in CT enclosures are used to meter the service entrance where the service-entrance conductors are too large to terminate to the terminals in the meter base. Refer to the local guidelines of the utility or the local city electrical ordinance to determine the location and height of meter bases. See Figure 5-13.

■ *Master Electrician's Test Tip.* Some examination boards require persons taking tests to know the location and heights of meter bases per local codes.■

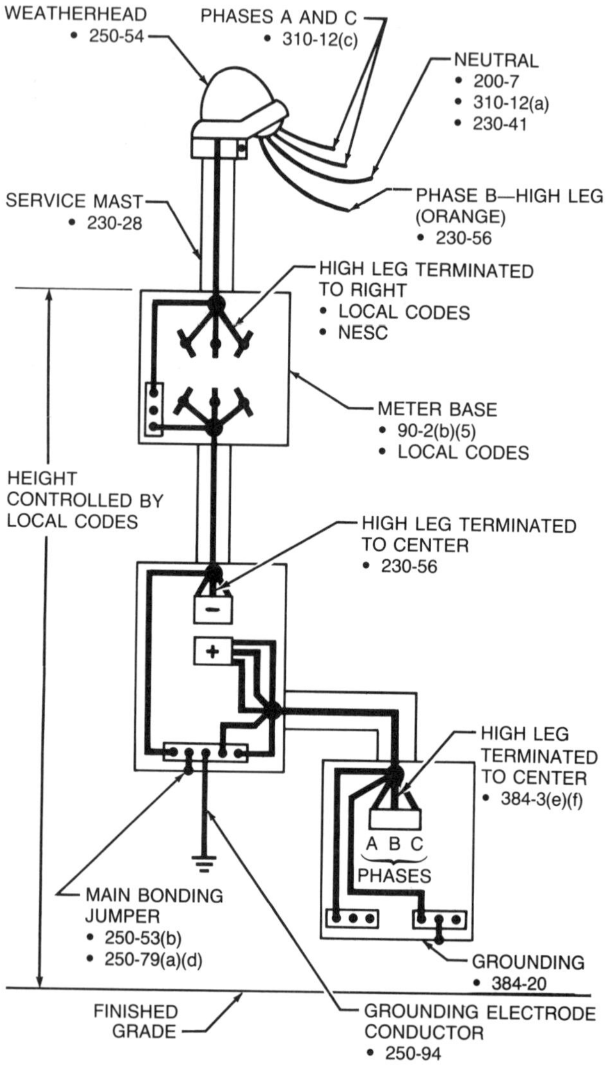

WYE SERVICE		
PHASES	TERMINAL	COLOR CODE RECOMMENDED
A	Left	Low V – Black High V – Brown
B	Center	Low V – Red High V – Orange
C	Right	Low V – Blue High V – Yellow
DELTA SERVICE		
A	Left	Black
B High Leg	Center	Orange
C	Right	Blue

Note: See 310-12.
Part A – Grounded conductors
Part B – EGC
Part C – Ungrounded conductors

Figure 5-13. The maximum height of meter bases is regulated by local codes. The high leg of an open- or closed-delta system shall be terminated to phase C (right leg) in a self-contained meter base. The high leg shall be terminated to phase B (center leg) at the service equipment.

WIRING METHODS—230-43

Section 230-43 lists the types of conductors that can be used as service-entrance conductors to supply power from the meter base or CT can to the service disconnecting means in the service equipment. Service-entrance conductors terminated to the lugs of a panelboard or switchgear shall have enough ampacity to supply the total calculated load. However, if the service-entrance conductors are connected to the terminals of a main overcurrent protection device or fusible disconnect, they shall be large enough to supply the rating of that overcurrent protection device. An in-line main is the only proper way to protect a panelboard or switchgear from overload conditions.

Open Wiring on Insulators

Open wiring on insulators is an exposed wiring method that shall not be concealed by the structure, finish, or the building. Where exposed to weather the insulation shall be approved for wet locations. Such types of insulation are RHW, THW, TW, THWN, and XHHW per Table 310-16 and Table 310-13. Section 310-8(a) lists the types of insulation approved for general-use conductors in wet locations. Open wiring on insulators is routed on the outside of a building where taps are made to a weatherhead for the service-entrance conductors to be routed through a riser to a meter base and from the meter base to the service equipment disconnecting means on the outside or inside of the building.

Rigid Metal Conduit—230-43

Rigid metal conduit is the oldest and most reliable wiring method of all the wiring methods listed in 230-43. It can be installed for the service-entrance conductors where other wiring methods are prohibited because of physical damage that could occur. Table 346-12 requires ½″ to ¾″ rigid metal conduit to be supported at intervals not exceeding 10′-0″ and within 3′-0″ of every junction box and fitting. One inch conduit shall be supported at intervals not exceeding 12′-0″ and also within 3′-0″ of each outlet box, junction box, cabinet, or fitting.

Rigid metal conduit of 1¼″ to 1½″ is supported at intervals not exceeding 14′-0″ with the 3′-0″ minimum also required at each outlet box, junction box, cabinet, or fitting. For sizes 2″ through 2½″, supports shall be provided at intervals not exceeding 16′-0″ and for 3″ and larger, supports shall be provided at intervals not exceeding 20′-0″. Supports shall be provided within 3′-0″ of each outlet box, junction box, cabinet, or fitting. Some utilities and local city codes only allow rigid metal conduit to be used as a riser per 230-28.

IMC—230-43

Intermediate metal conduit (IMC) is similar to rigid metal conduit, except it has a thinner wall and is lighter in weight. IMC can usually be distinguished from rigid metal conduit because it has a bright finish. IMC is supported per the requirements of Table 346-12. It can be used in the same manner as rigid metal conduit except as a riser to house service-entrance conductors according to some local codes.

EMT—230-43

Electrical metallic tubing (EMT) can be used to house service-entrance conductors but cannot be used as a riser. In most locations it cannot be installed underground because soil conditions can deteriorate the galvanized coating, and short circuits can occur between phase conductors and the conduit or earth. EMT cannot be exposed to physical damage without proper protection. For example, EMT cannot be run down a wall and connected to PVC underground by a driveway without guideposts. EMT shall be supported within 3′-0″ of every outlet box, junction box, cabinet, and fitting and at intervals not exceeding 10′-0″ in length. See Figure 5-14.

SE Cable—230-43

Service-entrance cable (SE) is a single conductor or multiple conductor assembly with or without an overall covering. It can be used as a service riser but not as a service mast. Type *USE* service-entrance cable can be used underground where identified for such use. Service-entrance cable shall be supported by straps or other approved means within 12″ of every service head. Additional supports shall be provided along the service-entrance cable at intervals not exceeding 30″ and within 12″ of the meter per 230-51.

Wireways—230-43

Wireways are sheet metal enclosures equipped with hinged or movable covers in which conductors can be laid in place after the wireway has been installed as a complete wiring method. They are manufactured in lengths of 1′ to 10′ and in various widths and depths.

Couplings, elbows, end plates, and accessories can be purchased to connect the wireway together to route it from point to point. Wireways shall be supported at intervals not exceeding 5′-0″ unless specifically approved for support at greater intervals. The distance between supports shall never exceed 10′-0″. Section 362-8 requires wireways extended through walls to be unbroken lengths with no removable covers. See Figure 5-15.

Busways—230-43

Busways are grounded metal enclosures containing factory-mounted bare insulated conductors, which are usually copper or aluminum bars. They are used to supply power to switchgears and larger pieces of equipment where service-entrance conductors from Table 310-16 are not feasible or economical. Busways shall

SIZING CONDUIT USING SAME SIZE CONDUCTORS AND INSULATION

What size EMT is required for five #12 THWN Cu conductors?

Step 1. *Table 3B, Ch. 9:*
½″ EMT holds ten #12 THWN Cu

Answer. **EMT = ½″**

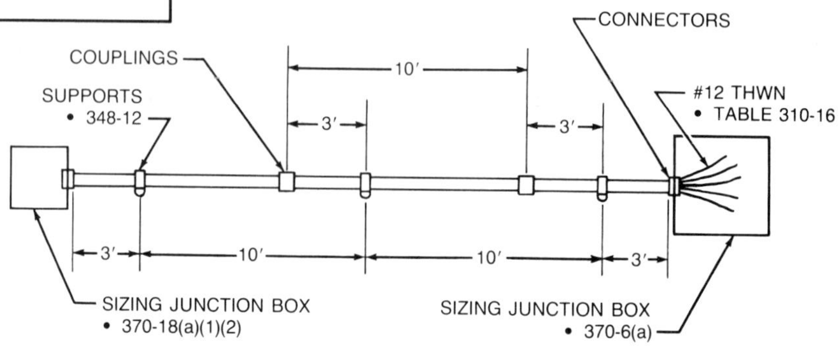

Figure 5-14. Conduit size for conductors of the same size with the same type of insulation is selected from Table 3A, 3B, or 3C, Ch. 9. Conduit for conductors with different types of insulation is selected from Tables 4 and 5, Ch. 9 based on square inch areas.

SIZING WIREWAYS USING COMBINATION CONDUCTORS WITH DIFFERENT INSULATION

What size wireway is required to contain four #1/0 THWN, four #2/0 THWN, four #3/0 XHHW, six #4/0 THW, and three #1 THW Cu conductors?

Step 1. *Table 5, Ch. 9:*

#1/0 THWN =
.1893 sq. in. × 4 = .7572 sq. in.
#2/0 THHN =
.2265 sq. in. × 4 = .906 sq. in.
#3/0 XHHW =
.2715 sq. in. × 4 = 1.086 sq. in.
#4/0 THW =
.3904 sq. in. × 6 = 2.3424 sq. in.
#1 THW =
.2027 sq. in. × 3 = .6081 sq. in.
Total = 5.6997 sq. in.

Step 2. *362-5:* Multiplier $= \frac{100}{20} = 5$

5.6997 sq. in. × 5 = 28.4985 sq. in.

Step 3. Checking multiplier.

$$\frac{5.6997 \text{ sq. in.}}{20\%} = 28.4985 \text{ sq. in.}$$

Step 4. *Chart:* 6″ × 6″ wireway = 36″

Step 5. Checking wireway.
36″ wireway × 20% fill = 7.2 sq. in.
7.2 sq. in. is greater than 5.6997 sq. in.

Answer. **Wireway = 6″ × 6″**

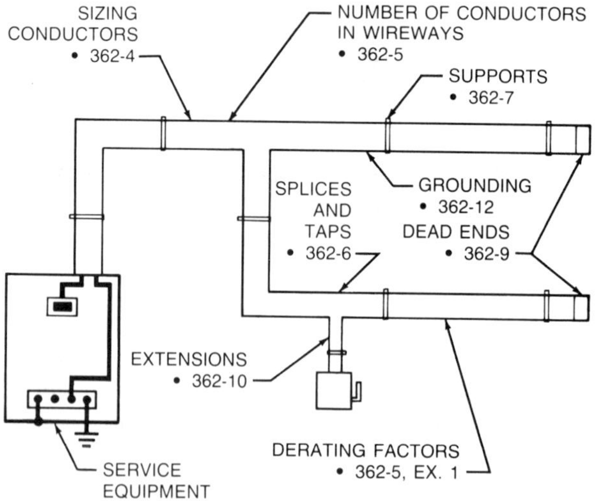

Figure 5-15. Wireways are sized from the square inch area of each conductor per Table 5, Ch. 9 based on the type of insulation of each conductor.

be supported at intervals not exceeding 5′-0″. Busways can be supported at greater intervals only if so marked by the manufacturer.

Auxiliary Gutters—230-43

Auxiliary gutters with associated fittings are identical to wireways. They are not permitted to extend more than 30′-0″ from the equipment to which they are connected per 374-2. Service-entrance conductors are routed to gutters when tapped and run to meter bases, service disconnecting switches, or other types of electrical equipment. Auxiliary gutters shall be supported at intervals not exceeding 5′-0″ in length. See Figure 5-16.

PVC—230-43

Two types of rigid nonmetallic conduit (PVC) are *Schedule 40* and *Schedule 80*. Schedule 40 cannot be exposed to physical damage. Table 347-8 lists the requirements for supporting PVC. It shall be supported within 3′-0″ of each box, cabinet, or other conduit termination.

PVC is supported at closer intervals than rigid metal conduit and IMC because when exposed to heat, the conduit tends to bow and will back out of its couplings and connectors. It will not reset when the conduit cools. See Figure 5-17.

MC Cable—230-43

Metal-clad cable (MC) is a factory assembly of one or more conductors, with each conductor individually

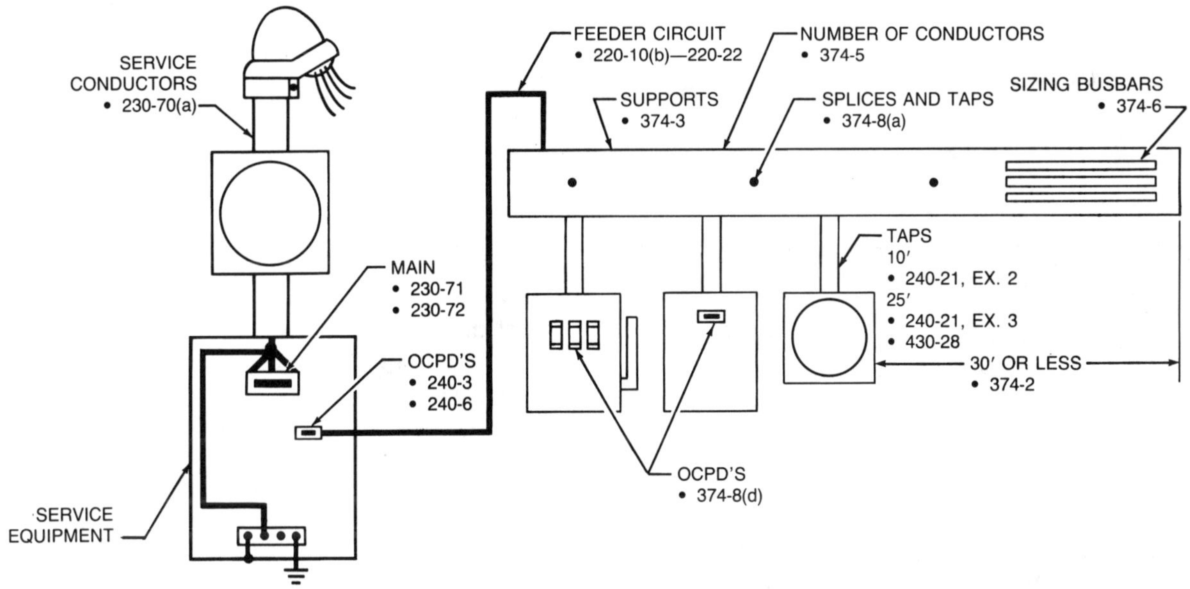

SIZING GUTTERS USING COMBINATION CONDUCTORS WITH DIFFERENT INSULATION

What size gutter is required to contain four #350 kcmil THHN, four #1/0 THW, four #2/0 XHHW, and four #3/0 RHW Cu conductors?

Step 1. *Table 5, Ch. 9:*

#350 kcmil THHN =
.5307 sq. in. × 4 = 2.1228 sq. in.

#1/0 THWN =
.2367 sq. in. × 4 = .9468 sq. in.

#2/0 XHHW =
.2265 sq. in. × 4 = 0.906 sq. in.

#3/0 RHW =
.4151 sq. in. × 4 = 1.6604 sq. in.

Total = 5.636 sq. in.

Step 2. *374-5:* Multiplier = $\frac{100}{20} = 5$

5.636 sq. in. × 5 = 28.18 sq. in.

Step 3. Checking multiplier.

$$\frac{5.636 \text{ sq. in.}}{20\%} = 28.18 \text{ sq. in.}$$

Step 4. *Chart:* 6″ × 6″ = 36″

Step 5. Checking gutter.
36″ gutter × 20% fill = 7.2 sq. in.
7.2 sq. in. is greater than 5.636 sq. in.

Answer. **Gutter = 6″ × 6″**

Figure 5-16. Gutters are sized by applying a multiplier times the total square inch area of the conductors. The square inch area of each conductor is determined per Table 5, Ch. 9.

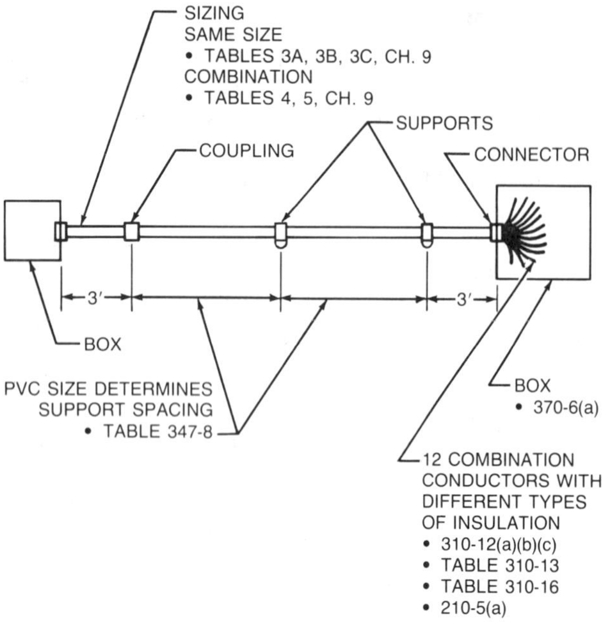

SIZING CONDUIT USING COMBINATION CONDUCTORS WITH DIFFERENT INSULATION

What size PVC is required for three #4 THWN, three #2 RHW, three #3 XHHW, and three #1 THW Cu conductors?

Step 1. *Table 5, Ch. 9:*

#4 THWN =
.0845 sq. in. × 3 = .2535 sq. in.
#2 RHW =
.2067 sq. in. × 3 = .6201 sq. in.
#3 XHHW =
.0995 sq. in. × 3 = .2985 sq. in.
#1 THW =
.2027 sq. in. × 3 = .6081 sq. in.
Total = 1.7802 sq. in.

Step 2. *Table 4, Ch. 9:*
1.780 sq. in. requires 1.92 sq. in. for 40% fill column

Answer. **PVC = 2½″**

Figure 5-17. Conduit size for combination conductors is selected from the total square inch area of the conductors per Table 5, Ch. 9.

insulated and enclosed in a metallic sheath of interlocking tape or a smooth or corrugated tube. MC cable is available in three designs: interlocking and metal tape, corrugated tube, and smooth tube. It is used for installation above ground unless marked for direct burial. MC cable suitable for use in cable trays, direct burial, or indirect sunlight is so marked.

Conductors in MC cables are type *CS* (copper sheath) and *ALS* (aluminum sheath). MC cable is rated for use up to 5000 V and listed in sizes #18 or larger for copper and #12 or larger for aluminum. It shall be supported and secured at intervals not exceeding 6′-0″ per 334-10(a). The armor of interlocking MC cable shall not be used as an equipment ground alone but may be used to supplement the interior equipment grounding conductor.

Flexible Metal Conduit (Greenfield)—230-43

Flexible metal conduit (Greenfield) can only be used in a service run not exceeding 6′-0″ in length. Used in this manner, a bonding jumper shall be provided on the outside of the Greenfield to bond each end to the conduits. Where this is not feasible, a bonding jumper may be pulled and routed inside the conduit out to the bonding point and inside the service to the bonding point. The size of the bonding jumper is determined by Table 250-94 based upon the largest phase conductor in the service run. Sections 250-79(d) and 250-79(f) list these requirements. A piece of Greenfield can be used to offset around an object when it is impractical to bend a saddle. The same requirements apply for liquidtight flexible nonmetallic conduit.

SERVICE EQUIPMENT—230, PART E

The service equipment is the heart of the distribution system. It can be compared to a tree. The service equipment, with its overcurrent protection devices protecting each load, serves as the trunk. The wiring that extends out from the overcurrent protection devices to the individual loads serves as the branches of the tree. The leaves that are connected to each branch extending out from the tree represent the loads.

The service equipment enclosure contains all disconnects that disconnect all conductors from the service-entrance conductors in a building or other type of structure. Feeder circuits with overcurrent protection devices installed in the service equipment can be routed throughout the building where taps can be made to the various loads that need to be served. Taps to various types of loads are either 10′, 25′, 100′, or 25′ including the primary tap transformer plus secondary conductors per 240-21, Ex. 1, 2, 3, 8, or 11.

Feeder circuits can also be used with overcurrent protection in the panelboard to feed subpanels at various locations in the building. From these subpanels, other feeders and branch circuits can supply loads in a central location to prevent routing feeder circuits or branch circuits back to the service equipment in one location.

Location—230-70(a)

The service disconnecting means shall be located either inside or outside the building at a readily accessible location nearest the point of entry of the service-entrance conductors. This provides easy access for the user of the service equipment to shut the service OFF in case of a problem. The NEC® does not list a maximum distance from the point of entry of the service-entrance conductors to a readily accessible location for the installation of the service disconnecting means. However, the length of the service-entrance conductors shall be kept to a minimum inside the building since power utilities provide limited overcurrent protection for the service-entrance conductors. A fault could occur in the service-entrance conductor that may ignite nearby combustible materials. The authority having jurisdiction shall determine this distance once the service-entrance conductors enter the building.

The entrance of the service-entrance conductors in a building can be from the roof, slab, or through the wall. When service-entrance conductors enter through the wall, the overcurrent protection device shall be placed as close as possible to the point of entry. When entering from the roof, the service-entrance conductors are run down the wall to a readily accessible location. Section 380-8(a) lists the maximum height of any overcurrent protection device as no higher than 6′-6″ from finished grade. This section also applies to service-entrance conductors entering through a slab. See Figure 5-18.

■ *Master Electrician's Test Tip.* By applying Section 230-6, the service-entrance conduit can have a covering of 2″ and the service-entrance conduit and its conductors would be considered outside of the building. With the service conduit covered by 2″ or more of concrete, there is no limitation on the distance service-entrance conductors can be run into a building. By placing an overcurrent protection device (fusible disconnect or main circuit breaker) beneath the meter base, the conductors run inside the building become a feeder circuit and can be run any distance and routed through the combustible interior parts of the building to the subpanel. The subpanel houses all breakers for branch circuit loads and feeders.■

Suitable—230-70(c)

Each service-entrance disconnecting means shall be suitable for the location in which it is installed. For example, if a service-entrance disconnecting means is located outside exposed to the weather, it shall be rated for damp and wet locations. If the service-entrance disconnecting means in equipment is placed inside the building and recessed into the wall, it shall be rated for recessed installation. If the service equipment is surface-mounted, the enclosure housing the service-entrance conductors and overcurrent protection devices shall be rated for exposed surface installations.

Marking—384-13, FPN; 230-70(b); 110-22

All service-entrance disconnects shall be plainly marked to identify the loads served. For example,

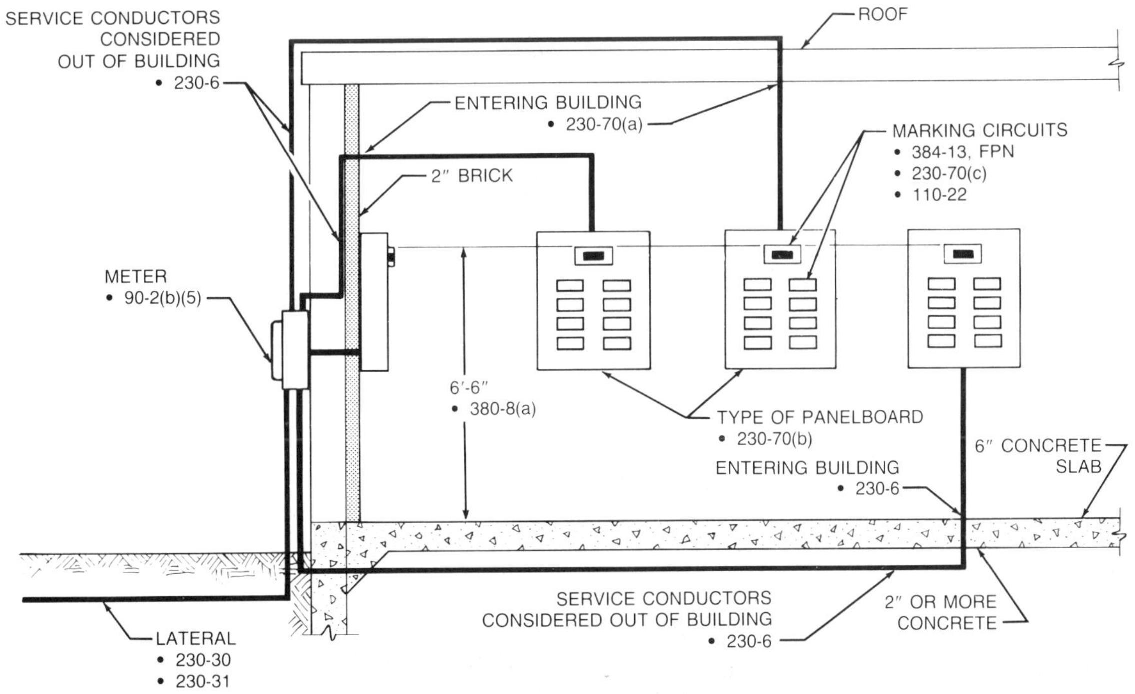

Figure 5-18. The service disconnecting means shall be installed as near the service conductor's point of entrance as possible per 230-70(a). Conductors are considered outside of a building per 230-6.

where two 240 V, 2-pole breakers or 3ϕ, 3-pole breakers are used to supply heating, air conditioning, water heaters, heating units, and similar types of loads, the disconnecting means shall be plainly marked so the user of these disconnects will know what loads are protected by each individual overcurrent protection device. Where 15 and 20 A overcurrent protection devices are used to supply lighting loads, appliance loads, and receptacle loads, 110-22 requires these branch circuits to be identified for the loads served.

Height—380-8(a)

The handle of the circuit breaker or disconnect switch, when used as a service equipment disconnecting means, shall be located so that it is readily accessible and operable from the floor at a height of no more than 6′-6″ in the ON position. Section 380-8(a) correlates with 240-24(a), which requires all overcurrent protection devices in panelboards to be readily accessible without having to climb over chairs or move furniture or other objects that would prevent ready access to the operation of the overcurrent protection device.

Section 230-91(a) also correlates with these sections. It requires the service overcurrent protection devices to be located at a readily accessible location nearest to the point of entry of the service-entrance conductors. If there is no readily accessible area and no available space for the service disconnecting means to be located ahead of the overcurrent protection devices, the authority having jurisdiction shall judge how far the service-entrance conductors can enter into the building before overcurrent protection shall be placed outside beneath the meter base or outside by the CT can per 230-70(a).

Numbering—230-71

One set of overhead or lateral service-entrance conductors is allowed to supply two to six disconnecting means instead of one single main disconnecting means. A single-occupancy building can be served by one main service disconnecting means or up to six disconnecting means that supply separate loads in the building. For two to six sets of fuses or circuit breakers, the total rating of the multiple-occupancy overcurrent protection devices is not required to match the ampacity of the service-entrance conductors per 230-80. Section 230-90(a), Ex. 3 permits not more than six circuit breakers or six sets of fuses to be considered as the overcurrent protection device. However, none of these individual overcurrent protection devices can have a rating or setting higher than the ampacity of the service-entrance conductors.

By applying these rules, it is possible for the total of the six overcurrent protection devices, fuses, or circuit breakers to be greater than the rating of the service-entrance conductors. Note that the size of the service-entrance conductors is required to be adequate for the connected load in each individual service disconnecting means. The service equipment shall be disconnected completely with six throws of the hand, whether there is one main overcurrent protection device in one enclosure or two to six overcurrent protection devices in the enclosure.

The service-entrance conductors can terminate in a gutter or SE can for six individual disconnects with fuses or six individual enclosures with circuit breakers, or there can be any combination of circuit breakers or fuses. In any case, the service equipment shall be disconnected with six throws of the hand or less. Section 230-82 has exceptions that allow taps ahead of the main where two to six more disconnecting means can be added for each particular exception. See Figure 5-19.

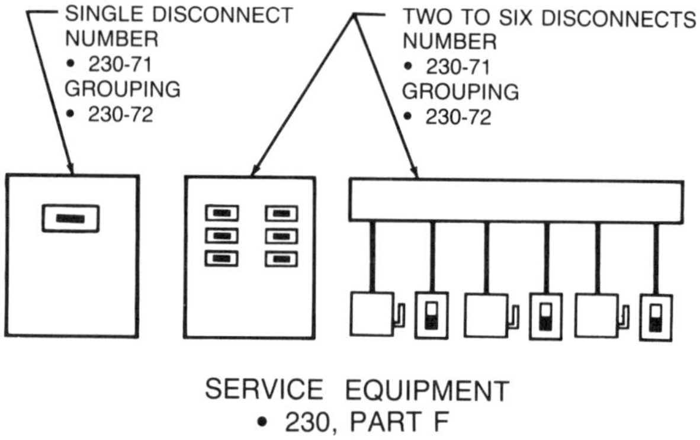

Figure 5-19. The service disconnecting means can be one to six mains in one enclosure or two to six separate disconnects grouped in one location.

Grouping—230-72

The two to six disconnects permitted in 230-71 are required to be grouped together so personnel will know where the disconnecting means is located for the service. Note that where a building is permitted to have more than one service drop or lateral per 230-2, each service shall have a service disconnecting means grouped in one location for each service. For this type of service, a directory shall be placed in each service location identifying the location of each service. This is for safety and to prevent someone from turning the disconnecting means OFF at one service location and then being surprised by a second service supplying power to equipment.

Access—240-24

Overcurrent protection devices for service-entrance conductors shall be an integral part of the service disconnecting means and located immediately adjacent to the service equipment. Sections 240-24(a) and 230-91(a) correlate to make this rule mandatory. *Readily accessible,* per 100, means capable of being quickly reached for operation, renewal, or inspection. The overcurrent protection device in the service equipment shall be accessible from the service equipment to the floor.

Single Occupancy—240-24(b); 230-91(a). For single occupancy, the service equipment shall be in a readily accessible location to the occupant of the building. The service equipment can be located beneath the meter base or CT can. It can also be in a garage, bedroom, living room, den, kitchen, or other location, providing ready access to the overcurrent protection devices. Section 230-70(b) requires that each main overcurrent protection device in the service equipment be identified for the loads served. Section 110-22 requires that all branch circuit overcurrent protection devices be identified for the loads served.

Multiple Occupancy—240-24(b), Ex.; 230-72(c), Ex.; 230-91(c), Ex. A multiple-occupancy building can have any number of dwelling units, offices, stores, or a combination of these independent of each other. Section 240-24(b) requires each occupant to have ready access to all overcurrent protection devices protecting the conductors supplying the occupancy. Section 240-24(b), Ex. allows the overcurrent protection device for a feeder or branch circuit to be placed in a location not accessible to each occupancy. To gain access, one shall go through management-supplied 24-hour per day personnel, or each occupancy shall have a key available for ready access to the service equipment.

Section 230-72(c), Ex. requires that each occupant have access to each occupancy service disconnecting means. This section correlates with 240-24(b), Ex. in requiring that each occupant have access to their overcurrent protection devices serving each individual load. Section 230-91(c), Ex. requires that individual occupants have access to their overcurrent protection devices. This exception also refers to 240-24(b) for the exception, which allows the overcurrent protection devices to be placed in one location. The management shall have personnel available to allow each occupant to have access. See Figure 5-20.

■ *Master Electrician's Test Tip.* The service disconnecting means, which can be six or fewer overcurrent protection devices located at each enclosure to supply individual loads, is required to be accessible to each occupant. If individual service-entrance conductors or feeders cannot be run to each occupancy per 230-6, the service disconnecting means with overcurrent protection devices can be installed in one location.■

Clearance—110-16(a); 110-34

Section 110-16(a) deals with service-entrance conductors rated at 600 V or less. Condition 1, 2, or 3 is applied in providing clearance in front of panelboards, switchgear, motor control centers, etc. Work space to the sides of service equipment shall not be less than 30″ wide in front of the electrical equipment. The work space shall also permit at least a 90° opening for the equipment doors or hinged panels. A problem occurs

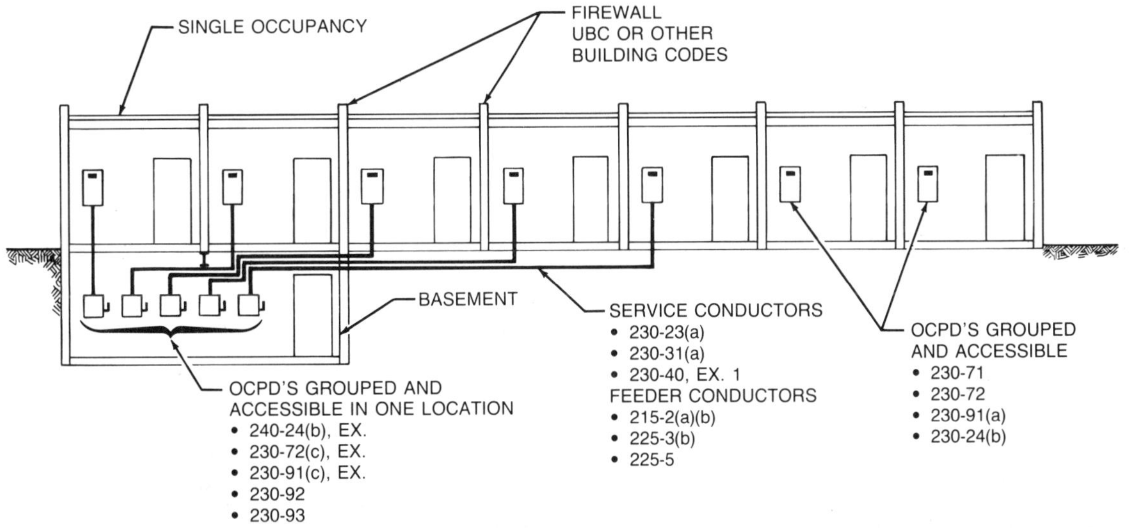

Figure 5-20. Overcurrent protection devices shall be grouped and accessible to each occupant or grouped in one location with each occupant having access. The overcurrent protection devices can be grouped in one location where management provides access for each occupant.

where the doors or hinged panels are wider than 30″. More than 30″ of space will be needed to provide enough room for the door to be opened at a 90° angle.

Condition 1 to Table 110-16(a) deals with the wall opposite the panelboard. This is an insulated wall such as drywall, wood, or other material that insulates the wall from the *grounded floor*. A grounded floor is connected by rebar to the foundation steel in the concrete on the grade-level floor. For example, where there is a wall of an insulated material that will insulate and isolate a person touching it from grounded objects, the minimum clearance in front of that panelboard for 0 V to 150 V is 3′-0″. Condition 2 is where the wall opposite the service is a grounded wall such as concrete, brick, or ceramic tile.

For 0 V to 150 V to ground, minimum clearance in front of the panelboard with a wall that is grounded or could connect a person's body touching it to ground is 3′-0″. Minimum clearance for 151 V to 600 V to ground is 3′-6″ for such an enclosure. For Condition 3 (operator between live parts on both sides of the work space), the minimum clearance for 0 V to 150 V to ground is 3′-0″. For 151 V to 600 V to ground, the minimum clearance is 4′-0″. See Figure 5-21.

Table 110-34(a) deals with service equipment for service-entrance conductors rated over 600 V. Condition 1, 2, or 3 also applies to service equipment with voltages over 600 V to ground. For example, a panelboard under Condition 1 with over 600 V to 2500 V to ground requires 3′-0″ clearance. A panelboard with over 25,000 V to 75,000 V to ground requires a 6′-0″ clearance per Condition 1. A panelboard with over 9000 V to 25,000 V to ground requires a clearance of 6′-0″ for Condition 2. Condition 3 requires a clearance of 12′-0″ for over 75,000 V to ground.

Installation—384-4; 710-9

Service equipment located in rooms or dedicated spaces shall have an exclusively dedicated space equivalent to the width and depth of the service equipment extending from the structural ceiling. No piping ducts or foreign equipment shall be installed in or passed through such a space or room. Four exceptions to 384-4 allow piping ducts and equipment to be located above the service equipment under certain conditions. Fine Print Note No. 2 to 384-4 denotes that it is not the intent of Section 384-4 to prohibit sprinkler protection for electrical installations.

Section 710-9 deals with the space above electrical equipment where the voltage of the service-entrance conductors is over 600 V. Pipes or ducts, foreign to the electrical equipment and that requires periodic maintenance or whose malfunction could endanger the operation of the electrical system, shall not be located above the service equipment. Protection shall be provided to avoid damage from condensation leaks and breaks in piping systems that are considered foreign systems to the electrical equipment. However, piping in other facilities shall not be considered foreign to the electrical equipment where they are provided for fire protection of the electrical installation and protection of the room or space. See Figure 5-22. *Note:* See 450-13 for installation of transformers. The dedicated space shall extend 25′ from the finished floor, or to the structural roof, whichever comes first.

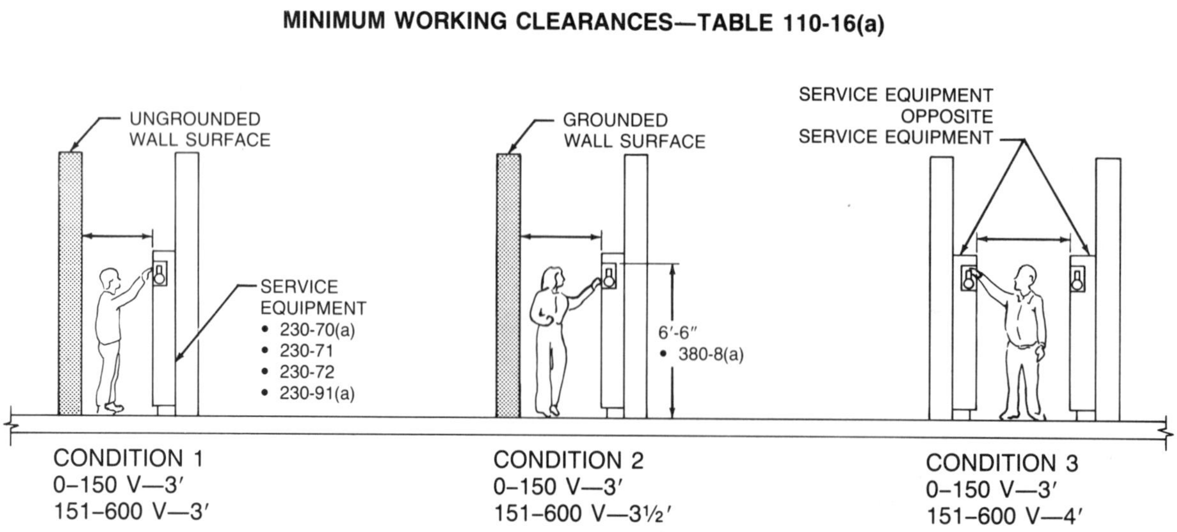

Figure 5-21. Clearances in front of service equipment are determined from Table 110-16(a) for 600 V or less and Table 110-34(a) for over 600 V.

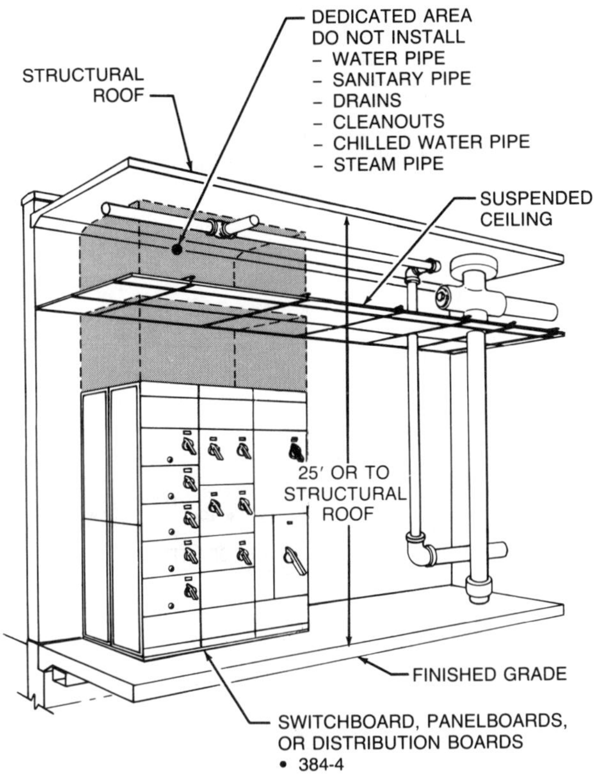

CLEARANCES	
IN FRONT OF EQUIPMENT	OVER TOP OF EQUIPMENT
600 V or less 110-16	600 V or less 384-4
Over 600 V 110-34	Over 600 V 710-9

Figure 5-22. Space above the switchboards, panelboards, and distribution boards shall be free from water pipes, sanitary pipes, steam pipes, and other pipes that can leak and deteriorate service enclosures.

Grounding Electrode System—250-81

Section 250-81 has four parts that shall be applied when connecting a grounding electrode system in a building or premise. See Figure 5-23. Part (a) deals with a metal underground water pipe. When there is at least 10′-0″ of metal water pipe in the ground, the metal underground water pipe shall be chosen as the grounding electrode. However, 250-80(a), which deals with metal water piping, requires the water piping system to be bonded with an additional grounding electrode. The additional grounding electrode can be building steel, a concrete-encased electrode, a ground ring or a driven rod, or a plate.

Service-entrance conductors are #250 kcmil THWN Cu. What size copper grounding electrode conductor is required to the metal underground water piping system that has 10′-0″ or more piping in the earth?

Step 1. *Table 310-16; 250-94:*
Service-entrance conductors = #250 kcmil THWN Cu

Step 2. *Table 250-94:*
Grounding electrode conductor = #2 Cu

Answer. **Grounding electrode conductor = #2 Cu**

What size supplementary copper grounding electrode conductor is required per 250-80(a) and 250-81(a) using a driven rod?

Step 1. *250-94, Ex. 1:*
Service-entrance conductors = #250 kcmil THWN Cu

Step 2. *250-94, Ex. 1:*
Grounding electrode conductor = #6 Cu

Answer. **Grounding electrode conductor = #6 Cu**

If there is not 10′-0″ of metal underground water pipe in the earth, the water piping system is no longer considered the grounding electrode. In most cases plastic water pipe is used to connect the water piping system in the premises to the utility system. After the PVC enters the building, copper or metal water piping is used to connect all fixtures throughout the premise. In this case, the metal water piping system shall not be chosen as the grounding electrode because it is isolated from the earth by the plastic conduit.

Should an electrical appliance of any type have a hot conductor connect to ground (ground fault) in any way to the metal water pipe isolated above grade, a person could receive an electrical shock or be severely hurt by an energized water line. Therefore, the metal water piping system shall be bonded to the grounding electrode system that is chosen per 250-81(b)(c)(d). The grounding electrode system shall be either metal water pipe, building steel, a concrete-encased electrode, or a ground ring. If these are not available, a driven rod is used. See Figure 5-24.

A plastic supply line is used to connect the utility's water system to the metal water piping in the building. Building steel is used as the main grounding electrode system. Service-entrance conductors are #400 kcmil THWN Cu. What size grounding electrode conductor, using copper, is required? What size supplementary ground is required to bond the water piping system into the grounding electrode system?

Grounding electrode conductor

Step 1. *250-80(a); Table 250-94:*
#400 kcmil THWN Cu requires #1/0 grounding electrode conductor

Answer. **Grounding electrode conductor = #1/0 Cu**

Supplementary ground

Step 1. *250-81(a); Table 250-94:*
#400 kcmil THWN Cu requires supplementary ground of #1/0

Answer. **Supplementary ground = #1/0 Cu**

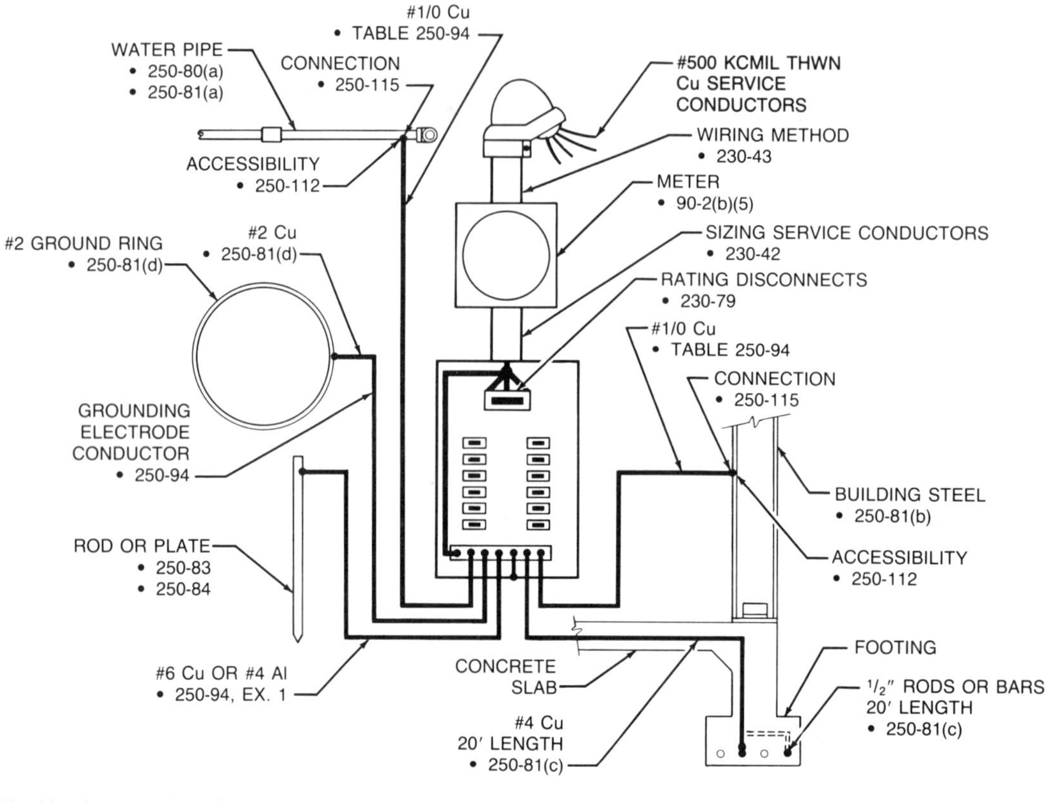

Figure 5-23. Metal water pipes, building steel, concrete-encased electrodes, ground rings, and driven rods or plates in a building structure shall be bonded together to form the grounding electrode system.

Part (b) to 250-81 pertains to metal frames of buildings used as grounding electrode systems. The grounding electrode conductor used to ground the service grounded terminal bar to building steel is sized from Table 250-94 based upon the largest phase conductor of the service entrance. No supplementary ground is required when the metal frame of the building is used as the grounding electrode. *Note:* All water piping is plastic.

Service-entrance conductors to a building are #3/0 THWN Cu. What size grounding electrode conductor is required to connect the grounded terminal bar in the service equipment to building steel?

Step 1. *250-81(b); Table 250-94:*
#3/0 THWN Cu conductors require #4 grounding electrode conductor

Answer. **Grounding electrode conductor = #4 Cu**

Part (c) of 250-81 deals with concrete-encased electrodes. A concrete-encased electrode can be ½″ rebar in 20′-0″ or more lengths either in one piece or spliced together or a #4 Cu conductor placed in the bottom or near the footing of the concrete foundation.

For concrete-encased electrodes, a #4 bare Cu conductor is the largest grounding electrode conductor required to connect the grounded terminal bar in the service equipment to the #4 bare conductor (grounding electrode) laid in the footing.

Part (d) to 250-81 deals with ground rings. A ground ring encircles a building or structure in direct contact with the earth at a depth of at least 2′-6″. The size of the ground ring conductor shall be at least #2 Cu. It shall circle the entire building and connect together. Where grounding electrode systems per 250-81(a), (b), (c), or (d) are not present in a building, a driven rod or plate is used per 250-83 and 250-84.

The largest grounding electrode conductor required to be connected to a driven rod or plate is #6 for copper and #4 for aluminum per 250-94, Ex. 1. A #6 Cu grounding electrode conductor is the largest required by the NEC® because the resistance to ground of a driven rod only has to be 25 ohms or less. By dividing 120 V to ground by 25 ohms, only 5 A will normally flow over the grounding electrode conductor. If 25 ohms or less is not obtained with a driven rod, a second rod shall be driven and located no closer than 6′-0″ from the first rod. Only two driven rods are required to be installed even if the resistance still measures over 25 ohms.

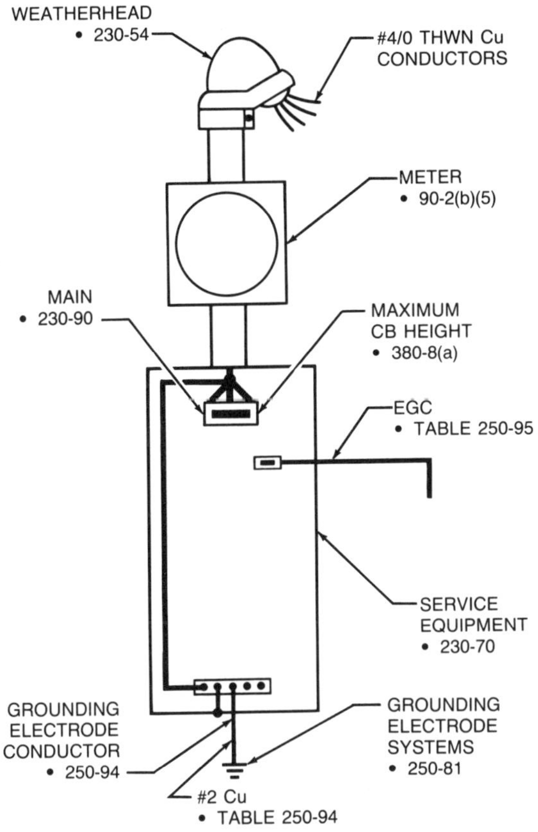

SIZING GROUNDING ELECTRODE CONDUCTOR WHERE KCMIL IS #1100 KCMIL OR LESS PER PHASE FOR COPPER

What size grounding electrode conductor is required to ground a service supplying #4/0 THWN Cu conductors? Use copper.

Step 1. *250-94:* Service-entrance conductors = #4/0 THWN Cu

Step 2. *Table 250-94:* #4/0 THWN requires #2 Cu

Answer. **Conductor = #2 Cu**

Figure 5-24. The grounding electrode conductor is sized from Table 250-94 based on the largest phase conductor of the service.

Grounding Electrode Conductor—250-94

The grounding electrode conductor is the grounding conductor that connects the grounded terminal of the service equipment to the grounding electrode or electrode system. Section 250-94 is used to size the grounding electrode conductor for connection to a water pipe system with at least 10′-0″ in the earth or to building steel connected to foundation steel. Section 250-81(c) requires the grounding electrode conductor to be at least #4 Cu when connected to a concrete-encased rebar or a #4 Cu conductor lying in the footing of the foundation.

Section 250-81(d) requires the grounding electrode conductor to be at least a #2 when connected to a #2 ground ring at least 2′-6″ in the earth surrounding the building. Section 250-92(a) lists the requirements for installing the grounding electrode conductor. A #4 or larger conductor, either copper or aluminum, used as a grounding electrode conductor shall be protected from physical damage where exposed to severe damage. A #6 grounding electrode conductor is also required to be protected from physical damage where exposed to only moderate physical damage. Section 250-91(a) requires the grounding electrode conductor to be unspliced from the service equipment to the connection with the grounding electrode. Exception 1 to 250-91(a) permits the grounding electrode conductor to be spliced to approved busbars.

Exception 2 to 250-91(a) allows the grounding electrode conductor to be spliced in a gutter, but from the grounded conductor spliced in the gutter, it shall be unspliced to the connection with the grounding electrode. Section 250-112 deals with splicing the grounding electrode conductor with a fitting to the grounding electrode. The connection of the grounding electrode conductor to a water pipe or building steel shall be accessible per 250-112 and be made in such a manner to provide a permanent and effective ground for the system.

The exception to 250-112 allows the connection of the grounding electrode conductor to the grounding electrode to be encased or buried when connected to a driven rod, concrete-encased electrode, or a ground ring. Note that ground clamps and other connections that are suitable for use are required to be used when embedded in concrete or the earth. Section 250-115 requires the connection of the grounding electrode conductor to the grounding electrode to be made with approved fittings, such as lugs, pressure connectors, listed clamps, or other approved means. Note that exothermic welding may be used. For example, a cadweld may be used to connect the grounding electrode conductor to the grounding electrode. See Figure 5-25.

Bonding—250-79(d)

Section 250-71(a) lists the items of the service equipment that shall be bonded together. These items are the service-entrance raceway, all service-entrance enclosures (junction boxes, etc.), meter bases, and other metal enclosures in the service equipment.

Section 250-72 lists the methods in which the items listed in 250-71(a) can be bonded. Bosses or threaded hubs into which rigid metal conduit or IMC can be threaded are considered an approved bonding means. Approved bonding locknuts and bonding bushings can be used.

Wedges with metal screws can be used to bond the conduit to the service equipment. Two locknuts and a bushing with a lug can be used with a bonding

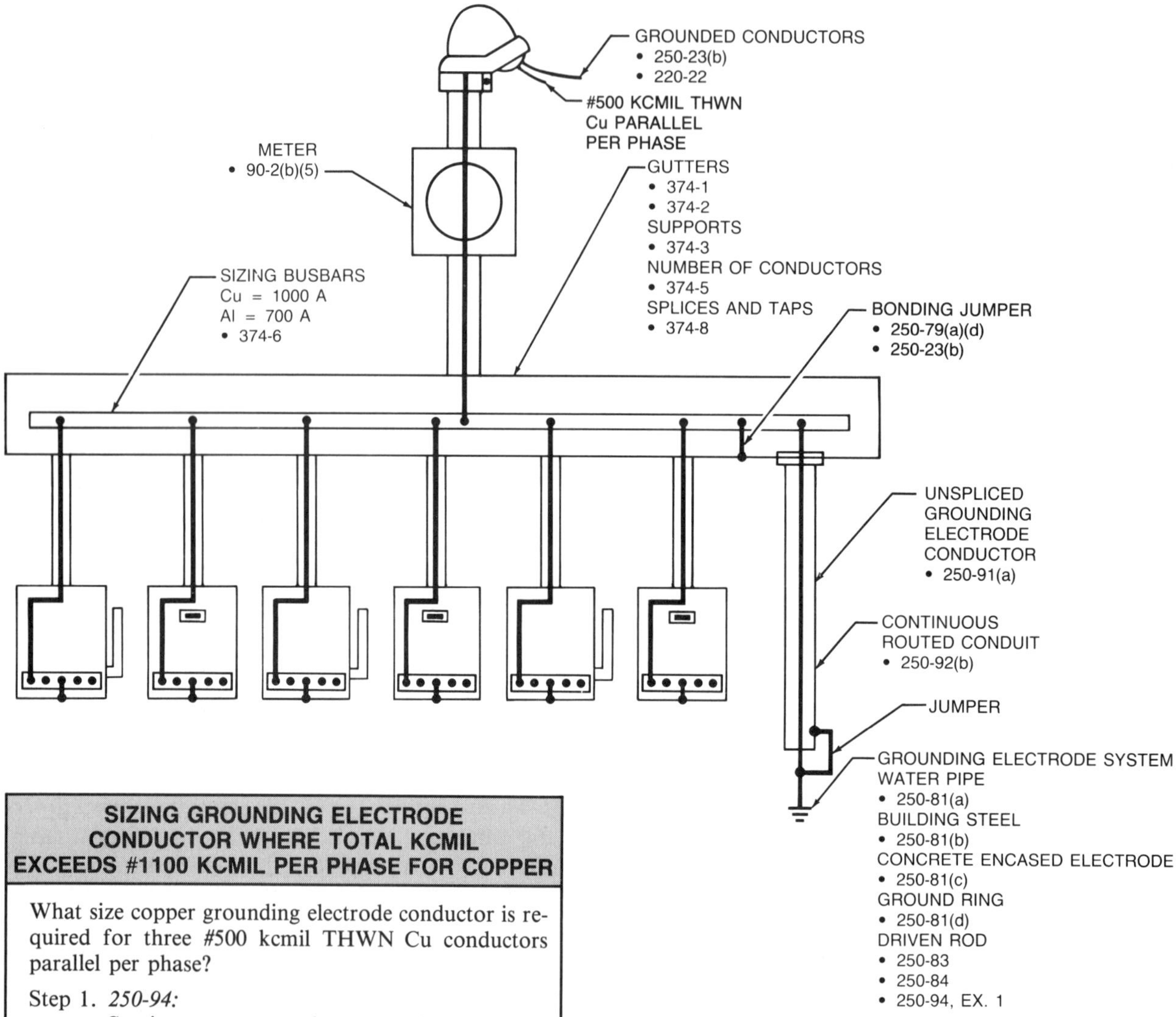

SIZING GROUNDING ELECTRODE CONDUCTOR WHERE TOTAL KCMIL EXCEEDS #1100 KCMIL PER PHASE FOR COPPER

What size copper grounding electrode conductor is required for three #500 kcmil THWN Cu conductors parallel per phase?

Step 1. *250-94:*
Service-entrance conductors = three #500 kcmil THWN Cu parallel per phase

Step 2. *250-94:* #500 kcmil × 3 = #1500 kcmil

Step 3. *Table 250-94:* #1500 kcmil requires #3/0 Cu

Answer. **Conductor = #3/0 Cu**

Figure 5-25. For phase conductors connected in parallel, the grounding electrode conductor is selected from Table 250-94 based on the phase with the greater kcmil rating.

jumper to ground the service conduit to the grounded service bar in the service equipment.

Section 250-79(d) requires the bonding jumper to be as large as the grounding electrode conductor up to #3/0 and where the service-entrance conductors on any one phase do not exceed #1100 kcmil. For service-entrance conductors larger than #1100 kcmil, the size of the bonding jumper shall be equivalent to 12½% of the kcmil rating of the largest phase conductor. See Figure 5-26.

Phases A, B, and C of a 4-wire, 3ϕ wye system are made up of five #500 kcmil THWN Cu conductors paralleled per phase. What size bonding jumper is required to ground five conduits entering a switchgear to the grounded busbar of the service equipment?

Step 1. *250-79(d):* #500 kcmil × 5 = 2500 kcmil

Step 2. *250-79(d):* 2500 kcmil × 12½% = 312.5 kcmil

Step 3. *Table 8, Ch. 9:* 312.5 kcmil requires #350 kcmil

Answer. **Bonding jumper = #350 kcmil**

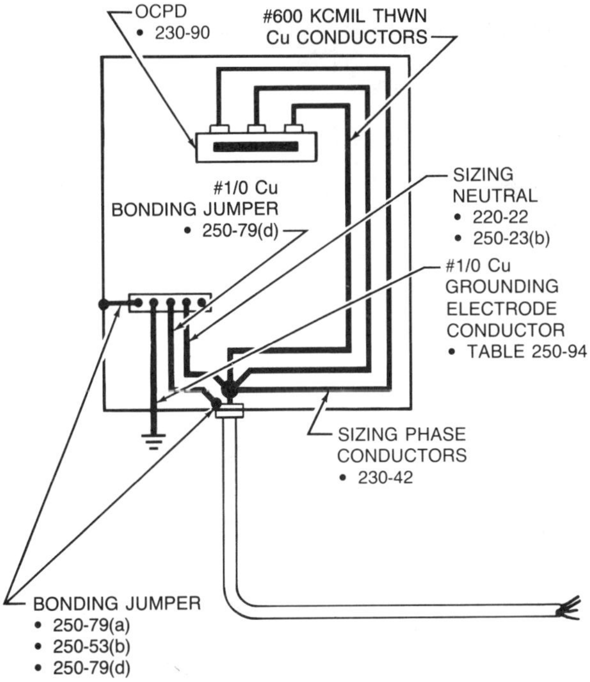

SIZING BONDING JUMPER WHERE TOTAL KCMIL IS #1100 KCMIL OR LESS PER PHASE

What size copper bonding jumper is required to bond a raceway with three #600 kcmil THWN Cu conductors and one neutral conductor?

Step 1. *250-94:* Service-entrance conductors = one #600 kcmil per phase

Step 2. *250-79(d); Table 250-94:* #600 kcmil THWN requires #1/0 Cu

Answer. **Bonding jumper = #1/0 Cu**

Figure 5-26. The bonding jumper used to ground the metal raceway of the service to the grounded neutral terminal in the service enclosure is selected from 250-79(d) and Table 250-94 based on the largest phase conductor.

What size bonding jumper is required to bond four conduits that enter a switchgear housing four #600 kcmil THWN Cu conductors parallel per phase?

Step 1. *250-79(d):* #600 kcmil × 4 = 2400 kcmil

Step 2. 2400 kcmil × 12½% = #300 kcmil

Step 3. *Table 8, Ch. 9:* #300 kcmil = #300 kcmil

Answer. **Bonding jumper = #300 kcmil**

It is easier to size a bonding jumper based upon the largest phase conductor in each conduit than to use one large bonding jumper to connect each conduit in series to the grounded terminal bus of the service equipment. See Figure 5-27.

Five conduits enter a switchgear. Each conduit contains four #400 kcmil conductors. What size bonding jumper is required to connect each conduit to the grounded terminal bus of the service equipment?

Step 1. *250-79(d):* Service-entrance per phase = #400 kcmil

Step 2. *Table 250-94:* #400 kcmil requires #1/0 Cu bonding jumper

Answer. **Cu bonding jumper = #1/0**

■ *Master Electrician's Test Tip.* If testing authorities require the bonding jumper to be selected on the larger phase conductor in any one conduit, size the bonding jumper from Table 250-94. If the bonding jumper is selected based upon the total kcmil rating of all conductors in the conduits, size the bonding jumper from Table 8, Ch. 9.■

SIZING BONDING JUMPER WHERE TOTAL KCMIL EXCEEDS #1100 KCMIL PER PHASE

What size copper bonding jumper is required to bond three raceways with one bonding jumper where each raceway contains four #700 kcmil conductors that are parallel per phase?

Step 1. *250-94:* Service-entrance conductors = three #700 kcmil parallel per phase

Step 2. *Table 8, Ch. 9:* #700 kcmil × 3 × .125 = 262.5 kcmil

Step 3. *Table 8, Ch. 9:* 262.5 kcmil requires #300 kcmil

Answer. **Bonding jumper = #300 kcmil**

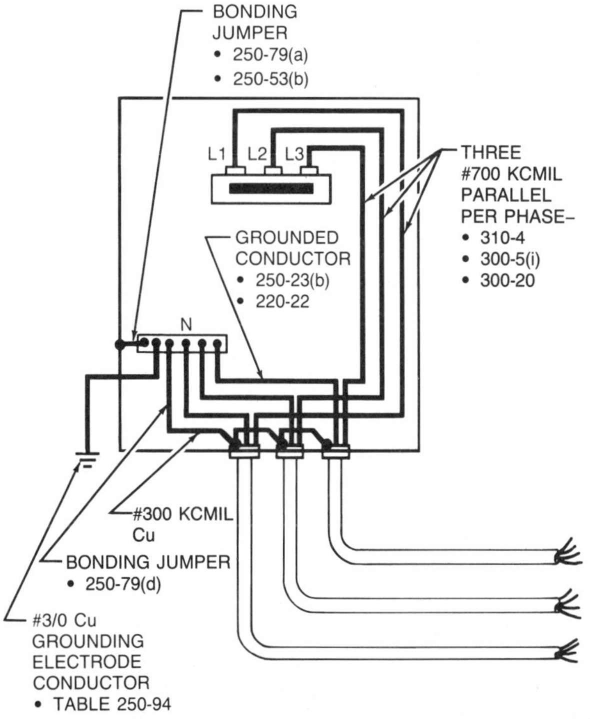

Figure 5-27. One bonding jumper used to connect all metal raceways of the service to the grounded terminal bar is sized at $12^1/_2$% of the largest kcmil rating of any one phase per 250-79(d) and Table 250-95.

Service Entrance and Grounding

Date ______________

Chapter 5—TEST

Name ______________________

______ **1.** What size conduit is required for twenty #12 THWN Cu conductors?

______ **2.** What size conduit is required for the following conductors having different types of insulation?
Ten #10 THWN Cu
Twelve #12 RHW Cu
Fifteen #6 XHHW Cu
Four #12 TW Cu

______ **3.** What size gutter is required for three #350 kcmil THW Cu feeder circuit conductors with the following tapped conductors?
Three #1 THWN Cu tapped to a switch
Three #4 XHHW Cu tapped to a panel
Three #3 RHW Cu tapped to motor control center

______ **4.** What size wireway is required for the following conductors?
Three #1/0 THW Cu
Four #2/0 THWN Al
Three #350 RHW Cu
Four #250 XHHW Al

______ **5.** What size OCPD is required to hold the LRC of a 125 HP, 460 V, 3ϕ fire pump motor?

______ **6.** What size THWN Cu conductors is required to supply power to a 150 HP, 460 V, 3ϕ fire pump motor?

______ **7.** What size kW generator set is required to start and run a 50 HP, 30 HP, 15 HP, and 10 HP, 480 V, 3ϕ group of motors? Use kW rating.

______ **8.** What size copper grounding electrode conductor is required to ground a 350 A service to a water pipe system? Service conductors are THWN Cu.

______ **9.** What size copper grounding electrode conductor is required to ground a 1650 A service paralleled five times per phase to building steel? Service conductors are THWN Cu.

______ **10.** What size copper grounding electrode conductor is required to ground a service with #1800 kcmil THWN Cu conductors per phase to a concrete-encased electrode?

______ **11.** What size copper bonding jumper is required to connect the metal raceways of a 690 A service to the grounded busbar in the service enclosure? The service is paralleled three times per phase with THWN Cu.

______ **12.** How many supports are required for 3″ rigid conduit routed 200′-0″ from point to point? Use rule-of-thumb method.

______ **13.** What is the clearance in front of a 277/480 V switchboard with an ungrounded wall directly opposite the switchboard?

______ **14.** What size copper grounding electrode conductor is required to ground a 225 A service supplied by THWN Cu conductors to a driven rod, and from the driven rod to a metal water pipe?

______ **15.** What size aluminum grounding electrode is required to ground a service at a common SE can for the following risers? The ground is building steel.
#4/0 THWN Al per phase
#3/0 THWN Al per phase
#2/0 THWN Al per phase
#1/0 THWN Al per phase
#250 kcmil THWN Al per phase

______ **16.** What is the current-carrying capacity of a ¼″ × 1″ Cu busbar?

______ **17.** What is the current-carrying capacity of a ½″ × 2″ Al busbar?

______ **18.** What is the normal current flow over a driven rod used to ground a 120/240 V, 600 A, 1ϕ service? The driven rod has a resistance of 25 ohms.

______ **19.** What size copper grounding electrode conductor is required to ground a 2000 A service to a ground ring?

______ **20.** What size copper grounding electrode conductor is required to ground an 800 A service to a ½″ rebar in a concrete foundation? The rebar is over 20′-0″ in length.

______ **21.** What size copper bonding jumper is required to connect the grounding terminal in a subpanel to metal water pipe isolated from the earth by a PVC water line? The feeder circuit is protected by 450 A fuses.

______ **22.** What size aluminum equipment grounding conductor is required for a piece of equipment protected by an 800 A main circuit breaker?

______ **23.** What size copper equipment grounding conductor is required for a piece of equipment protected by a 600 A main set of fuses? Each of three PVC conduits contains three #2/0 THWN Cu conductors paralleled per phase.

______ **24.** What is the current-carrying capacity of the following conductors installed in a wireway?
Four #6 THWN Cu
Four #8 THWN Cu
Six #4 THHN Al
Four #1 RHW Cu
Six #2 XHHW (75 °C) Cu
Four #1/0 THWN Cu
Four #3/0 THWN Al

______ **25.** What size junction box is required for a left wall lined with a 3″, 2½″, 2″, and 1″ conduit and a bottom wall lined with a 3″, 2½″, 2″, 2″, and 1″ conduit?

Chapter 6

Calculations—One-Family Dwellings

Calculations to determine the load in VA or amps for all dwelling units are done by applying either the standard or optional calculation. The standard calculation is the most difficult to apply. It produces larger VA and amperage ratings than the optional calculation. As a result, larger overcurrent protection devices and other components of the electrical service are required. The optional calculation is much easier to apply than the standard calculation. It produces smaller VA and amperage ratings than the standard calculation and is often used, as it requires smaller components for the service.

STANDARD CALCULATION—220, PART B

The standard calculation was the first method used to compute the load of dwelling units. Through the years the procedure has been modified to allow demand factors to be applied to certain types of loads. The percentages of demand that can be applied depend upon the condition of use of the equipment. When the standard calculation is used, the loads in the dwelling unit are divided into three groups:

1. General lighting loads
2. Small appliance loads
3. Special appliance loads

Four loads are selected from the three groups of loads. Demand factors can be applied to these four loads. The percentage of demand that can be applied to each load is based on the diversity of each load. Four loads to which demand factors can be applied are

1. General lighting load
2. Cooking equipment load
3. Dryer equipment load
4. Fixed appliance load

After the loads have been divided into three groups and demand factors applied to the four loads, two additional loads are calculated. The largest load between the heating and air conditioning load is selected. The final load is the largest motor load, which is selected from the air conditioning or the fixed appliance load.

General Lighting Loads—220-3(b); Table 220-3(b)

The general lighting load is found by multiplying the square footage of the dwelling unit by 3 VA per Table 220-3(b). The general lighting load supplies the *lighting loads* and the *general-purpose receptacle loads.* Lighting loads include all lighting outlets in the dwelling unit per 210-70(a). General-purpose receptacle loads include all general-purpose receptacles in the dwelling unit per Article 100 and 210-52(a).

General-purpose receptacle outlets include all outlets except those supplied by small appliance circuits or direct (individual) circuits per Article 100 and 220-16(a) and (b). For example, general-purpose receptacle outlets are located in bedrooms, dens, bathrooms, living rooms, garages, halls, and outside. Direct circuits serve only one load or piece of equipment that is cord- and plug-connected to a receptacle. The asterisk by the dwelling units in Table 220-3(b) refers to the first Footnote at the bottom of the Table. This includes the general lighting and general-purpose receptacle outlets in dwelling units and guest rooms of hotels and motels. See Figure 6-1.

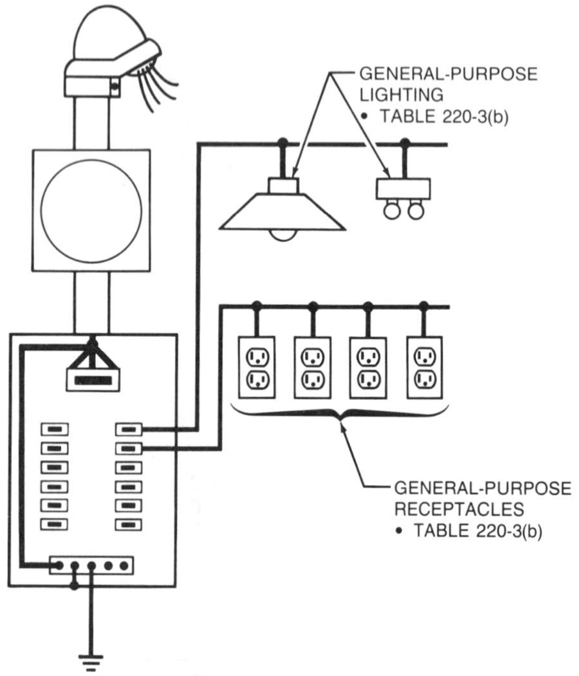

GENERAL-PURPOSE LIGHTING AND RECEPTACLE LOADS

What is the general-purpose lighting and receptacle load for an 1800 sq. ft. dwelling unit?

Step 1. *Table 220-3(b):* 1800 sq. ft. × 3 VA = 5400 VA

Answer. **Load = 5400 VA**

Figure 6-1. General-purpose lighting and receptacle loads are found by multiplying the square footage of the dwelling unit by 3 VA per Table 220-3(b). This VA rating supplies all general lighting in the dwelling unit plus all general-purpose receptacles located in the bedrooms, den, living room, hallways, etc.

What is the general lighting load for a 65′-0″ × 50′-0″ dwelling unit?

Step 1. *220-3(b):* 65′-0″ × 50′-0″ = 3250 sq. ft.

Step 2. *Table 220-3(b):*
3250 sq. ft. × 3 VA = 9750 VA

Answer. **General lighting load = 9750 VA**

Small Appliance Loads—220-16(a)

Small appliance loads supply receptacle outlets located in the kitchen, pantry, breakfast room, and dining room of a dwelling unit per 210-52(b)(1). Two small appliance circuits are required in a dwelling unit. No other outlets are permitted on the 20 A 2-wire circuits except those listed in 210-52(b), Ex. 1 and 2. One small appliance circuit is required to supply receptacle outlets located in the laundry room. Only the outlets used to serve laundry loads can be supplied by this circuit. The load for each small appliance circuit is computed at 1500 VA. This VA rating is used to size the service or feeder conductors and service equipment. See Figure 6-2.

What is the VA rating for the small appliance loads in a dwelling?

Step 1. *220-16(a)(b):* 1500 VA for each circuit

Step 2. *220-16(a)(b):* 1500 VA × 3 = 4500 VA

Answer. **VA rating = 4500 VA**

Special Appliance Loads—220-10(b)

Special appliance loads are loads such as heating units, water heaters, air conditioners, dishwashers, compactors, disposals, and motors. In most cases, special appliance loads are units that are supplied by direct circuits. These loads are not normally supplied by the general-purpose circuits. Special appliance loads are determined by the VA rating of each unit. The VA rating of each unit is figured at 125% for continuous duty and 100% for noncontinuous duty, with demand factors applied to certain loads. Special appliance loads are usually fed by direct circuits. See Figure 6-3.

What is the service load in VA for a 25 kW heating unit supplied by a 240 V, 1ϕ circuit?

Step 1. *Text:* $VA = kW \times 1000$
$VA = 25$ kW × 1000 = 25,000 VA

Step 2. *220-15:* 25,000 VA × 100% = 25,000 VA

Answer. **Load = 25,000 VA**

Note: Section 220-17, Ex. does not allow a demand factor to be applied to this special appliance load.

Applying Demand Factors—220, Part B

Demand factors can be applied to the following four loads:

1. General lighting loads. *Table 220-3(b)*
2. Cooking equipment loads. *220-19; Table 220-19*
3. Dryer loads. *220-18; Table 220-18*
4. Fixed appliance loads. *220-17*

Except for the general lighting loads, the cooking equipment, dryer, and fixed appliance loads are selected from the special appliance loads. Demand factors are then applied based on the number and rating of each unit. Demand factors are permitted due to the *diversity* of each load. Diversity is where the full load of the unit is not used continuously. For example, the elements in an oven are turned on to full kW to heat the oven. The heat setting is then adjusted to a lower

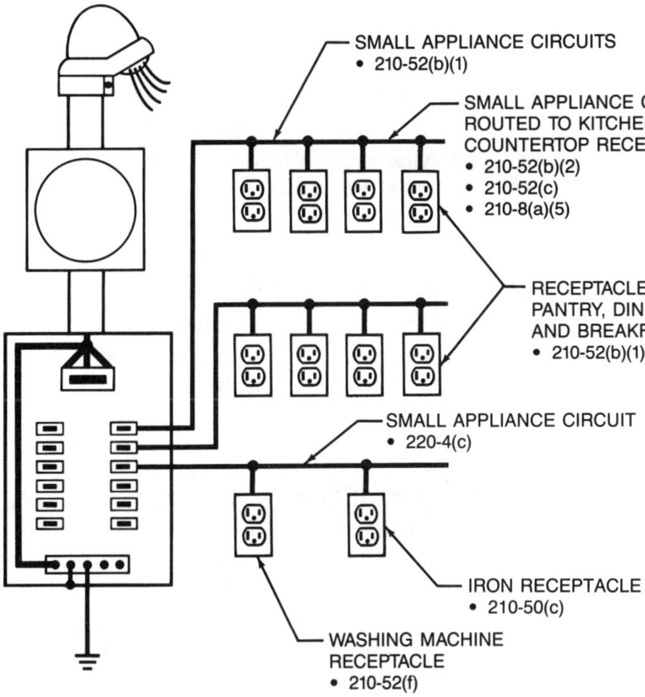

SMALL APPLIANCE LOADS

What is the small appliance load for a dwelling unit?

Step 1. *220-16(a):* 1500 VA × 2 = 3000 VA

Step 2. *220-16(b):* 1500 VA × 1 = 1500 VA

Answer. **Load = 4500 VA**

Figure 6-2. Two 20 A small appliance circuits are required to supply receptacle outlets located in the kitchen, pantry, dining room, and breakfast room. One 20 A small appliance circuit is required for the laundry room. The small appliance circuits are computed at 1500 VA each for sizing the elements of the service equipment.

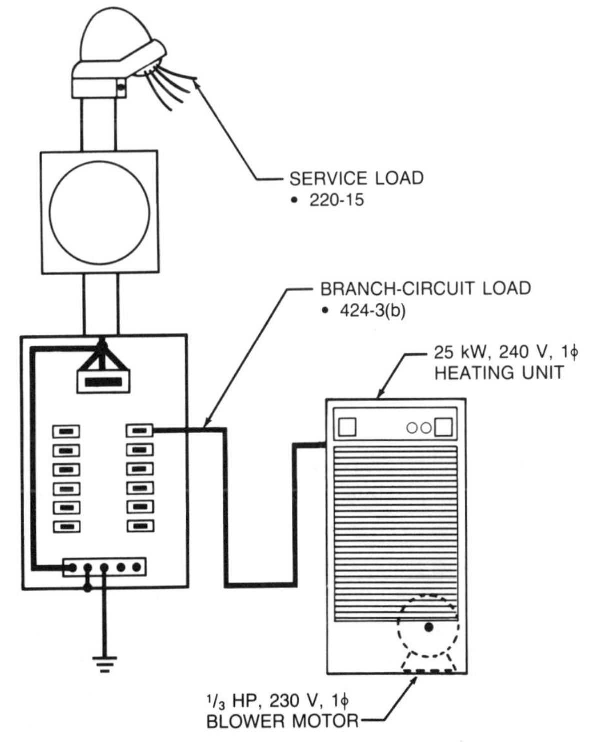

SPECIAL APPLIANCE LOADS

What is the heating load used for the service load and the load used for the branch circuit load?

Service load

Step 1. *220-15:* 25 kW × 1000 × 100% = 25,000 VA

Answer. **Load = 25,000 VA**

Branch circuit load

Step 1. *424-3(b); Table 430-148:*

Heating load =	25,000 VA
Motor load (3.6 A × 240 V) =	864 VA
Total =	25,864 VA

Step 2. *424-3(b):*
25,864 VA × 125% = 32,330 VA

Answer. **Load = 32,330 VA**

Figure 6-3. Special appliance loads are figured at 125% for continuous duty and 100% for noncontinuous duty with demand factors applied to certain loads.

temperature to cook the food. The load on the circuit is less than the full kW rating of the range. See Figure 6-4.

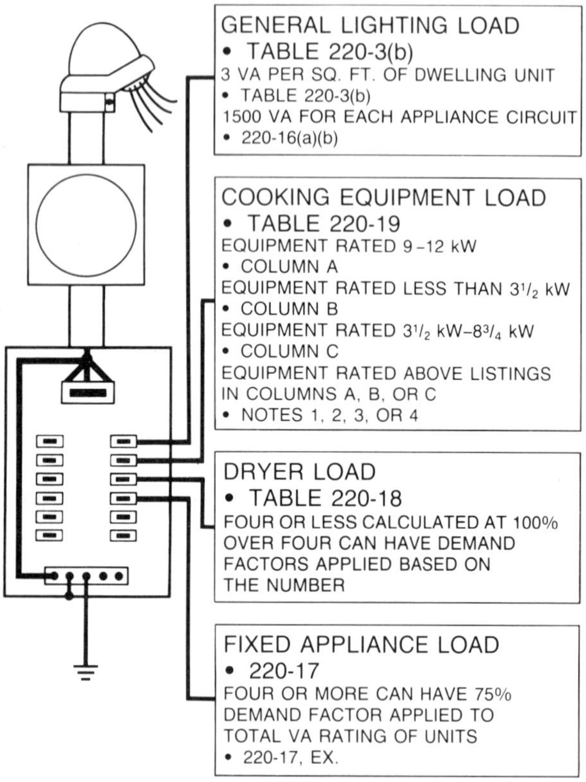

Figure 6-4. Demand factors can be applied to four loads in a dwelling.

General Lighting Loads—Table 220-11 and Small Appliance Loads—220-16(a)(b). Section 220-3(b) requires 3 VA per square foot to be selected from Table 220-3(b) to determine the general lighting load for a dwelling unit. The asterisk by the 3 VA per square foot refers to the Footnotes at the bottom of Table 220-3(b), which allow the general-purpose receptacles to be figured in this calculation.

Table 220-11 permits a demand factor to be applied to the general lighting loads, and 220-16(a)(b) allows the small appliance loads to be included in this computation.

What is the general lighting load, including the small appliance load, for a 50′-0″ × 50′-0″ dwelling unit?

Step 1. General lighting loads. *Table 220-3(b):*
50′-0″ × 50′-0″ × 3 VA = 7500 VA

Step 2. Small appliance loads. *220-16(a)(b):*
1500 VA × 3 VA = 4500 VA

Step 3. Total loads.
General lighting loads = 7500 VA
Small appliance loads = 4500 VA
Total loads = 12,000 VA

Step 4. Applying demand factors. *Table 220-11:*
First 3000 VA × 100% = 3000 VA
Next 9000 VA × 35% = 3150 VA
Total load = 6150 VA

Answer. **Total load = 6150 VA**

The general-purpose lighting and receptacle loads plus the small appliance loads are computed at 6150 VA when sizing the service-entrance equipment and conductors. See Figure 6-5.

Cooking Equipment Loads—220-19; Table 220-19. The demand load for cooking equipment is determined from Table 220-19 and Footnotes based on the kW rating and number of units. The units of cooking equipment in the Table are based on ranges, wall-mounted ovens, and counter-mounted cooktops. A range consists of both the cooktop and oven elements, which are considered one unit. Table 220-19, Column A is used for cooking equipment of over 8¾ kW to 12 kW. Column B is used for cooking equipment loads rated less than 3½ kW. Column C is used for cooking equipment loads rated at 3½ kW to 8¾ kW. See Figure 6-6.

Column A gives the demand of cooking equipment loads in kW. These are already computed and it is not necessary to multiply the kW rating of the unit by a percentage.

What is the demand load for a 10.5 kW range?

Step 1. *Table 220-19, Column A:*
10.5 kW = 8 kVA demand load

Answer. **Demand load = 8 kVA**

Column B lists a percentage based on the number of cooking units times the total kW to obtain the demand load.

What is the demand load for a piece of cooking equipment rated at 3 kW?

Step 1. *Table 220-19, Column B:*
3 kW × 80% = 2.4 kVA

Answer. **Demand load = 2.4 kVA**

Column C lists a percentage based on the number of cooking units times the total kW to determine the demand load.

What is the demand load for a piece of cooking equipment rated at 8 kW?

Step 1. *Table 220-19, Column C:*
8 kW × 80% = 6.4 kVA

Answer. **Demand load = 6.4 kVA**

Note 1 requires the demand in Column A to be increased 5% for each kW that exceeds 12 kW for pieces of cooking equipment rated over 12 kW to 27 kW.

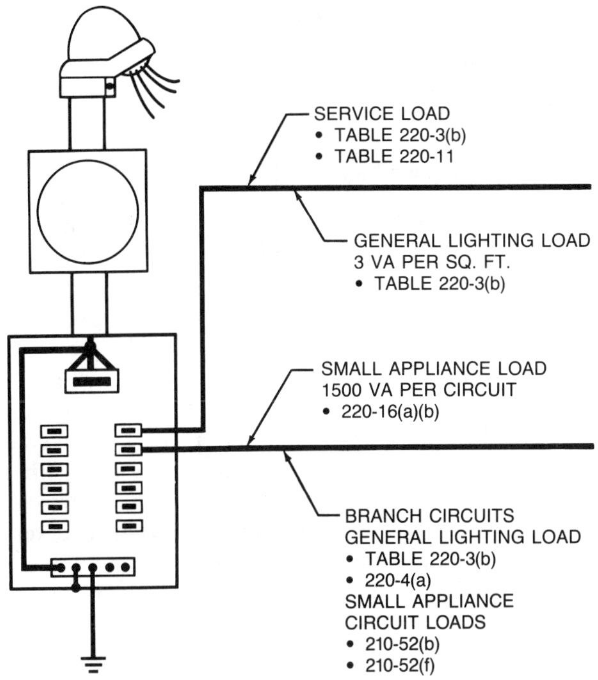

GENERAL LIGHTING LOADS

What is the demand load for the general lighting load of a 3000 sq. ft. dwelling unit?

Step 1. *Table 220-3(b):* 3000 sq. ft. × 3 VA = 9000 VA

Step 2. *220-16(a)(b):* 1500 VA × 3 = 4500 VA

Step 3. Total loads.

General lighting load =	9000 VA
Small appliance load =	4500 VA
Total =	13,500 VA

Step 4. *Table 220-11:*

First 3000 VA × 100% =	3000 VA
Next 10,500 VA × 35% =	3675 VA
Total =	6675 VA

Answer. **Load = 6675 VA**

Figure 6-5. General lighting loads for dwelling units are calculated at 3 VA per sq. ft. plus small appliance loads at 1500 VA each. The total of these loads is calculated at 100% of the first 3000 VA and 35% of the remaining VA up to 120,000 VA. See Table 220-11.

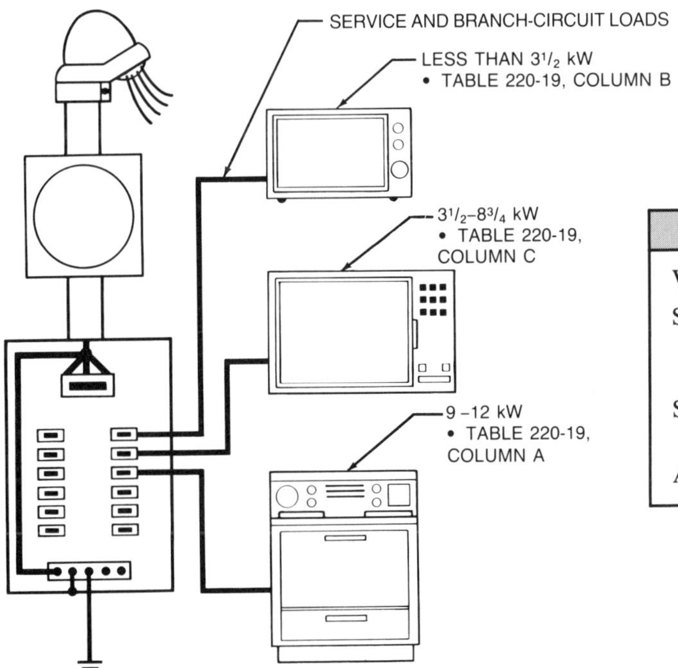

COOKING EQUIPMENT LOADS

What is the demand load for an 18 kW range?

Step 1. *Note 1 to Table 220-19:*
18 kW − 12 kW = 6 kW
6 kW × 5% = 30%

Step 2. *Table 220-19, Column A:*
8 kW × 130% = 10.4 kVA

Answer. **Demand load = 10.4 kVA**

NOTES TO TABLE 220-19
NOTE 1: OVER 12–27 kW
SAME kW RATING
NOTE 2: OVER 12–27 kW
UNEQUAL RATINGS
NOTE 3: 1¾–8¾ kW
NOTE 4: BRANCH-CIRCUIT LOAD

Figure 6-6. Demand loads for cooking equipment are found by the demand factors listed in Table 220-19, Column A, B, or C based on the kW rating of the equipment. Note 1, 2, 3, or 4 is applied if the kW rating is greater than that listed in Column A, B, or C.

What is the demand load for a freestanding range rated at 20 kW?

Step 1. *Note 1 to Table 220-19:*
20 kW − 12 kW = 8 kW
8 kW × 5% = 40%

Step 2. *Table 220-19, Column A:*
8 kW × 140% = 11.2 kVA

Answer. **Demand load = 11.2 kVA**

Note 2 to Table 220-19 applies to cooking equipment of unequal ratings over 8¾ kW to 27 kW. The average rating is determined by adding the kW ratings of all units and dividing by the number of units. Each range below 12 kW is figured at 12 kW. After the average rating is obtained, the demand in Column A for the number of units is increased 5% for each kW exceeding 12 kW.

What is the demand load for four pieces of cooking equipment rated at 10 kW, 16 kW, 18 kW, and 22 kW?

Step 1. *Note 2 to Table 220-19:*
Total kW = 12 kW + 16 kW + 18 kW + 22 kW = 68 kW

$$\text{Average rating} = \frac{68 \text{ kW}}{4 \text{ kW}} = 17 \text{ kW}$$

17 kW − 12 kW = 5 kW
5 kW × 5% = 25%

Step 2. *Table 220-19, Column A:*
17 kW × 125% = 21.25 kVA

Answer. **Demand load = 21.25 kVA**

Note 4 to Table 220-19 is applied to find the demand load for a branch circuit supplying one cooktop with one or two wall-mounted ovens. Each piece of cooking equipment is served by a tap from the branch circuit per 240-21, Ex. 5 and 210-19(b), Ex. 2. The pieces of cooking equipment shall be installed in the same room in order to apply Note 4 to Table 220-19. A cooktop and oven located in the same room provides the same electrical situation as a freestanding range containing a cooktop and oven.

The average rating for the branch circuit is determined by adding the kW rating of each unit and taking 5% of the kW exceeding 12 kW. The demand load is found by multiplying the kW rating of one unit per Table 220-19, Column A by the average rating. Note that one counter-mounted cooktop and one or two wall-mounted ovens are considered one appliance (unit) in Table 220-19, Column A. See Figure 6-7.

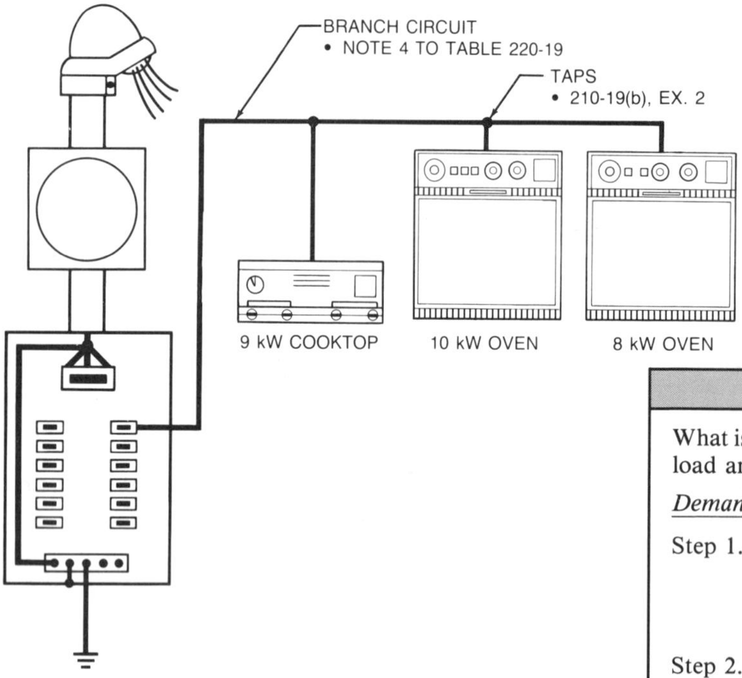

COOKING EQUIPMENT LOADS

What is the demand load in kVA for the branch-circuit load and the load in kW for the 10 kW oven?

Demand load

Step 1. *Note 4 to Table 220-19:*
9 kW + 10 kW + 8 kW = 27 kW
27 kW − 12 kW = 15 kW
15 kW × 5% = 75%

Step 2. *Table 220-19, Column A:*
8 kW × 175% = 14 kVA

Answer. **Demand load = 14 kVA**

10 kW load tapped from branch circuit

Step 1. *Note 4 to Table 220-19:*
10 kW × 100% = 10 kVA

Answer. **Demand load = 10 kVA**

Figure 6-7. Branch circuit demand load is determined by adding the kW of the units together and taking 5% of all kW exceeding 12 kW. The demand in Table 220-19 is selected from Column A based on one range and multiplied by the percentage produced by 5% times the kW exceeding 12 kW. The load for each unit is calculated at 100%.

What is the demand load for a branch circuit supplying a 10 kW cooktop and two 8 kW ovens located in the kitchen of a dwelling unit?

Step 1. *Note 4 to Table 220-19:*
Total kW = 10 kW + 8 kW + 8 kW = 26 kW
26 kW − 12 kW = 14 kW
14 kW × 5% = 70%

Step 2. *Table 220-19; Column A:*
8 kW × 170% = 13.6 kVA

Answer. **Demand load = 13.6 kVA**

■ *Master Electrician's Test Tip.* Some testing authorities allow the demand load for service-entrance conductors and equipment to be selected from Note 4 to Table 220-19. Most testing authorities only permit the demand load for the branch circuit to be computed by applying the provisions of Note 4 to Table 220-19.■

Dryer Loads—220-18; Table 220-18. The demand load for dryer equipment is determined per Table 220-18. The load for a single dryer is figured at 5000 VA or the nameplate rating, whichever is greater.

What is the demand load in kVA for a 4 kW dryer used in a dwelling unit?

Step 1. *220-18; Table 220-18:*
4 kW = 5 kVA demand load

Answer. **Demand load = 5 kVA**

The demand load for four or fewer dryers is calculated at 100% of the nameplate rating. Note that 5000 VA or the nameplate rating, whichever is greater, shall be used when applying Table 220-18. A percentage based on the number of dryers can be applied when there are five or more dryers. See Figure 6-8.

What is the demand load for 10 dryers rated at 5500 VA each?

Step 1. *220-18; Table 220-18:* 10 dryers = 50% demand

Step 2. *Table 220-18:*
5500 VA × 10 × 50% = 27,500 VA

Answer. **Demand load = 27,500 VA**

Fixed Appliance Loads—220-17. Fixed appliance loads that cannot be selected from the special appliance loads and have demand factors applied are heating equipment loads, air conditioning loads, cooking equipment loads, and dryer equipment loads. Section 220-17, Ex. prohibits these loads from having a 75% demand factor applied. After separating these fixed appliance loads, all other fixed appliances grouped in the special appliance loads are considered fixed appliance loads and can usually have the 75% demand factor applied. See Figure 6-9.

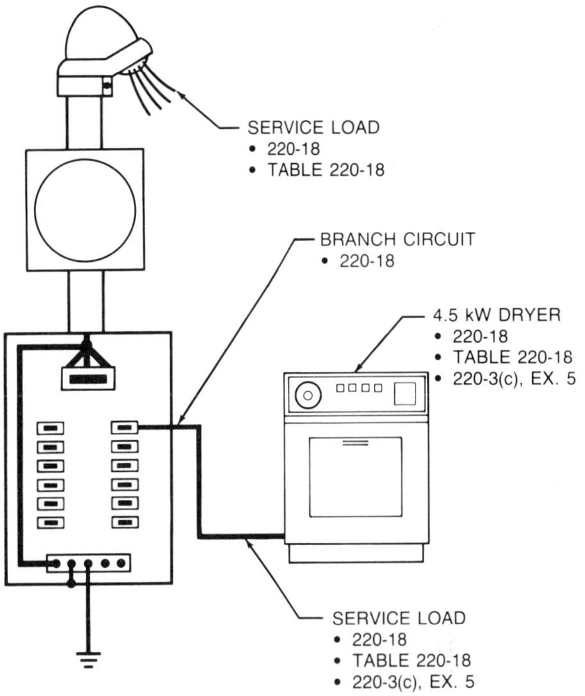

DRYER LOADS

What is the demand load for a 4.5 kW dryer?

Step 1. *220-18:* 4.5 kW = 5 kW

Step 2. *Table 220-18:* Four or fewer dryers = 100%
5 kW × 100% = 5 kVA

Answer. **Demand load = 5 kVA**

Figure 6-8. Demand loads for dryers are calculated at 5000 VA or the nameplate rating, whichever is greater. Four or fewer dryers are calculated at 100% of the nameplate rating or 5000 VA, whichever is greater.

What is the demand load for the following fixed appliances?

15,000 VA heating unit .... 240 V, 1ϕ
5040 VA A/C .... 240 V, 1ϕ
5500 VA water heater .... 240 V, 1ϕ
10,000 VA cooktop .... 240 V, 1ϕ
8500 VA oven .... 240 V, 1ϕ
2880 VA water pump .... 240 V, 1ϕ
2160 VA sump pump .... 240 V, 1ϕ
1320 VA compactor .... 120 V, 1ϕ
1200 VA disposal .... 120 V, 1ϕ
1800 VA dishwasher .... 120 V, 1ϕ
780 VA attic fan .... 120 V, 1ϕ
1200 VA microwave .... 120 V, 1ϕ
5000 VA dryer .... 240 V, 1ϕ
720 VA blower motor .... 240 V, 1ϕ

Step 1. Special appliance loads. *220-17, Ex.:*
Pull out the following loads:
Heating load
Air conditioning load
Cooking equipment load
Dryer load

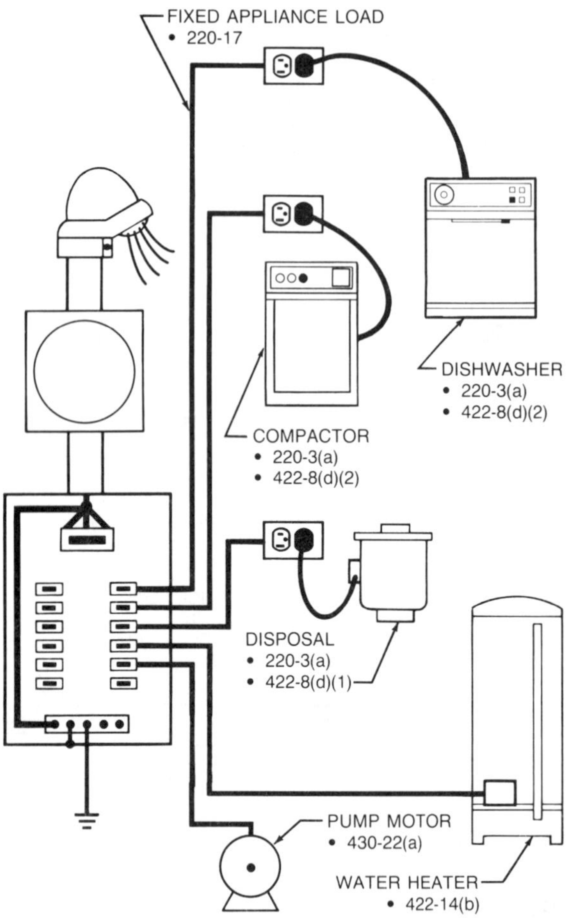

FIXED APPLIANCE LOADS

What is the demand load for the fixed appliance loads?

20 kW heating unit	240 V, 1ϕ
5.5 kW A/C unit	240 V, 1ϕ
6 kW water heater	240 V, 1ϕ
1.2 kW compactor	120 V, 1ϕ
1.6 kW dishwasher	120 V, 1ϕ
1.5 kW disposal	120 V, 1ϕ
6.6 kW attic fan	120 V, 1ϕ
1.3 kW microwave	120 V, 1ϕ
1.6 kW pump motor	120 V, 1ϕ
12 kW range	240 V, 1ϕ

Step 1. *220-17:*

Water heater =	6 kVA
Compactor =	1.2 kVA
Dishwasher =	1.6 kVA
Disposal =	1.5 kVA
Attic fan =	6.6 kVA
Microwave =	1.3 kVA
Pump motor =	1.6 kVA
Total =	19.8 kVA

Step 2. *220-17:* 19.8 kVA × 75% = 14.85 kVA

Answer. **Demand load = 14.85 kVA**

Figure 6-9. A 75% demand factor can be applied to fixed appliance loads for four or more fixed appliances in a dwelling unit per 220-17.

Step 2. Total VA rating. *220-17:*

Water heater load =	5500 VA
Water pump load =	2880 VA
Sump pump load =	2160 VA
Compactor load =	1320 VA
Disposal load =	1200 VA
Dishwasher load =	1800 VA
Attic fan load =	780 VA
Microwave load =	1200 VA
Blower motor load =	720 VA
Total =	17,560 VA

Step 2. Applying demand factors. *220-17:*
17,560 VA × 75% = 13,170 VA

Answer. **Demand load = 13,170 VA**

■ *Master Electrician's Test Tip.* Some examining boards consider the blower motor as part of the heating unit load per 220-15. Also, some examination boards consider the attic fan as a noncoincident load because it does not operate when the heat or A/C is ON.■

Heating or Air Conditioning Load—220-21. The largest load between the heating and A/C load is used per 220-21. The heating load, at 100%, is compared to 100% of the A/C load and the smaller of the two loads is dropped. See Figure 6-10.

What is the load to be used between a 20 kW, 1ϕ heating unit and a 23 A, 240 V, 1ϕ A/C unit?

Step 1. Computed rating of loads. *220-21:*

Heating unit load

VA = 20 kW × 1000 × 100% = 20,000 VA

A/C load

VA = 23 A × 240 V × 100% = 5520 VA

Step 2. Selecting larger load. *220-21:*
20,000 VA is greater than 5520 VA

Answer. **Load = 20,000 VA**

■ *Master Electrician's Test Tip.* Some examining boards calculate the heating load at 100% per 220-15 and the A/C load at 125% per 440-34, and then select the largest load between the two.■

Largest Motor Load—220-14. The largest motor load is taken out of the special appliance loads and added to the total calculation at 25% of the motor's full-load current rating. Section 220-14 refers to 430-25, which references 430-22(a) for calculating the load for one motor with other loads. See Figure 6-11.

■ *Master Electrician's Test Tip.* Some examining boards consider the A/C load eligible as the largest motor load, even when the A/C is dropped due to the heating unit being the largest load. Some examining boards do not consider the blower motor in the heating unit eligible as the largest motor load.■

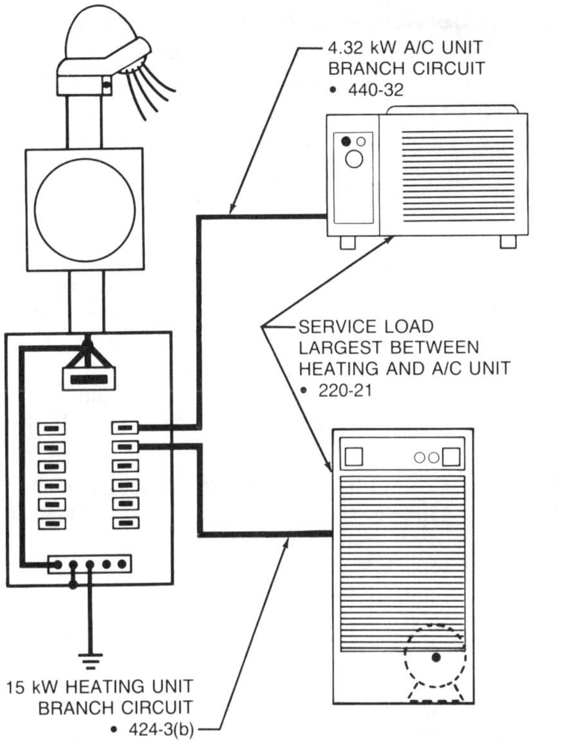

LARGEST LOAD BETWEEN HEATING AND A/C LOAD

A dwelling unit has a 15 kW heating load and a 4.32 kW A/C load. Which load is used to size the service?

Step 1. *220-21:* Heating load =
15 kW × 100% = 15 kVA
A/C load = 4.32 kW × 100% = 4.32 kVA

Answer. **Load = 15 kVA**

Figure 6-10. The largest load between the heating and A/C is determined by selecting the larger of the two loads.

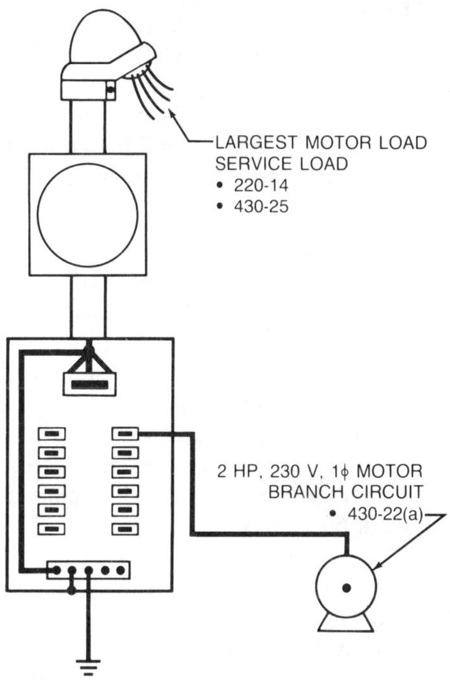

LARGEST MOTOR LOAD

What is the VA load for the largest motor?

Step 1. *Table 430-148:* 2 HP = 12 A

Step 2. *220-14; 430-22(a); 430-25:*
12 A × 125% = 15 A

Step 3. *Text:* 15 A × 240 V = 3600 VA

Answer. **Load = 3600 VA**

Figure 6-11. The largest motor load is determined by selecting the motor with the largest rating. The largest motor chosen shall be eligible after other demands have been applied.

What is the load for a 5 HP, 230 V, 1ϕ motor with 16,000 VA of other loads?

Step 1. *430-6(a); Table 430-148:*
Motor load = 5 HP = 28 A
28 A × 240 V = 6720 VA
Other loads = 16,000 VA
Total = 22,720 VA

Step 2. *220-14; 430-25; 430-22(a):*
Total loads = 22,720 VA
Largest motor load =
6720 VA × 25% = 1680 VA
Total = 24,400 VA

Answer. **Load = 24,400 VA**

What is the load for a 3 HP and a 2 HP, 230 V, 1ϕ group of motors with 15,000 VA of other loads?

Step 1. *430-6(a); Table 430-148:*
Motor load = 3 HP = 17 A
17 A × 240 V = 4080 VA
Motor load = 2 HP =
12 A × 240 V = 2880 VA
Other loads = 15,000 VA

Step 2. *220-14; 430-25; 430-24:*
Motor load = 4080 VA
Motor load = 2880 VA
Other loads = 15,000 VA
Largest motor load =
4080 VA × 25% = 1020 VA
Total = 22,980 VA

Answer. **Load = 22,980 VA**

EXAMPLES

The size of loads and their methods of use determine the total calculated load in a dwelling unit. Many arrangements of loads can produce different computations for sizing and selecting conductors, overcurrent protection devices, and other elements of the service equipment.

Minimum Size Service—230-42(b)

The minimum size service for a dwelling unit is 100 A for a computed load of 10,000 VA or more, or where there are six or more 2-wire branch circuits. The calculated load of 10,000 VA or more is after the demand factors have been applied and all loads have been totaled. If the computed load does not exceed 10,000 VA or the number of 2-wire branch circuits does not exceed five, a 60 A service can be used.

60 A Service—230-42(b)(3). A 60 A service can be installed in a dwelling unit if the computed load does not exceed 10,000 VA or if the number of 2-wire branch circuits does not exceed five. See Figure 6-12.

60 A SERVICE

What size service is required for a 1200 sq. ft. dwelling unit without other loads?

Step 1. General lighting load. *Table 220-3(b):*

1200 sq. ft. × 3 VA =	3600 VA
Small appliance loads. *220-16(a):*	
1500 VA × 2 circuits =	3000 VA
Laundry load. *220-16(b):*	
1500 VA × 1 =	1500 VA
Total =	8100 VA
Applying demand factors. *Table 220-11:*	
First 3000 VA × 100% =	3000 VA
Next 5100 VA × 35% =	1785 VA
Total =	4785 VA

Step 2. *230-42(b)(2):* Total service load = 4785 VA
4785 VA is less than 10,000 VA

Step 3. *230-42(b)(1); 220-4(a):* $No. = \frac{VA}{OCPD \times V}$

$$No. = \frac{3600\ \text{VA}}{15\ \text{A CB} \times 120\ \text{V}} = 2$$

Step 4. Total no. of 2-wire circuits.

General lighting load =	2
Small appliance loads =	3
Total =	5 (2-wire circuits)

Answer. **Service = 60 A**

Figure 6-12. A 60 A service can be installed in a dwelling unit if the computed load does not exceed 10,000 VA or if there are no more than six 2-wire branch circuits.

100 A Service—230-42(b)(1)(2). A 100 A service shall be installed in a dwelling unit if the computed load is 10,000 VA or more or if the number of 2-wire branch circuits is six or more. See Figure 6-13.

100 A SERVICE

A 1500 sq. ft. dwelling unit has the following loads:
5000 VA dryer .... 240 V, 1ϕ
5500 VA water heater .... 240 V, 1ϕ
1200 VA dishwasher .... 120 V, 1ϕ
What size service is required?

Step 1. General lighting load. *Table 220-3(b):*

1500 sq. ft. × 3 VA =	4500 VA
Small appliance loads. *220-16(a):*	
1500 VA × 2 circuits =	3000 VA
Laundry load. *220-16(b):*	
1500 VA × 1 =	1500 VA
Total =	9000 VA
Applying demand factors. *Table 220-11:*	
First 3000 VA × 100% =	3000 VA
Next 6000 VA × 35% =	2100 VA
Total =	5100 VA

Step 2. Adding loads.

General lighting load =	5100 VA
Dryer load =	5000 VA
Water heater load =	5500 VA
Dishwasher load =	1200 VA
Total load =	16,800 VA

Step 3. *230-42(b)(2):*
16,800 VA is greater than 10,000 VA

Step 4. *230-42(b)(1); 220-4(a):* $No. = \frac{VA}{OCPD \times V}$

$$No. = \frac{4500\ \text{VA}}{15\ \text{A CB} \times 120\ \text{V}} = 2.5 = 3$$

Step 5. Total no. of 2-wire circuits.
General lighting load = 3 (15 A CB)

Dryer load =

$$\frac{5000\ \text{VA}}{240\ \text{V}} = 20.8\ \text{A} = 1\ (30\ \text{A CB})$$

Water heater load =

$$\frac{5500\ \text{VA} \times 125\%}{240\ \text{V}} = 29\ \text{A} = 1\ (30\ \text{A CB})$$

Dishwasher load =

$$\frac{1200\ \text{VA}}{120\ \text{V}} = 10\ \text{A} = 1\ (15\ \text{A CB})$$

Small appliance load = 3 (20 A CB)
Total = 9 (2-wire circuits)

Answer. **Service = 100 A**

Figure 6-13. A 100 A service can be installed in a dwelling unit if the computed load exceeds 10,000 VA or if there are six or more 2-wire circuits.

■ *Master Electrician's Test Tip.* The size of the service-entrance conductors can be selected from Table 310-16 based on a 100 A service which requires #3 THWN Cu conductors. This is considered the maximum rating. However, Note 3 to Tables 310-16–310-19 allows a #4 THWN Cu conductor to be used as a 100 A service. This is considered the minimum rating. ■

Service with Gas Heat

A 2400 sq. ft. dwelling has an 85,000 Btu gas heating unit, gas cooktop, gas oven, and loads as follows. The service conductors are THWN Al. What size conductors is required for phases A and B and the neutral? See Figure 6-14.

4 A blower motor .... 120 V, 1ϕ
5500 VA water heater .... 240 V, 1ϕ
23 A A/C unit .... 240 V, 1ϕ
5000 VA dryer .... 240 V, 1ϕ
1400 VA dishwasher .... 120 V, 1ϕ
1000 VA compactor .... 120 V, 1ϕ
980 VA disposal .... 120 V, 1ϕ
1440 VA microwave .... 120 V, 1ϕ
840 VA attic fan .... 120 V, 1ϕ

(1) What size OCPD is required for the A/C unit using the minimum setting?

Step 1. *440-22(a):* 23 A × 175% = 40.25 A

Step 2. *240-3, Ex. 6; 240-6:* 40 A is a standard rating

Answer. **OCPD = 40 A**

(2) What size nonmetallic-sheathed cable is required to supply the A/C unit?

Step 1. *440-32:* 23 A × 125% = 28.75 A

Step 2. *Table 310-16:* #10 NMC = 30 A

Answer. **NMC = #10-2 w/ground**

(3) What size nonmetallic-sheathed cable is required to supply the dishwasher?

Step 1. *210-23(a); 220-3(a):* $I = \frac{1400 \text{ VA}}{120 \text{ V}} = 11.7 \text{ A}$

Step 2. *210-23(a); 220-3(a):* 11.7 A × 100% = 11.7 A

Step 3. *Table 310-16:* #14 NMC = 15 A

Answer. **NMC = #14-2 w/ground**

(4) What size nonmetallic-sheathed cable is required to supply the water heater?

Step 1. *Text:* $I = \frac{VA}{V}$: $I = \frac{5500 \text{ VA}}{240 \text{ V}} = 22.9 \text{ A}$

Step 2. *422-14(b):* 22.9 A × 125% = 28.6 A

Step 3. *Table 310-16:* #10 NMC = 30 A

Answer. **NMC = #10-2 w/ground**

(5) What size OCPD is required for the water heater?

Step 1. *Text:* $I = \frac{VA}{V}$: $I = \frac{5500 \text{ VA}}{240 \text{ V}} = 22.9 \text{ A}$

Step 2. *422-14(b):* 22.9 A × 125% = 28.6 A

Step 3. *240-3, Ex. 4; 240-6:*
30 A is the next higher standard size

Answer. **OCPD = 30 A**

Service with Electric Heat

A 3500 sq. ft. dwelling has the following loads. What size conductors is required for phases A and B and the neutral? The service conductors are THWN Al. See Figure 6-15.

35 kW heating unit .... 240 V, 1ϕ
8400 VA A/C unit .... 240 V, 1ϕ
6000 VA water heater .... 240 V, 1ϕ
10 kW cooktop .... 240 V, 1ϕ
Two 9 kW ovens .... 240 V, 1ϕ
5500 VA dryer .... 240 V, 1ϕ
1440 VA blower motor .... 240 V, 1ϕ
1650 VA dishwasher .... 120 V, 1ϕ
1200 VA disposal .... 120 V, 1ϕ
1080 VA compactor .... 120 V, 1ϕ
960 VA attic fan .... 120 V, 1ϕ
1440 VA microwave .... 120 V, 1ϕ
1960 VA water pump .... 120 V, 1ϕ

(1) What size aluminum grounding electrode conductor is required to ground the service to the metal water pipe?

Step 1. *Table 250-94:* #500 kcmil Al requires #1/0 Al

Answer. **Conductor = #1/0 Al**

(2) What size aluminum conductor is required for a driven rod to act as a supplemental ground?

Step 1. *250-80(a); 250-81(a); 250-94, Ex. 1a; 250-94:*
Driven rod requires #4 Al

Answer. **Conductor = #4 Al**

(3) What size overcurrent protection device is required for the service equipment?

Step 1. *Table 310-16:* #500 kcmil THWN Al = 310 A

Step 2. *240-3, Ex. 4; 230-90(a); 240-6:*
350 A is the next higher standard size

Answer. **OCPD = 350 A**

(4) What size panelboard is required for the service equipment?

Step 1. *384-13; 384-16(a):* OCPD = 350 A

Step 2. *Chart:* 350 A OCPD requires a 400 A panelboard

Answer. **Panelboard = 400 A**

Service with Electric Heat and Heat Pump

A 3000 sq. ft. dwelling has the following loads. What size conductors is required for phases A and B and the neutral? The service conductors are THWN Cu. See Figure 6-16.

30 kW heating unit .... 240 V, 1ϕ
6720 VA A/C unit (heat pump) ...... 240 V, 1ϕ
5550 VA water heater .... 240 V, 1ϕ
10.5 kW cooktop .... 240 V, 1ϕ
9 kW oven .... 240 V, 1ϕ
5000 VA dryer .... 240 V, 1ϕ
1260 VA blower motor .... 240 V, 1ϕ
1600 VA dishwasher .... 120 V, 1ϕ
1050 VA disposal .... 120 V, 1ϕ
980 VA compactor .... 120 V, 1ϕ
950 VA garage door motor .... 120 V, 1ϕ
1200 VA sump pump .... 120 V, 1ϕ

(1) How many 15 A 2-wire circuits are required for the general lighting and receptacle loads?

STANDARD CALCULATION—ONE-FAMILY DWELLING

1. GENERAL LIGHTING LOADS—Table 220-3(b)

			PHASES		NEUTRAL
2400 sq. ft. × 3 VA =		7200 VA			
Small appliance loads—220-16(a)					
1500 VA × 2 circuits =		3000 VA			
Laundry load—220-16(b)					
1500 VA × 1 =		1500 VA			
		11,700 VA			
Applying demand factors—Table 220-11					
First 3000 VA × 100% =	3000 VA				
Next 8700 VA × 35% =	3045 VA				
Remaining — VA × 25% =	— VA				
(total)	6045 VA		6045 VA		6045 VA

2. COOKING EQUIPMENT LOADS—Table 220-19; Notes

Col. A — VA demand =	— VA			
Col. B — VA × — % =	— VA			
Col. C — VA × — % =	— VA			
(total)	— VA	— VA	× 70% =	— VA

3. DRYER LOAD—220-18; Table 220-18

5000 VA × 100 % =	5000 VA	5000 VA	× 70% =	3500 VA

4. FIXED APPLIANCE LOADS—220-17

Dishwasher =	1400 VA			
Disposal =	980 VA			
Compactor =	1000 VA			
Water heater =	5500 VA			
Blower motor =	480 VA			
Microwave =	1440 VA			
Attic fan =	840 VA			
— =	— VA			(120 V Loads × 75%)
(total)	11,640 VA × 75% =	8730 VA	8730 VA	4605 VA

5. HEATING OR A/C LOAD—220-21

Heating unit = — VA × 100% =	— VA		
A/C unit = 5520 VA × 100% =	5520 VA		
Heat pump = — VA × 100% =	— VA		
(largest load)	5520 VA	5520 VA	— VA

6. LARGEST MOTOR LOAD—220-14

ϕ 5520 VA × 25% =	1380 VA	1380 VA	
N 1000 VA × 25% =	250 VA		250 VA
	TOTAL =	26,675 VA	14,400 VA

1ϕ service: $I = \frac{VA}{V}$

PHASES $I = \frac{26{,}675\ VA}{240\ V} = 111\ A$

NEUTRAL $I = \frac{14{,}400\ VA}{240\ V} = 60\ A$

If over 200 A—220-22; First 200 A × 100% = 200 A

Remaining ______ A × 70% = ______ A

______ A

SIZING SERVICE CONDUCTORS

Step 1. Phases A and B
Table 310-16 = #1/0 THWN Al
Note 3 to Table 310-16
#1/0 THWN Al

Step 2. Neutral
Table 310-16 = 6 THWN Al
Note 3 to Table 310-16
#2 THWN Al

Figure 6-14. The standard calculation can be used when determining the load in VA and A to select the elements of the service equipment with gas heat.

STANDARD CALCULATION—ONE-FAMILY DWELLING

1. GENERAL LIGHTING LOADS—Table 220-3(b)

3500 sq. ft. × 3 VA = 10,500 VA

Small appliance loads—220-16(a)

1500 VA × 2 circuits = 3000 VA

Laundry load—220-16(b)

1500 VA × 1 = 1500 VA

15,000 VA

Applying demand factors—Table 220-11

First 3000 VA × 100% = 3000 VA

Next 12,000 VA × 35% = 4200 VA

Remaining — VA × 25% = — VA

(total) 7200 VA | PHASES 7200 VA | NEUTRAL 7200 VA

2. COOKING EQUIPMENT LOADS—Table 220-19; Notes

Col. A 28,000 VA demand = 14,000 VA

Col. B — VA × — % = — VA

Col. C — VA × — % = — VA

(total) 14,000 VA | PHASES 14000 VA × 70% = NEUTRAL 9800 VA

3. DRYER LOAD—220-18; Table 220-18

5500 VA × 100 % = 5500 VA | PHASES 5500 VA × 70% = NEUTRAL 3850 VA

4. FIXED APPLIANCE LOADS—220-17

Dishwasher = 1650 VA

Disposal = 1200 VA

Compactor = 1080 VA

Water heater = 6000 VA

Blower motor = 1440 VA

Attic fan = 960 VA

Microwave = 1440 VA

Waterpump = 1960 VA

(total) 15,730 VA × 75% = 11,797.5 VA | PHASES 11,797.5 VA | NEUTRAL (120 V Loads × 75%) 6217.5 VA

5. HEATING OR A/C LOAD—220-21

Heating unit = 35,000 VA × 100% = 35,000 VA

A/C unit = 8400 VA × 100% = 8400 VA

Heat pump = — VA × 100% = — VA

(largest load) 35,000 VA | PHASES 35,000 VA | NEUTRAL — VA

6. LARGEST MOTOR LOAD—220-14

ϕ 1960 VA × 25% = 490 VA | PHASES 490 VA

N — VA × 25% = — VA | NEUTRAL 490 VA

TOTAL = PHASES 73,987.5 VA | NEUTRAL 27,557.5 VA

1ϕ service: $I = \frac{VA}{V}$

PHASES $I = \frac{73{,}987.5\ VA}{240\ V} = 308\ A$

NEUTRAL $I = \frac{27{,}557.5\ VA}{240\ V} = 114.8\ A$

If over 200 A—220-22; First 200 A × 100% = 200 A

Remaining ______ A × 70% = ______ A

______ A

SIZING SERVICE CONDUCTORS	
Phases A and B	= #500 kcmil THWN Al (Note 3 = #500 kcmil)
Neutral	= #1/0 THWN Al (Note 3 = #300)

Figure 6-15. The standard calculation can be used when determining the load in VA and A to select the elements of the service equipment with electric heat.

STANDARD CALCULATION—ONE-FAMILY DWELLING

1. GENERAL LIGHTING LOADS—Table 220-3(b)

3000 sq. ft. × 3 VA = 9000 VA

Small appliance loads—220-16(a)

1500 VA × 2 circuits = 3000 VA

Laundry load—220-16(b)

1500 VA × 1 = 1500 VA

13,500 VA

Applying demand factors—Table 220-11

First 3000 VA × 100% = 3000 VA

Next 10,500 VA × 35% = 3675 VA

Remaining ______ VA × 25% = ______ VA

(total) 6675 VA — PHASES: 6675 VA — NEUTRAL: 6675 VA

2. COOKING EQUIPMENT LOADS—Table 220-19; Notes

Col. A 19,500 VA demand = 11,000 VA

Col. B — VA × — % = — VA

Col. C — VA × — % = — VA

(total) 11,000 VA — PHASES: 11,000 VA × 70% = NEUTRAL: 7700 VA

3. DRYER LOAD—220-18; Table 220-18

5000 VA × 100 % = 5000 VA — PHASES: 5000 VA × 70% = NEUTRAL: 3500 VA

4. FIXED APPLIANCE LOADS—220-17

Dishwasher = 1600 VA

Disposal = 1050 VA

Compactor = 980 VA

Water heater = 5550 VA

Blower motor = 1260 VA

Garage door mtr. = 950 VA

Sump pump = 1200 VA

— = — VA

(total) 12,590 VA × 75% = 9442.5 VA — PHASES: 9,442.5 VA — NEUTRAL (120 V Loads × 75%): 4335 VA

5. HEATING OR A/C LOAD—220-21

Heating unit = 30,000 VA × 100% = 30,000 VA

A/C unit = ______ VA × 100% = ______ VA

Heat pump = 6720 VA × 100% = 6720 VA

(largest load) 36,720 VA — PHASES: 36,720 VA — NEUTRAL: — VA

6. LARGEST MOTOR LOAD—220-14

ϕ 6720 VA × 25% = 1680 VA — PHASES: 1680 VA

N 1200 VA × 25% = 300 VA — NEUTRAL: 300 VA

TOTAL = PHASES: 70,517.5 VA — NEUTRAL: 22,510 VA

1ϕ service: $I = \frac{VA}{V}$

PHASES $I = \frac{70{,}517.5\ VA}{240\ V}$ = 293.8 A

NEUTRAL $I = \frac{22{,}510\ VA}{240\ V}$ = 93.8 A

If over 200 A—220-22; First 200 A × 100% = 200 A

Remaining ______ A × 70% = ______ A

______ A

SIZING SERVICE CONDUCTORS

Phases A and B = #350 kcmil THWN Cu (Note 3 = #250 kcmil)

Neutral = #3 THWN Cu

250-23(b) = #2 THWN Cu (Note 3 = #3/0)

Figure 6-16. The standard calculation can be used when determining elements of the service equipment with electric heat and heat pumps.

Step 1. *Table 220-3(b):* 3000 sq. ft. × 3 VA = 9000 VA

Step 2. *220-4(a):* $No. = \frac{VA}{CB \times V}$

$$No. = \frac{9000\ VA}{15\ A\ CB} \times 120\ V = 5$$

Answer. **Circuits = 5**

(2) What size nonmetallic-sheathed cable is required to supply the branch circuit for the cooktop and oven per Note 4 to Table 220-19?

Step 1. Finding total kW.

Cooktop =	10.5 kW
Oven =	9 kW
Total =	19.5 kW

Step 2. *Note 4 to Table 220-19:*
19.5 kW − 12 kW = 7.5 kW
7.5 kW × 5% = 37.5%

Step 3. *Table 220-19, Column A:*
8 kW × 137.5% = 11 kW

Step 4. *Text:* $I = \frac{kVA \times 1000}{V}$

$$I = \frac{11\ kVA \times 1000}{240\ V} = 45.8\ A$$

Step 5. *Table 310-16, 60°C Column:* #6 NMC = 55 A

Answer. **NMC = #6-2 w/ground**

(3) What size overcurrent protection device is required for the branch circuit to the cooktop and oven?

Step 1. *Table 310-16:* #6 NMC = 55 A

Step 2. *240-3, Ex. 4; 240-6; 210-19(b), Ex. 2:*
60 A is the next higher standard size

Answer. **OCPD = 60 A**

Note: Apply 210-19(b), Ex. 2 where smaller taps are made to supply each piece of cooking equipment.

Service with 120 V A/C Window Unit

A 1600 sq. ft., two-story dwelling has the following loads. What size conductors is required for phases A and B and the neutral? The service conductors are THWN Cu. See Figure 6-17.

10 kW heating unit (downstairs)	240 V, 1ϕ
10 kW heating unit (upstairs)	240 V, 1ϕ
18 A A/C central unit (downstairs)	240 V, 1ϕ
18 A A/C central unit (upstairs)	240 V, 1ϕ
16 A A/C window unit	120 V, 1ϕ
12 kW range	240 V, 1ϕ
1800 VA dishwasher	120 V, 1ϕ
1250 VA compactor	120 V, 1ϕ
1000 VA disposal	120 V, 1ϕ
5500 VA water heater	240 V, 1ϕ
Two 720 VA blower motors	240 V, 1ϕ
5000 VA dryer	240 V, 1ϕ
720 VA stereo/intercom	120 V, 1ϕ

(1) What size overcurrent protection device is required for the service-entrance conductors (maximum size)?

Step 1. *Table 310-16, 75°C Column:*
#3/0 THWN Cu = 200 A

Step 2. *230-90(a); 240-3, Ex. 4; 240-6:*
200 A is a standard size

Answer. **OCPD = 200 A**

(2) What size 3-wire, 1ϕ panelboard is required?

Step 1. *Chart:* 200 A requires a 200 A OCPD

Answer. **Panelboard = 200 A**

(3) What size nonmetallic-sheathed cable is required for the feeder circuit to a panelboard supplying the 10 kW heating unit and 18 A A/C central unit upstairs?

Step 1. *220-10(b); 220-15; 220-21; 440-34:*
Heating unit load =
10,000 VA × 100% = 10,000 VA
A/C unit load =
18 A × 240 V × 100% = 4320 VA

Step 2. *220-21:* 10,000 VA is the largest load

Step 3. *424-3(b):*

Heating unit load =	
10,000 VA × 125% =	12,500 VA
Blower motor load =	
720 VA × 125% =	900 VA
Total =	13,400 VA

Step 4. *Text:* $I = \frac{VA}{V}$

$$I = \frac{13{,}400\ VA}{240\ V} = 55.8\ A$$

Step 5. *Table 310-16, 60°C Column; 240-3, Ex. 4; 240-6:*
#4 NMC Cu = 70 A

Answer. **NMC = #4-2 w/ground**

(4) What size overcurrent protection device is required for the feeder circuit to the 10 kW heating unit and the feeder circuit to the 18 A A/C unit upstairs? Size the overcurrent protection device located in the panelboard by the heating unit to serve the A/C unit.

Heating unit

Step 1. *Table 310-16, 60°C Column:* #4 NMC Cu = 70 A

Step 2. *240-3, Ex. 4; 240-6:* 70 A is a standard size

Answer. **OCPD = 70 A**

A/C unit

Step 1. *440-22(a); 240-3, Ex. 6; 240-6:*
A/C unit load = 18 A × 175% = 31.5 A
30 A is the next lower standard size

Answer. **OCPD = 30 A**

Service Using THWN on 75°C Lugs

An 1800 sq. ft. dwelling has the following loads. What size conductors are required for phases A and B and the neutral? The service conductors are THWN Cu and terminate to 75°C lugs. See Figure 6-18.

20 kW heating unit	240 V, 1ϕ
6000 VA central A/C unit	240 V, 1ϕ
8 kW cooktop	240 V, 1ϕ
6 kW oven	240 V, 1ϕ
6500 VA water heater	240 V, 1ϕ
5500 VA dryer	240 V, 1ϕ
840 VA blower motor	240 V, 1ϕ
1650 VA dishwasher	120 V, 1ϕ
1280 VA compactor	120 V, 1ϕ
1340 VA disposal	120 V, 1ϕ
1200 VA microwave	120 V, 1ϕ

STANDARD CALCULATION—ONE-FAMILY DWELLING

1. GENERAL LIGHTING LOADS—Table 220-3(b)

1600 sq. ft. × 3 VA = 4800 VA

Small appliance loads—220-16(a)

1500 VA × 2 circuits = 3000 VA

Laundry load—220-16(b)

1500 VA × 1 = 1500 VA

9300 VA

Applying demand factors—Table 220-11

First 3000 VA × 100% = 3000 VA

Next 6300 VA × 35% = 2205 VA

Remaining — VA × 25% = — VA

(total) 5205 VA

	PHASES		NEUTRAL
1. General lighting	5205 VA		5205 VA

2. COOKING EQUIPMENT LOADS—Table 220-19; Notes

Col. A 12,000 VA demand = 8000 VA

Col. B — VA × — % = — VA

Col. C — VA × — % = — VA

(total) 8000 VA — PHASES 8000 VA × 70% = NEUTRAL 5600 VA

3. DRYER LOAD—220-18; Table 220-18

5000 VA × 100 % = 5000 VA — PHASES 5000 VA × 70% = NEUTRAL 3500 VA

4. FIXED APPLIANCE LOADS—220-17

Dishwasher =	1800 VA
Disposal =	1000 VA
Compactor =	1250 VA
Water heater =	5500 VA
Blower motor =	720 VA
Stereo/intercom	720 VA
Blower motor =	720 VA
— =	— VA

(total) 11,710 VA × 75% = 8782.5 VA — PHASES 8782.5 VA — NEUTRAL (120 V Loads × 75%) 3577.5 VA

5. HEATING OR A/C LOAD—220-21

Heating unit = 20,000 VA × 100% = 20,000 VA

A/C unit = 10,560 VA × 100% = 10,560 VA

Heat pump = — VA × 100% = — VA

(largest load) 20,000 VA — PHASES 20,000 VA — NEUTRAL 1920 VA

6. LARGEST MOTOR LOAD—220-14

ϕ 1920 VA × 25% = 480 VA — PHASES 480 VA

N 1920 VA × 25% = 480 VA — NEUTRAL 480 VA

TOTAL = PHASES 47,467.5 VA — NEUTRAL 20,282.5 VA

1ϕ service: $I = \frac{VA}{V}$

PHASES $I = \frac{47{,}467.5\ VA}{240\ V} = 197.8\ A$

NEUTRAL $I = \frac{20{,}282.5\ VA}{240\ V} = 84.5\ A$

SIZING SERVICE CONDUCTORS	
Phases A and B	= #3/0 THWN Cu (Note 3 = #2/0)
Neutral	= #4 THWN Cu (Note 3 = #1)

If over 200 A—220-22; First 200 A × 100% = 200 A

Remaining ______ A × 70% = ______ A

______ A

Figure 6-17. The standard calculation can be used when determining elements of the service equipment with 120 V, A/C window units.

STANDARD CALCULATION—ONE-FAMILY DWELLING

1. GENERAL LIGHTING LOADS—Table 220-3(b)

		PHASES		NEUTRAL
1800 sq. ft. × 3 VA =	5400 VA			
Small appliance loads—220-16(a): 1500 VA × 2 circuits =	3000 VA			
Laundry load—220-16(b): 1500 VA × 1 =	1500 VA			
	9900 VA			
Applying demand factors—Table 220-11				
First 3000 VA × 100% =	3000 VA			
Next 6900 VA × 35% =	2415 VA			
Remaining — VA × 25% =	— VA			
(total)	5415 VA	5415 VA		5415 VA

2. COOKING EQUIPMENT LOADS—Table 220-19; Notes

		PHASES		NEUTRAL
Col. A — VA demand =	— VA			
Col. B 14,000 VA × 65 % =	9100 VA			
Col. C — VA × — % =	— VA			
(total)	9100 VA	9100 VA	× 70% =	6370 VA

3. DRYER LOAD—220-18; Table 220-18

		PHASES		NEUTRAL
5500 VA × 100 % =	5500 VA	5500 VA	× 70% =	3850 VA

4. FIXED APPLIANCE LOADS—220-17

			PHASES	NEUTRAL
Dishwasher =	1650 VA			
Disposal =	1340 VA			
Compactor =	1280 VA			
Water heater =	6500 VA			
Blower motor =	840 VA			
Microwave =	1200 VA			
— =	— VA			
— =	— VA			(120 V Loads × 75%)
(total)	12,810 VA × 75% =	9607.5 VA	9607.5 VA	4102.5 VA

5. HEATING OR A/C LOAD—220-21

		PHASES	NEUTRAL
Heating unit = 20,000 VA × 100% =	20,000 VA		
A/C unit = 6000 VA × 100% =	6000 VA		
Heat pump = — VA × 100% =	— VA		
(largest load)	20,000 VA	20,000 VA	— VA

6. LARGEST MOTOR LOAD—220-14

		PHASES	NEUTRAL
ϕ 1340 VA × 25% =	335 VA	335 VA	
N 1340 VA × 25% =	— VA		335 VA
	TOTAL =	49,957.5 VA	20,072.5 VA

1ϕ service: $I = \frac{VA}{V}$

PHASES $I = \frac{49{,}957.5\ VA}{240\ V}$ = 208 A

NEUTRAL $I = \frac{20{,}072.5\ VA}{240\ V}$ = 83.6 A

SIZING SERVICE CONDUCTORS

Table 310-16: Phases A and B = #4/0 THWN Cu (Note 3 = #3/0)

Table 310-16: Neutral = #4 THWN Cu (Note 3 = #1/0)

If over 200 A—220-22; First 200 A × 100% = 200 A

Remaining ______ A × 70% = ______ A

______ A

Figure 6-18. The standard calculation can be used when determining elements of the service equipment using 75°C lugs to terminate service conductors.

(1) What size THWN Cu conductors is required to supply the 20 kW heating unit?

Step 1. *Text:* $I = \frac{VA}{V}$

$$I = \frac{20{,}000\text{ VA}}{240\text{ V}} = 83\text{ A}$$

$$I = \frac{840\text{ VA}}{240\text{ V}} = 3.5\text{ A}$$

Step 2. *424-3(b):*
Heating element load =
83 A × 125% = 103.8 A
Blower motor load =
3.5 A × 125% = 4.4 A
Total = 108.2 A

Step 3. *Table 310-16, 75°C Column:*
#2 THWN Cu = 115 A

Answer. **Conductors = #2 THWN Cu**

(2) What size overcurrent protection device is required for the 20 kW heating unit?

Step 1. *424-3(b); 240-3, Ex. 4; 240-6:*
#2 THWN Cu = 115 A
125 A is the next higher standard size

Answer. **OCPD = 125 A**

(3) What size EMT is required for the circuit supplying the 20 kW heating unit?

Step 1. *Table 3B, Ch. 9:* Two #2 THWN Cu = 1″ EMT

Answer. **EMT = 1″**

Dividing VA on Line and Neutral

A 1200 sq. ft. dwelling has the following loads. What is the total VA and amps for phases A and B and the neutral? All loads are assumed to be balanced. See Figure 6-19.

15 kW heating unit .... 240 V, 1ϕ
20 A central A/C unit .... 240 V, 1ϕ
6000 VA water heater .... 240 V, 1ϕ
10,000 VA cooktop .... 240 V, 1ϕ
9000 VA oven .... 240 V, 1ϕ
4500 VA dryer .... 240 V, 1ϕ
1680 VA dishwasher .... 120 V, 1ϕ
980 VA disposal .... 120 V, 1ϕ
1200 VA compactor .... 120 V, 1ϕ
1200 VA microwave .... 120 V, 1ϕ
300 VA blower motor .... 240 V, 1ϕ

(1) What size nonmetallic-sheathed cable (Romex) is required to supply power to the branch circuit for the dryer?

Step 1. *220-3(c), Ex. 5; 220-18; Text:* $I = \frac{VA}{V}$

$$I = \frac{5000\text{ VA}}{240\text{ V}} = 20.8\text{ A}$$

Step 2. *336-26; Table 310-16, 60°C Column:*
#10 NMC Cu = 30 A

Answer. **NMC = #10-2 w/ground**

(2) What size overcurrent protection device is required for the dryer?

Step 1. *Table 310-16, Footnotes:*
#10 NMC requires a 30 A OCPD

Answer. **OCPD = 30 A**

(3) What size nonmetallic-sheathed cable is required for the branch circuit to the central A/C unit?

Step 1. *440-32:* 20 A × 125% = 25 A

Step 2. *Table 310-16, 60°C Column, Footnotes:*
#12 NMC Cu = 25 A

Answer. **NMC = #12-2 w/ground**

(4) What size overcurrent protection device is required for the central A/C unit? Use the maximum rating.

Step 1. *440-22(a):* 20 A × 225% = 45 A

Step 2. *240-3, Ex. 4; 240-6:* 45 A is a standard size

Answer. **OCPD = 45 A**

(5) What size nonmetallic-sheathed cable is required for the branch circuit supplying the cooktop?

Step 1. *Table 220-19, Column A:* 10 kW = 8 kW demand

Step 2. *Text:* $I = \frac{kW \times 1000}{V}$

$$I = \frac{8\text{ kW} \times 1000}{240\text{ V}} = 33\text{ A}$$

Step 3. *336-26; Table 310-16, 60°C Column:*
#8 NMC Cu = 40 A

Answer. **NMC = #8-2 w/ground**

(6) What size overcurrent protection device is required for the cooktop?

Step 1. *Table 310-16, 60°C Column:* #8 NMC Cu = 40 A

Step 2. *240-3, Ex. 4; 240-6:* 40 A is a standard size

Answer. **OCPD = 40 A**

Dividing Amps on Line and Neutral

The following is a step-by-step procedure for dividing the load in amps on the line and neutral. Note that for simplicity all loads are assumed to be balanced.

A 1750 sq. ft. dwelling has the following loads. Balance phases A and B and the neutral. See Figure 6-20.

20 kW heating unit .... 240 V, 1ϕ
22 A central A/C unit .... 240 V, 1ϕ
960 VA blower motor .... 240 V, 1ϕ
6500 VA water heater .... 240 V, 1ϕ
8 kW cooktop .... 240 V, 1ϕ
7 kW oven .... 240 V, 1ϕ
5 kW dryer .... 240 V, 1ϕ
1800 VA dishwasher .... 120 V, 1ϕ
1250 VA disposal .... 120 V, 1ϕ
1450 VA compactor .... 120 V, 1ϕ
780 VA attic fan .... 120 V, 1ϕ

(1) What size service-entrance conductors, using THWN Cu, is required?

Step 1. Largest phase (line) conductor.
209.59 A is larger than 207.31 A

Step 2. *Table 310-16, 75°C Column:*
#4/0 THWN Cu = 230 A

Answer. **Conductors = #4/0 THWN Cu**

(2) What size overcurrent protection device is required in the panelboard for the service-entrance conductors?

STANDARD CALCULATION—ONE-FAMILY DWELLING 1ϕ (CARRYING PHASES AND NEUTRAL IN VA)

1. GENERAL LIGHTING LOADS—Table 220-3(b)

1200 sq. ft. × 3 VA = 3600 VA

Small appliance loads—220-16(a)

1500 VA × 1 circuits = 3000 VA

Laundry load—220-16(b)

1500 VA × 1 = 1500 VA

8100 VA

Applying demand factors—Table 220-11

First 3000 VA × 100% = 3000 VA

Next 5100 VA × 35% = 1785 VA

Remaining — VA × 25% = — VA

(total) 4785 VA ÷ 2 =

Item	PHASE A	PHASE B		NEUTRAL
1. General lighting (total) 4785 VA ÷ 2 =	2392.5 VA	2392.5 VA		2392.5 VA
2. COOKING EQUIPMENT LOADS—Table 220-19, Notes				
Col. A — units = 19,000 VA demand =	11,000 VA	11000 VA	× 70% =	7700 VA
Col. B — VA × — units × — % =	— VA	— VA	× 70% =	— VA
Col. C — VA × — units × — % =	— VA	— VA	× 70% =	— VA
3. DRYER LOAD—220-18; Table 220-18				
5000 VA × 1 units × 100 % =	5000 VA	5000 VA	× 70% =	3500 VA
4. FIXED APPLIANCE LOADS—220-17				
Dishwasher = 1680 VA × 75% ÷ — =	1260 VA	— VA		1260 VA
Disposal = 980 VA × 75% ÷ — =	735 VA	— VA		735 VA
Compactor = 1200 VA × 75% ÷ — =	— VA	900 VA		900 VA
Water heater = 6000 VA × 75% ÷ — =	4500 VA	4500 VA		— VA
Microwave = 1200 VA × 75% ÷ — =	— VA	900 VA		900 VA
Blower motor = 300 VA × 75% ÷ — =	225 VA	225 VA		— VA
— = — VA × 75% ÷ — =	— VA	— VA		— VA
— = — VA × 75% ÷ — =	— VA	— VA		— VA
— = — VA × 75% ÷ — =	— VA	— VA		— VA
5. HEATING OR A/C LOAD—220-21				
Heating unit = 20,000 VA × 100% × — =	20,000 VA	20,000 VA		— VA
A/C unit = 6000 VA × 100% × — =	— VA	— VA		— VA
Heat pump = — VA × 100% × — =	— VA	— VA		— VA
6. LARGEST MOTOR LOAD—220-14				
ϕ 1200 VA × 25% =	300 VA	— VA		
N 1200 VA × 25% =	— VA	— VA		300 VA
TOTAL =	45,412.5 VA	44,917.5 VA		17,687.5 VA

1ϕ service: I = VA / V

PHASES I = 45,412.5 VA / 240 V = 189 A

NEUTRAL I = 17,687.5 VA / 240 V = 73.7 A

If over 200 A—220-22; First 200 A × 100% = 200 A

Remaining ______ A × 70% = ______ A

______ A

Figure 6-19. The standard calculation can be used when balancing the load in VA on the phases and neutral.

STANDARD CALCULATION—ONE-FAMILY DWELLING 1ϕ (CARRYING PHASES AND NEUTRAL IN AMPS)

1. GENERAL LIGHTING LOADS—Table 220-3(b)

1750 sq. ft. × 3 VA = 5250 VA

Small appliance loads—220-16(a)

1500 VA × 2 circuits = 3000 VA

Laundry load—220-16(b)

1500 VA × 1 = 1500 VA

9750 VA

Applying demand factors—Table 220-11

First 3000 VA × 100% = 3000 VA

Next 6750 VA × 35% = 2362.5 VA

Remaining — VA × 25% = — VA

	PHASE A	PHASE B	NEUTRAL
(total) 5362.5 VA ÷ 240 V =	22.34 A	22.34 A	22.34 A

2. COOKING EQUIPMENT LOADS—Table 220-19, Notes

	PHASE A	PHASE B		NEUTRAL
Col. A — VA demand ÷ — V =	— A	— A	× 70% =	— A
Col. B — VA × — % ÷ — V =	— A	— A	× 70% =	— A
Col. C 15,000 VA × 65 % ÷ 240 V =	40.63 A	40.63 A	× 70% =	28.44 A

3. DRYER LOAD—220-18; Table 220-18

	PHASE A	PHASE B		NEUTRAL
5000 VA × 100 % ÷ 240 V =	20.83 A	20.83 A	× 70% =	14.58 A

4. FIXED APPLIANCE LOADS—220-17

		PHASE A	PHASE B	NEUTRAL
Dishwasher =	1800 VA × 75% ÷ 120 V =	11.25 A	— A	11.25 A
Disposal =	1250 VA × 75% ÷ 120 V =	— A	7.81 A	7.81 A
Compactor =	1450 VA × 75% ÷ 120 V =	— A	9.06 A	9.06 A
Water heater =	6500 VA × 75% ÷ 240 V =	20.31 A	20.31 A	— A
Attic fan =	780 VA × 75% ÷ 120 V =	4.88 A	— A	4.88 A
Blower motor =	960 VA × 75% ÷ 240 V =	3 A	3 A	— A
— =	— VA × 75% ÷ — V =	— A	— A	— A
— =	— VA × 75% ÷ — V =	— A	— A	— A
— =	— VA × 75% ÷ — V =	— A	— A	— A

5. HEATING OR A/C LOAD—220-21

	PHASE A	PHASE B	NEUTRAL
Heating unit = 20,000 VA × 100% ÷ 240 V =	83.33 A	83.33 A	— A
A/C unit = 5280 VA × 100% ÷ 240 V =	— A	— A	— A
Heat pump = — VA × 100% ÷ — V =	— A	— A	— A

6. LARGEST MOTOR LOAD—220-14

	PHASE A	PHASE B	NEUTRAL
ϕ 1450 VA × 25% ÷ 120 V =	3.02 A	— A	
N 1450 VA × 25% ÷ 120 V =			3.02 A
TOTAL =	209.59 A	207.31 A	101.38 A

If over 200 A—220-22; First 200 A × 100% = 200 A

Remaining ______ A × 70% = ______ A

______ A

Figure 6-20. The standard calculation can be used when balancing the load in amps on the phases and neutral.

Step 1. *Table 310-16, 75°C Column:*
#4/0 THWN Cu = 230 A (Note 3 = #3/0)

Step 2. *240-3, Ex. 4; 240-6:*
250 A is the next higher standard size

Answer. **OCPD = 250 A**

(3) What size 3-wire, 1ϕ panelboard is required for the service?

Step 1. *Chart:*
300 A panelboard is required for a 250 A OCPD

Answer. **Panelboard = 300 A**

(4) What size copper grounding electrode conductor is required to ground the service equipment to a metal water pipe?

Step 1. *250-80(a); 250-81(a); Table 250-94:*
#4/0 THWN Cu requires #2 Cu (Note 3 = #4)

Answer. **Conductor = #2 Cu**

OPTIONAL CALCULATION: NEW OR EXISTING—220-30

The optional calculation provides a simpler method of computing the load of dwelling units. The loads are divided into two groups. The first group of loads is *other loads,* which consist of all loads except the heating and A/C loads. The second group of loads is either the *heating load* or the *A/C load.* The two groups of loads are added to find the total load. This total load is then used to select the elements of the service equipment.

Percentages used in the optional calculation are derived from demand factors. The demand factors are derived from monitoring the total kVA used in dwelling units in different locations in the United States. Table 220-30 was developed from these studies. It is used to calculate the load in dwelling units based on these percentages. The optional method listed in Table 220-30 also applies to dwelling units that are separate units located in a multifamily dwelling. The service conductors shall have an ampacity of at least 100 A for the optional method to be applied. The optional calculation can be applied to new or existing dwelling units.

Other Loads—220-30(b)

Other loads consist of the 3 VA per square foot which serve all the general lighting and general-purpose receptacle loads per 220-30(b)(2). Small appliance loads are computed at 1500 VA for each 20 A circuit per 220-30(b)(1) and 220-16(a)(b). These are then added to the other loads column. Special appliance loads are added to the other loads except for the heating and A/C loads per 220-30(b)(3). Motor loads are included in the other loads column per 220-30(b)(4). Note that the largest motor load at 25% is not added to the other loads per 220-14. All loads in the other loads column are computed at 100% of their nameplate ratings. Demand factors are applied at 100% for the first 10,000 VA of the other loads total and the remaining VA is computed at 40% per Table 220-30(4).

What is the VA rating for a 2500 sq. ft. dwelling unit with the following loads? See Figure 6-21.

25 kW heating unit .... 240 V, 1ϕ
25 A A/C unit .... 240 V, 1ϕ
6 kW water heater .... 240 V, 1ϕ
5 kW dryer .... 240 V, 1ϕ
12 kW range .... 240 V, 1ϕ
1600 VA dishwasher .... 120 V, 1ϕ
1200 VA compactor .... 120 V, 1ϕ
1000 VA disposal .... 120 V, 1ϕ
1680 VA water pump .... 120 V, 1ϕ
2400 VA swimming pool pump .... 240 V, 1ϕ
1200 VA microwave .... 120 V, 1ϕ

Heating or A/C Loads—Table 220-30(1)(2)(3)(4)

The heating or A/C load, which is the second group of loads, is selected by the following procedures:

1. The heating load for three or fewer units is calculated at 65% of the total kW rating.
2. The heating load for four or more units is calculated at 40% of the total kW rating.

The calculated load is then compared to 100% of the total VA rating of the heating unit or A/C load and the smaller load is dropped per Table 220-30.

A dwelling has a 35 kW heating unit and a 36.5 A A/C unit. The service is 240 V, 1ϕ. What is the load in VA for the group?

Step 1. Larger of heating or A/C load. *Table 220-30(1)(2)(3)(4):*
Heating = 35,000 VA × 65% = 22,750 VA
A/C = 240 V × 36.5 A =
8760 VA × 100% = 8760 VA

Step 2. *Table 220-30(1)(2)(3)(4); 220-21:*
Heating load is the largest load

Answer. **Load = 22,750 VA**

OPTIONAL CALCULATION—EXISTING—220-31

The optional calculation can be used to size the elements of the service equipment in existing dwelling units per 220-31 to determine if additional loads can be added. The service is required to be supplied by a 120/240 V or 120/208 V, 3-wire, 1ϕ system. The loads in the dwelling unit are divided into two groups. The first group is all loads in the dwelling figured at 100% of their nameplate ratings plus 3 VA per sq. ft. per 220-31 and the small appliance loads per 220-31. The second group of loads is the heating or A/C unit load computed at 100% and the smaller load dropped per 220-31. Note that the heating unit loads are figured at 100% for four or fewer units.

■ *Master Electrician's Test Tip.* Most examining boards do not permit the optional calculation per 220-31 to be applied for more than four electric space heating units.■

OPTIONAL CALCULATION—ONE-FAMILY DWELLING (NEW OR EXISTING)

1. OTHER LOADS—220-30(b)

General lighting loads—220-30(b)(2)

2500 sq. ft. × 3 VA = 7500 VA

Small appliance and laundry loads—220-30(b)(1)

1500 VA × 3 circuits = 4500 VA

Special loads—220-30(b)(3)(4)

Dishwasher =	1600 VA
Disposal =	1000 VA
Compactor =	1200 VA
Water heater =	6000 VA
Dryer =	5000 VA
Range =	12,000 VA
Water pump =	1680 VA
Pool pump =	2400 VA
Microwave =	1200 VA
— =	— VA
(total)	44,080 VA

Applying demand factors—Table 220-30(4)

First 10,000 VA × 100% = 10,000 VA

Remaining 34,080 VA × 40% = 13,632 VA

(total) 23,632 VA

PHASES 23,632 VA

2. HEATING OR A/C LOAD—Table 220-30(1)(2)(3)

Heating units (3 or less) = 25,000 VA × 65% = 16,250 VA

Heating units (4 or more) = — VA × 40% = — VA

A/C unit = 6000 VA × 100% = 6000 VA

Heat pump = — VA × 100% = — VA

(largest load) 16,250 VA — 16,250 VA

TOTAL = 39,882 VA

NEUTRAL

(Loads from Standard Calculation)

1. General lighting = 6150 VA
2. Cooking equipment = 5600 VA
3. Dryer = 3500 VA
4. Fixed appliance = 5010 VA
5. Heating or A/C = — VA
6. Largest motor = 420 VA

TOTAL = 20,680 VA

86 A

1φ service: $I = \frac{VA}{V}$

PHASES $I = \frac{39{,}882 \text{ VA}}{240 \text{ V}} = 166$ A

NEUTRAL $I = \frac{20{,}680 \text{ VA}}{240 \text{ V}} = 86$ A

166 A

Figure 6-21. The optional calculation can be used to find the service load in VA for a new or existing dwelling unit.

Existing Loads—220-31

The existing loads in a dwelling unit are determined by adding 3 VA per sq. ft. to the small appliance loads plus the nameplate ratings of each special appliance load. This VA rating is then reduced by demand factors in 220-31 to find the existing load. The *existing load* is classified as Group 1 of the two loads.

What is the existing load in VA in an 1100 sq. ft. dwelling unit with the following loads? See Figure 6-22.

5500 VA dryer .... 240 V, 1ϕ
1200 VA dishwasher .... 120 V, 1ϕ
900 VA disposal .... 120 V, 1ϕ
800 VA compactor .... 120 V, 1ϕ
11 kW range .... 240 V, 1ϕ
5000 VA water heater .... 240 V, 1ϕ

Added Load—220-31

The *added load,* which is Group 2 loads, is found by adding the total VA rating of the load at 100%. The added load is usually an air conditioner, dryer, heating unit, or other special appliance load. The VA rating of the added load is added to the existing loads to obtain the total VA rating of the existing dwelling.

What is the total VA rating of a 5 kW, 240 V, 1ϕ water heater?

Step 1. *220-31:* 5000 VA × 100% = 5000 VA

Answer. **VA = 5000 VA**

EXAMPLES

The optional calculation is the most commonly used method to determine the total load to be used in selecting the elements of the service equipment. When used for existing dwelling units, it produces a lower total VA rating than the standard calculation. The following examples illustrate the procedure for applying the optional calculation to find the computed loads of existing dwelling units.

Adding an A/C Load

An existing 23′-0″ × 45′-0″ dwelling is supplied by a 3-wire, #4 THWN Cu service and has the following loads. Can a 4320 W central A/C unit be added? See Figure 6-23.

10 kW range .... 240 V, 1ϕ
5 kW dryer .... 240 V, 1ϕ
980 VA compactor .... 120 V, 1ϕ
820 VA disposal .... 120 V, 1ϕ
1200 VA dishwasher .... 120 V, 1ϕ

(1) What size THWN Cu conductors is required to supply power to the A/C unit?

Step 1. *Text:* $I = \frac{VA}{V}$

$$I = \frac{4320\ \text{VA}}{240\ \text{V}} = 18\ \text{A}$$

Step 2. *440-32:* 18 A × 125% = 22.5 A

Step 3. *Table 310-16, 75°C Column:*
#12 THWN Cu = 25 A

Answer. **Conductors = #12 THWN Cu**

(2) What size overcurrent protection device is required for the branch circuit to the A/C unit? Use the maximum rating.

Step 1. *440-22(a):* 18 A × 225% = 40.5 A

Step 2. *Table 310-16, Footnotes; 240-3, Ex. 6; 240-6:*
40 A is the next lower standard size

Answer. **OCPD = 40 A**

(3) What size nonmetallic-sheathed cable is required to supply power to the A/C unit?

Step 1. *336-26; Table 310-16, 60°C Column:*
#12 NMC = 25 A

Answer. **NMC = #12-2 w/ground**

Adding 120 V A/C Window Units

Can four 15 A, 120 V A/C window units be added to an existing 25′-0″ × 40′-0″ dwelling unit having the following loads? The dwelling unit is supplied by a 3-wire, #2 THWN Cu service. See Figure 6-24.

9.5 kW range .... 240 V, 1ϕ
4.5 kW dryer .... 240 V, 1ϕ
1080 VA blower motor (gas heat) .... 120 V, 1ϕ
1600 VA dishwasher .... 120 V, 1ϕ
1000 VA disposal .... 120 V, 1ϕ
780 VA attic fan .... 120 V, 1ϕ
1500 VA microwave .... 120 V, 1ϕ
1440 VA sump pump .... 240 V, 1ϕ

(1) Will a #2 THWN Cu conductor be large enough after adding the four 120 V A/C window units?

Step 1. *Table 310-16:* #2 THWN Cu = 115 A

Step 2. *220-31:* Existing load = 98 A

Answer. **Yes**

APPLYING STANDARD AND OPTIONAL CALCULATIONS

Some testing authorities require the load of a dwelling unit to be calculated and the totals carried to columns that list the values for the standard and optional calculations along with the neutral. Those taking the test should be familiar with this method of determining total ratings.

Standard Calculation—220-10(b)–220-21

The first column of loads lists the VA ratings for the phase conductors using the standard calculation. Six loads are calculated to determine the VA rating of a dwelling unit.

Neutral—220-10(b)–220-22

The second column of loads lists the VA ratings for the neutral conductor (based upon the standard calculation). Six loads may be required to determine the VA rating. Only five loads are required when the heating or A/C is 240 V, 1ϕ.

OPTIONAL CALCULATION—ONE-FAMILY DWELLING (EXISTING ONLY)

1. OTHER LOADS—220-31

General lighting loads—220-31

1100 sq. ft. × 3 VA = 3300 VA

Small appliance and laundry loads—220-31

1500 VA × 3 circuits = 4500 VA

Special loads—220-31

Dishwasher =	1200	VA
Disposal =	900	VA
Compactor =	800	VA
Water heater =	5000	VA
Range =	11,000	VA
Dryer =	5500	VA
— =	—	VA
— =	—	VA
— =	—	VA
— =	—	VA
(total)	32,200	VA

Applying demand factors—220-31

First 8000 VA × 100% =	8000	VA
Remaining 24,200 VA × 40% =	—	VA
Added appliance = — VA × 100% =	9680	VA
(total)	17,680	VA

PHASES

VA

A

1φ service: $I = \frac{VA}{V}$

I = — VA / V = — A

(*Note:* Total ampacity of service shall not exceed service conductor ampacity. Existing service conductor ampacity = — A.)

NEUTRAL

(Loads from Standard Calculation)

1. General lighting = — VA
2. Cooking equipment = — VA
3. Dryer = — VA
4. Fixed appliance = — VA
5. Heating or A/C = — VA
6. Largest motor = — VA

TOTAL = — VA

— A

Figure 6-22. The optional calculation can be used to find the service load in VA for existing loads.

OPTIONAL CALCULATION—ONE-FAMILY DWELLING (EXISTING ONLY)

1. OTHER LOADS—220-31

General lighting loads—220-31

1035 sq. ft. × 3 VA = 3105 VA

Small appliance and laundry loads—220-31

1500 VA × 3 circuits = 4500 VA

Special loads—220-31

Dishwasher =	1200 VA
Disposal =	820 VA
Compactor =	980 VA
Water heater =	— VA
Dryer =	5000 VA
Range =	10,000 VA
— =	— VA
— =	— VA
— =	— VA
— =	— VA
(total)	25,605 VA

Applying demand factors—220-31

		PHASES
First 8000 VA × 100% =	8000 VA	
Remaining 17,605 VA × 40% =	7042 VA	
Added appliance = 4320 VA × 100% =	4320 VA	
(total)	19,362 VA	19,362 VA

1ϕ service: $I = \frac{VA}{V}$

$I = \frac{19{,}362 \text{ VA}}{240 \text{ V}} = 80.7 \text{ A}$

PHASES: 80.7 A

(*Note:* Total ampacity of service shall not exceed service conductor ampacity. Existing service conductor ampacity = 85 A.)

NEUTRAL

(Loads from Standard Calculation)

1. General lighting =	4611.8 VA
2. Cooking equipment =	5600 VA
3. Dryer =	3500 VA
4. Fixed appliance =	3000 VA
5. Heating or A/C =	— VA
6. Largest motor =	245 VA
TOTAL =	16,956.8 VA
	70.7 A

Yes ... appliance can be added.

Figure 6-23. The optional calculation can be used to determine if an A/C central unit can be added to an existing service.

OPTIONAL CALCULATION—ONE-FAMILY DWELLING (EXISTING ONLY)

1. OTHER LOADS—220-31

General lighting loads—220-31

1000 sq. ft. × 3 VA = 3000 VA

Small appliance and laundry loads—220-31

1500 VA × 3 circuits = 4500 VA

Special loads—220-31

Dishwasher =	1600 VA
Disposal =	1000 VA
Compactor =	— VA
Water heater =	— VA
Range =	9500 VA
Dryer =	4500 VA
Blower motor =	1080 VA
Attic fan =	780 VA
Microwave =	1500 VA
Sump pump =	1440 VA
(total)	28,900 VA

Applying demand factors—220-31

First 8000 VA × 100% = 8000 VA

Remaining 20,900 VA × 40% = 8360 VA

Added appliance = 7200 VA × 100% = 7200 VA

(total) 23,560 VA

PHASES

23,560 VA

98 A

1ϕ service: $I = \frac{VA}{V}$

$I = \frac{23{,}560 \text{ VA}}{240 \text{ V}} = 98$ A

(*Note:* Total ampacity of service shall not exceed service conductor ampacity. Existing service conductor ampacity = 115 A.)

Yes — load can be added —

NEUTRAL

(Loads from Standard Calculation)

1. General lighting = 4575 VA
2. Cooking equipment = 5600 VA
3. Dryer = 3500 VA
4. Fixed appliance = 4470 VA
5. Heating or A/C = — VA
6. Largest motor = 270 VA

TOTAL = 18,415 VA

76.7 A

Figure 6-24. The optional calculation can be used to determine if an appliance can be added to an existing service.

Optional Calculation—220-30; Table 220-30

The third column of loads lists the VA ratings for the phase conductors (based upon the optional calculation). Two loads are required to determine the VA rating of the dwelling unit.

EXAMPLES

The procedure for determining the loads in VA for each column of loads is to apply the standard calculation. This calculation produces the larger load. The third column of loads is determined by applying the optional calculation, which produces the smaller load of the two methods. When the optional calculation is used, the neutral can be based on the phase conductors instead of the VA rating of neutral-related loads that are figured by the standard calculation.

Carrying Standard and Optional Calculations with Neutral

A 2600 sq. ft. dwelling has a 240 V, 1ϕ service and the following loads. Find the amperage of phases A and B and the neutral. See Figure 6-25.

20 kW heating unit .... 240 V, 1ϕ
4680 VA A/C unit .... 240 V, 1ϕ
6500 VA water heater .... 240 V, 1ϕ
10 kW cooktop .... 240 V, 1ϕ
1104 VA blower motor .... 240 V, 1ϕ
4500 VA dryer .... 240 V, 1ϕ
12 kVA oven .... 240 V, 1ϕ
900 VA attic fan .... 120 V, 1ϕ
1428 VA microwave .... 120 V, 1ϕ
1586 VA disposal .... 120 V, 1ϕ
1556 VA compactor .... 120 V, 1ϕ
2400 VA dishwasher .... 120 V, 1ϕ
1200 VA hydromassage tub .... 120 V, 1ϕ

(1) What size service conductors is required using THWN Cu?

Step 1. *Table 310-16, 75°C Column:*
Standard calculation.
229.8 A = #4/0 THWN Cu (Note 3 = #3/0)
Optional calculation. 171.6 A = #2/0 THWN Cu
Note 3 allows #1/0 THWN Cu

Answer. Standard calculation.
Conductors = **#4/0 THWN Cu** (Note 3 = #3/0)
Optional calculation.
Conductors = **#2/0 or #1/0 THWN Cu**

(2) What size main OCPD is required for the service panel?

Step 1. *Table 310-16, 75°C Column:*
Standard calculation. #4/0 THWN Cu = 230 A
Optional calculation. #2/0 THWN Cu = 175 A

Step 2. *240-3, Ex. 4; 240-6:*
230 A = 250 A OCPD
175 A = 175 A OCPD

Answer. Standard calculation. **OCPD** = **250 A**
Optional calculation. **OCPD** = **175 A**
(Note 3 to Tables 310-16–310-19 allows a 175 A OCPD with #1/0 THWN Cu conductors.)

(3) What size service panel is required?

Step 1. *Chart:*
Standard calculation.
250 A OCPD = 300 A panel
Optional calculation.
175 A OCPD = 225 A panel

Answer. Standard calculation. **Service panel** = **300 A**
Optional calculation. **Service panel** = **225 A**

(4) What size grounding electrode conductor is required? Use copper.

Step 1. *Table 250-94:*
Standard calculation. #4/0 = #2 Cu
(#3/0 = #4 Cu)
Optional calculation.
#2/0 = #4 Cu; #1/0 = #6 Cu

Answer. Standard calculation. **Conductor** = **#2 Cu or #4 Cu**
Optional calculation. **Conductor** = **#4 Cu**
(Note 3 to Tables 310-16–310-19 requires #6 Cu.)

STANDARD AND OPTIONAL CALCULATIONS—ONE-FAMILY DWELLING

	STANDARD PHASES	STANDARD NEUTRAL	OPTIONAL PHASES
1. GENERAL LIGHTING LOADS—Table 220-3(b)			
2600 sq. ft. × 3 VA =	7800 VA	7800 VA	7800 VA
Small appliance loads—220-16(a)			
1500 VA × 2 circuits =	3000 VA	3000 VA	3000 VA
Laundry load—220-16(b)			
1500 VA × 1 =	1500 VA	1500 VA	1500 VA
	12,300 VA	12,300 VA	
Applying demand factors—Table 220-11			
First 3000 VA × 100% =	3000 VA	3000 VA	
Next 9300 VA × 35% =	3255 VA	3255 VA	
Remaining — VA × 25% =	— VA	— VA	
2. COOKING EQUIPMENT LOADS—Table 220-19; Notes			
Col. A 22,000 VA demand =	11,000 VA × 70% =	7700 VA	22,000 VA
Col. B — VA × — % =	— VA × 70% =	— VA	— VA
Col. C — VA × — % =	— VA × 70% =	— VA	— VA
3. DRYER LOAD—220-18; Table 220-18			
5000 VA × 100 % =	5000 VA × 70% =	3500 VA	4500 VA
4. FIXED APPLIANCE LOADS—220-17			
Dishwasher = 2400 VA × 75% =	1800 VA	1800 VA	2400 VA
Disposal = 1586 VA × 75% =	1189.5 VA	1189.5 VA	1586 VA
Compactor = 1556 VA × 75% =	1167 VA	1167 VA	1556 VA
Water heater = 6500 VA × 75% =	4875 VA	— VA	6500 VA
Blower motor = 1104 VA × 75% =	828 VA	— VA	1104 VA
Attic fan = 900 VA × 75% =	675 VA	675 VA	900 VA
Microwave = 1428 VA × 75% =	1071 VA	1071 VA	1428 VA
Hydro tub = 1200 VA × 75% =	900 VA	900 VA	1200 VA
		(total) =	55,474 VA
5. HEATING OR A/C LOAD—220-21			
Heating unit = 20,000 VA × 100% =	20,000 VA	— VA × 65 % =	13,000 VA
A/C unit = 4680 VA × 100% =	— VA	— VA	— VA
Heat pump = — VA × 100% =	— VA	— VA	— VA
6. LARGEST MOTOR LOAD—220-14			
ϕ 1586 VA × 25% =	396.5 VA		
N 1586 VA × 25% =		396.5 VA	
TOTAL =	55,157 VA	24,654 VA	

1ϕ service: I = VA / V		
STANDARD PHASES	I = 55,157 VA / 240 V =	229.8 A
OPTIONAL PHASES	I = 41,189.6 VA / 240 V =	171.6 A
NEUTRAL	I = 24,654 VA / 240 V =	102.7 A

Applying demand factors—Table 220-30(4)	
First 10,000 VA × 100% =	10,000 VA
Remaining 45,474 VA × 40% =	18,189.6 VA
Heating load =	13,000 VA
TOTAL =	41,189.6 VA

If over 200 A—220-22; First 200 A × 100% =	200 A
Remaining ______ A × 70% =	______ A
	A

Figure 6-25. The standard and optional calculations can be used to carry the phases and neutral when calculating the service load.

Calculations— One-Family Dwellings

Date ____________________

Chapter 6—TEST

Name ______________________________

1. What is the general lighting and receptacle load for a 40′-0″ × 60′-0″ one-family dwelling unit with a 120/240 V, 1ϕ service?

2. How many 15 A, 2-wire circuits are required for a 2400 sq. ft. one-family dwelling unit?

3. What is the demand load for a 2800 sq. ft. one-family dwelling unit with four small appliance circuits? The laundry circuit is included.

4. What is the demand load in VA for a 22 kW freestanding electric range?

5. What is the demand load in kW for a 4500 VA electric clothes dryer?

6. A one-family dwelling supplied by a 120/240 V, 1ϕ service has a 25 kW heating unit and a 21 A, 240 V, 1ϕ A/C central unit. Find the largest load in VA.

7. A one-family dwelling unit supplied by a 240 V, 1ϕ service has the following fixed appliance loads:
   - 10 kW heating unit .... 240 V, 1ϕ
   - 4560 VA A/C unit .... 240 V, 1ϕ
   - 6 kW water heater .... 240 V, 1ϕ
   - 12 kW range .... 240 V, 1ϕ
   - 760 VA blower motor .... 240 V, 1ϕ
   - 5000 VA clothes dryer .... 240 V, 1ϕ
   - 2880 VA water pump .... 240 V, 1ϕ
   - 720 VA sump pump .... 120 V, 1ϕ
   - 1200 VA microwave .... 120 V, 1ϕ
   - 785 VA disposal .... 120 V, 1ϕ
   - 960 VA compactor .... 120 V, 1ϕ
   - 1920 VA dishwasher .... 120 V, 1ϕ

7-1. What is the demand load in VA?

7-2. What is the largest motor load in VA?

8. What is the load in VA for a 7.5 kW cooktop and an 8 kW wall-mounted oven?

9. A 2500 sq. ft. one-family dwelling is supplied by a 120/240 V, 1ϕ service and has the following loads:
   - 25 kW heating unit .... 240 V, 1ϕ
   - 6000 VA A/C unit .... 240 V, 1ϕ
   - 6500 VA water heater .... 240 V, 1ϕ
   - 10.5 kW cooktop .... 240 V, 1ϕ
   - 1320 VA blower motor .... 240 V, 1ϕ
   - 5000 VA clothes dryer .... 240 V, 1ϕ
   - 2880 VA water pump .... 240 V, 1ϕ
   - 12 kW oven .... 240 V, 1ϕ
   - 12 kW oven .... 240 V, 1ϕ
   - 660 VA overhead grarage door .... 120 V, 1ϕ
   - 1200 VA microwave .... 120 V, 1ϕ
   - 1280 VA disposal .... 120 V, 1ϕ
   - 1470 VA compactor .... 120 V, 1ϕ
   - 1660 VA dishwasher .... 120 V, 1ϕ

__________ **9-1.** What is the fixed appliance load? Use the standard calculation.

__________ **9-2.** What is the demand load for other loads? Use the optional calculation per 220-30.

__________ **9-3.** What is the demand load for other loads? Use the optional calculation per 220-31.

__________ **9-4.** What is the heating load in kW? Use the optional calculation per 220-30.

__________ **9-5.** What is the VA rating of the largest motor? Use the standard calculation.

__________ **10.** What is the heating load in kW for a 20 kW, 240 V, 1ϕ heating unit? Use the optional calculation per 220-30.

__________ **11.** What is the heating load in kW for three 8 kW, 240 V, 1ϕ heating units? Use the optional calculation per 220-30.

__________ **12.** What is the heating load in kW for five 6 kW, 240 V, 1ϕ heating units? Use the optional calculation per 220-30.

__________ **13.** A one-family dwelling unit has 32,850 VA of other loads. What is the demand load in VA? Use the optional calculation per 220-30.

__________ **14.** A one-family dwelling unit supplied by a 120/240 V, 1ϕ service has a 10 kW heating unit, three 1920 VA A/C window units, and a 19.5 A, 240 V A/C central unit. Which VA load is the largest? Use the optional calculation.

__________ **15.** A 1680 sq. ft. one-family dwelling unit supplied by a 120/240 V, 1ϕ service has the following loads:

20 kW heating unit .... 240 V, 1ϕ
3816 VA A/C unit .... 240 V, 1ϕ
5500 VA water heater .... 240 V, 1ϕ
10 kW cooktop .... 240 V, 1ϕ
600 VA blower motor .... 240 V, 1ϕ
5000 VA clothes dryer .... 240 V, 1ϕ
9.5 kW oven .... 240 V, 1ϕ
828 VA attic fan .... 120 V, 1ϕ
1200 VA microwave .... 120 V, 1ϕ
1380 VA disposal .... 120 V, 1ϕ
1500 VA compactor .... 120 V, 1ϕ
1700 VA dishwasher .... 120 V, 1ϕ

__________ **15-1.** What is the total VA of other loads? Use the optional calculation per 220-30.

__________ **15-2.** Find the largest load between the heating and A/C loads. Use the optional calculation per 220-30.

__________ **16.** Find the other loads for an existing 1080 sq. ft. dwelling unit with a 120/240 V, 1ϕ service having the following loads. Use the optional calculation per 220-31.

10 kW range .... 240 V, 1ϕ
5 kW clothes dryer .... 240 V, 1ϕ
1.6 kW dishwasher .... 240 V, 1ϕ
900 VA disposal .... 240 V, 1ϕ

__________ **17.** Find the heating unit load in VA for an existing dwelling unit with two 6 kW heating units. Use the optional calculation per 220-31.

_______ **18.** A 42′-0″ × 42′-0″ one-family dwelling is supplied by a 120/240 V, 1ϕ service and has the following loads:

10 kW heating unit . . . . 240 V, 1ϕ
4560 VA A/C unit . . . . 240 V, 1ϕ
6 kW water heater . . . . 240 V, 1ϕ
12 kW range . . . . 240 V, 1ϕ
760 VA blower motor . . . . 240 V, 1ϕ
5000 VA clothes dryer . . . . 240 V, 1ϕ
2880 VA water pump . . . . 240 V, 1ϕ
720 VA sump pump . . . . 240 V, 1ϕ
1200 VA microwave . . . . 120 V, 1ϕ
785 VA disposal . . . . 120 V, 1ϕ
960 VA compactor . . . . 120 V, 1ϕ
1920 VA dishwasher . . . . 120 V, 1ϕ

Use the form STANDARD CALCULATION—ONE-FAMILY DWELLING to find the following:

_______ **18-1.** Service rating in VA.

_______ **18-2.** Service rating in A.

_______ **18-3.** Size maximum and minimum THWN Cu conductors for phases A and B.

_______ **18-4.** Size THWN Cu conductor for the neutral.

_______ **18-5.** Size main for the service equipment (based on maximum).

_______ **18-6.** Size of the service panelboard.

_______ **18-7.** Size of the grounding electrode conductor for the maximum service. Use copper.

_______ **19.** A 50′-0″ × 50′-0″ one-family dwelling is supplied by a 120/240 V, 1ϕ service and has the following loads:

25 kW heating unit . . . . 240 V, 1ϕ
6 kW A/C unit . . . . 240 V, 1ϕ
5.5 kW water heater . . . . 240 V, 1ϕ
11 kW oven . . . . 240 V, 1ϕ
9 kW cooktop . . . . 240 V, 1ϕ
5 kW clothes dryer . . . . 240 V, 1ϕ
1000 VA water pump . . . . 120 V, 1ϕ
890 VA attic fan . . . . 240 V, 1ϕ
980 VA compactor . . . . 120 V, 1ϕ
1550 VA dishwasher . . . . 120 V, 1ϕ
1200 VA disposal . . . . 120 V, 1ϕ

Use the form STANDARD CALCULATION—ONE-FAMILY DWELLING to find the following:

_______ **19-1.** Service rating in VA for the phases.

_______ **19-2.** Service rating in A for the neutral.

______ **20.** A 2650 sq. ft. one-family dwelling is supplied by a 120/240 V, 1ϕ service and has the following loads:

30 kW heating unit .... 240 V, 1ϕ
6500 VA A/C unit .... 240 V, 1ϕ
6 kW water heater .... 240 V, 1ϕ
8 kW oven .... 240 V, 1ϕ
6 kW cooktop .... 240 V, 1ϕ
780 VA blower motor .... 240 V, 1ϕ
5000 VA clothes dryer .... 240 V, 1ϕ
1560 VA water pump .... 120 V, 1ϕ
1440 VA sump pump .... 120 V, 1ϕ
1200 VA microwave .... 120 V, 1ϕ
1500 VA compactor .... 120 V, 1ϕ
1800 VA dishwasher .... 120 V, 1ϕ
1320 VA disposal .... 120 V, 1ϕ

Use the form STANDARD CALCULATION—ONE-FAMILY DWELLING to find the following:

______ **20-1.** Service rating in VA for the phases.

______ **20-2.** Service rating in A for the phases.

______ **20-3.** Neutral load in A.

______ **21.** A 3600 sq. ft. one-family dwelling is supplied by a 120/240 V, 1ϕ service and has the following loads:

20 kW heating unit .... 240 V, 1ϕ
3816 VA A/C unit .... 240 V, 1ϕ
5500 VA water heater .... 240 V, 1ϕ
9.5 kW oven .... 240 V, 1ϕ
9 kW oven .... 240 V, 1ϕ
10 kW cooktop .... 240 V, 1ϕ
5000 VA clothes dryer .... 240 V, 1ϕ
600 VA blower motor .... 240 V, 1ϕ
828 VA attic fan .... 120 V, 1ϕ
1200 VA microwave .... 120 V, 1ϕ
1500 VA compactor .... 120 V, 1ϕ
1700 VA dishwasher .... 120 V, 1ϕ
1380 VA disposal .... 120 V, 1ϕ
½ HP circulating pump .... 120 V, 1ϕ
Five small appliance circuits for portable appliances .... 120 V, 1ϕ

Use the form STANDARD CALCULATION—ONE-FAMILY DWELLING to find the following:

______ **21-1.** Service load in A for phases.

______ **21-2.** Size maximum THWN Cu conductors for phases A and B.

______ **21-3.** Size maximum THWN Cu conductor for the neutral.

______ **21-4.** Size of the service panelboard.

______ **21-5.** Size main for the service equipment.

______ **21-6.** Size of the grounding electrode conductor. Use copper.

______ **22.** A 1075 sq. ft. one-family dwelling is supplied by a 130 A, 120/240 V, 1ϕ service and has the following loads. Can a 10 kW, 240 V, 1ϕ heating unit and a 28 A, 240 V, 1ϕ heat pump operating together be added to the existing service? Use the form OPTIONAL CALCULATION—ONE-FAMILY DWELLING (EXISTING ONLY).

10 kW cooktop .... 240 V, 1ϕ
8 kW oven .... 240 V, 1ϕ
5 kW clothes dryer .... 240 V, 1ϕ
5.5 kW water heater .... 240 V, 1ϕ
1.2 kW dishwasher .... 120 V, 1ϕ
1140 VA microwave .... 120 V, 1ϕ

______ **23.** A 1680 sq. ft. one-family dwelling is supplied by a 120/240 V, 1ϕ service and has the following loads:

20 kW for one heating unit .... 240 V, 1ϕ
3816 VA A/C unit .... 240 V, 1ϕ
5500 VA water heater .... 240 V, 1ϕ
9.5 kW oven .... 240 V, 1ϕ
10 kW cooktop .... 240 V, 1ϕ
5000 VA clothes dryer .... 240 V, 1ϕ
600 VA blower motor .... 240 V, 1ϕ
828 VA attic fan .... 120 V, 1ϕ
1200 VA microwave .... 120 V, 1ϕ
1500 VA compactor .... 120 V, 1ϕ
1700 VA dishwasher .... 120 V, 1ϕ
1380 VA disposal .... 120 V, 1ϕ

Use the form OPTIONAL CALCULATION—ONE-FAMILY DWELLING (NEW OR EXISTING) to find the following:

______ **23-1.** Service load in VA for phases.

______ **23-2.** Service load in A for phases.

______ **23-3.** Service conductors for phases A and B. Use THWN Cu.

______ **23-4.** Size THWN Cu conductor for the neutral. Use the form STANDARD CALCULATION—ONE-FAMILY DWELLING.

______ **24.** A 2100 sq. ft. one-family dwelling is supplied by a 120/240 V, 1ϕ service and has the following loads:

25 kW for two heating units .... 240 V, 1ϕ
6 kW A/C unit with heat pump (operating separately) .... 240 V, 1ϕ
6500 VA water heater .... 240 V, 1ϕ
Two 6 kW ovens .... 240 V, 1ϕ
8 kW cooktop .... 240 V, 1ϕ
4500 VA clothes dryer .... 240 V, 1ϕ
1080 VA blower motors (total load) .... 240 V, 1ϕ
660 VA attic fan .... 120 V, 1ϕ
1500 VA microwave .... 120 V, 1ϕ
1260 VA compactor .... 120 V, 1ϕ
1850 VA dishwasher .... 120 V, 1ϕ
1140 VA disposal .... 120 V, 1ϕ

Use the form OPTIONAL CALCULATION—ONE-FAMILY DWELLING (NEW OR EXISTING) to find the following:

______ **24-1.** Other loads in VA.

______ **24-2.** Largest of heating or A/C.

______ **24-3.** Service load in VA for phases.

______ **24-4.** Service load in A for phases.

__________ **24-5.** Service conductors for phases A and B. Use THWN Cu (size maximum and minimum).

__________ **24-6.** Neutral conductor. Use THWN Cu. Use the form STANDARD CALCULATION--ONE-FAMILY DWELLING.

__________ **25.** A 2350 sq. ft. one-family dwelling is supplied by a 120/240 V, 1ϕ service and has the following loads:

25 kW for four heating units . . . . 240 V, 1ϕ
5040 VA A/C unit . . . . 240 V, 1ϕ
5500 VA water heater . . . . 240 V, 1ϕ
10 kW oven . . . . 240 V, 1ϕ
10 kW oven . . . . 240 V, 1ϕ
9 kW cooktop . . . . 240 V, 1ϕ
5000 VA clothes dryer . . . . 240 V, 1ϕ
960 VA blower motors (total load) . . . . 240 V, 1ϕ
900 VA attic fan . . . . 120 V, 1ϕ
1260 VA microwave . . . . 120 V, 1ϕ
1020 VA compactor . . . . 120 V, 1ϕ
1800 VA dishwasher . . . . 120 V, 1ϕ
1176 VA disposal . . . . 120 V, 1ϕ

Use the form OPTIONAL CALCULATION—ONE-FAMILY DWELLING (NEW OR EXISTING) to find the following:

__________ **25-1.** Service load in A for phases.

__________ **25-2.** Service conductors for phases A and B. Use THWN Cu (maximum size).

__________ **25-3.** Neutral conductor. Use THWN Cu. Use the form STANDARD CALCULATION—ONE-FAMILY DWELLING.

__________ **25-4.** SE cable to supply the heating units. Use TW Cu.

__________ **25-5.** OCPD for heating units.

__________ **25-6.** OCPD for A/C (minimum).

__________ **25-7.** NMC for A/C.

STANDARD CALCULATION—ONE-FAMILY DWELLING

Use for Problem 18

1. GENERAL LIGHTING LOADS—Table 220-3(b)

______ sq. ft. × 3 VA = ______ VA

Small appliance loads—220-16(a)

______ VA × ______ circuits = ______ VA

Laundry load—220-16(b)

______ VA × 1 = ______ VA

______ VA

Applying demand factors—Table 220-11

First 3000 VA × 100% = 3000 VA

Next ______ VA × 35% = ______ VA

Remaining ______ VA × 25% = ______ VA

	PHASES		NEUTRAL
(total) ______ VA	VA		VA

2. COOKING EQUIPMENT LOADS—Table 220-19; Notes

Col. A ______ VA demand = ______ VA

Col. B ______ VA × ______% = ______ VA

Col. C ______ VA × ______% = ______ VA

	PHASES		NEUTRAL
(total) ______ VA	VA	× 70% =	VA

3. DRYER LOAD—220-18; Table 220-18

	PHASES		NEUTRAL
______ VA × ______% = ______ VA	VA	× 70% =	VA

4. FIXED APPLIANCE LOADS—220-17

Dishwasher = ______ VA

Disposal = ______ VA

Compactor = ______ VA

Water heater = ______ VA

______ = ______ VA

______ = ______ VA

______ = ______ VA

______ = ______ VA

	PHASES	NEUTRAL (120 V Loads × 75%)
(total) ______ VA × 75% = ______ VA	VA	VA

5. HEATING OR A/C LOAD—220-21

Heating unit = ______ VA × 100% = ______ VA

A/C unit = ______ VA × 100% = ______ VA

Heat pump = ______ VA × 100% = ______ VA

	PHASES	NEUTRAL
(largest load) ______ VA	VA	VA

6. LARGEST MOTOR LOAD—220-14

	PHASES	NEUTRAL
ϕ ______ VA × 25% = ______ VA	VA	
N ______ VA × 25% = ______ VA	______	VA
TOTAL =	VA	VA

1ϕ service: I = $\frac{VA}{V}$

PHASES I = $\frac{\_\_\_\_\_\_ VA}{V}$ = ______ A

NEUTRAL I = $\frac{\_\_\_\_\_\_ VA}{V}$ = ______ A

If over 200 A—220-22; First 200 A × 100% = 200 A

Remaining ______ A × 70% = ______ A

______ A

Use for Problem 19

STANDARD CALCULATION—ONE-FAMILY DWELLING

1. GENERAL LIGHTING LOADS—Table 220-3(b)

_______ sq. ft. × 3 VA = _______ VA

Small appliance loads—220-16(a)

_______ VA × _______ circuits = _______ VA

Laundry load—220-16(b)

_______ VA × 1 = _______ VA

_______ VA

Applying demand factors—Table 220-11

First 3000 VA × 100% = 3000 VA

Next _______ VA × 35% = _______ VA

Remaining _______ VA × 25% = _______ VA

(total) _______ VA — PHASES: VA — NEUTRAL: VA

2. COOKING EQUIPMENT LOADS—Table 220-19; Notes

Col. A _______ VA demand = _______ VA

Col. B _______ VA × _______% = _______ VA

Col. C _______ VA × _______% = _______ VA

(total) _______ VA — PHASES: VA × 70% = NEUTRAL: VA

3. DRYER LOAD—220-18; Table 220-18

_______ VA × _______% = _______ VA — PHASES: VA × 70% = NEUTRAL: VA

4. FIXED APPLIANCE LOADS—220-17

Dishwasher = _______ VA

Disposal = _______ VA

Compactor = _______ VA

Water heater = _______ VA

_______ = _______ VA

_______ = _______ VA

_______ = _______ VA

_______ = _______ VA

(total) _______ VA × 75% = _______ VA — PHASES: VA — NEUTRAL (120 V Loads × 75%): VA

5. HEATING OR A/C LOAD—220-21

Heating unit = _______ VA × 100% = _______ VA

A/C unit = _______ VA × 100% = _______ VA

Heat pump = _______ VA × 100% = _______ VA

(largest load) _______ VA — PHASES: VA — NEUTRAL: VA

6. LARGEST MOTOR LOAD—220-14

ϕ _______ VA × 25% = _______ VA — PHASES: VA

N _______ VA × 25% = _______ VA — NEUTRAL: VA

TOTAL = PHASES: VA — NEUTRAL: VA

1ϕ service: $I = \frac{VA}{V}$

PHASES $I = \frac{\_\_\_\_\_\_\ VA}{V}$ = _______ A

NEUTRAL $I = \frac{\_\_\_\_\_\_\ VA}{V}$ = _______ A

If over 200 A—220-22; First 200 A × 100% = 200 A

Remaining _______ A × 70% = _______ A

A

Use for Problem 20

STANDARD CALCULATION—ONE-FAMILY DWELLING

1. GENERAL LIGHTING LOADS—Table 220-3(b)

_____ sq. ft. × 3 VA = _____ VA

Small appliance loads—220-16(a)

_____ VA × _____ circuits = _____ VA

Laundry load—220-16(b)

_____ VA × 1 = _____ VA

_____ VA

Applying demand factors—Table 220-11

First 3000 VA × 100% = 3000 VA

Next _____ VA × 35% = _____ VA

Remaining _____ VA × 25% = _____ VA

		PHASES		NEUTRAL
(total)	_____ VA	VA		VA

2. COOKING EQUIPMENT LOADS—Table 220-19; Notes

Col. A _____ VA demand = _____ VA

Col. B _____ VA × _____% = _____ VA

Col. C _____ VA × _____% = _____ VA

(total) _____ VA | VA | × 70% = | VA

3. DRYER LOAD—220-18; Table 220-18

_____ VA × _____% = _____ VA | VA | × 70% = | VA

4. FIXED APPLIANCE LOADS—220-17

Dishwasher = _____ VA

Disposal = _____ VA

Compactor = _____ VA

Water heater = _____ VA

_____ = _____ VA

_____ = _____ VA

_____ = _____ VA

_____ = _____ VA

(total) _____ VA × 75% = _____ VA | VA | (120 V Loads × 75%) VA

5. HEATING OR A/C LOAD—220-21

Heating unit = _____ VA × 100% = _____ VA

A/C unit = _____ VA × 100% = _____ VA

Heat pump = _____ VA × 100% = _____ VA

(largest load) _____ VA | VA | VA

6. LARGEST MOTOR LOAD—220-14

ϕ _____ VA × 25% = _____ VA | VA

N _____ VA × 25% = _____ VA | | VA

TOTAL = VA | VA

1ϕ service: I = $\frac{VA}{V}$

PHASES I = $\frac{\_\_\_\_\_ VA}{V}$ = _____ A

NEUTRAL I = $\frac{\_\_\_\_\_ VA}{V}$ = _____ A

If over 200 A—220-22; First 200 A × 100% = 200 A

Remaining _____ A × 70% = _____ A

A

STANDARD CALCULATION—ONE-FAMILY DWELLING

Use for Problem 21

1. GENERAL LIGHTING LOADS—Table 220-3(b)

_______ sq. ft. × 3 VA = _______ VA

Small appliance loads—220-16(a)

_______ VA × _______ circuits = _______ VA

Laundry load—220-16(b)

_______ VA × 1 = _______ VA

_______ VA

Applying demand factors—Table 220-11

First 3000 VA × 100% = 3000 VA

Next _______ VA × 35% = _______ VA

Remaining _______ VA × 25% = _______ VA

		PHASES	NEUTRAL
(total)	_______ VA	VA	VA

2. COOKING EQUIPMENT LOADS—Table 220-19; Notes

Col. A _______ VA demand = _______ VA

Col. B _______ VA × _______% = _______ VA

Col. C _______ VA × _______% = _______ VA

(total) _______ VA | VA × 70% = | VA

3. DRYER LOAD—220-18; Table 220-18

_______ VA × _______% = _______ VA | VA × 70% = | VA

4. FIXED APPLIANCE LOADS—220-17

Dishwasher = _______ VA

Disposal = _______ VA

Compactor = _______ VA

Water heater = _______ VA

_______ = _______ VA

_______ = _______ VA

_______ = _______ VA

_______ = _______ VA

(total) _______ VA × 75% = _______ VA | VA | (120 V Loads × 75%) VA

5. HEATING OR A/C LOAD—220-21

Heating unit = _______ VA × 100% = _______ VA

A/C unit = _______ VA × 100% = _______ VA

Heat pump = _______ VA × 100% = _______ VA

(largest load) _______ VA | VA | VA

6. LARGEST MOTOR LOAD—220-14

ϕ _______ VA × 25% = _______ VA | VA

N _______ VA × 25% = _______ VA | | VA

TOTAL = VA | VA

1ϕ service: $I = \frac{VA}{V}$

PHASES $I = \frac{\_\_\_\_\_\_ VA}{V}$ = _______ A

NEUTRAL $I = \frac{\_\_\_\_\_\_ VA}{V}$ = _______ A

If over 200 A—220-22; First 200 A × 100% = 200 A

Remaining _______ A × 70% = _______ A

A

Use for Problem 22

OPTIONAL CALCULATION—ONE-FAMILY DWELLING (EXISTING ONLY)

1. OTHER LOADS—220-31

General lighting loads—220-31

__________ sq. ft. × 3 VA = __________ VA

Small appliance and laundry loads—220-31

__________ VA × ______ circuits = __________ VA

Special loads—220-31

Dishwasher = __________ VA

Disposal = __________ VA

Compactor = __________ VA

Water heater = __________ VA

____________________ = __________ VA

____________________ = __________ VA

____________________ = __________ VA

____________________ = __________ VA

____________________ = __________ VA

____________________ = __________ VA

(total) __________ VA

Applying demand factors—220-31

First 8000 VA × 100% = 8000 VA

Remaining __________ VA × 40% = __________ VA

Added appliance = __________ VA × 100% = __________ VA

(total) __________ VA

PHASES

__________ **VA**

__________ **A**

1ϕ service: $I = \frac{VA}{V}$

$I = \frac{\_\_\_\_\_\_ VA}{V}$ = __________ **A**

(*Note:* Total ampacity of service shall not exceed service conductor ampacity. Existing service conductor ampacity = __________ A.)

NEUTRAL

(Loads from Standard Calculation)

1. General lighting = __________ VA
2. Cooking equipment = __________ VA
3. Dryer = __________ VA
4. Fixed appliance = __________ VA
5. Heating or A/C = __________ VA
6. Largest motor = __________ VA

TOTAL = __________ **VA**

__________ **A**

OPTIONAL CALCULATION—ONE-FAMILY DWELLING (NEW OR EXISTING)

Use for Problem 23

1. **OTHER LOADS—220-30(b)**

General lighting loads—220-30(b)(2)

_______ sq. ft. × 3 VA = _______ VA

Small appliance and laundry loads—220-30(b)(1)

_______ VA × _______ circuits = _______ VA

Special loads—220-30(b)(3)(4)

Dishwasher = _______ VA

Disposal = _______ VA

Compactor = _______ VA

Water heater = _______ VA

_______ = _______ VA

_______ = _______ VA

_______ = _______ VA

_______ = _______ VA

_______ = _______ VA

_______ = _______ VA

(total) _______ VA

Applying demand factors—Table 220-30(4)

First 10,000 VA × 100% = 10,000 VA

Remaining _______ VA × 40% = _______ VA

(total) _______ VA

PHASES

VA

2. **HEATING OR A/C LOAD—Table 220-30(1)(2)(3)**

Heating units (3 or less) = _______ VA × 65% = _______ VA

Heating units (4 or more) = _______ VA × 40% = _______ VA

A/C unit = _______ VA × 100% = _______ VA

Heat pump = _______ VA × 100% = _______ VA

(largest load) _______ VA | VA

TOTAL = VA

NEUTRAL

(Loads from Standard Calculation)

1. General lighting = _______ VA
2. Cooking equipment = _______ VA
3. Dryer = _______ VA
4. Fixed appliance = _______ VA
5. Heating or A/C = _______ VA
6. Largest motor = _______ VA

TOTAL = VA

A

1ϕ service: $I = \frac{VA}{V}$

PHASES $I = \frac{\_\_\_\_\_\_ VA}{V}$ = _______ A

NEUTRAL $I = \frac{\_\_\_\_\_\_ VA}{V}$ = _______ A

A

STANDARD CALCULATION—ONE-FAMILY DWELLING

Use for Problem 23-4

1. GENERAL LIGHTING LOADS—Table 220-3(b)

_______ sq. ft. × 3 VA = _______ VA

Small appliance loads—220-16(a)

_______ VA × _______ circuits = _______ VA

Laundry load—220-16(b)

_______ VA × 1 = _______ VA

_______ VA

Applying demand factors—Table 220-11

First 3000 VA × 100% = 3000 VA

Next _______ VA × 35% = _______ VA

Remaining _______ VA × 25% = _______ VA

	PHASES		NEUTRAL
(total) _______ VA	VA		VA

2. COOKING EQUIPMENT LOADS—Table 220-19; Notes

Col. A _______ VA demand = _______ VA

Col. B _______ VA × _______% = _______ VA

Col. C _______ VA × _______% = _______ VA

(total) _______ VA — VA × 70% = VA

3. DRYER LOAD—220-18; Table 220-18

_______ VA × _______% = _______ VA — VA × 70% = VA

4. FIXED APPLIANCE LOADS—220-17

Dishwasher = _______ VA

Disposal = _______ VA

Compactor = _______ VA

Water heater = _______ VA

_______ = _______ VA

_______ = _______ VA

_______ = _______ VA

_______ = _______ VA

(total) _______ VA × 75% = _______ VA — VA — (120 V Loads × 75%) VA

5. HEATING OR A/C LOAD—220-21

Heating unit = _______ VA × 100% = _______ VA

A/C unit = _______ VA × 100% = _______ VA

Heat pump = _______ VA × 100% = _______ VA

(largest load) _______ VA — VA — VA

6. LARGEST MOTOR LOAD—220-14

ϕ _______ VA × 25% = _______ VA — VA

N _______ VA × 25% = _______ VA — _______ — VA

TOTAL = VA — VA

1ϕ service: I = $\frac{VA}{V}$

PHASES I = $\frac{\_\_\_\_\_\_\_ VA}{V}$ = _______ A

NEUTRAL I = $\frac{\_\_\_\_\_\_\_ VA}{V}$ = _______ A

If over 200 A—220-22; First 200 A × 100% = 200 A

Remaining _______ A × 70% = _______ A

A

OPTIONAL CALCULATION—ONE-FAMILY DWELLING (NEW OR EXISTING)

Use for Problem 24

1. OTHER LOADS—220-30(b)

General lighting loads—220-30(b)(2)

________ sq. ft. × 3 VA = ________ VA

Small appliance and laundry loads—220-30(b)(1)

________ VA × ________ circuits = ________ VA

Special loads—220-30(b)(3)(4)

Dishwasher = ________ VA

Disposal = ________ VA

Compactor = ________ VA

Water heater = ________ VA

________________ = ________ VA

________________ = ________ VA

________________ = ________ VA

________________ = ________ VA

________________ = ________ VA

________________ = ________ VA

(total) ________ VA

Applying demand factors—Table 220-30(4)

First 10,000 VA × 100% = 10,000 VA

Remaining ________ VA × 40% = ________ VA

(total) ________ VA

PHASES

VA

2. HEATING OR A/C LOAD—Table 220-30(1)(2)(3)

Heating units (3 or less) = ________ VA × 65% = ________ VA

Heating units (4 or more) = ________ VA × 40% = ________ VA

A/C unit = ________ VA × 100% = ________ VA

Heat pump = ________ VA × 100% = ________ VA

(largest load) ________ VA VA

TOTAL = VA

A

NEUTRAL

(Loads from Standard Calculation)

1. General lighting = ________ VA
2. Cooking equipment = ________ VA
3. Dryer = ________ VA
4. Fixed appliance = ________ VA
5. Heating or A/C = ________ VA
6. Largest motor = ________ VA

TOTAL = VA

A

1ϕ service: $I = \frac{VA}{V}$

PHASES $I = \frac{\_\_\_\_\ VA}{V}$ = ________ A

NEUTRAL $I = \frac{\_\_\_\_\ VA}{V}$ = ________ A

STANDARD CALCULATION—ONE-FAMILY DWELLING

Use for Problem 24-6

1. GENERAL LIGHTING LOADS—Table 220-3(b)

______ sq. ft. × 3 VA = ______ VA

Small appliance loads—220-16(a)

______ VA × ______ circuits = ______ VA

Laundry load—220-16(b)

______ VA × 1 = ______ VA

______ VA

Applying demand factors—Table 220-11

First 3000 VA × 100% = 3000 VA

Next ______ VA × 35% = ______ VA

Remaining ______ VA × 25% = ______ VA

	PHASES	NEUTRAL
(total) ______ VA	VA	VA

2. COOKING EQUIPMENT LOADS—Table 220-19; Notes

Col. A ______ VA demand = ______ VA

Col. B ______ VA × ______% = ______ VA

Col. C ______ VA × ______% = ______ VA

(total) ______ VA | VA × 70% = | VA

3. DRYER LOAD—220-18; Table 220-18

______ VA × ______% = ______ VA | VA × 70% = | VA

4. FIXED APPLIANCE LOADS—220-17

Dishwasher = ______ VA

Disposal = ______ VA

Compactor = ______ VA

Water heater = ______ VA

______ = ______ VA

______ = ______ VA

______ = ______ VA

______ = ______ VA

(120 V Loads × 75%)

(total) ______ VA × 75% = ______ VA | VA | VA

5. HEATING OR A/C LOAD—220-21

Heating unit = ______ VA × 100% = ______ VA

A/C unit = ______ VA × 100% = ______ VA

Heat pump = ______ VA × 100% = ______ VA

(largest load) ______ VA | VA | VA

6. LARGEST MOTOR LOAD—220-14

ϕ ______ VA × 25% = ______ VA | VA

N ______ VA × 25% = ______ VA | ______ | VA

TOTAL = | VA | VA

1ϕ service: $I = \dfrac{VA}{V}$

PHASES $I = \dfrac{\_\_\_\_\_\_ VA}{V} =$ ______ A

NEUTRAL $I = \dfrac{\_\_\_\_\_\_ VA}{V} =$ ______ A

If over 200 A—220-22; First 200 A × 100% = 200 A

Remaining ______ A × 70% = ______ A

A

Use for Problem 25

OPTIONAL CALCULATION—ONE-FAMILY DWELLING (NEW OR EXISTING)

1. OTHER LOADS—220-30(b)

General lighting loads—220-30(b)(2)

________ sq. ft. × 3 VA = ________ VA

Small appliance and laundry loads—220-30(b)(1)

________ VA × ________ circuits = ________ VA

Special loads—220-30(b)(3)(4)

Dishwasher = ________ VA

Disposal = ________ VA

Compactor = ________ VA

Water heater = ________ VA

________ = ________ VA

________ = ________ VA

________ = ________ VA

________ = ________ VA

________ = ________ VA

________ = ________ VA

(total) ________ VA

Applying demand factors—Table 220-30(4)

First 10,000 VA × 100% = 10,000 VA

Remaining ________ VA × 40% = ________ VA

(total) ________ VA

PHASES ________ VA

2. HEATING OR A/C LOAD—Table 220-30(1)(2)(3)

Heating units (3 or less) = ________ VA × 65% = ________ VA

Heating units (4 or more) = ________ VA × 40% = ________ VA

A/C unit = ________ VA × 100% = ________ VA

Heat pump = ________ VA × 100% = ________ VA

(largest load) ________ VA ________ VA

TOTAL = ________ VA

NEUTRAL

(Loads from Standard Calculation)

1. General lighting = ________ VA
2. Cooking equipment = ________ VA
3. Dryer = ________ VA
4. Fixed appliance = ________ VA
5. Heating or A/C = ________ VA
6. Largest motor = ________ VA

TOTAL = ________ VA

________ A

1ϕ service: $I = \frac{VA}{V}$

PHASES I = ________ VA / V = ________ A

NEUTRAL I = ________ VA / V = ________ A

________ A

STANDARD CALCULATION—ONE-FAMILY DWELLING

Use for Problem 25-3

1. GENERAL LIGHTING LOADS—Table 220-3(b)

_____ sq. ft. × 3 VA = _____ VA

Small appliance loads—220-16(a)

_____ VA × _____ circuits = _____ VA

Laundry load—220-16(b)

_____ VA × 1 = _____ VA

_____ VA

Applying demand factors—Table 220-11

First 3000 VA × 100% = 3000 VA

Next _____ VA × 35% = _____ VA

Remaining _____ VA × 25% = _____ VA

(total) _____ VA — PHASES: _____ VA — NEUTRAL: _____ VA

2. COOKING EQUIPMENT LOADS—Table 220-19; Notes

Col. A _____ VA demand = _____ VA

Col. B _____ VA × _____% = _____ VA

Col. C _____ VA × _____% = _____ VA

(total) _____ VA — PHASES: _____ VA × 70% = NEUTRAL: _____ VA

3. DRYER LOAD—220-18; Table 220-18

_____ VA × _____% = _____ VA — PHASES: _____ VA × 70% = NEUTRAL: _____ VA

4. FIXED APPLIANCE LOADS—220-17

Dishwasher – _____ VA

Disposal = _____ VA

Compactor = _____ VA

Water heater = _____ VA

_____ = _____ VA

_____ = _____ VA

_____ = _____ VA

_____ = _____ VA

(total) _____ VA × 75% = _____ VA — PHASES: _____ VA — NEUTRAL (120 V Loads × 75%): _____ VA

5. HEATING OR A/C LOAD—220-21

Heating unit = _____ VA × 100% = _____ VA

A/C unit = _____ VA × 100% = _____ VA

Heat pump = _____ VA × 100% = _____ VA

(largest load) _____ VA — PHASES: _____ VA — NEUTRAL: _____ VA

6. LARGEST MOTOR LOAD—220-14

ϕ _____ VA × 25% = _____ VA — PHASES: _____ VA

N _____ VA × 25% = _____ VA — NEUTRAL: _____ VA

TOTAL = PHASES: _____ VA — NEUTRAL: _____ VA

1ϕ service: $I = \frac{VA}{V}$

PHASES $I = \frac{\_\_\_\_\ VA}{V}$ = _____ A

NEUTRAL $I = \frac{\_\_\_\_\ VA}{V}$ = _____ A

If over 200 A—220-22; First 200 A × 100% = 200 A

Remaining _____ A × 70% = _____ A

_____ A

Calculations—Multifamily Dwellings

Chapter 7

Calculations to find VA and amps of the service for dwellings are based on Article 220. The standard calculation is applied per 220, Part B and the optional calculation is applied per 220, Part C. The standard calculation always produces larger VA and amp ratings than the optional calculation; therefore, the optional calculation is usually applied so that smaller elements of the service equipment can be selected.

Tables are used to find general lighting loads based on the square footage of occupancies. Demand factors are applied for dryers and ranges based on the number of appliances. Standardized forms facilitate calculations. Other elements of the service are based on the size of the service.

STANDARD CALCULATION—220, PART B

The load for a number of dwelling units located in an apartment complex, condominium complex, or townhouse complex is determined by multiplying the square footage of each dwelling unit by 3 VA per square foot. The VA of each unit is then multiplied by the number of units. The small appliance circuits are computed at 1500 VA each times the number of dwelling units. Each special appliance load is computed at its nameplate rating and demand factors are applied where permitted. These values are multiplied by the number of dwelling units.

The loads in each dwelling unit are divided into three groups:

1. General lighting loads
2. Small appliance loads
3. Special appliance loads

Four loads are selected from the three groups of loads. Demand factors can then be applied. The loads to which demand factors can be applied are

1. General lighting loads
2. Cooking equipment loads
3. Dryer equipment loads
4. Fixed appliance loads

Two additional loads are computed. They are

1. Largest between heating or A/C unit load
2. Largest motor load

General Lighting Loads—220-3(b)

The general lighting load is found by selecting 3 VA per square foot from Table 220-3(b) for a dwelling unit. The square footage of each dwelling in the complex is multiplied by 3 VA per square foot times the number of units to find the total VA rating for the general lighting load.

What is the total VA rating of fifteen 990 sq. ft. dwelling units located in an apartment complex?

Step 1. *Table 220-3(b):*
990 sq. ft. × 3 VA × 15 = 44,550 VA

Answer. **VA rating = 44,550 VA**

Most apartment complexes have dwelling units with varying square footages. In these cases, each dwelling with the same square footage is multiplied by 3 VA per square foot and the totals are added to obtain the VA rating for the apartment complex. See Figure 7-1.

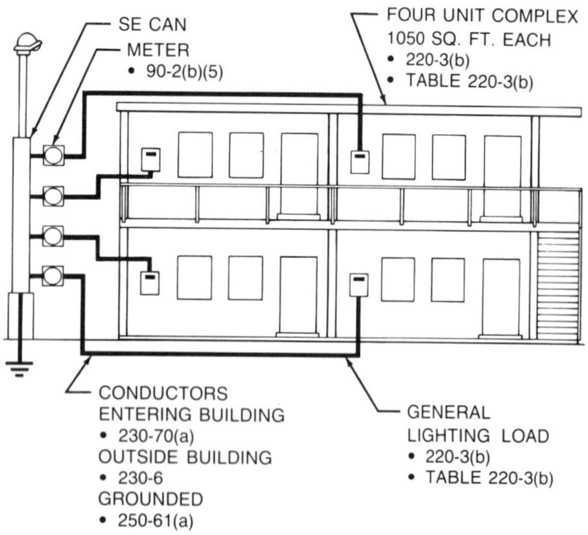

GENERAL LIGHTING AND RECEPTACLE LOADS

What are the general lighting and receptacle loads for the four unit complex?

Step 1. *Table 220-3(b):*
1050 sq. ft. × 3 VA × 4 units = 12,600 VA

Answer. **Loads = 12,600 VA**

Figure 7-1. The general lighting and receptacle loads are found by multiplying the square footage of each unit by 3 VA. For units with the same square footage and loads, the total VA rating is multiplied by the number of units.

What is the VA rating for five 880 sq. ft. dwellings, five 900 sq. ft. dwellings, five 950 sq. ft. dwellings, and five 1050 sq. ft. dwellings in an apartment complex?

Step 1. *Table 220-3(b):*

880 sq. ft. × 3 VA × 5 =	13,200 VA
900 sq. ft. × 3 VA × 5 =	13,500 VA
950 sq. ft. × 3 VA × 5 =	14,250 VA
1050 sq. ft. × 3 VA × 5 =	15,750 VA
Total =	56,700 VA

Answer. **VA rating = 56,700 VA**

Small Appliance Loads—220-16(a)

Each dwelling unit located in a complex of apartments, condominiums, or townhouses shall have at least two small appliance circuits rated at 20 A, each installed for receptacle outlets located in the kitchen, pantry, dining room, and breakfast room. The VA rating used for each circuit to compute the load used to select the elements of the service equipment is 1500 VA per circuit. To determine the small appliance load for a number of dwelling units in a complex, 1500 VA times 2 is multiplied by the number of units.

What is the small appliance load in VA for 20 units in an apartment complex?

Step 1. *220-16 (a):* 1500 VA × 2 × 20 = 60,000 VA

Answer. **Small appliance load = 60,000 VA**

One 20 A small appliance circuit is required for the laundry room to serve receptacle outlets related to the laundry equipment.

What is the load in VA for the small appliance circuits used for the laundry equipment in twenty dwelling units located in an apartment complex?

Step 1. *220-16(b):* 1500 VA × 1 × 20 = 30,000 VA

Answer. **Load = 30,000 VA**

Section 210-52(f), Ex. 1 does not require a 20 A appliance circuit to be installed in each dwelling unit in an apartment complex for laundry facilities available to all tenants. Where the laundry facilities are furnished for each tenant, the load for the laundry equipment is connected to the house meter of the apartment complex. See Figure 7-2.

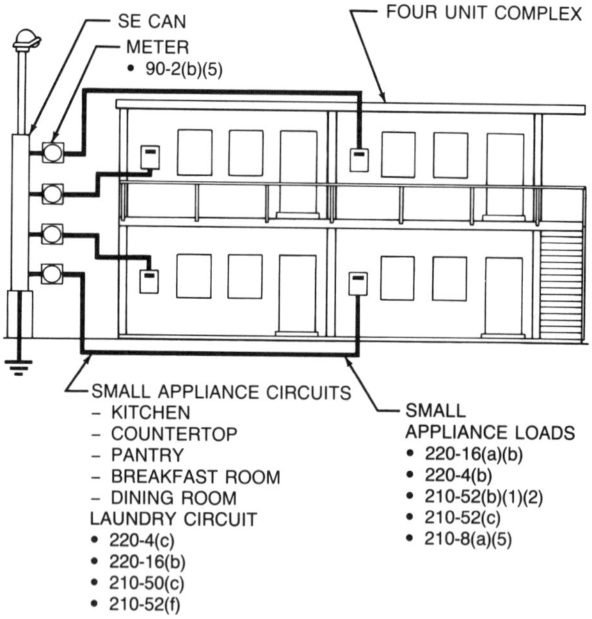

SMALL APPLIANCE LOADS

What is the small appliance load for the four unit complex?

Step 1. *220-16(a)(b):*
1500 VA × 3 × 4 units = 18,000 VA

Answer. **Load = 18,000 VA**

Figure 7-2. Two small appliance circuits at 1500 VA each and one small appliance circuit (for laundry loads) are computed at 1500 VA for a minimum of 4500 VA for the small appliance loads.

Twenty pieces of laundry equipment are used at continuous duty in the laundry facilities in an apartment complex. What is the load in VA for the laundry equipment with a connected load of 18 kW?

Step 1. *210-52(f), Ex. 1; 220-10(b):*
18 kW × 125% = 22.5 kW

Step 2. *Text:* 22.5 kW × 1000 = 22,500 VA

Answer. **Load** = 22,500 VA

Special Appliance Loads—220-3(a); 220-10(b)

Special appliance loads are loads such as heating units, A/C units, water heaters, dishwashers, disposals, and compactors. These loads are usually served by direct circuits. However, special appliance loads under certain requirements and conditions of use can be served by general-purpose circuits per 210-23(a), 430-42, and 430-53. The VA rating of each special appliance load not classified as a fixed appliance is multiplied by the number of dwelling units in the complex. For four or more units, a demand factor of 75% is permitted per 220-17.

What is the load in VA for twenty 10 kW heating units in dwelling units located in an apartment complex? Apply the standard calculation.

Step 1. *220-15:* 10 kW × 100% × 20 = 200 kW

Step 2. *Text:* 200 kW × 1000 = 200,000 VA

Answer. **Load** = **200,000 VA**

Applying Demand Factors—220 Part B

Demand factors can be applied to four loads. These are

1. General lighting loads. *Table 220-3(b); 220-16(a)(b)*
2. Cooking equipment loads. *Table 220-19*
3. Dryer loads. *Table 220-18*
4. Fixed appliance loads. *220-17*

The number of dwelling units in the complex determines the percentage of demand factors that can be applied.

General Lighting Loads and Small Appliance Loads—Table 220-11. The VA rating for the general lighting load is determined by multiplying the square footage of each dwelling unit by 3 VA per square foot times the number of units in the complex per Table 220-3(b).

The small appliance loads are determined by multiplying 1500 VA by 3 times the number of dwelling units to derive the VA rating per 220-16(a)(b). The general lighting loads and small appliance loads are added together and the demand factors of Table 220-11 are applied. The demand load is used for the general lighting loads and the small appliance loads to size the elements of the service equipment. See Figure 7-3.

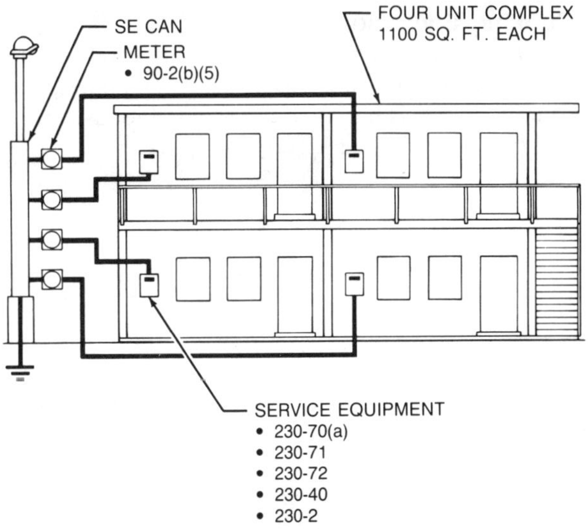

APPLYING DEMAND FACTORS TO GENERAL LIGHTING AND RECEPTACLE LOADS

What is the demand load for the general lighting and receptacle loads plus the small appliance loads for the four unit complex?

Step 1. *Table 220-3(b):*
1100 sq. ft. × 3 VA × 4 = 13,200 VA

Step 2. *220-16(a)(b):* 1500 VA × 3 × 4 = 18,000 VA

Step 3. Total load.
13,200 VA + 18,000 VA = 31,200 VA

Step 4. Applying demand factors. *Table 220-11:*
First 3000 VA × 100% = 3000 VA
Next 28,200 VA × 35% = 9870 VA
Total = 12,870 VA

Answer. **Load** = **12,870 VA**

Figure 7-3. The general lighting and receptacle loads are found by multiplying the square footage of each dwelling unit by 3 VA, adding 1500 VA for each small appliance circuit, and multiplying by the number of units. Demand factors are applied per Table 220-11.

What is the demand load in VA for twenty 990 sq. ft. dwelling units with three small appliance circuits located in an apartment complex?

Step 1. *Table 220-3(b):*
990 sq. ft. × 3 VA × 20 = 59,400 VA

Step 2. *220-16(a)(b):* 1500 VA × 3 × 20 = 90,000 VA

Step 3. Totaling loads. *Text:*
General lighting load = 59,400 VA
Small appliance load = 90,000 VA
Total = 149,400 VA

Step 4. Applying demand factors. *Table 220-11:*
First 3000 VA × 100% = 3000 VA
Next 117,000 VA × 35% = 40,950 VA
Remaining 29,400 VA × 25% = 7350 VA
Total = 51,300 VA

Answer. **Demand load** = **51,300 VA**

Cooking Equipment Loads—Table 220-19. The demand load for cooking equipment for dwelling units located in a complex of apartments, condominiums, or townhouses is based on the demand loads listed in Column B or C. See Figure 7-4.

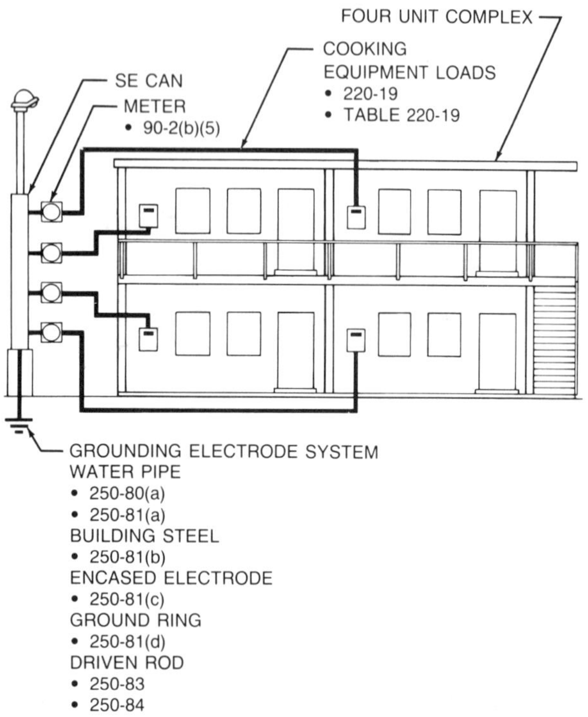

DEMAND LOADS FOR COOKING EQUIPMENT

What is the demand load for four 8 kW ranges installed in a dwelling unit in a complex?

Step 1. *Table 220-19, Column C:*
8 kW × 4 × 50% = 16 kVA

Answer. **Load = 16 kVA**

Figure 7-4. Demand loads for cooking equipment are based on the kW rating of each piece of cooking equipment.

What is the demand load in kVA for twenty 9 kW ranges located in dwelling units in an apartment complex?

Step 1. *Table 220-19, Column A:*
Twenty 9 kW ranges = 35 kW

Answer. **Demand load = 35 kVA**

Table 220-19, Column B is applied for ranges, counter-mounted cooktops, and wall-mounted ovens with kW ratings less than $3\frac{1}{2}$ kW.

What is the demand load in kVA for twenty 3 kW cooktops and twenty 3 kW ovens?

Step 1. *Table 220-19, Column B:*
Twenty 3 kW cooktops
Twenty 3 kW ovens
40 total = 30% demand factor

Step 2. Applying demand factors.
3 kVA × 40 × 30% = 36 kVA

Answer. **Demand load = 36 kVA**

Where the kW rating of each piece of cooking equipment is between $3\frac{1}{2}$ kW to $8\frac{3}{4}$ kW, Column C is used to determine the demand load. The percentage of the demand factor is selected based upon the number of units.

What is the demand load in kVA for ten 8.5 kW ovens and ten 7.5 kW cooktops installed in ten dwelling units of a townhouse complex?

Step 1. *Table 220-19, Column C:*
Ten 8.5 kW ovens
Ten 7.5 kW cooktops
20 total = 28% demand factor

Step 2. Applying demand factors.
8.5 kVA × 10 × 28% = 23.8 kVA
7.5 kVA × 10 × 28% = 21 kVA
Total = 44.8 kVA

Answer. **Demand load = 44.8 kVA**

Where the kW rating of each unit does not fall in Column B or C but falls in both Columns B and C, each Column is used to determine demand factors based upon the number of units.

What is the demand load in kVA for fifteen 2.5 kW cooktops and fifteen 5 kW ovens in the dwelling units of a condominium building?

Step 1. *Table 220-19, Columns B and C:*
Fifteen 2.5 kW cooktops = 40% demand factor
Fifteen 5 kW ovens = 32% demand factor

Step 2. Applying demand factors.
2.5 kVA × 15 × 40% = 15 kVA
5 kVA × 15 × 32% = 24 kVA
Total = 39 kVA

Answer. **Demand load = 39 kVA**

■*Master Electrician's Test Tip.* Some examination boards allow all pieces of cooking equipment rated over $1\frac{3}{4}$ kW and up to $8\frac{3}{4}$ kW to be grouped and the percentages in Column B or C to be used to find the demand load.■

What is the demand load in kVA of thirty 10 kW ranges located in an apartment complex?

Step 1. *Table 220-19, Column A:*
30 ranges =
15 kW + 1 kW per range demand factor

Step 2. Applying demand factors.
30 ranges = 15 kW
1 kW per range = 30 kW
Total = 45 kW

Answer. **Demand load = 45 kVA**

Dryer Loads—Table 220-18. There shall be over four dryers before demand factors are applied to the number of dryers located in dwelling units in a complex. The standard calculation requires each dryer to be computed at 5000 VA or the nameplate rating, whichever is greater per 220-3(c), Ex. 5 and 220-18. For example, a dryer with a nameplate rating of 4500 VA is calculated at 5000 VA and a dryer with a nameplate rating of 6500 VA is calculated at 6500 VA. See Figure 7-5.

Twelve 5000 W dryers are installed in an apartment complex. What is the demand load in VA?

Step 1. *Table 220-18:*
Twelve 5000 VA dryers = 45% demand factor

Step 2. 5000 VA × 12 × 45% = 27,000 VA

Answer. **Demand load = 27,000 VA**

DEMAND LOADS FOR DRYERS

What is the demand load in VA for one 5000 VA dryer?

Step 1. *220-18; Table 220-18:*
Fewer than 4 dryers = 100%

Step 2. *220-18; Table 220-18:*
5000 VA × 100% = 5000 VA

Answer. **Load = 5000 VA**

What is the demand load in VA for fifteen 5500 VA dryers?

Step 1. *220-18; Table 220-18:*
5500 VA × 15 × 40% = 33,000 VA

Answer. **Load = 33,000 VA**

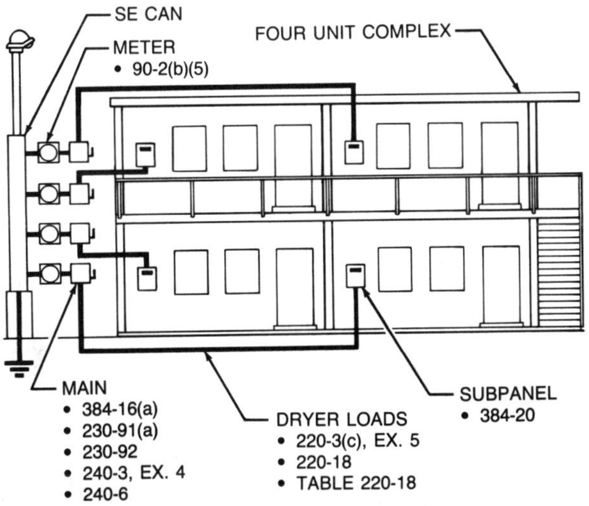

Figure 7-5. Demand loads for dryers are found by multiplying the percentages in Table 220-18 by the VA rating of each dryer.

Fixed Appliance Loads—220-17. Each fixed appliance load is taken at nameplate rating and multiplied by the number of units. If the number of units is four or more, a demand factor of 75% is permitted per 220-17.

The exception to 220-17 will not permit the 75% demand factor to be applied to the heating unit, air conditioning equipment, cooking equipment, or dryer equipment. Fixed appliance loads can be supplied by an individual branch circuit or be connected to a general-purpose branch circuit. Portable appliances connected to the small appliance circuits are not considered fixed appliances. The blower motor is considered a fixed appliance because it can be used during the heating or air conditioning cycle, or as a circulating fan. See Figure 7-6.

■*Master Electrician's Test Tip.* Some testing authorities require the blower motor to be calculated as part of the heating load and added to the service load at 100% per 220-15.■

What is the demand load in VA for eighteen 6 A, 120 V, 1ϕ disposals in dwelling units located in an apartment complex?

Step 1. *Text: VA = V × A*
120 V × 6 A = 720 VA

Step 2. Applying demand factors. *220-17:*
720 VA × 18 × 75% = 9720 VA

Answer. **Demand load = 9720 VA**

If the disposal were the only fixed appliance load in the dwelling unit, as there are 18 dwelling units in the complex, four or more fixed appliances are present. The 75% demand factor multiplied by the number of dwelling units times the VA rating of the fixed appliance is permitted.

■*Master Electrician's Test Tip.* Some examination boards interpret 220-17 to require four or more fixed appliances for each dwelling unit in a complex in order to apply the 75% demand factor.■

In each dwelling unit having fixed appliances with different nameplate ratings, each appliance load is multiplied by 75% to derive the demand load for each fixed appliance.

What is the demand load in VA for four dwelling units with the following fixed appliances?

Unit 1

7.5 kW heating unit	240 V, 1ϕ
3000 VA A/C unit	240 V, 1ϕ
5500 water heater	240 V, 1ϕ
10 kW cooktop	240 V, 1ϕ
600 VA blower motor	240 V, 1ϕ
5000 VA dryer	240 V, 1ϕ
9.5 kW oven	240 V, 1ϕ
828 VA attic fan	120 V, 1ϕ
1200 VA microwave	120 V, 1ϕ
1380 VA disposal	120 V, 1ϕ
1500 VA compactor	120 V, 1ϕ
1700 VA dishwasher	120 V, 1ϕ

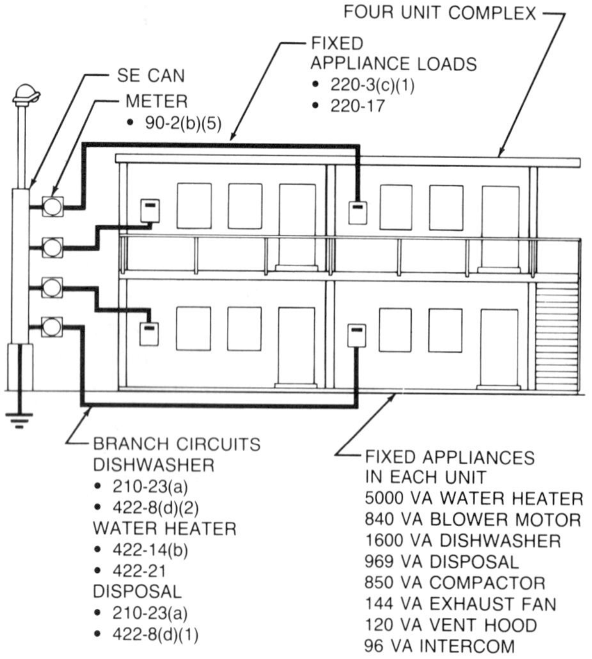

APPLYING DEMAND FACTORS FOR FIXED APPLIANCE LOADS

What is the demand load for fixed appliances in the four unit complex?

Step 1. Applying demand factors. *220-17:*

5000 VA water heater × 4 units × 75% =	15,000 VA
840 VA blower motor × 4 units × 75% =	2520 VA
1600 VA dishwasher × 4 units × 75% =	4800 VA
969 VA disposal × 4 units × 75% =	2907 VA
850 VA compactor × 4 units × 75% =	2550 VA
144 VA exhaust fan × 4 units × 75% =	432 VA
120 VA vent hood × 4 units × 75% =	360 VA
96 VA intercom × 4 units × 75% =	288 VA
Total =	28,857 VA

Answer. **Load = 28,857 VA**

Figure 7-6. Demand loads for fixed appliances are found by multiplying the VA rating of each appliance by 75% and using the total for the service load.

Unit 2

8 kW heating unit .... 240 V, 1ϕ
2880 VA A/C unit .... 240 V, 1ϕ
4500 VA water heater .... 240 V, 1ϕ
8 kW cooktop .... 240 V, 1ϕ
600 VA blower motor .... 240 V, 1ϕ
5000 VA dryer .... 240 V, 1ϕ
8 kW oven .... 240 V, 1ϕ
1200 VA microwave .... 120 V, 1ϕ
680 VA disposal .... 120 V, 1ϕ
720 VA compactor .... 120 V, 1ϕ
1600 VA dishwasher .... 120 V, 1ϕ

Unit 3

8 kW heating unit .... 240 V, 1ϕ
3120 VA A/C unit .... 240 V, 1ϕ
4500 water heater .... 240 V, 1ϕ
8.5 kW cooktop .... 240 V, 1ϕ
696 VA blower motor .... 240 V, 1ϕ
5000 VA dryer .... 240 V, 1ϕ
7 kW oven .... 240 V, 1ϕ
1656 VA attic fan .... 120 V, 1ϕ
1260 VA microwave .... 120 V, 1ϕ
864 VA disposal .... 120 V, 1ϕ
1176 VA compactor .... 120 V, 1ϕ
1650 VA dishwasher .... 120 V, 1ϕ

Unit 4

10 kW heating unit .... 240 V, 1ϕ
4536 VA A/C unit .... 240 V, 1ϕ
5000 VA water heater .... 240 V, 1ϕ
10 kW cooktop .... 240 V, 1ϕ
936 VA blower motor .... 240 V, 1ϕ
5000 VA dryer .... 240 V, 1ϕ
12 kW oven .... 240 V, 1ϕ
1500 VA microwave .... 120 V, 1ϕ
948 VA disposal .... 120 V, 1ϕ
1020 VA compactor .... 120 V, 1ϕ
1680 VA dishwasher .... 120 V, 1ϕ

Unit 1

Step 1. *220-17:*

Water heater =	5500 VA
Blower motor =	600 VA
Attic fan =	828 VA
Microwave =	1200 VA
Disposal =	1380 VA
Compactor =	1500 VA
Dishwasher =	1700 VA

Unit 2

Step 2. *220-17:*

Water heater =	4500 VA
Blower motor =	600 VA
Microwave =	1200 VA
Disposal =	680 VA
Compactor =	720 VA
Dishwasher =	1600 VA

Unit 3

Step 3. *220-17:*

Water heater =	4500 VA
Blower motor =	696 VA
Attic fan =	1656 VA
Microwave =	1260 VA

Disposal =	864 VA
Compactor =	1176 VA
Dishwasher =	1650 VA

Unit 4

Step 4. *220-17:*

Water heater =	5000 VA
Blower motor =	936 VA
Microwave =	1500 VA
Disposal =	948 VA
Compactor =	1020 VA
Dishwasher =	1680 VA
Total =	44,894 VA

Step 5. Applying demand factors. *220-17:*
44,894 VA × 75% = 33,670.5 VA

Answer. **Demand load = 33,670.5 VA**

Largest Load Between Heating and A/C Load—220-21. The largest load between the heating and A/C load is selected and added to the other five loads to determine the service elements. The heating load is figured at 100% and compared to 100% of the A/C load. The largest load is selected. The smallest load is dropped and not used again, per 220-21, until the branch circuits are computed. See Figure 7-7.

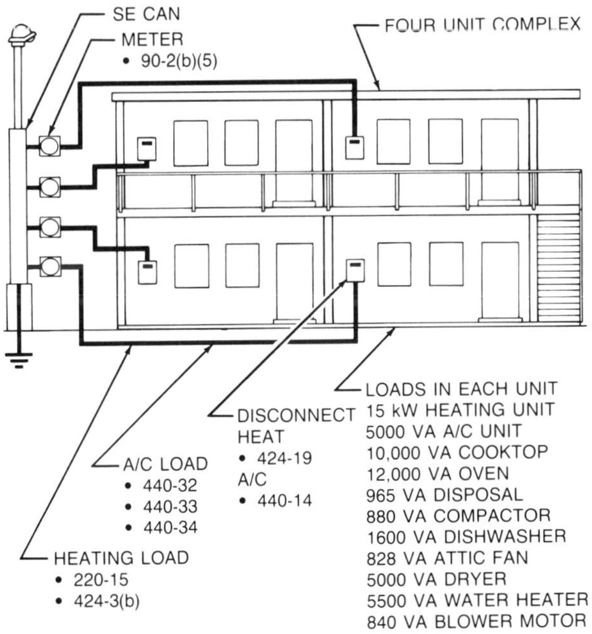

LARGEST OF HEATING AND A/C LOAD

What is the largest load between the heating and A/C load?

Step 1. *220-21:*
Heating load =
15 kW × 1000 × 4 = 60,000 VA
A/C load = 5000 VA × 4 = 20,000

Answer. **Heating load = 60,000 VA**

Figure 7-7. The largest of the heating or A/C load is used per 220-21.

What is the largest load for eight dwelling units located in an apartment complex? Each apartment has a 10 kW, 240 V, 1ϕ heating unit and a 4.5 kVA, 240 V, 1ϕ A/C load.

Step 1. Computing loads. *220-21:*
Heating load = 10 kW × 1000 = 10,000 VA
A/C load = 4.5 kVA × 1000 = 4500 VA

Step 2. Selecting largest motor load. *220-21:*
10,000 VA × 8 = 80,000 VA
4500 VA × 8 = 36,000 VA

Answer. **Largest load = 80,000 VA**

Note: Where the heating unit load is greater, the heating elements are calculated at 100% and multiplied by the number of dwelling units in the complex. Where each heating unit in a dwelling unit of a complex has a different rating, each heating unit is calculated at 100% and multiplied by the number of units.

■*Master Electrician's Test Tip.* Some examination boards multiply the heating load by 100% and the A/C load by 125%. The largest load between the two ratings is then selected.■

Largest Motor Load—220-14. The largest motor in any one dwelling unit is calculated at 25% of its full-load current rating per 220-14, 430-17, and 430-25. The full-load current ratings are selected from Table 430-148 for 1ϕ motors per 430-6(a). Section 220-14 refers to 430-24 for two or more motors in a group; however, this requirement pertains only to motors in a group. Section 220-14 refers to 430-25, which refers to 430-22(a) for one motor with other loads and 430-24 for two or more motors with other loads. See Figure 7-8.

What is the load in VA for a 3 HP, 230 V, 1ϕ motor, which is the largest motor, in an apartment complex with 225,000 VA for other motors and loads?

Step 1. *430-6(a); Table 430-148:*

Motor load = 17 A × 240 V =	4080 VA
Other loads =	225,000 VA
Total =	229,080 VA

Step 2. *220-14; 430-25; 430-24:*

Total loads =	229,080 VA
Largest motor load = 4080 VA × 25% =	1020 VA
Total =	230,100 VA

Answer. **Load = 230,100 VA**

Each 980 sq. ft. apartment in a 24 unit apartment complex has the following loads:

7.5 kW heating unit	240 V, 1ϕ
3816 VA A/C unit	240 V, 1ϕ
5500 VA water heater	240 V, 1ϕ
10 kW range	240 V, 1ϕ
600 VA blower motor	240 V, 1ϕ
5000 VA dryer	240 V, 1ϕ
828 VA attic fan	120 V, 1ϕ
1200 VA microwave	120 V, 1ϕ
1380 VA disposal	120 V, 1ϕ
1500 VA compactor	120 V, 1ϕ
1700 VA dishwasher	120 V, 1ϕ

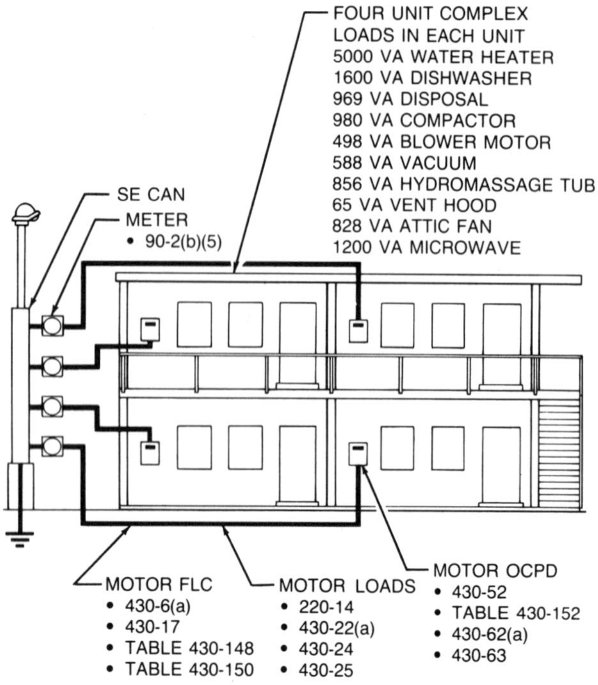

LARGEST MOTOR LOAD

What is the largest motor load?

Step 1. *220-14; 430-24; 430-25:*
Compactor = 980 VA × 25% = 245 VA

Answer. **Load = 245 VA**

Figure 7-8. The largest motor load is found by multiplying the VA rating of the largest motor by 25%.

The total loads for the apartment complex are

General lighting loads	58,590 VA
Cooking equipment loads	39,000 VA
Dryer loads	42,000 VA
Fixed appliance loads	228,744 VA
Electric heating loads	180,000 VA

What is the largest motor load?

Step 1. *220-14; 430-25; 430-24:*
1500 VA × 25% = 375 VA

Answer. **Largest motor load = 375 VA**

▪*Master Electrician's Test Tip.* Some examination boards require the A/C load to be calculated at 125%, compared to 100% of the heating load, and the smaller load dropped per 220-21. Some examination boards require the A/C load, even when dropped per 220-21, to be considered a possibility for the largest motor load per 220-14.▪

EXAMPLES USING STANDARD CALCULATION—220, PART B

The service load in VA or amps will vary in rating due to the square footage and number of dwelling units and nameplate ratings of electrical equipment times the number of units. All loads are multiplied by the number of dwelling units in the complex and demand factors applied where applicable.

Service with Units Having Same Square Footage—220-10(b)–220-22

Sixteen 1050 sq. ft. dwelling units in an apartment complex have the following loads and are served by a 120/240 V, 1ϕ supply. Calculate the service load for phases A and B and the neutral. See Figure 7-9.

10 kW heating unit	240 V, 1ϕ
4920 VA A/C unit	240 V, 1ϕ
5500 VA water heater	240 V, 1ϕ
864 VA hydromassage tub	120 V, 1ϕ
6 kW cooktop	240 V, 1ϕ
8.5 kW oven	240 V, 1ϕ
864 VA blower motor	240 V, 1ϕ
4500 VA dryer	240 V, 1ϕ
1200 VA disposal	120 V, 1ϕ
696 VA compactor	120 V, 1ϕ
1600 VA dishwasher	120 V, 1ϕ

(1) What size nonmetallic-sheathed cable (Romex) is required to supply the hydromassage tub?

Step 1. *430-22(a):* 864 VA × 125% = 1080 VA

Step 2. *Text:* $I = \frac{A}{V}$

$$I = \frac{1080\text{ VA}}{120\text{ V}} = 9\text{ A}$$

Step 3. *336-26; Table 310-16, 60°C Column:*
9 A requires #14 Cu

Answer. **Nonmetallic-sheathed cable = #14-2 w/ground**

(2) What size overcurrent protection device and type circuit is required for the hydromassage tub?

OCPD

Step 1. *Text:* $I = \frac{VA}{V}$

$$I = \frac{864\text{ VA}}{120\text{ V}} = 7.2\text{ A}$$

Step 2. *Table 430-152:* 7.2 A × 250% = 18 A

Step 3. *240-3, Ex. 6; 430-52, Ex. 1; 240-6:*
20 A is the next higher standard size

Answer. **OCPD = 20 A**

Circuit

Step 1. *680-70:* GFCI circuit required

Answer. **Circuit = GFCI**

(3) What size service-entrance conductors, using THWN Cu paralleled seven times per phase, is required?

Step 1. *Text:* $\frac{1735.7\text{ A}}{7} = 247.9\text{ A}$

Step 2. *310-4; Table 310-16, 75°C Column:*
247.9 A requires #250 kcmil THWN Cu

Answer. **Conductors = #250 kcmil THWN Cu**

STANDARD CALCULATION—MULTIFAMILY DWELLING

1. GENERAL LIGHTING LOADS—Table 220-3(b)

1050 sq. ft. × 3 VA × 16 units = 50,400 VA

— sq. ft. × 3 VA × — units = — VA

— sq. ft. × 3 VA × — units = — VA

50,400 VA

Small appliance loads—220-16(a)

1500 VA × 2 circuits × 16 units = 48,000 VA

Laundry load—220-16(b)

1500 VA × 1 × 16 units = 24,000 VA

122,400 VA

Applying demand factors—Table 220-11

First 3000 VA × 100% = 3000 VA

Next 117,000 VA × 35% = 40,950 VA

Remaining 2400 VA × 25% = 600 VA

(total) 44,550 VA

	PHASES		NEUTRAL
1. General lighting loads	44,550 VA		44,550 VA

2. COOKING EQUIPMENT LOADS—Table 220-19, Notes

Col. A — units = — VA demand = — VA

Col. B — VA × — units × — % = — VA

Col. C 14,500 VA × 16 units × 22 % = 51,040 VA

(total) 51,040 VA — PHASES 51,040 VA × 70% = NEUTRAL 35,728 VA

3. DRYER LOAD—220-18; Table 220-18

5000 VA × 16 units × 40 % = 32,000 VA — PHASES 32,000 VA × 70% = NEUTRAL 22,400 VA

4. FIXED APPLIANCE LOADS—220-17

Dishwasher = 1600 VA

Disposal = 1200 VA

Compactor = 696 VA

Water heater = 5500 VA

Hydro tub = 864 VA

Blowermotor = 864 VA

(total) 10,724 VA × 16 units × 75% = 128,688 VA — PHASES 128,688 VA

NEUTRAL (120 V Loads × units × 75%) 52,320 VA

5. HEATING OR A/C LOAD—220-21

Heating unit = 10,000 VA × 100% × 16 units = 160,000 VA

A/C unit = 4920 VA × 100% × 16 units = 78,720 VA

Heat pump = — VA × 100% × — units = — VA

(largest load) 160,000 VA — PHASES 160,000 VA — NEUTRAL — VA

6. LARGEST MOTOR LOAD—220-14

ϕ 1200 VA × 25% = 300 VA — PHASES 300 VA

N 1200 VA × 25% = 300 VA — NEUTRAL 300 VA

TOTAL = PHASES 416,578 VA — NEUTRAL 155,298 VA

1ϕ service: $I = \frac{VA}{V}$

PHASES $I = \frac{416{,}578\ VA}{240\ V}$ = 1735.7 A

NEUTRAL $I = \frac{155{,}298\ VA}{240\ V}$ = 647 A

3ϕ service: $I = \frac{VA}{V \times \sqrt{3}}$

$I = \frac{\_\_\_\ VA}{V \times \sqrt{3}}$ = ____ A

$I = \frac{\_\_\_\ VA}{V \times \sqrt{3}}$ = ____ A

If over 200 A—220-22; First 200 A × 100% = 200 A

Remaining 447 A × 70% = 312.9 A

512.9 A

Figure 7-9. The service load in VA and amps is found by using the form, STANDARD CALCULATION—MULTIFAMILY DWELLING.

(4) What size neutral conductor, using THWN Cu paralleled seven times, is required?

Step 1. *Text:* $\frac{512.9 \text{ A}}{7} = 73 \text{ A}$

Step 2. *310-4; Table 310-16, 75°C Column:*
#1/0 is the smallest conductor to be connected in parallel

Answer. **Conductor = #1/0 THWN Cu**

(5) What size conduit is required for each multiple run?

Step 1. *Table 5, Ch. 9:*

#250 kcmil THWN =	
.4026 sq. in. × 2 =	.8052 sq. in.
#1/0 THWN =	
.1893 sq. in. × 1 =	.1893 sq. in.
Total =	.9945 sq. in.

Step 2. *Table 4, Ch. 9:*
.9945 sq. in. area requires 1.34 sq. in. area

Answer. **Conduit = 2″**

Service with Units Having Different Square Footages—220-10(b)–220-22

Four different size dwelling units in an apartment complex have varying loads as follows. The service is 120/240 V, 1ϕ. What is the service load for phases A and B and the neutral? See Figure 7-10.

Unit 1—890 sq. ft.

8 kW heating unit .... 240 V, 1ϕ
3600 VA A/C .... 240 V, 1ϕ
950 VA disposal .... 120 V, 1ϕ
1550 VA dishwasher .... 120 V, 1ϕ
10 kW range .... 240 V, 1ϕ
4500 VA water heater .... 240 V, 1ϕ
5000 VA dryer .... 240 V, 1ϕ
696 VA blower motor .... 240 V, 1ϕ

Unit 2—920 sq. ft.

10 kW heating unit .... 240 V, 1ϕ
4680 VA A/C unit .... 240 V, 1ϕ
12 kW range .... 240 V, 1ϕ
5500 VA water heater .... 240 V, 1ϕ
5000 VA dryer .... 240 V, 1ϕ
1050 VA disposal .... 120 V, 1ϕ
975 VA compactor .... 120 V, 1ϕ
1600 VA dishwasher .... 120 V, 1ϕ
10.8 VA vent hood .... 120 V, 1ϕ
864 VA blower motor .... 240 V, 1ϕ

Unit 3—975 sq. ft.

10 kW heating unit .... 240 V, 1ϕ
4680 VA A/C unit .... 240 V, 1ϕ
9 kW cooktop .... 240 V, 1ϕ
10 kW oven .... 240 V, 1ϕ
850 VA disposal .... 120 V, 1ϕ
910 VA compactor .... 120 V, 1ϕ
1400 VA dishwasher .... 120 V, 1ϕ
12 VA vent fan .... 120 V, 1ϕ
5000 VA dryer .... 240 V, 1ϕ
6500 VA water heater .... 240 V, 1ϕ
864 VA blower motor .... 240 V, 1ϕ

Unit 4—975 sq. ft.

15 kW heating unit .... 240 V, 1ϕ
5040 VA A/C unit .... 240 V, 1ϕ
6500 VA water heater .... 240 V, 1ϕ
5000 VA dryer .... 240 V, 1ϕ
1600 VA dishwasher .... 120 V, 1ϕ
950 VA disposal .... 120 V, 1ϕ
10 kW cooktop .... 240 V, 1ϕ
12 kW oven .... 240 V, 1ϕ
1176 VA blower motor .... 240 V, 1ϕ
11.8 VA vent hood .... 120 V, 1ϕ

(1) What size THWN Cu conductors, paralleled four times per phase, is required for phases A and B?

Step 1. *Text:* $\frac{522 \text{ A}}{4} = 130.5 \text{ A}$

Step 2. *310-4; Table 310-16, 75°C Column:*
130.5 A requires #1/0 THWN Cu

Answer. **Phases A and B conductors = #1/0 THWN Cu**

(2) What size THWN Cu conductors, paralleled four times, is required for the neutral?

Step 1. *Text:* $\frac{206 \text{ A}}{4} = 51.5 \text{ A}$

Step 2. *310-4; Table 310-16, 75°C Column:*
#1/0 is the smallest conductor to be connected in parallel

Answer. **Neutral = #1/0 THWN Cu**

(3) What minimum size feeder circuit conductors, using THWN Cu, is required for Unit 4? See Figure 7-11.

Step 1. *Note 3 to Tables 310-16–310-19:*
181 A requires #2/0 THWN Cu

Answer. **Conductors = #2/0 THWN Cu**

(4) What size time delay fuses is required for the disconnect under the meter base?

Step 1. *240-3, Ex. 4; 240-6:* #2/0 requires 200 A TDFs

Answer. **TDFs = 200 A**

(5) What size disconnect is required to house the time delay fuses?

Step 1. *Chart:* 200 A TDFs requires 200 A disconnect

Answer. **Disconnect = 200 A**

Service with Units Having Gas Heat—220-10; 220-20

Eight 950 sq. ft. dwelling units in an apartment complex are equipped with the same loads as follows. The complex is supplied by a 120/240 V, 1ϕ service. Calculate the service load for phases A and B and the neutral. See Figure 7-12.

Gas heating unit
4680 VA A/C unit .... 240 V, 1ϕ
5000 VA water heater .... 240 V, 1ϕ
Gas cooktop
Gas oven
5000 VA dryer .... 240 V, 1ϕ
696 VA blower motor .... 120 V, 1ϕ
1800 VA dishwasher .... 120 V, 1ϕ
690 VA disposal .... 120 V, 1ϕ
750 VA compactor .... 120 V, 1ϕ
180 VA exhaust fan .... 120 V, 1ϕ

STANDARD CALCULATION—MULTIFAMILY DWELLING

1. GENERAL LIGHTING LOADS—Table 220-3(b)

890 sq. ft. × 3 VA × 1 units = 2670 VA

920 sq. ft. × 3 VA × 1 units = 2760 VA

975 sq. ft. × 3 VA × 2 units = 5850 VA

11,280 VA

Small appliance loads—220-16(a)

1500 VA × 2 circuits × 4 units = 12,000 VA

Laundry load—220-16(b)

1500 VA × 1 × 4 units = 6000 VA

29,280 VA

Applying demand factors—Table 220-11

First 3000 VA × 100% = 3000 VA

Next 26,280 VA × 35% = 9198 VA

Remaining — VA × 25% = — VA

(total) 12,198 VA | PHASES 12,198 VA | NEUTRAL 12,198 VA

2. COOKING EQUIPMENT LOADS—Table 220-19, Notes

Col. A 6 units = 21,000 VA demand = 21,000 VA

Col. B — VA × — units × — % = — VA

Col. C — VA × — units × — % = — VA

(total) 21,000 VA | 21,000 VA | × 70% = 14,700 VA

3. DRYER LOAD—220-18; Table 220-18

5000 VA × 4 units × 100 % = 20,000 VA | 20,000 VA | × 70% = 14,000 VA

4. FIXED APPLIANCE LOADS—220-17

Dishwasher = 6150 VA

Disposal = 3800 VA

Compactor = 1885 VA

Water heater = 23,000 VA

Blower motor = 3600 VA

Vent hood = 34.6 VA

(total) 38,469.6 VA × 1 units × 75% = 28,852.2 VA | 28,852.2 VA | (120 V Loads × units × 75%) 8902.2 VA

5. HEATING OR A/C LOAD—220-21

Heating unit = 43,000 VA × 100% × 1 units = 43,000 VA

A/C unit = 18,000 VA × 100% × 1 units = 18,000 VA

Heat pump = — VA × 100% × — units = — VA

(largest load) 43,000 VA | 43,000 VA | — VA

6. LARGEST MOTOR LOAD—220-14

ϕ 1176 VA × 25% = 294 VA | 294 VA

N 1050 VA × 25% = 262.5 VA | 262.5 VA

TOTAL = 125,344.2 VA | 50,062.7 VA

1ϕ service: $I = \frac{VA}{V}$

PHASES $I = \frac{125{,}344.2 \text{ VA}}{240 \text{ V}}$ = 522 A

NEUTRAL $I = \frac{50{,}062.7 \text{ VA}}{240 \text{ V}}$ = 208.6 A

3ϕ service: $I = \frac{VA}{V \times \sqrt{3}}$

$I = \frac{VA}{V \times \sqrt{3}}$ = ____ A

$I = \frac{VA}{V \times \sqrt{3}}$ = ____ A

If over 200 A—220-22; First 200 A × 100% = 200 A

Remaining 8.6 A × 70% = 6 A

206 A

Figure 7-10. The service load in VA and amps for apartment units with different square footages is found by using the form, STANDARD CALCULATION—MULTIFAMILY DWELLING.

STANDARD CALCULATION—ONE-FAMILY DWELLING

1. GENERAL LIGHTING LOADS—Table 220-3(b)

975 sq. ft. × 3 VA = 2925 VA

Small appliance loads—220-16(a)

1500 VA × 2 circuits = 3000 VA

Laundry load—220-16(b)

1500 VA × 1 = 1500 VA

7425 VA

Applying demand factors—Table 220-11

First 3000 VA × 100% = 3000 VA

Next 4425 VA × 35% = 1548.8 VA

Remaining — VA × 25% = — VA

(total) 4548.8 VA

	PHASES		NEUTRAL
1. General lighting loads	4548.8 VA		4648.8 VA
2. Cooking equipment loads	11,000 VA	× 70% =	7700 VA
3. Dryer load	5000 VA	× 70% =	3500 VA
4. Fixed appliance loads	7678.4 VA	(120 V Loads × 75%)	2561.8 VA
5. Heating or A/C load	15,000 VA		— VA
6. Largest motor load	294 VA		237.5 VA
TOTAL =	43521.2 VA		18,548.1 VA

2. COOKING EQUIPMENT LOADS—Table 220-19; Notes

Col. A 22,000 VA demand = 11,000 VA

Col. B — VA × — % = — VA

Col. C — VA × — % = — VA

(total) 11,000 VA

3. DRYER LOAD—220-18; Table 220-18

5000 VA × 100 % = 5000 VA

4. FIXED APPLIANCE LOADS—220-17

Dishwasher = 1600 VA

Disposal = 950 VA

Compactor = — VA

Water heater = 6500 VA

Vent hood = 11.8 VA

Blower motor = 1176 VA

— = — VA

— = — VA

(total) 10,237.8 VA × 75% = 7678.4 VA

5. HEATING OR A/C LOAD—220-21

Heating unit = 15,000 VA × 100% = 15,000 VA

A/C unit = 5040 VA × 100% = 5040 VA

Heat pump = — VA × 100% = — VA

(largest load) 15,000 VA

6. LARGEST MOTOR LOAD—220-14

ϕ 1176 VA × 25% = 294 VA

N 950 VA × 25% = 237.5 VA

1ϕ service: $I = \frac{VA}{V}$

PHASES $I = \frac{43521.2\ VA}{240\ V}$ = 181 A

NEUTRAL $I = \frac{18,548.1\ VA}{240\ V}$ = 77 A

If over 200 A—220-22; First 200 A × 100% = 200 A

Remaining ______ A × 70% = ______ A

______ A

Figure 7-11. The feeder circuit load in VA and amps for a single dwelling unit in an apartment complex is found by using the form, STANDARD CALCULATION—MULTIFAMILY DWELLING. Conductors are sized from the total amperage.

STANDARD CALCULATION—MULTIFAMILY DWELLING

1. GENERAL LIGHTING LOADS—Table 220-3(b)

950 sq. ft. × 3 VA × 8 units = 22,800 VA
— sq. ft. × 3 VA × — units = — VA
— sq. ft. × 3 VA × — units = — VA
22,800 VA

Small appliance loads—220-16(a)
1500 VA × 2 circuits × 8 units = 24,000 VA

Laundry load—220-16(b)
1500 VA × 1 × 8 units = 12,000 VA
58,800 VA

Applying demand factors—Table 220-11
First 3000 VA × 100% = 3000 VA
Next 55,800 VA × 35% = 19,530 VA
Remaining — VA × 25% = — VA
(total) 22,530 VA

	PHASES		NEUTRAL
1. General lighting	22,530 VA		22,530 VA

2. COOKING EQUIPMENT LOADS—Table 220-19, Notes

Col. A — units = — VA demand = — VA
Col. B — VA × — units × — % = — VA
Col. C — VA × — units × — % = — VA
(total) — VA — PHASES: — VA × 70% = — VA NEUTRAL

3. DRYER LOAD—220-18; Table 220-18

6000 VA × 8 units × 60 % = 24,000 VA — PHASES: 24,000 VA × 70% = 16,800 VA NEUTRAL

4. FIXED APPLIANCE LOADS—220-17

Dishwasher = 1800 VA
Disposal = 690 VA
Compactor = 750 VA
Water heater = 5000 VA
Blower motor = 696 VA
Exhaust fan = 180 VA
(total) 9116 VA × 8 units × 75% = 54,696 VA — PHASES: 54,696 VA — NEUTRAL (120 V Loads × units × 75%): 24,696 VA

5. HEATING OR A/C LOAD—220-21

Heating unit = — VA × 100% × — units = — VA
A/C unit = 4680 VA × 100% × 8 units = 37,440 VA
Heat pump = — VA × 100% × — units = — VA
(largest load) 37,440 VA — PHASES: 37,440 VA — NEUTRAL: — VA

6. LARGEST MOTOR LOAD—220-14

ϕ 4680 VA × 25% = 1170 VA — PHASES: 1170 VA
N 750 VA × 25% = 187.5 VA — NEUTRAL: 187.5 VA

TOTAL = PHASES: 139,836 VA — NEUTRAL: 64,213.5 VA

1ϕ service: $I = \frac{VA}{V}$

PHASES $I = \frac{139{,}836\ VA}{240\ V} = 582.7\ A$

NEUTRAL $I = \frac{64{,}213.5\ VA}{240\ V} = 267.6\ A$

3ϕ service: $I = \frac{VA}{V \times \sqrt{3}}$

$I = \frac{VA}{V \times \sqrt{3}} =$ ______ A

$I = \frac{VA}{V \times \sqrt{3}} =$ ______ A

If over 200 A—220-22; First 200 A × 100% = 200 A
Remaining 67.6 A × 70% = 47 A
247 A

Figure 7-12. The service load in VA and amps for apartment units with the same loads is found by using the form, STANDARD CALCULATION—MULTIFAMILY DWELLING.

(1) What size THWN Cu conductors, paralleled three times per phase, is required for phases A and B?

Step 1. *Text:* $\frac{582.7 \text{ A}}{3} = 194 \text{ A}$

Step 2. *310-4; Table 310-16, 75°C Column:*
194 A requires #3/0 THWN Cu

Answer. **Phases A and B conductors = #3/0 THWN Cu**

(2) What size THWN Cu conductors, paralleled three times per phase, is required for the neutral?

Step 1. *Text:* $\frac{247 \text{ A}}{3} = 82 \text{ A}$

Step 2. *310-4; Table 310-16, 75°C Column:*
#1/0 is the smallest conductor that can be connected in parallel

Answer. **Neutral = #1/0 THWN Cu**

(3) What size conduit is required for the parallel conductors in each run? There are three runs.

Step 1. *Table 5, Ch. 9:*

#3/0 THWN =		
	.2715 sq. in. × 2 =	.543 sq. in.
#1/0 THWN =		
	.1893 sq. in. × 1 =	.1893 sq. in.
	Total =	.7323 sq. in.

Step 2. *Table 4, Ch. 9:* .7323 sq. in. requires .82 sq. in.

Answer. **Conduit = 1½″**

(4) What size grounding electrode conductor, using copper, is required? Use water pipe as a grounding electrode.

Step 1. *250-94; Table 8, Ch. 9:*

$$\#3/0 = \frac{167{,}800 \text{ CM}}{1000} \times 3 = 503.4 \text{ kcmil}$$

Step 2. *Table 250-94:* 503.4 kcmil = #1/0 Cu

Answer. **Grounding electrode conductor = #1/0 Cu**

(5) What is the maximum size overcurrent protection device required for the air conditioner?

Step 1. *Text:* $I = \frac{VA}{V}$

$$I = \frac{4680 \text{ VA}}{240 \text{ V}} = 19.5 \text{ A}$$

Step 2. *440-32:* 19.5 A × 225% = 43.9 A

Step 3. *240-3, Ex. 6; 240-6:*
40 A is the next lower standard size

Answer. **OCPD = 40 A**

Service with 120 V A/C Window Units

Six 680 sq. ft. dwelling units in an apartment complex have the same loads as follows. The complex is supplied by a 120/240 V, 1ϕ service. What is the total VA and amp ratings of the service for phases A and B and the neutral? See Figure 7-13.

8 kW heating unit .... 240 V, 1ϕ
4500 VA water heater .... 240 V, 1ϕ
10 kW range .... 240 V, 1ϕ
5000 VA dryer .... 240 V, 1ϕ
696 VA blower motor .... 240 V, 1ϕ
1500 VA dishwasher .... 120 V, 1ϕ
690 VA disposal .... 120 V, 1ϕ
720 VA compactor .... 120 V, 1ϕ
120 VA intercom .... 120 V, 1ϕ
1920 VA A/C window unit .... 120 V, 1ϕ
1800 VA A/C window unit .... 120 V, 1ϕ
1740 VA A/C window unit .... 120 V, 1ϕ

(1) What size THWN Cu conductors is required for the service? The conductors are paralleled three times per phase.

Step 1. *Text:* $\frac{596.6 \text{ A}}{3} = 198.9 \text{ A}$

Step 2. *310-4; Table 310-16, 75°C Column:*
198.9 A requires #3/0 THWN Cu

Answer. **Conductors = #3/0 THWN Cu**

(2) What size THWN Cu conductors is required for the neutral? The conductors are paralleled three times per phase.

Step 1. *Text:* $\frac{328 \text{ A}}{3} = 109 \text{ A}$

Step 2. *310-4; Table 310-16, 75°C Column:*
#1/0 THWN Cu is the smallest conductor that can be connected in parallel

Answer. **Neutral = #1/0 THWN Cu**

(3) What size CB is required for the service equipment?

Step 1. *240-3, Ex. 4:*
#3/0 THWN Cu = 200 A × 3 = 600 A

Step 2. *240-6:* 600 A is a standard size

Answer. **CB = 600 A**

(4) What size panelboard is required for the service?

Step 1. *384-13:* 600 A CB = 600 A panelboard

Answer. **Panelboard = 600 A**

(5) How many 15 A, 2-wire circuits are required for the general lighting and receptacle loads?

Step 1. *220-4(a):*

$$\text{No. of CBs} = \frac{\text{sq. ft.} \times 3 \text{ VA} \times 6}{\text{CB size} \times 120 \text{ V}}$$

$$\text{No. of CBs} = \frac{680 \text{ sq. ft.} \times 3 \text{ VA} \times 6}{15 \text{ A} \times 120 \text{ V}} = 6.8$$

Answer. **2-wire circuits = seven per dwelling unit**

Service with Heat Pumps

Four dwelling units in an apartment complex have square footages and loads as follows. Heat pumps operate with the heating units. The service is 120/240 V, 1ϕ. Calculate the service for phases A and B and the neutral. See Figure 7-14.

Units 1 and 2—1000 sq. ft.

10 kW heating unit .... 240 V, 1ϕ
4000 VA heat pump .... 240 V, 1ϕ
690 VA disposal .... 120 V, 1ϕ
1600 VA dishwasher .... 120 V, 1ϕ
12 kW range .... 240 V, 1ϕ
5000 VA water heater .... 240 V, 1ϕ
5000 VA dryer .... 240 V, 1ϕ
200 VA vent hood .... 120 V, 1ϕ
628 VA blower motor .... 240 V, 1ϕ
928 VA compactor .... 120 V, 1ϕ

STANDARD CALCULATION—MULTIFAMILY DWELLING

1. GENERAL LIGHTING LOADS—Table 220-3(b)

680 sq. ft. × 3 VA × 6 units =	12,240 VA	
— sq. ft. × 3 VA × — units =	— VA	
— sq. ft. × 3 VA × — units =	— VA	
	12,240 VA	
Small appliance loads—220-16(a): 1500 VA × 2 circuits × 6 units =	18,000 VA	
Laundry load—220-16(b): 1500 VA × 1 × 6 units =	9000 VA	
	39,240 VA	

Applying demand factors—Table 220-11

		PHASES	NEUTRAL
First 3000 VA × 100% =	3000 VA		
Next 36,240 VA × 35% =	12,684 VA		
Remaining — VA × 25% =	— VA		
(total)	15,684 VA	15,684 VA	15,684 VA

2. COOKING EQUIPMENT LOADS—Table 220-19, Notes

Col. A 6 units = 60,000 VA demand =	21,000 VA		
Col. B — VA × — units × — % =	— VA		
Col. C — VA × — units × — % =	— VA		
(total)	21,000 VA	21,000 VA	× 70% = 14,700 VA

3. DRYER LOAD—220-18; Table 220-18

5000 VA × 6 units × 70 % =	21,000 VA	21,000 VA	× 70% = 14,700 VA

4. FIXED APPLIANCE LOADS—220-17

Dishwasher =	1500 VA
Disposal =	690 VA
Compactor =	720 VA
Water heater =	4500 VA
Blower motor =	696 VA
Intercom =	120 VA

(total) 8226 VA × 6 units × 75% = 37,017 VA — Phases: 37,017 VA — Neutral (120 V Loads × units × 75%): 13,635 VA

5. HEATING OR A/C LOAD—220-21

Heating unit = 8000 VA × 100% × 6 units =	48,000 VA		
A/C unit = 5460 VA × 100% × 6 units =	32,760 VA		
Heat pump = — VA × 100% × — units =	— VA		
(largest load)	48,000 VA	48,000 VA	32,760 VA

6. LARGEST MOTOR LOAD—220-14

ϕ 1920 VA × 25% =	480 VA	480 VA	
N 1920 VA × 25% =	480 VA		480 VA
TOTAL =		143,181 VA	91,959 VA

1ϕ service: $I = \frac{VA}{V}$

PHASES $I = \frac{143{,}181\ VA}{240\ V} = 596.6$ A

NEUTRAL $I = \frac{91{,}959\ VA}{240\ V} = 383$ A

3ϕ service: $I = \frac{VA}{V \times \sqrt{3}}$

$I = \frac{\_\_\_\ VA}{V \times \sqrt{3}} = \_\_\_$ A

$I = \frac{\_\_\_\ VA}{V \times \sqrt{3}} = \_\_\_$ A

If over 200 A—220-22; First 200 A × 100% = 200 A

Remaining 183 A × 70% = 128 A

328 A

Figure 7-13. The service load in VA and amps for apartment units with A/C window units is found by using the form, STANDARD CALCULATION—MULTIFAMILY DWELLING.

STANDARD CALCULATION—MULTIFAMILY DWELLING

1. GENERAL LIGHTING LOADS—Table 220-3(b)

1000 sq. ft. × 3 VA × 2 units = 6000 VA
1100 sq. ft. × 3 VA × 2 units = 6600 VA
— sq. ft. × 3 VA × — units = — VA
12,600 VA

Small appliance loads—220-16(a)
1500 VA × 2 circuits × 4 units = 12,000 VA

Laundry load—220-16(b)
1500 VA × 1 × 4 units = 6000 VA
30,600 VA

Applying demand factors—Table 220-11
First 3000 VA × 100% = 3000 VA
Next 27,600 VA × 35% = 9660 VA
Remaining — VA × 25% = — VA
(total) 12,660 VA

	PHASES		NEUTRAL
1. General lighting	12,660 VA		12,660 VA

2. COOKING EQUIPMENT LOADS—Table 220-19, Notes

Col. A 4 units = 44,000 VA demand = 17,000 VA
Col. B — VA × — units × — % = — VA
Col. C — VA × — units × — % = — VA
(total) 17,000 VA — PHASES 17,000 VA × 70% = NEUTRAL 11,900 VA

3. DRYER LOAD—220-18; Table 220-18

5000 VA × 4 units × 100 % = 20,000 VA — PHASES 20,000 VA × 70% = NEUTRAL 14,000 VA

4. FIXED APPLIANCE LOADS—220-17

Dishwasher = 6400 VA
Disposal = 2760 VA
Compactor = 3226 VA
Water heater = 20,000 VA
Blower motor = 2648 VA
Vent hood = 760 VA
(total) 35,794 VA × — units × 75% = 26,845.5 VA — PHASES 26,845.5 VA — NEUTRAL (120 V Loads × units × 75%) 9859.5 VA

5. HEATING OR A/C LOAD—220-21

Heating unit = 10,000 VA × 100% × 4 units = 40,000 VA
A/C unit = — VA × 100% × — units = — VA
Heat pump = 4000 VA × 100% × 4 units = 16,000 VA
(largest load) 56,000 VA — PHASES 56,000 VA — NEUTRAL — VA

6. LARGEST MOTOR LOAD—220-14

ϕ 4000 VA × 25% = 1000 VA — PHASES 1000 VA
N 928 VA × 25% = 232 VA — NEUTRAL 232 VA

TOTAL = PHASES 133,505.5 VA — NEUTRAL 48,651.5 VA

1ϕ service: $I = \frac{VA}{V}$

PHASES $I = \frac{133{,}605.5\ VA}{240\ V}$ = 556 A

NEUTRAL $I = \frac{48{,}651.5\ VA}{240\ V}$ = 202.7 A

3ϕ service: $I = \frac{VA}{V \times \sqrt{3}}$

$I = \frac{\_\_\_\ VA}{\_\_\_\ V \times \sqrt{3}}$ = ___ A

$I = \frac{\_\_\_\ VA}{\_\_\_\ V \times \sqrt{3}}$ = ___ A

If over 200 A—220-22; First 200 A × 100% = 200 A
Remaining 2.7 A × 70% = 1.9 A
201.9 A

Figure 7-14. The service load in VA and amps for apartment units with heating units and heat pumps is found by using the form, STANDARD CALCULATION—MULTIFAMILY DWELLING. Heating and heat pump loads are added together at 100%.

Units 3 and 4—1100 sq. ft.

10 kW heating unit .... 240 V, 1ϕ
4000 VA A/C unit .... 240 V, 1ϕ
10 kW range .... 240 V, 1ϕ
5000 VA water heater .... 240 V, 1ϕ
4500 VA dryer .... 240 V, 1ϕ
690 VA disposal .... 120 V, 1ϕ
685 VA compactor .... 120 V, 1ϕ
1600 VA dishwasher .... 120 V, 1ϕ
180 VA vent hood .... 120 V, 1ϕ
696 VA blower motor .... 240 V, 1ϕ

(1) What size THHN Al conductors, paralleled three times, is required for phases A and B using 75 °C lugs?

Step 1. *Text:* $\frac{556 \text{ A}}{4} = 139 \text{ A}$

Step 2. *310-4; Table 310-16, 90°C Column:* 139 A = #3/0 THHN Al

Step 3. Rating of Terminals. *Table 310-16; General Information for Electrical Construction by UL:*
90 °C = #2/0 THHN Al
75 °C = #3/0 THHN Al

Answer. **Phases A and B conductors = #3/0 THHN Al**

(2) What size THHN Al conductors, paralleled four times, is required for the neutral?

Step 1. *Text:* $\frac{201.9 \text{ A}}{4} = 50.5 \text{ A}$

Step 2. *310-4; Table 310-16, 90°C Column:* #1/0 is the smallest size conductor to be connected in parallel

Answer. **Neutral = #1/0 THHN Al**

(3) What size conduit is required for four runs? Use maximum size aluminum conductors.

Step 1. *Table 5, Ch. 9:*
#3/0 THHN = .2715 sq. in. × 2 = .543 sq. in.
#1/0 THHN = .1893 sq. in. × 1 = .1893 sq. in.
Total = .7323 sq. in.

Step 2. *Table 4, Ch. 9:* .7323 requires .82 sq. in. area

Answer. **Conduit = 1½″**

(4) What size branch circuit conductors, using nonmetallic-sheathed cable (Romex), is required to supply the 12 kW range?

Step 1. *210-19(b); Table 220-19, Column A:* 12 kW = 8 kW demand

Step 2. *Text:* $I = \frac{kW \times 1000}{V}$

$$I = \frac{8 \text{ kW} \times 1000}{240 \text{ V}} = 33 \text{ A}$$

Step 3. *336-26; Table 310-16, 60°C Column:* 33 A = #8

Answer. **Conductors = #8-2 w/ground**

(5) Are all three conductors required to be insulated?

Step 1. *250-61(b), Ex. 1; 250-60(c):* Yes

Answer. **Yes**

(6) What size branch circuit conductors, using nonmetallic-sheathed cable (Rope), is required to supply the 690 VA disposal?

Step 1. *Text:* $I = \frac{VA}{V}$

$$I = \frac{690 \text{ VA}}{120 \text{ V}} = 5.8 \text{ A}$$

Step 2. *430-6(a); 430-22(a):* 5.8 A × 125% = 7.25 A

Step 3. *336-26; Table 310-16, 60°C Column:* 7.25 A = #14

Answer. **Conductors = #14-2 w/ground**

(7) What size circuit breaker is required to allow the attic fan to start and run?

Step 1. *430-6(a); Table 430-152:* 7.25 A × 250% = 9 A

Step 2. *240-3, Ex. 6; 430-52, Ex. 1; 240-6:* 15 A is a standard size OCPD

Answer. **CB = 15 A**

Dividing Volt Amps on Line and Neutral

▪*Master Electrician's Test Tip:* Some examination boards require that the VA rating of each load be listed on one form for phases A, B, C, and the neutral when figuring the service loads.▪

Fifteen 900 sq. ft. dwelling units in an apartment complex have identical loads as follows. The service is 120/208 V, 3ϕ, 4-wire. The loads are considered balanced per phase. Calculate the service in VA for phases A, B, C, and the neutral. See Figure 7-15.

10 kW heating unit .... 208 V, 1ϕ
5040 VA A/C unit .... 208 V, 1ϕ
5000 VA water heater .... 208 V, 1ϕ
12 kW range .... 208 V, 1ϕ
5000 VA dryer .... 208 V, 1ϕ
696 VA blower motor .... 208 V, 1ϕ
1450 VA dishwasher .... 120 V, 1ϕ
690 VA disposal .... 120 V, 1ϕ
750 VA compactor .... 120 V, 1ϕ
1920 VA bathroom heater (ceiling mounted) .... 120 V, 1ϕ

(1) What is the service load in VA for the ranges? Use the 120/208 V, 3ϕ, 4-wire supply.

Step 1. *220-19:* $\frac{15 \text{ ranges}}{3\phi} = 5$

Step 2. 5 ranges per phase × 2 = 10 ranges

Step 3. *Table 220-19, Column A:* 10 ranges = 25 kW

Step 4. *220-19:* 25 kW × 1000 = 25,000 VA

Step 5. *Service load:* $\frac{25{,}000 \text{ VA}}{2 \text{ phases}} \times 3\phi = 37{,}500 \text{ VA}$

Answer. **Service load = 37,500 VA**

(2) What size branch circuit conductors, using nonmetallic-sheathed cable (Romex), is required to supply the dishwasher?

Step 1. *Text:* $I = \frac{VA}{V}$

$$I = \frac{1450 \text{ VA}}{120 \text{ V}} = 12 \text{ A}$$

Step 2. *336-26; Table 310-16, 60°C Column:* 12 A = #14

Answer. **Conductors = #14-2 w/ground**

STANDARD CALCULATION—MULTIFAMILY DWELLING 3ϕ (CARRYING PHASES AND NEUTRAL IN VA)

1. GENERAL LIGHTING LOADS—Table 220-3(b)

900 sq. ft. × 3 VA × 15 units = 40,500 VA
— sq. ft. × 3 VA × — units = — VA
— sq. ft. × 3 VA × — units = — VA
40,500 VA

Small appliance loads—220-16(a)
1500 VA × 2 circuits × 15 units = 45,000 VA

Laundry load—220-16(b)
1500 VA × 1 × 15 units = 22,500 VA
108,000 VA

Applying demand factors—Table 220-11
First 3000 VA × 100% = 3000 VA
Next 105,000 VA × 35% = 36,750 VA
Remaining — VA × 25% = — VA

	PHASE A	PHASE B	PHASE C		NEUTRAL
(total) 39,750 VA ÷ 3 =	13,250 VA	13,250 VA	13,250 VA		13,250 VA
2. COOKING EQUIPMENT LOADS—Table 220-19, Notes					
Col. A 15 units = 180,000 VA demand = 37,500 VA ÷ 3 =	12,500 VA	12,500 VA	12,500 VA	× 70% =	8750 VA
Col. B — VA × — units × — % = — VA ÷ 3 =	— VA	— VA	— VA	× 70% =	— VA
Col. C — VA × — units × — % = — VA ÷ 3 =	— VA	— VA	— VA	× 70% =	— VA
3. DRYER LOAD—220-18; Table 220-18					
5000 VA × 15 units × 40 % ÷ 3 =	10,000 VA	10,000 VA	10,000 VA	× 70% =	7000 VA
4. FIXED APPLIANCE LOADS—220-17					
Dishwasher = 1450 VA × 15 units × 75% ÷ 3 =	5437.5 VA	5437.5 VA	5437.5 VA		5437.5 VA
Disposal = 690 VA × 15 units × 75% ÷ 3 =	2587.5 VA	2587.5 VA	2587.5 VA		2587.5 VA
Compactor = 750 VA × 15 units × 75% ÷ 3 =	2812.5 VA	2812.5 VA	2812.5 VA		2812.5 VA
Water heater = 5000 VA × 15 units × 75% ÷ 3 =	18,750 VA	18,750 VA	18,750 VA		— VA
Blower = 696 VA × 15 units × 75% ÷ 3 =	2610 VA	2610 VA	2610 VA		— VA
Bath htr. = 1920 VA × 15 units × 75% ÷ 3 =	7200 VA	7200 VA	7200 VA		7200 VA
— = — VA × — units × 75% ÷ 3 =	— VA	— VA	— VA		— VA
— = — VA × — units × 75% ÷ 3 =	— VA	— VA	— VA		— VA
— = — VA × — units × 75% ÷ 3 =	— VA	— VA	— VA		— VA
5. HEATING OR A/C LOAD—220-21					
Heating unit = 10,000 VA × 100% × 15 units ÷ 3 =	50,000 VA	50,000 VA	50,000 VA		— VA
A/C unit = 5040 VA × 100% × 15 units ÷ 3 =	— VA	— VA	— VA		— VA
Heat pump = — VA × 100% × — units ÷ 3 =	— VA	— VA	— VA		— VA
6. LARGEST MOTOR LOAD—220-14					
ϕ 750 VA × 25% =	187.5 VA	— VA	— VA		
N 750 VA × 25% =					187.5 VA
TOTAL =	125,335 VA	125,147.5 VA	125,147.5 VA		47,225 VA

3ϕ service: $I = \dfrac{VA}{V \times \sqrt{3}}$

$I = \dfrac{\_\_\_\_ VA}{\_\_\_\_ V \times \sqrt{3}} = \_\_\_\_ A$

$I = \dfrac{\_\_\_\_ VA}{\_\_\_\_ V \times \sqrt{3}} = \_\_\_\_ A$

If over 200 A—220-22; First 200 A × 100% = 200 A
Remaining ____ A × 70% = ____ A
____ A

Figure 7-15. The service load in VA can be easily divided on phases A, B, and C and the neutral by using the form, STANDARD CALCULATION—MULTIFAMILY DWELLING.

(3) Can the dishwasher be cord- and plug-connected with a 42″ SO cord supplied by a 15 A branch circuit?

Step 1. *210-23(a):* 15 A × 80% = 12 A

Step 2. 12 A = 12 A load

Step 3. *422-8(d)(2)a:* 3′-0″ to 4′-0″ cord

Answer. **Yes**

Dividing Amps on Line and Neutral

The following is a step-by-step procedure for calculating the service load in amps for phases A, B, and C and the neutral.

Twenty-one 860 sq. ft. dwelling units in an apartment complex are supplied by a 120/208 V, 4-wire, 3ϕ system. Each dwelling unit has the following loads. All loads are considered balanced per phase. Calculate the service in amps for phases A, B, and C and the neutral. See Figure 7-16.

8 kW heating unit .... 208 V, 1ϕ
3120 kW A/C unit .... 208 V, 1ϕ
4500 VA water heater .... 208 V, 1ϕ
9 kW cooktop .... 208 V, 1ϕ
10 kW oven .... 208 V, 1ϕ
5000 VA dryer .... 208 V, 1ϕ
582 VA blower motor .... 208 V, 1ϕ
1500 VA dishwasher .... 120 V, 1ϕ
650 VA disposal .... 120 V, 1ϕ
725 VA compactor .... 120 V, 1ϕ
108 VA exhaust fan .... 120 V, 1ϕ

(1) What is the service load in amps for the cooktops and ovens using the 120/208 V, 4-wire, 3ϕ system?

Step 1. $\dfrac{\text{42 cooktops and ovens}}{3\phi} = 14$

Step 2. 14 cooktops and ovens per phase × 2 = 28

Step 3. *Table 220-19, Column A:*
15 kW + 28 kW = 43 kW

Step 4. *Table 220-19, Column A:*
43 kW × 1000 = 43,000 VA

Step 5. Service load. $\dfrac{43{,}000\text{ VA}}{2\text{ phases}} \times 3\phi = 64{,}500\text{ VA}$

Step 6. *Text:* $I = \dfrac{VA}{V \times \sqrt{3}}$

$I = \dfrac{64{,}500\text{ VA}}{208\text{ V} \times 1.732} = 179\text{ A}$

Answer. **Service load = 179 A**

(2) What size THWN Cu conductors, paralleled four times per phase, is required for phases A, B, and C?

Step 1. *Text:* $\dfrac{909.5\text{ A}}{4} = 227\text{ A}$

Step 2. *310-4; Table 310-16, 75°C Column:*
208 A = #4/0 THWN Cu

Answer. **Conductors = *#4/0* THWN Cu**

(3) What size THWN Cu conductors is required for the neutral?

Step 1. *220-22:*

First 200 A × 100% =	200 A
Next 312.9 A × 70% =	219 A
Total =	419 A

Step 2. *Text:* $\dfrac{419\text{ A}}{4} = 104.8\text{ A}$

Step 3. *310-4; Table 310-16, 75°C Column:*
#1/0 is the smallest conductor that can be connected in parallel

Answer. **Conductors = *#1/0* THWN Cu**

(4) What size conduit is required for all conductors routed through one conduit?

Step 1. *Table 5, Ch. 9:*

#4/0 THWN = .3278 sq. in. × 12 =	3.9336 sq. in.
#1/0 THWN = .1893 sq. in. × 4 =	.7572 sq. in.
Total =	4.6908 sq. in.

Step 2. *Table 4, Ch. 9:* 4.6908 sq. in. requires 5.09 sq. in.

Answer. **Conduit = 4″**

(5) Can the service load of 909.5 A be supplied by the 16 conductors pulled in one conduit?

Step 1. *Table 310-16, 75°C Column:*
#4/0 THWN Cu = 230 A

Step 2. *Note 8 to Tables 310-16–310-19:*
230 A × 70% = 161 A

Step 3. *Text:* 161 A × 4 = 644 A

Step 4. Service load. 644 A is less than 909.5 A

Answer. **No**

OPTIONAL CALCULATION—220-32

The service load for dwelling units located in a multifamily complex can be computed using the optional calculation. The optional calculation requires that a demand factor be selected from Table 220-32. The percentage of the demand factor is based upon the number of units in the complex. Dwelling units shall be equipped with electric cooking and either electric heating or air conditioning in order to use the optional calculation. With both electric heating and air conditioning present in each dwelling unit, the smaller load is dropped. Each dwelling unit shall be supplied by an individual feeder. The neutral (grounded conductor) can be computed with the standard calculation by applying Sections 220-10(b)–220-22. Before the size of the neutral is selected by the actual calculated load, 250-23(b) shall be complied with. The house loads are computed using the standard calculation.

Other Loads (Connected)—220-32(c)

Other loads are connected loads such as the 3 VA per square foot of each dwelling unit in the complex per 220-32(c)(2).

What is the general lighting and receptacle load for twenty-four 890 sq. ft. dwelling units?

Step 1. *220-32(c)(2):*
890 sq. ft. × 3 VA × 24 = 64,080 VA

Answer. **Load = 64,080 VA**

STANDARD CALCULATION—MULTIFAMILY DWELLING 3ϕ (CARRYING PHASES AND NEUTRAL IN AMPS)

1. GENERAL LIGHTING LOADS—Table 220-3(b)

860 sq. ft. × 3 VA × 21 units = 54,180 VA
— sq. ft. × 3 VA × — units = — VA
— sq. ft. × 3 VA × — units = — VA
54,180 VA

Small appliance loads—220-16(a)
1500 VA × 2 circuits × 21 units = 63,000 VA

Laundry load—220-16(b)
1500 VA × 1 × 21 units = 31,500 VA
148,680 VA

Applying demand factors—Table 220-11
First 3000 VA × 100% = 3000 VA
Next 117,000 VA × 35% = 40,950 VA
Remaining 28,680 VA × 25% = 7170 VA

	PHASE A	PHASE B	PHASE C		NEUTRAL
(total) 51,120 VA ÷ 120 V ÷ 3 =	142 A	142 A	142 A		142 A
2. COOKING EQUIPMENT LOADS—Table 220-19, Notes					
Col. A 42 units = 399,000 VA demand = 64,500 VA ÷ 120 V ÷ 3 =	179.2 A	179.2 A	179.2 A	× 70% =	125.4 A
Col. B — VA × — units × — % = — VA ÷ — V ÷ 3 =	— A	— A	— A	× 70% =	— A
Col. C — VA × — units × — % = — VA ÷ — V ÷ 3 =	— A	— A	— A	× 70% =	— A
3. DRYER LOAD—220-18; Table 220-18					
5000 VA × 21 units × 35 % ÷ 208 V ÷ 3 =	58.9 A	58.9 A	58.9 A	× 70% =	41.2 A
4. FIXED APPLIANCE LOADS—220-17					
Dishwasher = 1500 VA × 21 units × 75% ÷ 120 V ÷ 3 =	65.6 A	65.6 A	65.6 A		65.6 A
Disposal = 650 VA × 21 units × 75% ÷ 120 V ÷ 3 =	28.4 A	28.4 A	28.4 A		28.4 A
Compactor = 725 VA × 21 units × 75% ÷ 120 V ÷ 3 =	31.7 A	31.7 A	31.7 A		31.7 A
Water heater = 4500 VA × 21 units × 75% ÷ 208 V ÷ 3 =	113.6 A	113.6 A	113.6 A		— A
Blower = 582 VA × 21 units × 75% ÷ 208 V ÷ 3 =	14.7 A	14.7 A	14.7 A		— A
Exhaust = 108 VA × 21 units × 75% ÷ 120 V ÷ 3 =	4.7 A	4.7 A	4.7 A		4.7 A
— = — VA × — units × 75% ÷ — V ÷ 3 =	— A	— A	— A		— A
— = — VA × — units × 75% ÷ — V ÷ 3 =	— A	— A	— A		— A
— = — VA × — units × 75% ÷ — V ÷ 3 =	— A	— A	— A		— A
5. HEATING OR A/C LOAD—220-21					
Heating unit = 8000 VA × 100% × 21 units ÷ 208 V ÷ 3 =	269.2 A	269.2 A	269.2 A		— A
A/C unit = 3120 VA × 100% × 21 units ÷ 208 V ÷ 3 =	— A	— A	— A		— A
Heat pump = — VA × 100% × — units ÷ — V ÷ 3 =	— A	— A	— A		— A
6. LARGEST MOTOR LOAD—220-14					
ϕ 725 VA × 25% ÷ 120 V =	1.5 A	— A	— A		
N 725 VA × 25% ÷ 120 V =					1.5 A
TOTAL =	909.5 A	908 A	908 A		512.9 A

If over 200 A—220-22; First 200 A × 100% = 200 A
Remaining 312.9 A × 70% = 219 A
419 A

Figure 7-16. The service load in amps can be easily divided on phases A, B, and C and the neutral by using the form, STANDARD CALCULATION—MULTIFAMILY DWELLING.

Small appliance circuits are computed at 1500 VA each. At least two small appliance circuits are required for the receptacles located in the kitchen, pantry, dining room, and breakfast room. One small appliance circuit is required for the laundry area per 210-52(f), Ex. 2 for laundry facilities not furnished by the management.

What is the small appliance load for an apartment complex with 24 dwellings?

Step 1. *220-32(c)(1):* 1500 VA × 3 × 24 = 108,000 VA

Answer. **Load = 108,000 VA**

Loads such as dishwashers, compactors, disposals, cooking equipment, and dryer equipment are figured at 100% of their nameplate rating. Motors are computed at nameplate ratings to derive the connected load.

What is the load in VA for a 10 kW oven and 8 kW cooktop in an apartment complex with 24 dwelling units?

Step 1. *220-32(c)(3):*
Cooktops =
10 kW × 1000 × 24 = 240,000 VA
Ovens =
8 kW × 1000 × 24 = 192,000 VA
Total = 432,000 VA

Answer. **Load = 432,000 VA**

What is the load in VA for twenty-four ¼ HP, 115 V, 1ϕ motors located in an apartment complex?

Step 1. *220-32(c)(4); 430-6(a); Table 430-148:*
¼ HP = 5.8 A

Step 2. *220-32(c)(4):* 5.8 A × 120 V × 24 = 16,704 VA

Answer. **Load = 16,704 VA**

Largest Load Between Heat and Air Conditioning—220-32(c)(5)

The largest load between the heating or air conditioning load is selected and the smaller load dropped per 220-32(c)(5). The largest load is then multiplied by 100% and the number of dwelling units to derive the total load for the selection of the service elements.

Heating and air conditioning is provided to 24 apartments by 10 kW heating units and 4800 VA A/C units. What is the load in VA for the largest load between the heating and A/C loads?

Step 1. *220-32(c)(5):*
Heating load = 10 kW × 1000 = 10,000 VA
A/C load = 4800 VA

Step 2. Largest load = 10,000 VA

Step 3. Total load =
10,000 VA × 100% × 24 = 240,000 VA

Answer. **Load = 240,000 VA**

House Loads—220-32(b)

House loads are the loads connected to the meter of the management and not the meters of each individual dwelling unit in the complex. Such loads are outside lighting, outside receptacles, laundry facilities, and other loads supplied by the house meter of the management.

What is the load in VA for the following house loads?

Outside receptacles . . . . . . 4500 VA (noncontinuous)
Outside receptacles . . . . . . 2160 VA (continuous)
Outside lighting . . . . 3600 VA (continuous)
Outside sign . . . . 2400 VA (continuous)

Step 1. *220-10(b):*
Receptacles =
4500 VA × 100% =
4500 VA (noncontinuous)
Receptacles =
2160 VA × 125% =
2700 VA (continuous)
Lighting =
3600 VA × 125% =
4500 VA (continuous)
Sign =
2400 VA × 125% =
3000 VA (continuous)
Total = 14,700 VA

Answer. **Load = 14,700 VA**

Note: House loads are not computed by Table 220-32. They are calculated per Sections 220-10(b)–220-22 using the standard calculation.

EXAMPLES USING OPTIONAL CALCULATION—220-32

The optional calculation is the simplest method to use in computing the load to select the service elements. Each load is figured at 100% of its nameplate rating. The small appliance loads are computed at 1500 VA each and the square footage of each dwelling is multiplied by 3 VA. The larger load between the heating or A/C is selected and the smaller load dropped. Demand factors, based on the number of dwelling units, per Table 220-32 are applied to the total load to determine service load.

Service with Units Having Same Square Footage—220-32; Table 220-32

Thirty 950 sq. ft. dwelling units in an apartment complex have a 120/208 V, 3ϕ, 4-wire service and the following loads. What is the service load in VA and amps for phases A, B, and C and the neutral? See Figure 7-17.

10 kW heating unit . . . . 208 V, 1ϕ
5016 VA A/C unit . . . . 208 V, 1ϕ
5000 VA water heater . . . . 208 V, 1ϕ
9 kW cooktop . . . . 208 V, 1ϕ
10 kW oven . . . . 208 V, 1ϕ
5000 VA dryer . . . . 208 V, 1ϕ
840 VA blower motor . . . . 208 V, 1ϕ
1500 VA dishwasher . . . . 120 V, 1ϕ
348 VA disposal . . . . 120 V, 1ϕ
450 VA compactor . . . . 120 V, 1ϕ
1200 VA microwave . . . . 120 V, 1ϕ
144 VA exhaust fan . . . . 120 V, 1ϕ

OPTIONAL CALCULATION—MULTIFAMILY DWELLING

1. OTHER LOADS—220-32

General lighting loads—220-32(c)(2) — **PHASES**

950 sq. ft. × 3 VA × 30 units = 85,500 VA — 85,500 VA
— sq. ft. × 3 VA × — units = — VA
— sq. ft. × 3 VA × — units = — VA
— sq. ft. × 3 VA × — units = — VA

Small appliance and laundry loads—220-32(c)(1)

1500 VA × 3 circuits × 30 units = 135,000 VA — 135,000 VA

Special loads—220-32(c)(3)

Dishwasher = 1500 VA
Disposal = 348 VA
Compactor = 450 VA
Water heater = 5000 VA
Dryer = 5000 VA
Cooktop = 9000 VA
Oven = 10,000 VA
Blower motor = 840 VA
Microwave = 1200 VA
Exhaust fan = 144 VA

(total) 33,482 VA × 30 units = 1,004,460 VA — 1,004,460 VA

2. HEATING OR A/C LOAD—220-32(5)

Heating unit = 10,000 VA × 100% × 30 units = 300,000 VA
A/C unit = 5016 VA × 100% × 30 units = 150,480 VA
Heat pump = — VA × 100% × — units = — VA

(largest load) 300,000 VA — 300,000 VA

TOTAL = 1,524,960 VA

Applying demand factors—Table 220-32

1,524,960 VA × 33 % = 503,236.8 VA

NEUTRAL

(Loads from Standard Calculation)

1. General lighting = 69,075 VA
2. Cooking equipment = 57,750 VA
3. Dryer = 31,500 VA
4. Fixed appliance = 81,945 VA
5. Heating or A/C = — VA
6. Largest motor = 112.5 VA

TOTAL = 240,382.5 VA

527 A

1ϕ service: $I = \frac{VA}{V}$

PHASES I = ____ VA / V = ____ A

NEUTRAL I = ____ VA / V = ____ A

3ϕ service: $I = \frac{VA}{V \times \sqrt{3}}$

$I = \frac{503{,}236.8 \text{ VA}}{208 \text{ V} \times \sqrt{3}} = 1397.9 \text{ A}$

$I = \frac{240{,}382.5 \text{ VA}}{208 \text{ V} \times \sqrt{3}} = 667.7 \text{ A}$

If over 200 A—220-22; First 200 A × 100% = 200 A
Remaining 467.7 A × 70% = 327 A
527 A

Figure 7-17. The service load in VA and amps for apartment units with the same square footage is found by using the form, OPTIONAL CALCULATION—MULTIFAMILY DWELLING. Total VA for 3ϕ service is divided by V × $\sqrt{3}$.

(1) What is the feeder circuit ampacity to one of the dwelling units? Use the optional calculation. See Figure 7-18.

(2) What size THWN Cu conductors is required for phases A and B and the neutral of the feeder circuit?

Phases A and B

Step 1. *Table 310-16, 75°C Column:*
139 A requires #1/0 THWN Cu

Answer. **Conductors = #1/0 THWN Cu**

Note: Note 3 to Tables 310-16–310-19 allows #1 THWN Cu conductors.

Neutral

Step 1. *Table 310-16, 75°C Column:*
89 A requires #3 THWN Cu

Answer. **Conductor = #3 THWN Cu**

(3) What size circuit breaker is required to protect the feeder circuit?

Step 1. *240-3, Ex. 4; 240-6:* 150 A = 150 A CB

Answer. **CB = 150 A**

(4) What size conduit is required for the feeder circuit? Use PVC.

Step 1. *Table 5, Ch. 9:*
#1/0 THWN =
.1893 sq. in. × 2 = .3786 sq. in.
#3 THWN =
.0995 sq. in. × 1 = .0995 sq. in.
Total = .4781 sq. in.

Step 2. *Table 5, Ch. 9:*
.4781 sq. in. area requires .60 sq. in. area

Answer. **Conduit = 1¼" PVC**

(5) What size copper equipment grounding conductor is required?

Step 1. *Table 250-95:* 150 A CB requires #6 Cu

Answer. **EGC = #6 Cu**

Service with Units Having Different Square Footage—220-32; Table 220-32

Four different sized dwelling units in an apartment complex have the following loads and are supplied by a 120/240 V, 1ϕ service. What is the total ampacity for phases A and B and the neutral? See Figure 7-19.

Unit 1—1200 sq. ft.

15 kW heating unit . . . . 240 V, 1ϕ
5040 VA A/C . . . . 240 V, 1ϕ
864 VA disposal . . . . 120 V, 1ϕ
1600 VA dishwasher . . . . 120 V, 1ϕ
12 kW range . . . . 240 V, 1ϕ
5500 VA water heater . . . . 240 V, 1ϕ
5000 VA dryer . . . . 240 V, 1ϕ
600 VA blower motor . . . . 240 V, 1ϕ

Unit 2—1100 sq. ft.

15 kW heating unit . . . . 240 V, 1ϕ
5040 VA A/C unit . . . . 240 V, 1ϕ
12 kW range . . . . 240 V, 1ϕ
5500 VA water heater . . . . 240 V, 1ϕ
5000 VA dryer . . . . 240 V, 1ϕ
864 VA disposal . . . . 120 V, 1ϕ
960 VA compactor . . . . 120 V, 1ϕ
1600 VA dishwasher . . . . 120 V, 1ϕ
118.8 VA vent hood . . . . 120 V, 1ϕ
600 VA blower motor . . . . 240 V, 1ϕ

Unit 3—1050 sq. ft.

10 kW heating unit . . . . 240 V, 1ϕ
4560 VA A/C unit . . . . 240 V, 1ϕ
9.5 kW cooktop . . . . 240 V, 1ϕ
10.5 kW oven . . . . 240 V, 1ϕ
696 VA disposal . . . . 120 V, 1ϕ
950 VA compactor . . . . 120 V, 1ϕ
1500 VA dishwasher . . . . 120 V, 1ϕ
5000 VA dryer . . . . 240 V, 1ϕ
5000 VA water heater . . . . 240 V, 1ϕ
456 VA blower motor . . . . 240 V, 1ϕ

Unit 4—1000 sq. ft.

10 kW heating unit . . . . 240 V, 1ϕ
4560 VA A/C unit . . . . 240 V, 1ϕ
5000 VA water heater . . . . 240 V, 1ϕ
4500 VA dryer . . . . 240 V, 1ϕ
1400 VA dishwasher . . . . 120 V, 1ϕ
696 VA disposal . . . . 120 V, 1ϕ
9 kW cooktop . . . . 240 V, 1ϕ
10 kW oven . . . . 240 V, 1ϕ
456 VA blower motor . . . . 240 V, 1ϕ
1200 VA microwave . . . . 120 V, 1ϕ
975 VA compactor . . . . 120 V, 1ϕ

Service with Units and House Loads—220-32; Table 220-32

Thirty-two 800 sq. ft. dwelling units are supplied by a 120/208 V, 3ϕ, 4-wire service. What is the amperage for phases A, B, and C and the neutral? See Figure 7-20.

House loads

Laundry loads . . . . 65,520 VA, 208 V, 1ϕ
(continuous and noncontinuous duty)
Outside lighting loads (electric
discharge) . 50 units of 180 VA each, 120 V, 1ϕ
(continuous duty)
Outside receptacle
loads . . . . 30 duplex receptacles, 120 V, 1ϕ
(noncontinuous duty)
Outside sign loads
(electric discharge) . . . . 1800 VA, 120 V, 1ϕ
(continuous duty)

Dwelling unit loads (with three small appliance circuits)

8 kW heating unit . . . . 208 V, 1ϕ
4.32 kW A/C unit . . . . 208 V, 1ϕ
4500 VA water heater . . . . 208 V, 1ϕ
6 kW cooktop . . . . 208 V, 1ϕ
8 kW oven . . . . 208 V, 1ϕ
5000 VA dryer . . . . 208 V, 1ϕ
600 VA blower motor . . . . 208 V, 1ϕ
1550 VA dishwasher . . . . 120 V, 1ϕ
864 VA disposal . . . . 120 V, 1ϕ
120 VA vent hood . . . . 120 V, 1ϕ
696 VA vacuum pump . . . . 120 V, 1ϕ
180 VA exhaust fan . . . . 120 V, 1ϕ

OPTIONAL CALCULATION—ONE-FAMILY DWELLING (NEW OR EXISTING)

1. OTHER LOADS—220-30(b)

General lighting loads—220-30(b)(2)

950 sq. ft. × 3 VA = 2850 VA

Small appliance and laundry loads—220-30(b)(1)

1500 VA × 3 circuits = 4500 VA

Special loads—220-30(b)(3)(4)

Dishwasher =	1500 VA
Disposal =	348 VA
Compactor =	450 VA
Water heater =	5000 VA
Cooktop =	9000 VA
Oven =	10,000 VA
Dryer =	5000 VA
Blower motor =	840 VA
Microwave =	1200 VA
Exhaust fan =	144 VA
(total)	40,832 VA

Applying demand factors—Table 220-30(4)

First 10,000 VA × 100% = 10,000 VA

Remaining 30,832 VA × 40% = 12,332.8 VA

(total) 22,332.8 VA

PHASES: 22,332.8 VA

2. HEATING OR A/C LOAD—Table 220-30(1)(2)(3)

Heating units (3 or less) = 10,000 VA × 65% = 6500 VA

Heating units (4 or more) = — VA × 40% = — VA

A/C unit = 5016 VA × 100% = 5016 VA

Heat pump = — VA × 100% = — VA

(largest load) 6500 VA — 6500 VA

TOTAL = 28,832.8 VA

NEUTRAL

(Loads from Standard Calculation)

1. General lighting = 4522.5 VA
2. Cooking equipment = 7700 VA
3. Dryer = 3500 VA
4. Fixed appliance = 2731.5 VA
5. Heating or A/C = — VA
6. Largest motor = 112.5 VA

TOTAL = 18,566.5 VA

89 A

1ϕ service: $I = \frac{VA}{V}$

PHASES $I = \frac{28{,}832.8\ VA}{208\ V}$ = 138.6 A

NEUTRAL $I = \frac{18{,}566.5\ VA}{208\ V}$ = 89 A

138.6 A

Figure 7-18. The service load in VA and amps for a feeder circuit to an individual dwelling in an apartment complex is found by using the form, OPTIONAL CALCULATION—ONE-FAMILY DWELLING (NEW OR EXISTING). Conductor size is based on this figure.

OPTIONAL CALCULATION—MULTIFAMILY DWELLING

1. OTHER LOADS—220-32

General lighting loads—220-32(c)(2)

		PHASES
1200 sq. ft. × 3 VA × 1 units =	3600 VA	13,050 VA
1100 sq. ft. × 3 VA × 1 units =	3300 VA	
1050 sq. ft. × 3 VA × 1 units =	3150 VA	
1000 sq. ft. × 3 VA × 1 units =	3000 VA	

Small appliance and laundry loads—220-32(c)(1)

1500 VA × 3 circuits × 4 units = 18,000 VA — 18,000 VA

Special loads—220-32(c)(3)

Dishwasher =	6100 VA
Disposal =	3120 VA
Compactor =	2885 VA
Water heater =	21,000 VA
Cooking eqp. =	63,000 VA
Dryer =	19,500 VA
Microwave =	1200 VA
Blower motor =	2112 VA
Vent fan =	118.8 VA
— =	— VA

(total) 119,035.8 VA × — units = 119,035.8 VA — 119,035.8 VA

2. HEATING OR A/C LOAD—220-32(c)(5)

Heating unit = 50,000 VA × 100% × — units =	50,000 VA	
A/C unit = 19,200 VA × 100% × — units =	19,200 VA	
Heat pump = — VA × 100% × — units =	— VA	
(largest load)	50,000 VA	50,000 VA
TOTAL		200,085.8 VA

Applying demand factors—Table 220-32

200,085.8 VA × 45 % = 90,038.6 VA

NEUTRAL

(Loads from Standard Calculation)

1. General lighting =	12,817.5 VA
2. Cooking equipment =	14,700 VA
3. Dryer =	14,000 VA
4. Fixed appliance =	10,067.9 VA
5. Heating or A/C =	— VA
6. Largest motor =	243.8 VA
TOTAL =	51,829.2 VA
	211 A

1ϕ service: $I = \frac{VA}{V}$

PHASES $I = \frac{90{,}038.6\ VA}{240\ V}$ = 375 A

NEUTRAL $I = \frac{51{,}829.2\ VA}{240\ V}$ = 215.9 A

3ϕ service: $I = \frac{VA}{V \times \sqrt{3}}$

$I = \frac{\_\_\_\ VA}{V \times \sqrt{3}}$ = ____ A

$I = \frac{\_\_\_\ VA}{V \times \sqrt{3}}$ = ____ A

If over 200 A—220-22; First 200 A × 100% = 200 A

Remaining 15.9 A × 70% = 11 A

211 A

Figure 7-19. The service load in VA and amps for apartment units with different square footages is found by using the form, OPTIONAL CALCULATION—MULTIFAMILY DWELLING.

OPTIONAL CALCULATION—MULTIFAMILY DWELLING

1. OTHER LOADS—220-32

General lighting loads—220-32(c)(2) — **PHASES**

800 sq. ft. × 3 VA × 32 units = 76,800 VA — 76,800 VA

— sq. ft. × 3 VA × — units = — VA

— sq. ft. × 3 VA × — units = — VA

— sq. ft. × 3 VA × — units = — VA

Small appliance and laundry loads—220-32(c)(1)

1500 VA × 3 circuits × 32 units = 144,000 VA — 144,000 VA

Special loads—220-32(c)(3)

Dishwasher = 1550 VA

Disposal = 864 VA

Compactor = — VA

Water heater = 4500 VA

Cooking loads = 14,000 VA

Dryer = 5000 VA

Blower mtr. = 600 VA

Vac. pump = 696 VA

Exh. fan = 180 VA

Vent hood = 120 VA

(total) 27,510 VA × 32 units = 880,320 VA — 880,320 VA

2. HEATING OR A/C LOAD—220-32(c)(5)

Heating unit = 8000 VA × 100% × 32 units = 256,000 VA

A/C unit = 4320 VA × 100% × 32 units = 138,240 VA

Heat pump = — VA × 100% × — units = — VA

(largest load) 256,000 VA — 256,000 VA

TOTAL = 1,357,120 VA

Applying demand factors—Table 220-32

1,357,120 VA × 31 % = 420,707.2 VA

NEUTRAL

(Loads from Standard Calculation)

1. General lighting = 69,150 VA
2. Cooking equipment = 50,176 VA
3. Dryer = 33,600 VA
4. Fixed appliance = 81,840 VA
5. Heating or A/C = — VA
6. Largest motor = 216 VA

TOTAL = 234,982 VA

516.9 A

1ϕ service: $I = \frac{VA}{V}$

PHASES $I = \frac{\_\_\_ VA}{\_\_\_ V}$ = ___ A

NEUTRAL $I = \frac{\_\_\_ VA}{\_\_\_ V}$ = ___ A

3ϕ service: $I = \frac{VA}{V \times \sqrt{3}}$

$I = \frac{420,707.2 \text{ VA}}{208 \text{ V} \times \sqrt{3}}$ = 1168.6 A

$I = \frac{234,982 \text{ VA}}{208 \text{ V} \times \sqrt{3}}$ = 652.7 A

If over 200 A—220-22; First 200 A × 100% = 200 A

Remaining 452.7 A × 70% = 316.9 A

516.9 A

Figure 7-20. The service load in VA and amps for apartment units with different square footages is found by using the form, OPTIONAL CALCULATION—MULTIFAMILY DWELLING. House loads are added to this figure.

(1) What is the ampacity of the house loads?

Step 1. Total house loads. *220-32(b):*

Laundry loads. *210-52(f), Ex. 1; 220-10(b):*
65,520 VA × 100% = 65,520 VA

Outside lighting loads. *210-22(b); 220-22:*
50 units × 180 VA × 125% = 11,250 VA

Outside receptacles. *220-3(c)(5); Table 220-13:*
30 plugs × 180 VA × 100% = 5400 VA

Outside sign load. *220-3, FPN; 600-6(c); 220-10(b):* 1800 VA × 125% = 2250 VA
65,520 VA + 11,250 VA + 5400 VA + 2250 VA = 84,420 VA

Step 2. *Text:* $I = \dfrac{VA}{V \times \sqrt{3}}$

$$I = \frac{84{,}420 \text{ VA}}{208 \text{ V} \times 1.732} = 234.5 \text{ A}$$

Answer. **House loads = 234.5 A**

(2) What is the service ampacity for phases A, B, and C and the neutral?

Phases

Step 1. *220-10–220-22:*

Apartment loads =	1168.6 A
House loads =	234.5 A
Total =	1403.1 A

Answer. **Phases A, B, and C = 1403.1 A**

Neutral

Step 1. *220-22:*

Apartment loads =	516.9 A
House loads =	52.5 A
Total =	569.4 A

Step 2. Applying demand factors. *220-22:*

First 200 A × 100% =	200 A
Next 452.7 A × 70% =	316.9 A
Total =	516.9 A

Plus electric discharge lighting
37.5 A × 100% = 37.5 A

Answer. **Neutral = 554.4 A**

Dividing Volt Amps on Line and Neutral

Some examination boards require that VA be divided on the line and the neutral using both calculations when computing the service load for a number of dwelling units in an apartment, condominium, or townhouse complex. VA ratings are carried with the optional calculation. The load for each piece of equipment is balanced as closely as possible to phases A and B when connected to the overcurrent protection devices in the panelboard. Some testing authorities require the amperage to be calculated and listed for the standard and optional calculation.

Ten 985 sq. ft. dwelling units in a townhouse complex are supplied by 120/240 V, 1ϕ services. Each dwelling unit is equipped with the following loads. Divide the VA on the line and neutral using the optional calculation. See Figure 7-21.

10 kW heating unit	240 V, 1ϕ
5 kW A/C unit	240 V, 1ϕ
5000 VA water heater	240 V, 1ϕ
10 kW range	240 V, 1ϕ
5000 VA dryer	240 V, 1ϕ
864 VA blower motor	240 V, 1ϕ
1320 VA dishwasher	120 V, 1ϕ
696 VA disposal	120 V, 1ϕ
785 VA compactor	120 V, 1ϕ
1500 VA microwave	120 V, 1ϕ
66 VA vent hood	120 V, 1ϕ
102 VA exhaust fan	120 V, 1ϕ

Twenty-one 950 sq. ft. dwelling units in a townhouse complex are supplied by 120/208 V, 3ϕ services. Each dwelling unit is equipped with the following loads. Divide the VA on the line and neutral using the optional calculation. See Figure 7-22.

10.5 kW heating unit	208 V, 1ϕ
4.5 kW A/C unit	208 V, 1ϕ
5000 VA water heater	208 V, 1ϕ
9.5 kW range	208 V, 1ϕ
5000 VA dryer	208 V, 1ϕ
864 VA blower motor	208 V, 1ϕ
1620 VA dishwasher	120 V, 1ϕ
696 VA disposal	120 V, 1ϕ
885 VA compactor	120 V, 1ϕ
1440 VA microwave	120 V, 1ϕ
66 VA vent hood	120 V, 1ϕ
120 VA exhaust fan	120 V, 1ϕ

OPTIONAL CALCULATION—MULTIFAMILY DWELLING 1ϕ (CARRYING PHASES AND NEUTRAL IN VA)

1. OTHER LOADS—220-32

	PHASE A	PHASE B	NEUTRAL (from Std. Calc.)
General lighting loads—220-32(c)(2)			
985 sq. ft. × 3 VA × 10 units ÷ 2 =	14,775 VA	14,775 VA	— VA
— sq. ft. × 3 VA × — units ÷ — =	— VA	— VA	— VA
— sq. ft. × 3 VA × — units ÷ — =	— VA	— VA	— VA
— sq. ft. × 3 VA × — units ÷ — =	— VA	— VA	— VA
Small appliance and laundry loads—220-32(c)(1)			
1500 VA × 3 circuits × 10 units ÷ 2 =	22,500 VA	22,500 VA	28,042.5 VA

Special loads—220-32(c)(3)

				PHASE A	PHASE B		NEUTRAL
Dishwasher =	1320 VA ×	10 units ÷	2 =	6600 VA	6600 VA	× 75% =	4950 VA
Disposal =	696 VA ×	10 units ÷	2 =	3480 VA	3480 VA	× 75% =	2610 VA
Compactor =	785 VA ×	10 units ÷	2 =	3925 VA	3925 VA	× 75% =	2943.8 VA
Water heater =	5000 VA ×	10 units ÷	— =	50,000 VA	50,000 VA	× 75% =	— VA
Cooktop/range =	10,000 VA ×	10 units ÷	— =	100,000 VA	100,000 VA	× 70% =	17,500 VA
Oven =	— VA ×	— units ÷	— =	— VA	— VA	× 70% =	— VA
Dryer =	5000 VA ×	10 units ÷	— =	50,000 VA	50,000 VA	× 70% =	17,500 VA
Blower motor =	864 VA ×	10 units ÷	— =	8640 VA	8640 VA	× 75% =	— VA
Microwave =	1500 VA ×	10 units ÷	2 =	7500 VA	7500 VA	× 75% =	5625 VA
Vent hood =	66 VA ×	10 units ÷	2 =	330 VA	330 VA	× 75% =	247.5 VA
Exhaust fan =	102 VA ×	10 units ÷	2 =	510 VA	510 VA	× 75% =	382.5 VA
— =	— VA ×	— units ÷	— =	— VA	— VA	× 75% =	— VA
— =	— VA ×	— units ÷	— =	— VA	— VA	× 75% =	— VA
— =	— VA ×	— units ÷	— =	— VA	— VA	× 75% =	— VA

2. HEATING OR A/C LOAD—220-32(c)(5)

	PHASE A	PHASE B	NEUTRAL
Heating unit = 10,000 VA × 100% × 10 units =	100,000 VA	100,000 VA	— VA
A/C unit = 5000 VA × 100% × 10 units =	— VA	— VA	— VA
Heat pump = — VA × 100% × — units =	— VA	— VA	— VA
TOTAL =	368,260 VA	368,260 VA	

Plus largest motor = 196.3 VA

79,997.6 VA

Applying demand factors—Table 220-32

_____ VA × _____% = _____ VA

1ϕ service: $I = \frac{VA}{V}$

PHASES $I = \frac{\_\_\_\_\_ VA}{V}$ = _____ A

NEUTRAL $I = \frac{\_\_\_\_\_ VA}{V}$ = _____ A

If over 200 A—220-22; First 200 A × 100% = 200 A

Remaining _____ A × 70% = _____ A

_____ A

Figure 7-21. The 1ϕ service load in VA can be easily divided on phases A, B, and the neutral by using the form, OPTIONAL CALCULATION—MULTIFAMILY DWELLING 1ϕ (CARRYING PHASES AND NEUTRAL IN VA).

OPTIONAL CALCULATION—MULTIFAMILY DWELLING 3ϕ (CARRYING PHASES AND NEUTRAL IN VA)

1. OTHER LOADS—220-32

	PHASE A	PHASE B	PHASE C		NEUTRAL (from Std. Calc.)
General lighting loads—220-32(c)(2)					
950 sq. ft. × 3 VA × 21 units ÷ 3 =	19950 VA	19,950 VA	19950 VA		— VA
— sq. ft. × 3 VA × — units ÷ — =	— VA	— VA	— VA		— VA
— sq. ft. × 3 VA × — units ÷ — =	— VA	— VA	— VA		— VA
— sq. ft. × 3 VA × — units ÷ — =	— VA	— VA	— VA		— VA
Small appliance and laundry loads—220-32(c)(1)					
1500 VA × 3 circuits × 21 units ÷ 3 =	31,500 VA	31,500 VA	31,500 VA		52,537.5 VA
Special loads—220-32(c)(3)					
Dishwasher = 1620 VA × 21 units ÷ 3 =	11,340 VA	11,340 VA	11,340 VA	× 75% =	8505 VA
Disposal = 696 VA × 21 units ÷ 3 =	4872 VA	4872 VA	4872 VA	× 75% =	3654 VA
Compactor = 885 VA × 21 units ÷ 3 =	6195 VA	6195 VA	6195 VA	× 75% =	4646.3 VA
Water heater = 5000 VA × 21 units ÷ 3 =	35,000 VA	35,000 VA	35,000 VA	× 75% =	— VA
Cooktop/range = 9500 VA × 21 units ÷ 3 =	66,500 VA	66,500 VA	66,500 VA	× 70% =	30,450 VA
Oven = — VA × — units ÷ — =	— VA	— VA	— VA	× 70% =	— VA
Dryer = 5000 VA × 21 units ÷ 3 =	35,000 VA	35,000 VA	35,000 VA	× 70% =	25,725 VA
Blower mtr. = 864 VA × 21 units ÷ 3 =	6048 VA	6048 VA	6048 VA	× 75% =	— VA
Microwave = 1440 VA × 21 units ÷ 3 =	10,080 VA	10,080 VA	10,080 VA	× 75% =	7560 VA
Venthood = 66 VA × 21 units ÷ 3 =	462 VA	462 VA	462 VA	× 75% =	346.5 VA
Exh. fan = 120 VA × 21 units ÷ 3 =	840 VA	840 VA	840 VA	× 75% =	630 VA
____ = ____ VA × ____ units ÷ ____ =	____ VA	____ VA	____ VA	× 75% =	____ VA
____ = ____ VA × ____ units ÷ ____ =	____ VA	____ VA	____ VA	× 75% =	____ VA
____ = ____ VA × ____ units ÷ ____ =	____ VA	____ VA	____ VA	× 75% =	____ VA

2. HEATING OR A/C LOAD—220-32(c)(5)

	PHASE A	PHASE B	PHASE C	NEUTRAL
Heating unit = 10,500 VA × 100% × 21 units ÷ 3 =	73500 VA	73,500 VA	73,500 VA	— VA
A/C unit = 4500 VA × 100% × 21 units ÷ — =	— VA	— VA	— VA	— VA
Heat pump = — VA × 100% × — units ÷ — =	— VA	— VA	— VA	— VA
TOTAL =	301,287 VA	301,287 VA	301,287 VA	
Plus largest motor =				221.3 VA
				134,275.6 VA

Applying demand factors—Table 220-32

________ VA × ________ % = ________ VA

3ϕ **service:** $I = \dfrac{VA}{V \times \sqrt{3}}$

$I = \dfrac{VA}{V \times \sqrt{3}} =$ ________ A

$I = \dfrac{VA}{V \times \sqrt{3}} =$ ________ A

If over 200 A—220-22; First 200 A × 100% = 200 A

Remaining ________ A × 70% = ________ A

________ A

Figure 7-22. The 3ϕ service load in VA can be easily divided on phases A, B, C, and the neutral by using the form, OPTIONAL CALCULATION—MULTIFAMILY DWELLING 3ϕ (CARRYING PHASES AND NEUTRAL IN VA).

Calculations— Multifamily Dwellings

Date ____________________

Chapter 7—TEST

Name ____________________________________

____________ 1. What are the general lighting and general-purpose receptacle loads in VA for a 40 unit apartment complex with each unit having 795 sq. ft. of space?

____________ 2. How many 15 A, 2-wire circuits are required to supply the general lighting and receptacle load in one of the dwelling units in question 1?

____________ 3. What are the small appliance circuit loads in VA for the kitchen, pantry, dining room, and breakfast room for the 40 dwelling units in question 1? Use standard method.

____________ 4. What are the small appliance loads in VA for the laundry circuits in the 40 dwelling units in question 1? Use standard method.

____________ 5. What is the demand load in VA for the phases and neutral for the general lighting loads in question 1? Use standard method.

____________ 6. What is the demand load in VA for the phases for forty 12 kW ranges installed in the dwelling units in question 1? Use standard method.

____________ 7. What is the demand load in VA for the phases for forty 8 kW ovens and forty 7 kW cooktops installed in the dwelling units in question 1? Use standard method.

____________ 8. What is the demand load in VA for the phases for forty 10 kW ovens and forty 8 kW cooktops installed in the dwelling units in question 1? Use optional method.

____________ 9. What is the demand load in VA for the phases for forty 3 kW ranges and forty 7 kW ovens installed in the dwelling units in question 1? Use standard method.

____________ 10. What is the demand load in VA for the phases for forty 5000 VA dryers installed in the dwelling units in question 1?

____________ 11. What is the demand load in VA for the phases for forty 4500 VA dryers installed in the dwelling units in question 1? Use the nameplate ratings of the dryers to compute the total load in VA per 220-32(c)(3).

____________ 12. What is the demand load in VA for the phases for the fixed appliance loads for forty 975 sq. ft. dwelling units with the following loads?

10 kW heating unit .... 240 V, 1ϕ
5000 VA A/C unit .... 240 V, 1ϕ
9000 VA cooktop .... 240 V, 1ϕ
10,000 VA oven .... 240 V, 1ϕ
960 VA disposal .... 120 V, 1ϕ
1020 VA compactor .... 120 V, 1ϕ
1600 VA dishwasher .... 120 V, 1ϕ
5000 VA dryer .... 240 V, 1ϕ
5500 VA water heater .... 240 V, 1ϕ
1728 VA blower motor .... 240 V, 1ϕ
2112 VA circulating pump .... 120 V, 1ϕ

__________ **13.** What is the load in VA for the phases for the fixed appliance load for forty 975 sq. ft. dwelling units supplied by a 120/240 V, 1ϕ service with the following loads? Three fixed appliances per unit × 40 units = 120 fixed appliances that the service conductors will carry per 220-17. Apply 75% per 220-17.

10 kW heating . . . . 240 V, 1ϕ
5 kVA A/C . . . . 240 V, 1ϕ
960 VA disposal . . . . 120 V, 1ϕ
1550 VA dishwasher . . . . 120 V, 1ϕ
12 kW range . . . . 240 V, 1ϕ
5500 VA water heater . . . . 240 V, 1ϕ
5000 VA dryer . . . . 240 V, 1ϕ

__________ **14.** What is the service load in VA for the phases between the heating and A/C load where the heating load is 10 kW and the A/C load is 4500 VA? There are 36 dwelling units in the complex. Use optional method.

__________ **15.** What is the service load in VA for the phases between the heating and A/C load for a 25 unit apartment complex? The heating load is 15 kW and the A/C load is a 6000 VA heat pump that can be used with the heating unit. Use optional method without applying demand factors.

__________ **16.** What is the percent of the demand factors used to determine the service load for a complex with 23 dwelling units?

__________ **17.** What is the service load in VA for the phases for the connected loads, other than the heat or A/C loads, in twenty-nine 1030 sq. ft. dwelling units in a complex with the following loads? The service is 120/240 V, 1ϕ.

10 kW heating unit . . . . 240 V, 1ϕ
4560 VA A/C unit . . . . 240 V, 1ϕ
5500 VA water heater . . . . 240 V, 1ϕ
10 kW cooktop . . . . 240 V, 1ϕ
10 kW oven . . . . 240 V, 1ϕ
4500 VA dryer . . . . 240 V, 1ϕ
969 VA blower motor . . . . 240 V, 1ϕ
8 kW oven . . . . 240 V, 1ϕ
1600 VA dishwasher . . . . 120 V, 1ϕ
864 VA disposal . . . . 120 V, 1ϕ
1200 VA microwave . . . . 120 V, 1ϕ
975 VA compactor . . . . 120 V, 1ϕ
250 VA bathroom ceiling heat lamp . . . . 120 V, 1ϕ

__________ **18.** What is the service load in VA for the phases for a 48 unit apartment complex after demand factors have been applied? The connected loads total 250,500 VA. Use optional method.

________ **19.** What is the service load in amps for the phases and neutral for twenty-eight 1010 sq. ft. dwelling units supplied by a 120/208 V, 3ϕ, 4-wire system in a complex with the following loads? The heat pump can operate with the heating unit. Use the form, STANDARD CALCULATION—MULTIFAMILY DWELLING.

10 kW heating unit .... 208 V, 1ϕ
5000 VA A/C and heat pump .... 208 V, 1ϕ
5500 VA water heater .... 208 V, 1ϕ
10 kW cooktop .... 208 V, 1ϕ
9 kW oven .... 208 V, 1ϕ
4500 VA dryer .... 208 V, 1ϕ
9 kW oven .... 208 V, 1ϕ
1500 VA dishwasher .... 120 V, 1ϕ
648 VA disposal .... 120 V, 1ϕ
750 VA compactor .... 120 V, 1ϕ
144 VA exhaust fan .... 120 V, 1ϕ
1200 VA bathroom ceiling heater .... 120 V, 1ϕ

Note: Heat pump can operate with the heating unit.

________ **20.** What is the service load in VA for the phases and neutral for six dwelling units in a complex supplied by a 120/240 V, 1ϕ service with the following loads? Use the form, STANDARD CALCULATION—MULTIFAMILY DWELLING.

Units 1 and 2—1200 sq. ft.

15 kW heating unit .... 240 V, 1ϕ
5400 VA A/C unit .... 240 V, 1ϕ
980 VA disposal .... 120 V, 1ϕ
980 VA compactor .... 120 V, 1ϕ
1600 VA dishwasher .... 120 V, 1ϕ
12 kVA range .... 240 V, 1ϕ
5500 VA water heater .... 240 V, 1ϕ
5000 VA dryer .... 240 V, 1ϕ
856 VA blower motor .... 240 V, 1ϕ
828 VA attic fan .... 120 V, 1ϕ

Units 3 and 4—1150 sq. ft.

15 kW heating unit .... 240 V, 1ϕ
5400 VA A/C unit .... 240 V, 1ϕ
12 kVA range .... 240 V, 1ϕ
5500 VA water heater .... 240 V, 1ϕ
5000 VA dryer .... 240 V, 1ϕ
980 VA disposal .... 120 V, 1ϕ
980 VA compactor .... 120 V, 1ϕ
1600 VA dishwasher .... 120 V, 1ϕ
856 VA blower motor .... 240 V, 1ϕ
828 VA attic fan .... 120 V, 1ϕ

Units 5 and 6—1100 sq. ft.

15 kW heating unit .... 240 V, 1ϕ
5400 VA A/C unit .... 240 V, 1ϕ
12 kW range .... 240 V, 1ϕ
980 VA disposal .... 120 V, 1ϕ
980 VA compactor .... 120 V, 1ϕ
1600 VA dishwasher .... 120 V, 1ϕ
828 VA attic fan .... 120 V, 1ϕ
5000 VA dryer .... 240 V, 1ϕ
5500 VA water heater .... 240 V, 1ϕ
856 VA blower motor .... 240 V, 1ϕ

______________ **21.** What is the service load in amps for the phases and neutral for four dwelling units in a complex supplied by a 120/240 V, 1ϕ service with the following loads? Use the form, OPTIONAL CALCULATION—MULTIFAMILY DWELLING.

Unit 1—750 sq. ft.

8 kW heating unit . . . . 240 V, 1ϕ
4056 VA A/C unit . . . . 240 V, 1ϕ
980 VA disposal . . . . 120 V, 1ϕ
1620 VA dishwasher . . . . 120 V, 1ϕ
12 kVA range . . . . 240 V, 1ϕ
5000 VA water heater . . . . 240 V, 1ϕ
5000 VA dryer . . . . 240 V, 1ϕ
696 VA blower motor . . . . 240 V, 1ϕ

Unit 2—875 sq. ft.

8 kW heating unit . . . . 240 V, 1ϕ
4056 VA A/C unit . . . . 240 V, 1ϕ
10.5 kVA range . . . . 240 V, 1ϕ
5500 VA water heater . . . . 240 V, 1ϕ
5000 VA dryer . . . . 240 V, 1ϕ
969 VA disposal . . . . 120 V, 1ϕ
1000 VA compactor . . . . 120 V, 1ϕ
1580 VA dishwasher . . . . 120 V, 1ϕ
696 VA vent hood . . . . 120 V, 1ϕ

Unit 3—965 sq. ft.

10 kW heating unit . . . . 240 V, 1ϕ
5160 VA A/C unit . . . . 240 V, 1ϕ
10 kVA cooktop . . . . 240 V, 1ϕ
12 VA oven . . . . 240 V, 1ϕ
969 VA disposal . . . . 120 V, 1ϕ
1000 VA compactor . . . . 120 V, 1ϕ
1600 VA dishwasher . . . . 120 V, 1ϕ
468 VA vent hood . . . . 120 V, 1ϕ
5000 VA dryer . . . . 240 V, 1ϕ
5000 VA water heater . . . . 240 V, 1ϕ
840 VA blower motor . . . . 120 V, 1ϕ

Unit 4—1020 sq. ft.

10 kW heating unit . . . . 240 V, 1ϕ
5160 VA A/C unit . . . . 240 V, 1ϕ
6500 VA water heater . . . . 240 V, 1ϕ
5000 VA dryer . . . . 240 V, 1ϕ
1450 VA dishwasher . . . . 120 V, 1ϕ
969 VA disposal . . . . 120 V, 1ϕ
10 kVA cooktop . . . . 240 V, 1ϕ
12 VA oven . . . . 240 V, 1ϕ
840 VA blower motor . . . . 240 V, 1ϕ
1200 VA microwave . . . . 120 V, 1ϕ

_______ **22.** What is the service load in amps for the phases and neutral for twenty-six 890 sq. ft. dwelling units in a complex supplied by a 120/208 V, 3ϕ, 4-wire service with the following loads? Use the forms, OPTIONAL CALCULATION—MULTIFAMILY DWELLING and STANDARD CALCULATION—MULTIFAMILY DWELLING.

8 kW heating unit . . . . 208 V, 1ϕ
4 kVA A/C unit . . . . 208 V, 1ϕ
Gas water heater
9.5 kW cooktop . . . . 208 V, 1ϕ
10.5 kW oven . . . . 208 V, 1ϕ
5000 VA dryer . . . . 208 V, 1ϕ
1450 VA dishwasher . . . . 120 V, 1ϕ
686 VA disposal . . . . 120 V, 1ϕ
792 VA compactor . . . . 120 V, 1ϕ
65 VA exhaust fan . . . . 120 V, 1ϕ
250 VA bathroom heat lamp . . . . 120 V, 1ϕ

_______ **23.** What is the service load in VA for fourteen 998 sq. ft. dwelling units in a complex with the following loads? The service is 120/240 V, 1ϕ. Use the forms, STANDARD CALCULATION—MULTIFAMILY DWELLING 1ϕ (CARRYING PHASES AND NEUTRAL IN VA) and OPTIONAL CALCULATION—MULTIFAMILY DWELLING 1ϕ (CARRYING PHASES AND NEUTRAL IN VA). Do not apply demand factors to the neutral for the optional calculation. Use only 120 V loads.

10 kW heating unit . . . . 240 V, 1ϕ
4.5 kW A/C unit . . . . 240 V, 1ϕ
5500 VA water heater . . . . 240 V, 1ϕ
8 kW cooktop . . . . 240 V, 1ϕ
6 kW oven . . . . 240 V, 1ϕ
6 kW oven . . . . 240 V, 1ϕ
5000 VA dryer . . . . 240 V, 1ϕ
600 VA blower motor . . . . 240 V, 1ϕ
1600 VA dishwasher . . . . 120 V, 1ϕ
864 VA disposal . . . . 120 V, 1ϕ
708 VA compactor . . . . 120 V, 1ϕ
60 VA intercom . . . . 120 V, 1ϕ

_______ **24.** What is the service in amps for twenty 965 sq. ft. dwelling units in a complex with the following loads? The service is 120/240 V, 1ϕ. Use the forms, OPTIONAL CALCULATION—MULTIFAMILY DWELLING 1ϕ (CARRYING PHASES AND NEUTRAL IN AMPS) and STANDARD CALCULATION—MULTIFAMILY DWELLING 1ϕ (CARRYING PHASES AND NEUTRAL IN AMPS). Carry the line and neutral loads.

8 kW heating unit . . . . 240 V, 1ϕ
4680 VA A/C unit . . . . 240 V, 1ϕ
5000 VA water heater . . . . 240 V, 1ϕ
9 kW cooktop . . . . 240 V, 1ϕ
10 kW oven . . . . 240 V, 1ϕ
5000 VA dryer . . . . 240 V, 1ϕ
969 VA blower motor . . . . 240 V, 1ϕ
1550 VA dishwasher . . . . 120 V, 1ϕ
864 VA disposal . . . . 120 V, 1ϕ
958 VA compactor . . . . 120 V, 1ϕ
108 VA exhaust fan . . . . 120 V, 1ϕ

______ **25.** What is the amps for the phases and the neutral for the feeder circuit of one dwelling unit in a 50 unit complex? Each unit contains 985 sq. ft. The supply voltage is 120/240 V, 1ϕ. Use the forms, STANDARD CALCULATION—ONE-FAMILY DWELLING and OPTIONAL CALCULATION—ONE-FAMILY DWELLING (NEW OR EXISTING).

10 kW heating unit . . . . 240 V, 1ϕ
4950 VA A/C unit . . . . 240 V, 1ϕ
5500 VA water heater . . . . 240 V, 1ϕ
9 kW cooktop . . . . 240 V, 1ϕ
10 kW oven . . . . 240 V, 1ϕ
10 kW oven . . . . 240 V, 1ϕ
5000 VA dryer . . . . 240 V, 1ϕ
925 VA blower motor . . . . 240 V, 1ϕ
1550 VA dishwasher . . . . 120 V, 1ϕ
985 VA disposal . . . . 120 V, 1ϕ
960 VA compactor . . . . 120 V, 1ϕ
1200 VA microwave . . . . 120 V, 1ϕ
1200 VA vacuum pump . . . . 120 V, 1ϕ

Use for Problem 19

STANDARD CALCULATION—MULTIFAMILY DWELLING

1. GENERAL LIGHTING LOADS—Table 220-3(b)

_______ sq. ft. × 3 VA × _______ units = _______ VA

_______ sq. ft. × 3 VA × _______ units = _______ VA

_______ sq. ft. × 3 VA × _______ units = _______ VA

_______ VA

Small appliance loads—220-16(a)

_______ VA × _______ circuits × _______ units = _______ VA

Laundry load—220-16(b)

_______ VA × 1 × _______ units = _______ VA

_______ VA

Applying demand factors—Table 220-11

First 3000 VA × 100% = 3000 VA

Next _______ VA × 35% = _______ VA

Remaining _______ VA × 25% = _______ VA

(total) _______ VA | PHASES: _______ VA | NEUTRAL: _______ VA

2. COOKING EQUIPMENT LOADS—Table 220-19, Notes

Col. A _______ units = _______ VA demand = _______ VA

Col. B _______ VA × _______ units × _______% = _______ VA

Col. C _______ VA × _______ units × _______% = _______ VA

(total) _______ VA | _______ VA × 70% = _______ VA

3. DRYER LOAD—220-18; Table 220-18

_______ VA × _______ units × _______% = _______ VA | _______ VA × 70% = _______ VA

4. FIXED APPLIANCE LOADS—220-17

Dishwasher = _______ VA

Disposal = _______ VA

Compactor = _______ VA

Water heater = _______ VA

_______ = _______ VA

_______ = _______ VA

(total) _______ VA × _______ units × 75% = _______ VA | _______ VA | (120 V Loads × units × 75%) _______ VA

5. HEATING OR A/C LOAD—220-21

Heating unit = _______ VA × 100% × _______ units = _______ VA

A/C unit = _______ VA × 100% × _______ units = _______ VA

Heat pump = _______ VA × 100% × _______ units = _______ VA

(largest load) _______ VA | _______ VA | _______ VA

6. LARGEST MOTOR LOAD—220-14

ϕ _______ VA × 25% = _______ VA | _______ VA

N _______ VA × 25% = _______ VA | _______ VA

TOTAL = _______ VA | _______ VA

1ϕ service: $I = \frac{VA}{V}$

PHASES $I = \frac{\_\_\_\_ VA}{\_\_\_\_ V} =$ _______ A

NEUTRAL $I = \frac{\_\_\_\_ VA}{\_\_\_\_ V} =$ _______ A

3ϕ service: $I = \frac{VA}{V \times \sqrt{3}}$

$I = \frac{\_\_\_\_ VA}{\_\_\_\_ V \times \sqrt{3}} =$ _______ A

$I = \frac{\_\_\_\_ VA}{\_\_\_\_ V \times \sqrt{3}} =$ _______ A

If over 200 A—220-22; First 200 A × 100% = 200 A

Remaining _______ A × 70% = _______ A

_______ A

Use for Problem 20

STANDARD CALCULATION—MULTIFAMILY DWELLING

1. GENERAL LIGHTING LOADS—Table 220-3(b)

_______ sq. ft. × 3 VA × _______ units = _______ VA

_______ sq. ft. × 3 VA × _______ units = _______ VA

_______ sq. ft. × 3 VA × _______ units = _______ VA

_______ VA

Small appliance loads—220-16(a)

_______ VA × _______ circuits × _______ units = _______ VA

Laundry load—220-16(b)

_______ VA × 1 × _______ units = _______ VA

_______ VA

Applying demand factors—Table 220-11

First 3000 VA × 100% = 3000 VA

Next _______ VA × 35% = _______ VA

Remaining _______ VA × 25% = _______ VA

(total) _______ VA

	PHASES	NEUTRAL
	VA	VA

2. COOKING EQUIPMENT LOADS—Table 220-19, Notes

Col. A _______ units = _______ VA demand = _______ VA

Col. B _______ VA × _______ units × _______% = _______ VA

Col. C _______ VA × _______ units × _______% = _______ VA

(total) _______ VA | VA × 70% = | VA

3. DRYER LOAD—220-18; Table 220-18

_______ VA × _______ units × _______% = _______ VA | VA × 70% = | VA

4. FIXED APPLIANCE LOADS—220-17

Dishwasher = _______ VA

Disposal = _______ VA

Compactor = _______ VA

Water heater = _______ VA

_______ = _______ VA

_______ = _______ VA

(total) _______ VA × _______ units × 75% = _______ VA | VA | (120 V Loads × units × 75%) VA

5. HEATING OR A/C LOAD—220-21

Heating unit = _______ VA × 100% × _______ units = _______ VA

A/C unit = _______ VA × 100% × _______ units = _______ VA

Heat pump = _______ VA × 100% × _______ units = _______ VA

(largest load) _______ VA | VA | VA

6. LARGEST MOTOR LOAD—220-14

ϕ _______ VA × 25% = _______ VA | VA

N _______ VA × 25% = _______ VA | | VA

TOTAL = | VA | VA

1ϕ service: I = $\frac{\text{VA}}{\text{V}}$

PHASES I = $\frac{\text{VA}}{\text{V}}$ = _______ A

NEUTRAL I = $\frac{\text{VA}}{\text{V}}$ = _______ A

3ϕ service: I = $\frac{\text{VA}}{\text{V} \times \sqrt{3}}$

I = $\frac{\text{VA}}{\text{V} \times \sqrt{3}}$ = _______ A

I = $\frac{\text{VA}}{\text{V} \times \sqrt{3}}$ = _______ A

If over 200 A—220-22; First 200 A × 100% = 200 A

Remaining _______ A × 70% = _______ A

A

Use for Problem 21

OPTIONAL CALCULATION—MULTIFAMILY DWELLING

1. OTHER LOADS—220-32

PHASES

General lighting loads—220-32(c)(2)

_____ sq. ft. × 3 VA × _____ units = _____ VA   VA

_____ sq. ft. × 3 VA × _____ units = _____ VA

_____ sq. ft. × 3 VA × _____ units = _____ VA

_____ sq. ft. × 3 VA × _____ units = _____ VA

Small appliance and laundry loads—220-32(c)(1)

_____ VA × _____ circuits × _____ units = _____ VA   VA

Special loads—220-32(c)(3)

Dishwasher = _____ VA

Disposal = _____ VA

Compactor = _____ VA

Water heater = _____ VA

_____ = _____ VA

_____ = _____ VA

_____ = _____ VA

_____ = _____ VA

_____ = _____ VA

_____ = _____ VA

(total) _____ VA × _____ units = _____ VA   VA

2. HEATING OR A/C LOAD—220-32(c)(5)

Heating unit = _____ VA × 100% × _____ units = _____ VA

A/C unit = _____ VA × 100% × _____ units = _____ VA

Heat pump = _____ VA × 100% × _____ units = _____ VA

(largest load) _____ VA   VA

TOTAL =   VA

Applying demand factors—Table 220-32

_____ VA × _____ % = _____ VA

NEUTRAL

(Loads from Standard Calculation)

1. General lighting = _____ VA
2. Cooking equipment = _____ VA
3. Dryer = _____ VA
4. Fixed appliance = _____ VA
5. Heating or A/C = _____ VA
6. Largest motor = _____ VA

TOTAL =   VA

   A

1ϕ service: $I = \frac{VA}{V}$

PHASES $I = \frac{\_\_\_\_\_\ VA}{V}$ = _____ A

NEUTRAL $I = \frac{\_\_\_\_\_\ VA}{V}$ = _____ A

3ϕ service: $I = \frac{VA}{V \times \sqrt{3}}$

$I = \frac{\_\_\_\_\_\ VA}{V \times \sqrt{3}}$ = _____ A

$I = \frac{\_\_\_\_\_\ VA}{V \times \sqrt{3}}$ = _____ A

If over 200 A—220-22; First 200 A × 100% = 200 A

Remaining _____ A × 70% = _____ A

   A

OPTIONAL CALCULATION—MULTIFAMILY DWELLING

Use for Problem 22

1. OTHER LOADS—220-32

General lighting loads—220-32(c)(2) — **PHASES**

_______ sq. ft. × 3 VA × _____ units = _______ VA

_______ sq. ft. × 3 VA × _____ units = _______ VA

_______ sq. ft. × 3 VA × _____ units = _______ VA

_______ sq. ft. × 3 VA × _____ units = _______ VA — VA

Small appliance and laundry loads—220-32(c)(1)

_______ VA × _____ circuits × _____ units = _______ VA — VA

Special loads—220-32(c)(3)

Dishwasher = _______ VA

Disposal = _______ VA

Compactor = _______ VA

Water heater = _______ VA

_______________ = _______ VA

_______________ = _______ VA

_______________ = _______ VA

_______________ = _______ VA

_______________ = _______ VA

_______________ = _______ VA

(total) _______ VA × _____ units = _______ VA — VA

2. HEATING OR A/C LOAD—220-32(c)(5)

Heating unit = _______ VA × 100% × _____ units = _______ VA

A/C unit = _______ VA × 100% × _____ units = _______ VA

Heat pump = _______ VA × 100% × _____ units = _______ VA

(largest load) _______ VA — VA

TOTAL = VA

Applying demand factors—Table 220-32

_______ VA × _______% = _______ VA

NEUTRAL

(Loads from Standard Calculation)

1. General lighting = _______ VA
2. Cooking equipment = _______ VA
3. Dryer = _______ VA
4. Fixed appliance = _______ VA
5. Heating or A/C = _______ VA
6. Largest motor = _______ VA

TOTAL = VA

A

1ϕ service: $I = \frac{VA}{V}$

PHASES $I = \frac{\_\_\_\_\_\_ VA}{V}$ = _______ A

NEUTRAL $I = \frac{\_\_\_\_\_\_ VA}{V}$ = _______ A

3ϕ service: $I = \frac{VA}{V \times \sqrt{3}}$

$I = \frac{\_\_\_\_\_\_ VA}{V \times \sqrt{3}}$ = _______ A

$I = \frac{\_\_\_\_\_\_ VA}{V \times \sqrt{3}}$ = _______ A

If over 200 A—220-22; First 200 A × 100% = 200 A

Remaining _______ A × 70% = _______ A

A

Use for Problem 22

STANDARD CALCULATION—MULTIFAMILY DWELLING

1. GENERAL LIGHTING LOADS—Table 220-3(b)

_____ sq. ft. × 3 VA × _____ units = _____ VA

_____ sq. ft. × 3 VA × _____ units = _____ VA

_____ sq. ft. × 3 VA × _____ units = _____ VA

_____ VA

Small appliance loads—220-16(a)

_____ VA × _____ circuits × _____ units = _____ VA

Laundry load—220-16(b)

_____ VA × 1 × _____ units = _____ VA

_____ VA

Applying demand factors—Table 220-11

First 3000 VA × 100% = 3000 VA

Next _____ VA × 35% = _____ VA

Remaining _____ VA × 25% = _____ VA

(total) _____ VA | PHASES: VA | NEUTRAL: VA

2. COOKING EQUIPMENT LOADS—Table 220-19, Notes

Col. A _____ units = _____ VA demand = _____ VA

Col. B _____ VA × _____ units × _____% = _____ VA

Col. C _____ VA × _____ units × _____% = _____ VA

(total) _____ VA | VA × 70% = VA

3. DRYER LOAD—220-18; Table 220-18

_____ VA × _____ units × _____% = _____ VA | VA × 70% = VA

4. FIXED APPLIANCE LOADS—220-17

Dishwasher = _____ VA

Disposal = _____ VA

Compactor = _____ VA

Water heater = _____ VA

_____ = _____ VA

_____ = _____ VA

(total) _____ VA × _____ units × 75% = _____ VA | VA | (120 V Loads × units × 75%) VA

5. HEATING OR A/C LOAD—220-21

Heating unit = _____ VA × 100% × _____ units = _____ VA

A/C unit = _____ VA × 100% × _____ units = _____ VA

Heat pump = _____ VA × 100% × _____ units = _____ VA

(largest load) _____ VA | VA | VA

6. LARGEST MOTOR LOAD—220-14

ϕ _____ VA × 25% = _____ VA | VA

N _____ VA × 25% = _____ VA | | VA

TOTAL = VA | VA

1ϕ service: $I = \frac{VA}{V}$

PHASES $I = \frac{\_\_\_\_\_ VA}{V}$ = _____ A

NEUTRAL $I = \frac{\_\_\_\_\_ VA}{V}$ = _____ A

3ϕ service: $I = \frac{VA}{V \times \sqrt{3}}$

$I = \frac{\_\_\_\_\_ VA}{V \times \sqrt{3}}$ = _____ A

$I = \frac{\_\_\_\_\_ VA}{V \times \sqrt{3}}$ = _____ A

If over 200 A—220-22; First 200 A × 100% = 200 A

Remaining _____ A × 70% = _____ A

A

Use for Problem 23

STANDARD CALCULATION—MULTIFAMILY DWELLING 1ϕ (CARRYING PHASES AND NEUTRAL IN VA)

1. GENERAL LIGHTING LOADS—Table 220-3(b)

_____ sq. ft. × 3 VA × _____ units = _____ VA

_____ sq. ft. × 3 VA × _____ units = _____ VA

_____ sq. ft. × 3 VA × _____ units = _____ VA

_____ VA

Small appliance loads—220-16(a)

_____ VA × _____ circuits × _____ units = _____ VA

Laundry load—220-16(b)

_____ VA × 1 × _____ units = _____ VA

_____ VA

Applying demand factors—Table 220-11

First 3000 VA × 100% = 3000 VA

Next _____ VA × 35% = _____ VA

Remaining _____ VA × 25% = _____ VA

	PHASE A	PHASE B		NEUTRAL
(total) _____ VA ÷ 2 =	_____ VA	_____ VA		_____ VA
2. COOKING EQUIPMENT LOADS—Table 220-19, Notes				
Col. A _____ units = _____ VA demand =	_____ VA	_____ VA	× 70% =	_____ VA
Col. B _____ VA × _____ units × _____% =	_____ VA	_____ VA	× 70% =	_____ VA
Col. C _____ VA × _____ units × _____% =	_____ VA	_____ VA	× 70% =	_____ VA
3. DRYER LOAD—220-18; Table 220-18				
_____ VA × _____ units × _____% =	_____ VA	_____ VA	× 70% =	_____ VA
4. FIXED APPLIANCE LOADS—220-17				
Dishwasher = _____ VA × _____ units × 75% ÷ _____ =	_____ VA	_____ VA		_____ VA
Disposal = _____ VA × _____ units × 75% ÷ _____ =	_____ VA	_____ VA		_____ VA
Compactor = _____ VA × _____ units × 75% ÷ _____ =	_____ VA	_____ VA		_____ VA
Water heater = _____ VA × _____ units × 75% ÷ _____ =	_____ VA	_____ VA		_____ VA
_____ = _____ VA × _____ units × 75% ÷ _____ =	_____ VA	_____ VA		_____ VA
_____ = _____ VA × _____ units × 75% ÷ _____ =	_____ VA	_____ VA		_____ VA
_____ = _____ VA × _____ units × 75% ÷ _____ =	_____ VA	_____ VA		_____ VA
_____ = _____ VA × _____ units × 75% ÷ _____ =	_____ VA	_____ VA		_____ VA
_____ = _____ VA × _____ units × 75% ÷ _____ =	_____ VA	_____ VA		_____ VA
5. HEATING OR A/C LOAD—220-21				
Heating unit = _____ VA × 100% × _____ units =	_____ VA	_____ VA		_____ VA
A/C unit = _____ VA × 100% × _____ units =	_____ VA	_____ VA		_____ VA
Heat pump = _____ VA × 100% × _____ units =	_____ VA	_____ VA		_____ VA
6. LARGEST MOTOR LOAD—220-14				
ϕ _____ VA × 25% =	_____ VA	_____ VA		
N _____ VA × 25% =				_____ VA
TOTAL =	VA	VA		VA

1ϕ service: $I = \frac{VA}{V}$

PHASES $I = \frac{\_\_\_\_\_ VA}{V}$ = _____ A

NEUTRAL $I = \frac{\_\_\_\_\_ VA}{V}$ = _____ A

If over 200 A—220-22; First 200 A × 100% = 200 A

Remaining _____ A × 70% = _____ A

A

Use for Problem 23

OPTIONAL CALCULATION—MULTIFAMILY DWELLING 1ϕ (CARRYING PHASES AND NEUTRAL IN VA)

1. OTHER LOADS—220-32	PHASE A	PHASE B	NEUTRAL (from Std. Calc.)
General lighting loads—220-32(c)(2)			
______ sq. ft. × 3 VA × ______ units ÷ ______ =	______ VA	______ VA	______ VA
______ sq. ft. × 3 VA × ______ units ÷ ______ =	______ VA	______ VA	______ VA
______ sq. ft. × 3 VA × ______ units ÷ ______ =	______ VA	______ VA	______ VA
______ sq. ft. × 3 VA × ______ units ÷ ______ =	______ VA	______ VA	______ VA
Small appliance and laundry loads—220-32(c)(1)			
______ VA × ______ circuits × ______ units ÷ ______ =	______ VA	______ VA	______ VA
Special loads—220-32(c)(3)			
Dishwasher = ______ VA × ______ units ÷ ______ =	______ VA	______ VA × 75% =	______ VA
Disposal = ______ VA × ______ units ÷ ______ =	______ VA	______ VA × 75% =	______ VA
Compactor = ______ VA × ______ units ÷ ______ =	______ VA	______ VA × 75% =	______ VA
Water heater = ______ VA × ______ units ÷ ______ =	______ VA	______ VA × 75% =	______ VA
Cooktop/range = ______ VA × ______ units ÷ ______ =	______ VA	______ VA × 70% =	______ VA
Oven = ______ VA × ______ units ÷ ______ =	______ VA	______ VA × 70% =	______ VA
Dryer = ______ VA × ______ units ÷ ______ =	______ VA	______ VA × 70% =	______ VA
______ = ______ VA × ______ units ÷ ______ =	______ VA	______ VA × 75% =	______ VA
______ = ______ VA × ______ units ÷ ______ =	______ VA	______ VA × 75% =	______ VA
______ = ______ VA × ______ units ÷ ______ =	______ VA	______ VA × 75% =	______ VA
______ = ______ VA × ______ units ÷ ______ =	______ VA	______ VA × 75% =	______ VA
______ = ______ VA × ______ units ÷ ______ =	______ VA	______ VA × 75% =	______ VA
______ = ______ VA × ______ units ÷ ______ =	______ VA	______ VA × 75% =	______ VA
______ = ______ VA × ______ units ÷ ______ =	______ VA	______ VA × 75% =	______ VA

2. HEATING OR A/C LOAD—220-32(c)(5)			
Heating unit = ______ VA × 100% × ______ units =	______ VA	______ VA	______ VA
A/C unit = ______ VA × 100% × ______ units =	______ VA	______ VA	______ VA
Heat pump = ______ VA × 100% × ______ units =	______ VA	______ VA	______ VA
TOTAL =	VA	VA	
		Plus largest motor =	______ VA
			VA

Applying demand factors—Table 220-32

______ VA × ______ % = ______ VA

1ϕ service: $I = \frac{VA}{V}$

PHASES $I = \frac{\text{\_\_\_\_\_\_ VA}}{V}$ = ______ A

NEUTRAL $I = \frac{\text{\_\_\_\_\_\_ VA}}{V}$ = ______ A

If over 200 A—220-22; First 200 A × 100% = 200 A

Remaining ______ A × 70% = ______ A

A

Use for Problem 24

OPTIONAL CALCULATION—MULTIFAMILY DWELLING 1ϕ (CARRYING PHASES AND NEUTRAL IN AMPS)

1. OTHER LOADS—220-32		PHASE A	PHASE B		NEUTRAL (from Std. Calc.)
General lighting loads—220-32(c)(2)					
____ sq. ft. × 3 VA × ____ units ÷ ____ V =		____ A	____ A		____ A
____ sq. ft. × 3 VA × ____ units ÷ ____ V =		____ A	____ A		____ A
____ sq. ft. × 3 VA × ____ units ÷ ____ V =		____ A	____ A		____ A
____ sq. ft. × 3 VA × ____ units ÷ ____ V =		____ A	____ A		____ A
Small appliance and laundry loads—220-32(c)(1)					
____ VA × ____ circuits × ____ units ÷ ____ V =		____ A	____ A		____ A
Special loads—220-32(c)(3)					
Dishwasher =	____ VA × ____ units ÷ ____ V =	____ A	____ A	× 75% =	____ A
Disposal =	____ VA × ____ units ÷ ____ V =	____ A	____ A	× 75% =	____ A
Compactor =	____ VA × ____ units ÷ ____ V =	____ A	____ A	× 75% =	____ A
Water heater =	____ VA × ____ units ÷ ____ V =	____ A	____ A	× 75% =	____ A
Cooktop/range =	____ VA × ____ units ÷ ____ V =	____ A	____ A	× 70% =	____ A
Oven =	____ VA × ____ units ÷ ____ V =	____ A	____ A	× 70% =	____ A
Dryer =	____ VA × ____ units ÷ ____ V =	____ A	____ A	× 70% =	____ A
____ =	____ VA × ____ units ÷ ____ V =	____ A	____ A	× 75% =	____ A
____ =	____ VA × ____ units ÷ ____ V =	____ A	____ A	× 75% =	____ A
____ =	____ VA × ____ units ÷ ____ V =	____ A	____ A	× 75% =	____ A
____ =	____ VA × ____ units ÷ ____ V =	____ A	____ A	× 75% =	____ A
____ =	____ VA × ____ units ÷ ____ V =	____ A	____ A	× 75% =	____ A
____ =	____ VA × ____ units ÷ ____ V =	____ A	____ A	× 75% =	____ A
____ =	____ VA × ____ units ÷ ____ V =	____ A	____ A	× 75% =	____ A

2. HEATING OR A/C LOAD—220-32(c)(5)				
Heating unit = ____ VA × 100% × ____ units ÷ ____ V =	____ A	____ A		____ A
A/C unit = ____ VA × 100% × ____ units ÷ ____ V =	____ A	____ A		____ A
Heat pump = ____ VA × 100% × ____ units ÷ ____ V =	____ A	____ A		____ A
TOTAL =	A	A		
			Plus largest motor =	____ A
				A

Applying demand factors—Table 220-32

____ A × ____ % = A

If over 200 A—220-22; First 200 A × 100% = 200 A

Remaining ____ A × 70% = ____ A

A

Use for Problem 24

STANDARD CALCULATION—MULTIFAMILY DWELLING 1ϕ (CARRYING PHASES AND NEUTRAL IN AMPS)

1. GENERAL LIGHTING LOADS—Table 220-3(b)

______ sq. ft. × 3 VA × ______ units = ______ VA

______ sq. ft. × 3 VA × ______ units = ______ VA

______ sq. ft. × 3 VA × ______ units = ______ VA

______ VA

Small appliance loads—220-16(a)

______ VA × ______ circuits × ______ units = ______ VA

Laundry load—220-16(b)

______ VA × 1 × ______ units = ______ VA

______ VA

Applying demand factors—Table 220-11

First 3000 VA × 100% = 3000 VA

Next ______ VA × 35% = ______ VA

Remaining ______ VA × 25% = ______ VA

	PHASE A	PHASE B		NEUTRAL
(total) ______ VA ÷ ______ V =	______ A	______ A		______ A
2. COOKING EQUIPMENT LOADS—Table 220-19, Notes				
Col. A ______ units = ______ VA demand = ______ VA ÷ ______ V =	______ A	______ A	× 70% =	______ A
Col. B ______ VA × ______ units × ______% = ______ VA ÷ ______ V =	______ A	______ A	× 70% =	______ A
Col. C ______ VA × ______ units × ______% = ______ VA ÷ ______ V =	______ A	______ A	× 70% =	______ A
3. DRYER LOAD—220-18; Table 220-18				
______ VA × ______ units × ______% ÷ ______ V =	______ A	______ A	× 70% =	______ A
4. FIXED APPLIANCE LOADS—220-17				
Dishwasher = ______ VA × ______ units × 75% ÷ ______ V =	______ A	______ A		______ A
Disposal = ______ VA × ______ units × 75% ÷ ______ V =	______ A	______ A		______ A
Compactor = ______ VA × ______ units × 75% ÷ ______ V =	______ A	______ A		______ A
Water heater = ______ VA × ______ units × 75% ÷ ______ V =	______ A	______ A		______ A
______ = ______ VA × ______ units × 75% ÷ ______ V =	______ A	______ A		______ A
______ = ______ VA × ______ units × 75% ÷ ______ V =	______ A	______ A		______ A
______ = ______ VA × ______ units × 75% ÷ ______ V =	______ A	______ A		______ A
______ = ______ VA × ______ units × 75% ÷ ______ V =	______ A	______ A		______ A
______ = ______ VA × ______ units × 75% ÷ ______ V =	______ A	______ A		______ A
5. HEATING OR A/C LOAD—220-21				
Heating unit = ______ VA × 100% × ______ units ÷ ______ V =	______ A	______ A		______ A
A/C unit = ______ VA × 100% × ______ units ÷ ______ V =	______ A	______ A		______ A
Heat pump = ______ VA × 100% × ______ units ÷ ______ V =	______ A	______ A		______ A
6. LARGEST MOTOR LOAD—220-14				
ϕ ______ VA × 25% ÷ ______ V =	______ A	______ A		______ A
N ______ VA × 25% ÷ 120 V =	______ A	______ A		______ A
TOTAL =	A	A		A

If over 200 A—220-22; First 200 A × 100% = 200 A

Remaining ______ A × 70% = ______ A

A

STANDARD CALCULATION—ONE-FAMILY DWELLING

Use for Problem 25

1. GENERAL LIGHTING LOADS—Table 220-3(b)

______ sq. ft. × 3 VA = ______ VA

Small appliance loads—220-16(a)

______ VA × ______ circuits = ______ VA

Laundry load—220-16(b)

______ VA × 1 = ______ VA

______ VA

Applying demand factors—Table 220-11

First 3000 VA × 100% = 3000 VA

Next ______ VA × 35% = ______ VA

Remaining ______ VA × 25% = ______ VA

	PHASES	NEUTRAL
(total) ______ VA	VA	VA

2. COOKING EQUIPMENT LOADS—Table 220-19; Notes

Col. A ______ VA demand = ______ VA

Col. B ______ VA × ______% = ______ VA

Col. C ______ VA × ______% = ______ VA

(total) ______ VA | VA × 70% = | VA

3. DRYER LOAD—220-18; Table 220-18

______ VA × ______% = ______ VA | VA × 70% = | VA

4. FIXED APPLIANCE LOADS—220-17

Dishwasher = ______ VA

Disposal = ______ VA

Compactor = ______ VA

Water heater = ______ VA

______ = ______ VA

______ = ______ VA

______ = ______ VA

______ = ______ VA

(total) ______ VA × 75% = ______ VA | VA | (120 V Loads × 75%) VA

5. HEATING OR A/C LOAD—220-21

Heating unit = ______ VA × 100% = ______ VA

A/C unit = ______ VA × 100% = ______ VA

Heat pump = ______ VA × 100% = ______ VA

(largest load) ______ VA | VA | VA

6. LARGEST MOTOR LOAD—220-14

ϕ ______ VA × 25% = ______ VA | VA

N ______ VA × 25% = ______ VA | ______ | VA

TOTAL = | VA | VA

1ϕ service: $I = \frac{VA}{V}$

PHASES $I = \frac{\_\_\_\_\_\_ VA}{V}$ = ______ A

NEUTRAL $I = \frac{\_\_\_\_\_\_ VA}{V}$ = ______ A

If over 200 A—220-22; First 200 A × 100% = 200 A

Remaining ______ A × 70% = ______ A

A

OPTIONAL CALCULATION—ONE-FAMILY DWELLING (NEW OR EXISTING)

Use for Problem 25

1. OTHER LOADS—220-30(b)

General lighting loads—220-30(b)(2)

________ sq. ft. × 3 VA = ________ VA

Small appliance and laundry loads—220-30(b)(1)

________ VA × ________ circuits = ________ VA

Special loads—220-30(b)(3)(4)

Dishwasher = ________ VA

Disposal = ________ VA

Compactor = ________ VA

Water heater = ________ VA

________ = ________ VA

________ = ________ VA

________ = ________ VA

________ = ________ VA

________ = ________ VA

________ = ________ VA

(total) ________ VA

Applying demand factors—Table 220-30(4)

First 10,000 VA × 100% = 10,000 VA

Remaining ________ VA × 40% = ________ VA

(total) ________ VA

PHASES

VA

2. HEATING OR A/C LOAD—Table 220-30(1)(2)(3)

Heating units (3 or less) = ________ VA × 65% = ________ VA

Heating units (4 or more) = ________ VA × 40% = ________ VA

A/C unit = ________ VA × 100% = ________ VA

Heat pump = ________ VA × 100% = ________ VA

(largest load) ________ VA   VA

TOTAL =   VA

A

NEUTRAL

(Loads from Standard Calculation)

1. General lighting = ________ VA
2. Cooking equipment = ________ VA
3. Dryer = ________ VA
4. Fixed appliance = ________ VA
5. Heating or A/C = ________ VA
6. Largest motor = ________ VA

TOTAL =   VA

A

1ϕ service: $I = \frac{VA}{V}$

PHASES I = $\frac{\text{\_\_\_\_\_\_\_\_ VA}}{\text{V}}$ = ________ A

NEUTRAL I = $\frac{\text{\_\_\_\_\_\_\_\_ VA}}{\text{V}}$ = ________ A

Calculations— Commercial and Industrial

Chapter 8

The National Electrical Code® recognizes all types of electrical installations based upon the type of occupancy or use. Businesses engaged in manufacturing, marketing, and distribution of goods are classified as either industrial or commercial locations. Loads are calculated based upon the type of occupancy and the electrical requirements.

The electrical service and elements of the service for industrial locations are based upon the demands of the electrical equipment used in the manufacture or repair of goods. These locations generally require greater ampacity ratings to select the service and elements of the equipment. Commercial locations such as offices, restaurants, and stores have diverse electrical requirements, which depend upon the size and type of equipment needed.

The standard calculation is used to size the service and elements of the service for both industrial and commercial locations. The optional calculation can be applied under certain conditions of use.

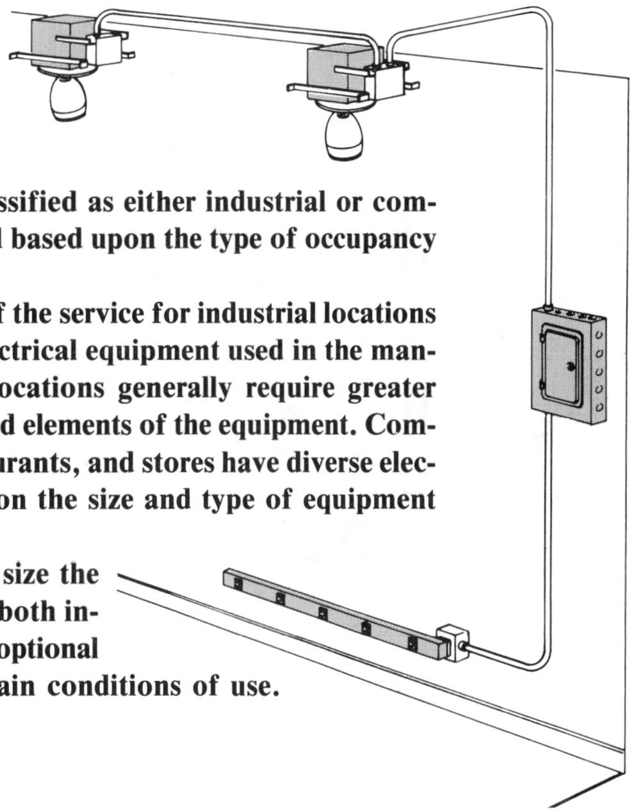

STANDARD CALCULATION

The standard calculation is used to compute the VA rating to select the elements of the service equipment. The selection of loads to apply the standard calculation is arranged differently for industrial and commercial loads than for the loads of one-family and multifamily dwelling units. There are six loads that shall be computed when the standard calculation is applied to derive the service load. Where permitted, demand factors can be applied to certain loads selected from the columns of six loads. The six loads are as follows:

Lighting loads. *220-3, Part B*
Receptacle loads. *220-3(c)(5)*
Special appliance loads. *220-10(b)*
Heating or A/C loads. *220-21*
Motor loads. *430-22(a); 430-24; 430-25*
Largest motor load. *220-14*

Different types of loads in each column of the six loads are computed. Each load is computed at continuous duty or noncontinuous duty. Demand factors are applied to the noncontinuous-duty loads by various sections of the NEC®.

Lighting Loads—220, Part B

Lighting loads are the first group of loads to be determined. Five lighting loads are computed to determine the total lighting load in industrial and commercial buildings. These five loads are

General lighting loads. *Table 220-3(b)*
Show window loads. *220-12*
Lighting track loads. *410-102*
Outside lighting loads. *220-10(b)*
Sign lighting loads. *600-6(c)*

Each lighting load is computed by the location and method in which it is used. Loads are calculated at continuous or noncontinuous duty and other lighting loads can have demand factors applied. Each lighting load should be computed and the total lighting loads determined before any other load is figured. See Figure 8-1.

General Lighting Loads—Table 220-3(b). General lighting loads are lighting units installed inside the building. This load is computed by the VA rating per sq. ft. from Table 220-3(b) based on the type of occupancy.

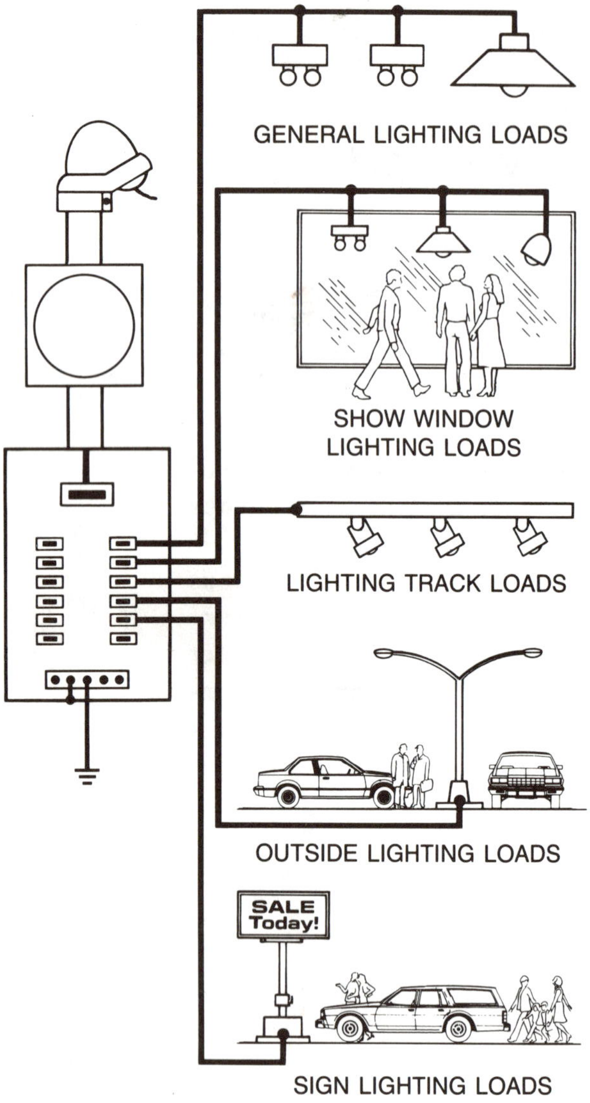

Figure 8-1. Five lighting loads are computed to determine the first group of loads, which is the lighting loads.

Table 220-3(b) is used to select the VA rating for listed occupancies. For example, a store building is figured at 3 VA per sq. ft. and a hospital at 2 VA per sq. ft. Other types of occupancies have different VA ratings per sq. ft. based on their operation and use. If an occupancy, such as a welding shop, is not listed in Table 220-3(b), the general lighting load shall be computed by one of the following:

Lamps for incandescent lighting. *220-3(c)(5)*

Lamps for recessed fixtures. *220-3(c)(3)*

Ballasts for electric discharge lighting. *210-22(b)*

Listed VA per sq. ft. or the unlisted VA rating is computed at 125% for continuous-duty loads and 100% for noncontinuous-duty loads per 220-10(b), with demand factors applied to certain lighting loads per Table 220-11. To determine the general lighting load, select the VA per sq. ft. from Table 220-3(b) and multiply by the square footage of the occupancy. Next, reference Table 220-11 to determine if a demand factor can be applied. If a demand factor cannot be applied, the lighting load is computed at 100% for noncontinuous-duty operation and 125% for continuous-duty operation per 220-10(b). See Figure 8-2.

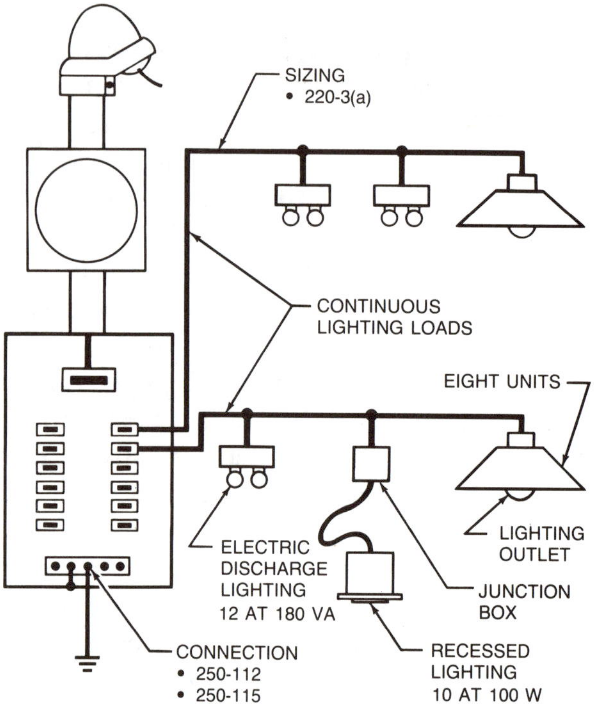

LISTED OCCUPANCY

What is the load in VA for the general lighting load in a combination barber shop and beauty parlor?

Step 1. *Table 220-3(b):*

1000 sq. ft. × 3 VA =	3000 VA
950 sq. ft. × 3 VA =	2850 VA
Total =	5850 VA

Step 2. *220-10(b):* 5850 VA × 125% = 7312.5 VA

Answer. **Load = 7312.5 VA**

UNLISTED OCCUPANCY

What is the load in VA for the general lighting load using fluorescent and incandescent lighting units?

Step 1.

210-22(b): 180 VA × 12 =	2160 VA
220-3(c)(3): 100 VA × 10 =	1000 VA
220-3(c)(5): 180 VA × 8 =	1440 VA
Total =	4600 VA

Step 2. *220-10(b):* 4600 VA × 125% = 5750 VA

Answer. **Load = 5750 VA**

Figure 8-2. In calculating the general lighting load for a listed or unlisted occupancy, the general lighting load is the first of the six loads to be determined.

What is the load in VA for a 2400 sq. ft. store building?

Step 1. *Table 220-3(b):* 2400 sq. ft. × 3 VA = 7200 VA

Step 2. *Table 220-11:* No demand factor allowed

Step 3. *220-10(b):* 7200 VA × 125% = 9000 VA

Answer. **Load = 9000 VA**

Four types of listed occupancies to which demand factors can be applied per Table 220-11 are dwelling units, hospitals, hotels and motels, and warehouses. All other types of listed or unlisted occupancies shall be computed at 100% for noncontinuous-duty operation and 125% for continuous-duty operation.

What is the lighting load in VA for a hospital with the patient bedrooms having an area of 200′-0″ × 225′-0″?

Step 1. *Table 220-3(b):*
200′-0″ × 225′-0″ × 2 VA = 90,000 VA

Step 2. Applying demand factors. *Table 220-11:*

First 50,000 VA × 40% =	20,000 VA
Next 40,000 VA × 20% =	8 000 VA
Total =	28,000 VA

Answer. **Load = 28,000 VA**

For occupancies not listed in Table 220-3(b), the VA rating of each lighting fixture is multiplied at 100% for noncontinuous-duty loads and 125% for continuous-duty loads.

What is the lighting load in VA for 75 lighting fixtures with each ballast rated at 1.5 A and supplied by 120 V?

Step 1. Finding VA. *Text:* $VA = V \times I$
VA = 120 V × 1.5 A = 180 VA

Step 2. *210-22(b):* 180 VA × 75 units = 13,500 VA

Step 3. *220-10(b):* 13,500 VA × 125% = 16,875 VA

Answer. **Load = 16,875 VA**

■*Master Electrician's Test Tip.* Some testing authorities require the general lighting calculated at 3 VA per sq. ft. to be compared to the total VA rating of the lighting fixtures and the larger load selected. This procedure is used where the listed occupancy and the number of fixtures with VA ratings are available.■

A store building has an 80′-0″ × 60′-0″ area with seventy-five 120 V lighting fixtures with each ballast rated at 180 VA to provide general lighting. What is the general lighting load?

Step 1. *Table 220-3(b):*
80′-0″ × 60′-0″ × 3 VA = 14,400 VA

Step 2. *Table 220-11:* No demand factor allowed

Step 3. *210-22(b):* 75 units × 180 VA = 13,500 VA

Step 4. *220-10(b):* 14,400 VA × 125% = 18,000 VA
13,500 VA × 125% = 16,875 VA

Answer. **Load = 18,000 VA**

The neutral is not considered a current-carrying conductor for lighting fixtures identified as incandescent. It is considered current-carrying for electric discharge lighting. It is not necessary to derate per Note 8 to Tables 310-16–310-19 for three (hot) ungrounded conductors and one (neutral) grounded conductor.

The connected lighting load of a store building is 68,000 VA with 50% or more electric discharge lighting. What size THWN Cu conductors is required for the 120/208 V, 3ϕ, 4-wire service?

Step 1. *220-10(b):* 68,000 VA × 125% = 85,000 VA

Step 2. *Text:* $I = \dfrac{VA}{V \times \sqrt{3}}$

$$I = \frac{85{,}000}{208\text{ V} \times 1.732} = 236\text{ A}$$

Step 3. *Table 310-16:*
236 A requires #250 kcmil THWN Cu

Step 4. Applying derating factors.
Note 8 to Tables 310 16 310-19:
#350 kcmil THWN Cu = 310 A
310 A × 80% = 248 A

Answer. **Conductors = #350 kcmil THWN Cu**

Show Window Loads—220-12. The lighting load in VA for the show window is determined by multiplying the linear feet of the show window by 200 VA per foot.

What is the lighting load in VA for a 60′-0″ show window?

Step 1. *220-12:* 60′-0″ × 200 VA = 12,000 VA

Answer. **Load = 12,000 VA**

■*Master Electrician's Test Tip.* It is not necessary to multiply 200 VA times the linear footage by 125%. Section 220-12 allows the linear footage of the show window to be multiplied by 200 VA whether the load is noncontinuous duty or continuous duty.■

If the number of lighting outlets is known, the VA rating of each fixture is multiplied by 125%. If the VA rating is not known, each outlet is computed at 180 VA times 125% to obtain the lighting load in VA for the show window. See Figure 8-3.

What is the lighting load in VA for a 50′-0″ show window with 42 lighting outlets?

Step 1. *220-12:* 50′-0″ × 200 VA = 10,000 VA

Step 2. Number of outlets. *220-3(c)(5):*
42 × 180 VA = 7560

Step 3. *220-10(b):* 7560 VA × 125% = 9450 VA

Answer. **Load = 10,000 VA**

Note: 210-62 requires a receptacle outlet for every 12 linear feet of show window space measured horizontally above the window. This receptacle outlet shall be installed directly above the show window area.

How many receptacle outlets are required above a show window with 60 linear feet measured horizontally?

Step 1. *210-62:* $No. = \dfrac{linear\ feet}{12'\text{-}0''}$

$$No. = \frac{60'\text{-}0''}{12'\text{-}0''} = 5$$

Answer. **Outlets = 5**

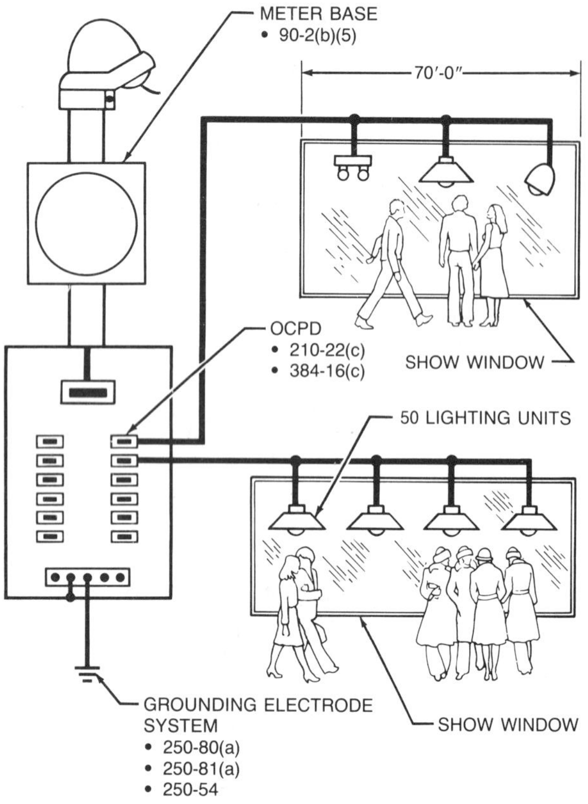

DETERMINING SHOW WINDOW LOAD

What is the load in VA for 70 linear feet of show window?

Step 1. *220-12:* 70′ × 200 VA = 14,000 VA

Answer. **Load = 14,000 VA**

What is the load in VA for 50 lighting outlets in a show window?

Step 1. *220-3(c)(5):* 180 VA × 50 = 9000 VA

Step 2. *220-10(b):* 9000 VA × 125% = 11,250 VA

Answer. **Load = 11,250 VA**

Figure 8-3. The load in VA for the show window lighting is determined by multiplying the linear footage by 200 VA or by multiplying each outlet by 180 VA times 125%.

Lighting Track Loads—410-102. The lighting load in VA for lighting track is determined by multiplying every 2′ of lighting track by 180 VA.

What is the lighting load in VA for 130′-0″ of lighting track in a store building?

Step 1. *410-102:* $VA = \frac{track\ length}{2'} \times 180\text{ VA}$

$$VA = \frac{130'\text{-}0''}{2'} \times 180\text{ VA} = 11{,}700\text{ VA}$$

Answer. **Load = 11,700 VA**

The number of 20 A, 2-wire circuits required to supply the lighting load in VA is determined by dividing the total VA by the amperage of the overcurrent protection device and multiplying by 120 V.

What is the number of 20 A, 2-wire circuits required to supply the 11,700 VA lighting track load?

Step 1. *220-4(a):* $No. = \frac{VA}{OCPD \times 120\text{ V}}$

$$No. = \frac{11{,}700\text{ VA}}{20\text{ A} \times 120\text{ V}} = 4.875$$

Answer. **Circuits = 5**

The number of lighting outlets permitted on a run of lighting track is determined by dividing the total VA of the track by 180 VA. See Figure 8-4.

What is the number of lighting outlets permitted on the 130′-0″ run of lighting track?

Step 1. *410-102:* $VA = \frac{track\ length}{2'} \times 180\text{ VA}$

$$VA = \frac{130'\text{-}0''}{2'} \times 180\text{ VA} = 11{,}700\text{ VA}$$

Step 2. *410-102:* $No. = \frac{VA\ rating\ of\ track}{180\text{ VA}}$

$$No. = \frac{11{,}700\text{ VA}}{180\text{ VA}} = 65$$

Answer. **Outlets = 65**

Outside Lighting Loads—220-10(b). The lighting load in VA for outside lighting loads is derived by multiplying the VA rating of each lighting unit by 100% for noncontinuous-duty loads and 125% for continuous-duty loads. See Figure 8-5.

What is the lighting load in VA for 25 continuous-duty lighting fixtures with a 180 VA ballast in each unit and 10 noncontinuous-duty lighting fixtures with each ballast having a rating of 225 VA?

Step 1. *210-22(b); 220-10(b):*

180 VA × 25 × 125% =	5625 VA
225 VA × 10 × 100% =	2250 VA
Total =	7875 VA

Answer. **Load = 7875 VA**

Sometimes the calculation requires amps instead of VA of each lighting fixture.

What is the lighting load in amps for 30 lighting fixtures used for outside lighting for six hours each night? Each ballast pulls 1.5 A.

Step 1. *210-22(b); 220-10(b):*
1.5 A × 30 × 125% = 56.25 A

Answer. **Load = 56 A**

Sign Lighting Loads—600-6(c). The lighting loads for signs are determined by the commercial occupancy or building having ground floor footage accessible to pedestrians. The problem occurs when a building has

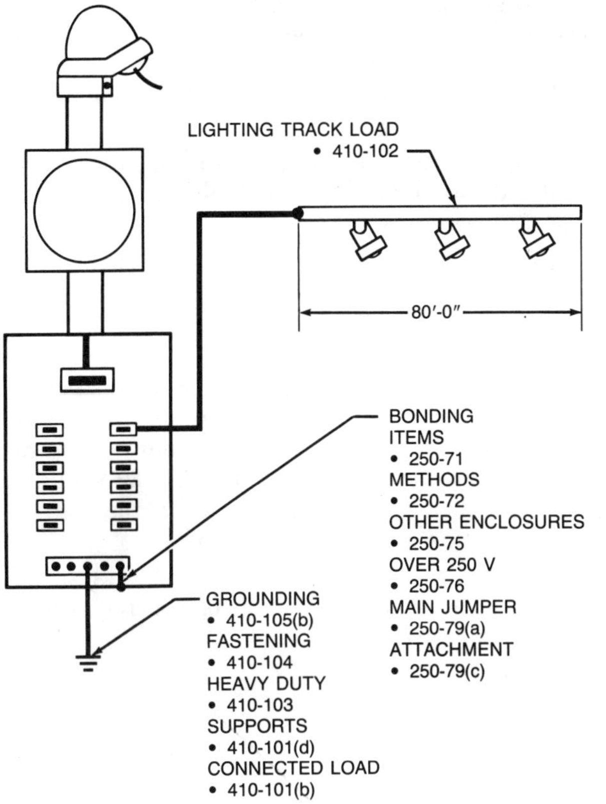

DETERMINING LIGHTING TRACK LOAD

What is the VA rating of 80′ of lighting track? How many outlets may be installed?

VA rating

Step 1. *410-102:*

$$VA = \frac{\text{track length}}{2'} \times 180 \text{ VA}$$

$$VA = \frac{80'}{2'} \times 180 \text{ VA} = 7200 \text{ VA}$$

Answer. **VA rating = 7200 VA**

No. of outlets

Step 1. $No. = \frac{VA}{180 \text{ VA}}$

$$No. = \frac{7200 \text{ VA}}{180 \text{ VA}} = 40$$

Answer. **Outlets = 40**

Figure 8-4. The load in VA for lighting track is determined by dividing the footage of the track by 2 and multiplying by 180 VA.

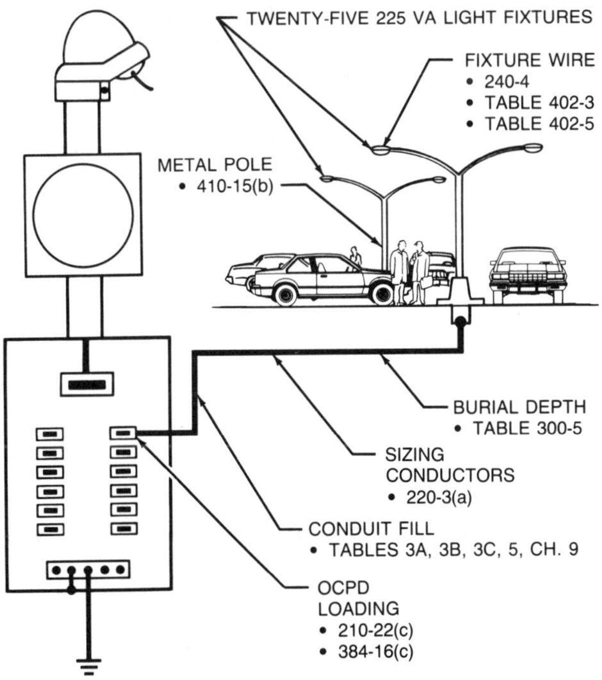

DETERMINING OUTSIDE LIGHTING LOAD

What is the lighting load in VA for 25 outside lighting fixtures pulling 225 VA each that will burn for five hours?

Step 1. Total VA. 25 × 225 VA = 5625 VA

Step 2. *220-10(b):* 5625 VA × 125% = 7031 VA

Answer. **Load = 7031 VA**

What is the load in VA for a 3840 VA outside lighting load that is used 2½ hours each night?

Step 1. *220-10(b):* 3840 VA × 100% = 3840 VA

Answer. **Load = 3840 VA**

Figure 8-5. The load in VA for the outside lighting load is derived by multiplying the VA of each lighting unit by the number times 125% for continuous-duty and 100% for noncontinuous-duty.

occupancies located above the first floor without grade-level access. Only those occupancies with grade-level access for pedestrians are required to have a minimum of 1200 VA provided for a sign lighting load. If the VA rating of the sign load exceeds 1200 VA, the VA rating is multiplied by 125%. See Figure 8-6.

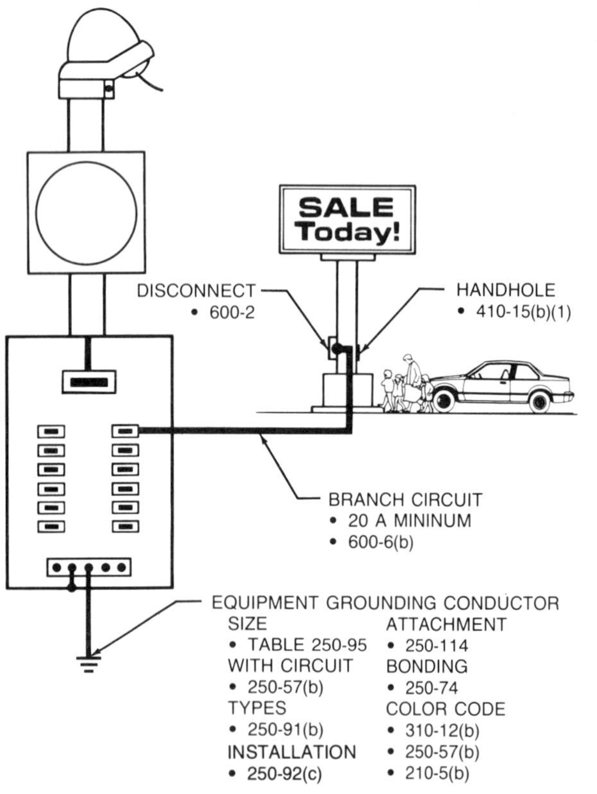

Figure 8-6. The load in VA for a sign is computed at 100% or 125% times the VA rating of the sign. The minimum VA is 1200 VA where the load is not known.

What is the VA rating for a sign lighting load of 1950 VA at continuous duty?

Step 1. *600-6(c); 220-10(b):*
1950 VA × 125% = 2437.5 VA

Answer. **VA rating = 2437.5 VA**

■*Master Electrician's Test Tip.* Some testing boards require each commercial building to have at least a 1200 VA sign lighting load regardless of ground-level access.■

Receptacle Loads

Receptacle loads are the second group of loads to be determined. They are divided into two subgroups: general-purpose receptacle outlets and multioutlet assemblies. Each subgroup is computed differently to select elements of the service equipment.

General-purpose Receptacle Loads—220-3(c)(5). The load in VA for the general-purpose receptacle outlets is determined by multiplying the number of outlets times 180 VA each times 125% for continuous-duty operation.

What is the load in VA for 52 general-purpose receptacles used to serve continuous-duty loads?

Step 1. *220-3(c)(5):* 52 × 180 VA = 9360 VA

Step 2. *220-10(b):* 9360 VA × 125% = 11,700 VA

Answer. **Load = 11,700 VA**

The number of 2-wire circuits required to supply power to continuous or noncontinuous-duty receptacle outlets is determined by one of two methods. In the first method, which is used for continuous-duty operation, multiply the number of outlets by 180 VA times 125% and divide the total by the size of the overcurrent protection device times the voltage.

How many 20 A, 2-wire circuits are required for 60 general-purpose receptacles used at continuous-duty in an office complex?

Step 1. *220-3(c)(5):* 60 × 180 VA = 10,800 VA

Step 2. *220-10(b):* 10,800 VA × 125% = 13,500 VA

Step 3. *220-4(a):* $No. = \frac{VA}{OCPD \times V}$

$$No. = \frac{13{,}500\ VA}{20\ A \times 120\ V} = 5.625$$

Answer. **Circuits = 6**

In the second method, which is used for noncontinuous-duty operation, multiply the number of receptacle outlets by 180 VA times 100% and divide the total by the size of the overcurrent protection device times 80% times the voltage. See Figure 8-7.

How many 2-wire circuits are required for 75 receptacles used at continuous-duty in an office building?

Step 1. *220-3(c)(5):* 75 × 180 VA = 13,500 VA

Step 2. *220-10(b); 220-4(a):*

$$No. = \frac{VA}{OCPD \times 80\% \times V}$$

$$No. = \frac{13{,}500\ VA}{20\ A \times 80\% \times 120\ V} = 7$$

Answer. **Circuits = 7**

■*Master Electrician's Test Tip.* Derating the current-carrying ability of the overcurrent protection devices by 80% is the same as multiplying the number of receptacle outlets by 180 VA times 125%. The 80% is derived by dividing 1 by 125% (1/125% = 80%).■

Receptacle outlets for cord- and plug-connected loads used at noncontinuous duty are computed per 220-3(c)(5) and Table 220-13. Noncontinuous-duty receptacles with a VA rating of 10,000 VA or less are figured at 100%. If the VA rating of the receptacle load exceeds 10,000 VA, a demand factor of 50% is applied to VA exceeding 10,000 VA per Table 220-13.

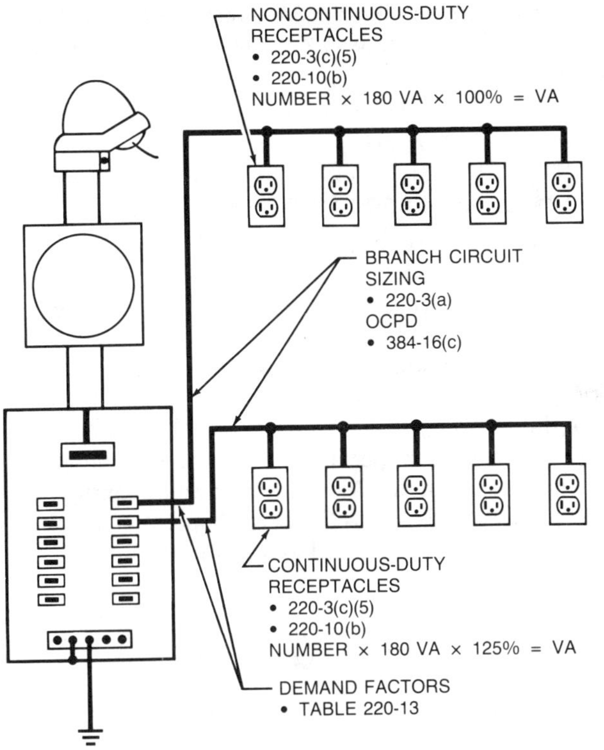

DETERMINING VA FOR CONTINUOUS- AND NONCONTINUOUS-DUTY RECEPTACLES

What is the load in VA for 55 continuous-duty receptacles?

Step 1. *220-3(c)(5):* 55 × 180 VA = 9900 VA

Step 2. *220-10(b):* 9900 VA × 125% = 12,375 VA

Answer. **Load = 12,375 VA**

What is the load in VA for 45 noncontinuous-duty receptacles?

Step 1. *220-3(c)(5):* 45 × 180 VA × 100% = 8100 VA

Answer. **Load = 8100 VA**

Figure 8-7. The load in VA for receptacles is determined by multiplying the number of receptacles by 180 A times 100% for noncontinuous-duty and 125% for continuous-duty. See Table 220-13 for the demand factors of the VA of noncontinuous-duty receptacles.

The number of branch circuits is determined by dividing the total VA by the size overcurrent protection device times 120 V. See Figure 8-8.

What is the VA rating for 55 general-purpose receptacle outlets used at noncontinuous-duty?

Step 1. *220-3(c)(5):* 55 × 180 VA = 9900 VA

Step 2. *220-10(b):* 9900 VA × 100% = 9900 VA

Answer. **VA rating = 9900 VA**

What is the VA rating for 116 general-purpose receptacle outlets used at noncontinuous-duty?

Step 1. *220-3(c)(5):* 116 × 180 VA = 20,880 VA

Step 2. *Table 220-13:*

First 10,000 VA × 100% =	10,000 VA
Next 10,880 VA × 50% =	5440 VA
Total =	15,440 VA

Answer. **VA rating = 15,440 VA**

How many 20 A, 2-wire circuits are required to supply a receptacle load of 20,880 VA?

Step 1. *220-4(a): No.* $= \dfrac{VA}{OCPD \times 120\ V}$

$$No. = \frac{20{,}880\ VA}{20\ V \times 120\ V} = 8.7\ VA$$

Answer. **No. = 9 VA**

■*Master Electrician's Test Tip.* The number of 2-wire circuits required to supply a receptacle load is determined by dividing the size of the overcurrent protection device times the voltage into the total VA rating before demand factors are applied.■

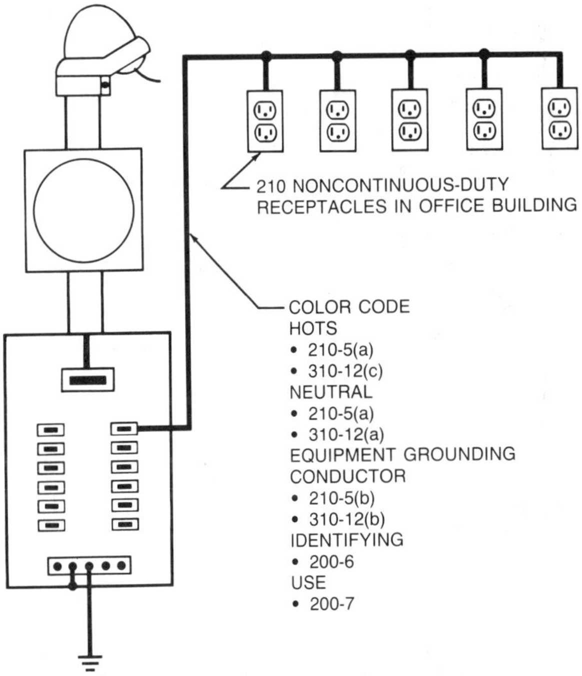

DETERMINING DEMAND FACTOR AND LOAD FOR RECEPTACLE LOADS

What is the load in VA for 210 receptacles used at noncontinuous duty?

Step 1. *220-3(c)(5):* 210 × 180 VA = 37,800 VA

Step 2. *Table 220-13:*

First 10,000 VA × 100% =	10,000 VA
Next 27,800 VA × 50% =	13,900 VA
Total =	23,900 VA

Answer. **Load = 23,900 VA**

Figure 8-8. The demand load in VA is derived by taking the first 10,000 VA of a receptacle load at 100%. All VA remaining is multiplied by 50%. The two loads in VA are added to determine the total VA.

Multioutlet Assemblies—220-3(c), Ex. 1. The VA rating used to calculate the load for a multioutlet assembly is based on the use of the assembly. For light-duty service, the VA rating is determined by dividing the length of the assembly by 5′-0″ and multiplying by 180 VA. For heavy-duty service, each foot of multi-outlet assembly is multiplied by 180 VA. See Figure 8-9.

What is the VA rating for 150′-0″ of multioutlet assembly used at light-duty in a TV shop?

Step 1. *220-3(c), Ex. 1:* $VA = \frac{length}{5'\text{-}0''} \times 180\ VA$

$$VA = \frac{150'\text{-}0''}{5'\text{-}0''} \times 180\ VA = 5400\ VA$$

Answer. **VA rating = 5400 VA**

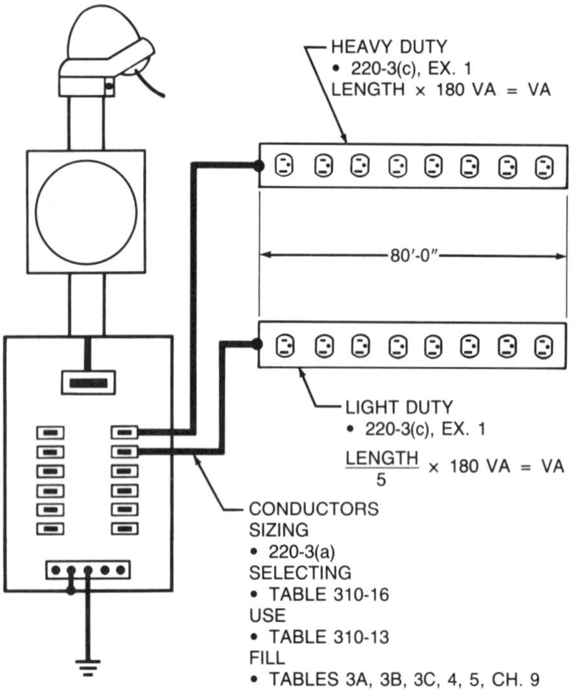

DETERMINING LOAD IN VA FOR MULTIOUTLET ASSEMBLY

What is the load in VA for 80′ of multioutlet assembly used at light duty?

Step 1. *220-3(c), Ex. 1:* $\frac{80'}{5'} \times 180\ VA = 2880\ VA$

Answer. **Load = 2880 VA**

What is the load in VA for 80′ of multioutlet assembly used at heavy duty?

Step 1. *220-3(c), Ex. 1:* 80′ × 180 VA = 14,400 VA

Answer. **Load = 14,400 VA**

Figure 8-9. The load in VA for multioutlet assembly is determined for light-duty by multiplying the length of the assembly by 180 VA. The load in VA for heavy-duty is derived by multiplying the assembly by 180 VA and dividing by 5.

What is the VA rating for 80′-0″ of multioutlet assembly used at heavy duty over workbenches in a motor shop?

Step 1. *220-3(c), Ex. 1: VA = length* × 180 VA
VA = 80′-0″ × 180 VA = 14,400 VA

Answer. **VA rating = 14,400 VA**

Special Appliance Loads—220-10(b)

Special appliance loads are the third group of loads to be determined. These loads, which include A/C, heating units, computers, processing machines, etc. are normally supplied by individual circuits. The load in VA for special appliance loads is determined by multiplying the VA rating of each load by 100% for non-continuous-duty operation and 125% for continuous-duty operation. A special appliance load operating for less than three hours is classified as a noncontinuous-duty load. A special appliance load operating for three hours or more is classified as a continuous-duty load. See Figure 8-10.

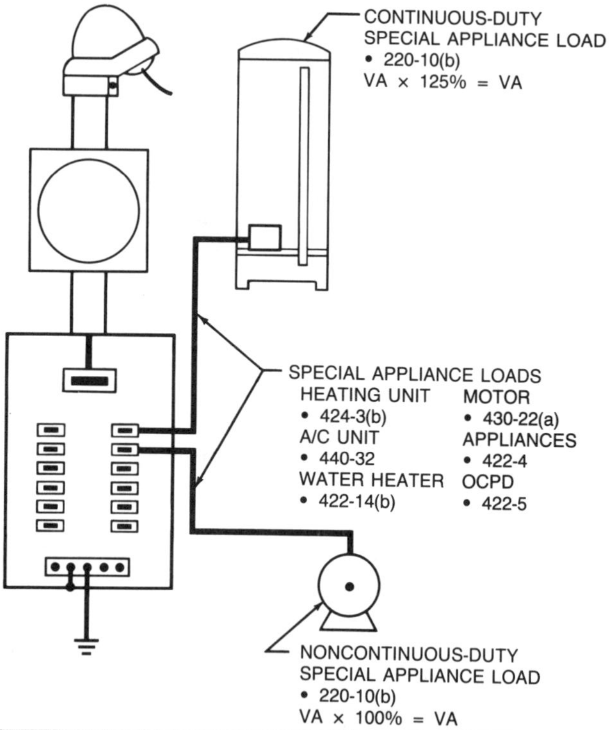

DETERMINING LOAD IN VA FOR SPECIAL APPLIANCE LOADS

What is the load in VA for a special appliance load rated at 14,880 VA and used at continuous duty?

Step 1. *220-10(b):* 14,880 VA × 125% = 18,600 VA

Answer. **Load = 18,600 VA**

What is the load in VA for a special appliance load rated at 12,950 VA and used at noncontinuous duty?

Step 1. *220-10(b):* 12,950 VA × 100% = 12,950 VA

Answer. **Load = 12,950 VA**

Figure 8-10. The load in VA foı a special appliance load is derived by multiplying the appliance load by 100% for noncontinuous-duty and 125% for continuous-duty.

What is the VA rating of a 208 V, 3ϕ, 50 A special appliance load operating for less than three hours and supplied by an individual circuit?

Step 1. *Text:* $VA = V \times \sqrt{3} \times A$
$VA = 208\text{ V} \times 1.732 \times 50\text{ A} = 18{,}000\text{ VA}$

Step 2. *220-10(b):* 18,000 VA × 100% = 18,000 VA

Answer. **VA rating = 18,000 VA**

What is the VA rating of a 480 V, 3ϕ, 35 A special appliance load operating for more than three hours and supplied by an individual circuit?

Step 1. *Text:* $VA = V \times \sqrt{3} \times A$
$VA = 480\text{ V} \times 1.732 \times 35\text{ A} = 29{,}085\text{ VA}$

Step 2. *220-10(b):* 29,085 VA × 125% = 36,356 VA

Answer. **VA rating = 36,356 VA**

Heating or A/C Loads—220-21

Heating or A/C loads is the fourth group of loads to be determined. The largest VA rating between the heating or A/C load is selected and the smaller of the two loads is dropped. It is not necessary to compute both loads to select the elements of the service equipment as the loads are not used simultaneously. The VA rating of each load is computed at 100% and the largest load of the two is selected. The load dropped is not used again except for sizing the elements of the branch circuit. See Figure 8-11.

A one-family dwelling has a 10 kW heating unit and a 5040 VA A/C unit. Which load is selected?

Step 1. *220-15:* 10 kW × 1000 × 100% = 10,000 VA
5040 VA × 100% = 5040 VA

Step 2. *220-21:* 10,000 VA is greater than 5040 VA

Answer. **Load = 10,000 VA**

■*Master Electrician's Test Tip.* Some examination boards require the heating load at 100% to be compared to 125% of the A/C load. If the A/C load is the largest motor per 220-14, it is used and not dropped per 220-21.■

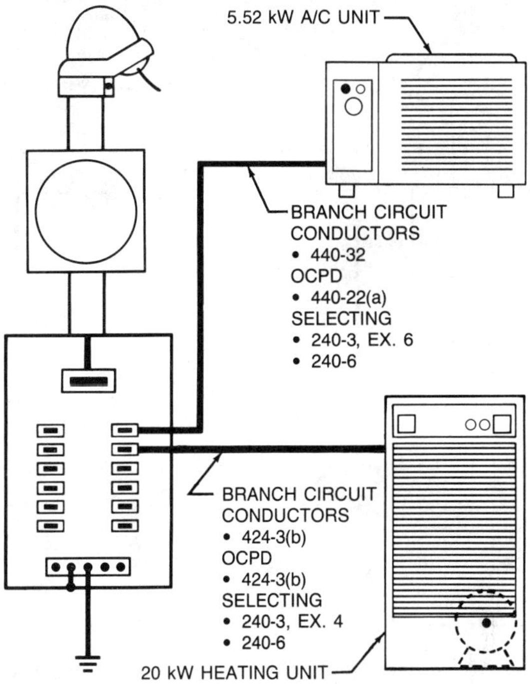

DETERMINING LARGEST LOAD BETWEEN HEATING AND A/C UNIT LOAD

What is the largest load between a 20 kW heating unit and a 5.52 kW A/C unit load?

Step 1. *220-15:* Heating load = 20 kW
A/C load = 5.52 kW

Step 2. *220-21:* 20 kW is greater than 5.52 kW

Answer. **Heating load = 20 kW**

Figure 8-11. The larger load between two dissimilar loads is used and the smaller load dropped when applying 220-21.

Motor Loads—430-22(a); 430-24; 430-25

Motor loads are the fifth group of loads to be determined. Motor loads are motors used as a single unit to drive a piece of equipment. Motors used in an approved assembly are usually not considered individual motor loads. The VA rating of motors is converted from amperage to VA by multiplying the amperage from Table 430-148 or Table 430-150 by the configuration of voltage. See Figure 8-12.

What is the VA rating for a group of 208 V, 3ϕ motors rated at 30 HP, 20 HP, and 15 HP?

Step 1. *430-6(a); Table 430-150, Footnotes:*
30 HP = 80 A × 110% = 88 A
20 HP = 54 A × 110% = 59.4 A
15 HP = 42 A × 110% = 46.2 A

Step 2. *Text:* $VA = V \times 1.732 \times A$

30 HP
$VA = 208\text{ V} \times 1.732 \times 88\text{ A} = 31{,}680\text{ VA}$

20 HP
$VA = 208\text{ V} \times 1.732 \times 59.4\text{ A} = 21{,}384\text{ VA}$

15 HP
$VA = 208\text{ V} \times 1.732 \times 46.2\text{ A} = 16{,}632\text{ VA}$

Step 3. Total VA. *Text:*
31,680 VA + 21,384 VA + 16,632 VA = 69,696 VA

Answer. **VA rating = 69,696 VA**

What is the largest load between a 25 kW heating unit and a 30 A A/C unit? The service is 240 V, 3ϕ.

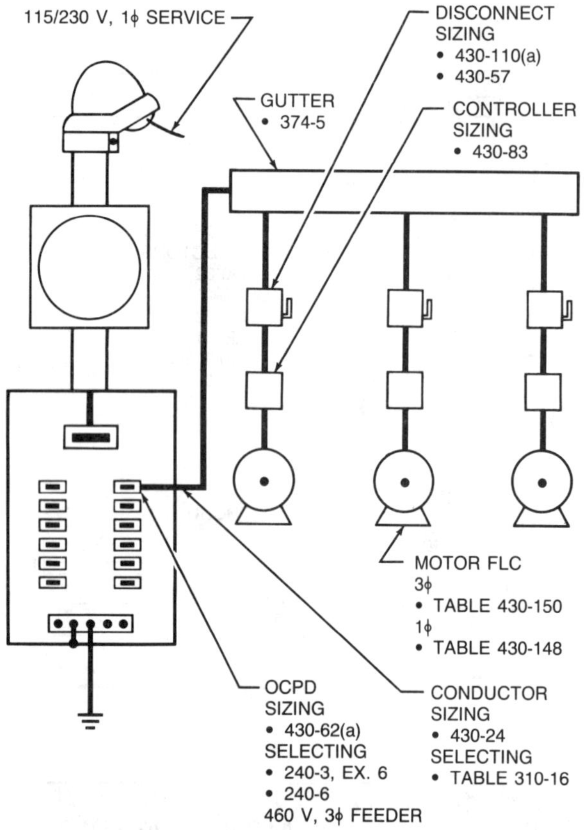

DETERMINING MOTOR LOAD IN AMPS AND VA

What is the load in amps and VA for a 3 HP, 5 HP, and 7.5 HP, 230 V, 1ϕ group of motors with a 122 A of other 1ϕ loads?

Step 1. *430-6(a); Table 430-148; 220-10(b):*
3 HP = 17 A
5 HP = 28 A
7.5 HP = 40 A

Step 2. Total motor load. 17 A + 28 + 40 A = 85 A

Answer. **Load = 85 A**

Step 3. *Text: VA = V × I*
VA = 230 V × 85 A = 19,550 VA

Answer. **Load = 19,550 VA**

Figure 8-12. The motor load in amps is selected per Table 430-148 for 1ϕ and Table 430-150 for 3ϕ. The VA is determined by multiplying the amps by the supply voltage.

Step 1. Selecting largest load. *220-21:*
Heating load =
25 kW × 1000 × 100% = 25,000 VA
A/C load =
240 V × 30 A × 1.732 = 12,480 VA

Answer. **Load = 25,000 VA**

Largest Motor Load—220-14

The largest motor load in VA between the heating unit load or the A/C load is the sixth load of the group of loads to be determined. The largest motor load is selected from one of the motors listed in the fifth loads, which is the group of motor loads. The VA rating of the largest motor is determined by multiplying the amperage of the unit by the voltage times 25% per 220-14. See Figure 8-13.

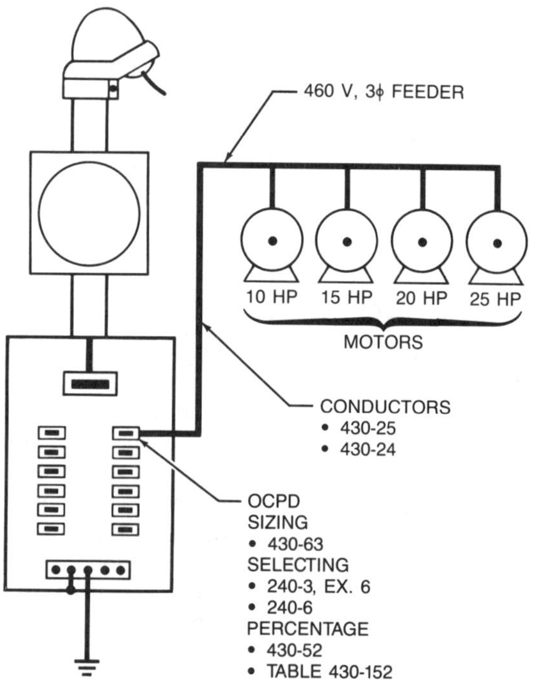

DETERMINING LARGEST MOTOR LOAD IN AMPS AND VA

What is the largest motor load for the group motors in amps and VA?

Step 1. *430-6(a); Table 430-150:*
10 HP = 14 A
15 HP = 21 A
20 HP = 27 A
25 HP = 34 A

Step 2. *430-25; 430-24:* 34 A × 25% = 8.5 A

Answer. **Load = 8.5 A**

Step 3. *Text: VA = E × 1.732 × I*
VA = 460 V × 1.732 × 8.5 A = 6774.5 VA

Answer. **Load = 6774.5 VA**

Figure 8-13. The largest motor load is determined by multiplying the largest motor from the group by 25%.

What is the largest motor load in VA of the following motors?

30 HP .... 240 V, 3ϕ
25 HP .... 240 V, 3ϕ
20 HP .... 240 V, 3ϕ
15 HP .... 240 V, 3ϕ
10 HP .... 240 V, 3ϕ

Step 1. *430-6(a); Table 430-150:*
30 HP = 80 A
25 HP = 68 A
20 HP = 54 A
15 HP = 42 A
10 HP = 28 A

Step 2. *430-17; 430-25:* $VA = V \times \sqrt{3} \times A \times 25\%$
$VA = 240\ V \times 1.732 \times 80\ A \times 25\% = 8320\ VA$

Answer. **Load = 8320 VA**

Note: Where the A/C load is greater than the heating load and the A/C is the largest motor load selected from the fourth or fifth loads, it is considered the largest motor load.

What is the largest motor load in VA of the following loads?

25 HP .... 80 V, 3ϕ = 34 A
20 HP .... 480 V, 3ϕ = 27 A
15 HP .... 480 V, 3ϕ = 21 A
10 HP .... 480 V, 3ϕ = 14 A
7.5 HP .... 480 V, 3ϕ = 11 A
A/C load .... 480 V, 3ϕ = 42 A

Step 1. *220-14; 440-33; 440-34:* 42 A × 25% = 10.5 A

Step 2. $VA = V \times \sqrt{3} \times I$
$VA = 480\ V \times 1.732 \times 10.5\ A = 8725.5\ VA$

Answer. **Load = 8725.5 VA**

■*Master Electrician's Test Tip.* Some testing agencies require the A/C load to be the largest motor load if it is greater in VA than the other motor loads. This is required even if the heating is greater in VA than the A/C load per 220-21.■

COMBINATION VOLTAGES

The service load in VA for combination voltages is determined by multiplying the amperage of each load by its respective voltage. The total amperage of each load multiplied by its voltage is divided by the higher voltage of the service to obtain the amperage to select the elements of the service equipment. The higher voltage is supplied by the utility company to the service equipment and transformers are installed to step down the voltage. Higher voltage to the service equipment allows smaller conductors and elements of the service equipment to be used. Equipment in commercial and industrial locations can require different voltage levels to operate. See Figure 8-14.

What is the service load in VA and amps for a building supplied by a 480 V, 3ϕ, 4-wire system with the following loads? All demand factors to loads have been applied and noncontinuous- and continuous-duty loads computed.

Lighting loads .... 65 A, 120 V, 1ϕ
Receptacle loads .... 55 A, 120 V, 1ϕ
Personal computer loads .... 35 A, 120 V, 1ϕ
Calculator loads .... 25 A, 120 V, 1ϕ
Lighting loads .... 21 A, 277 V, 1ϕ
A/C unit .... 18 A, 480 V, 3ϕ
Heating unit .... 48 A, 480 V, 3ϕ
Motor .... 21 A, 480 V, 3ϕ
Motor .... 27 A, 480 V, 3ϕ
Motor .... 34 A, 480 V, 3ϕ
Special loads .... 80 A, 208 V, 1ϕ
Special loads .... 110 A, 480 V, 3ϕ

Step 1. Lighting loads. *220, Part B:*
65 A × 120 V = 7800 VA
21 A × 277 V = 5817 VA
Total = 13,617 VA

Step 2. Receptacle loads. *220-3(c)(5):*
55 A × 120 V = 6600 VA
35 A × 120 V = 4200 VA
25 A × 120 V = 3000 VA
Total = 13,800 VA

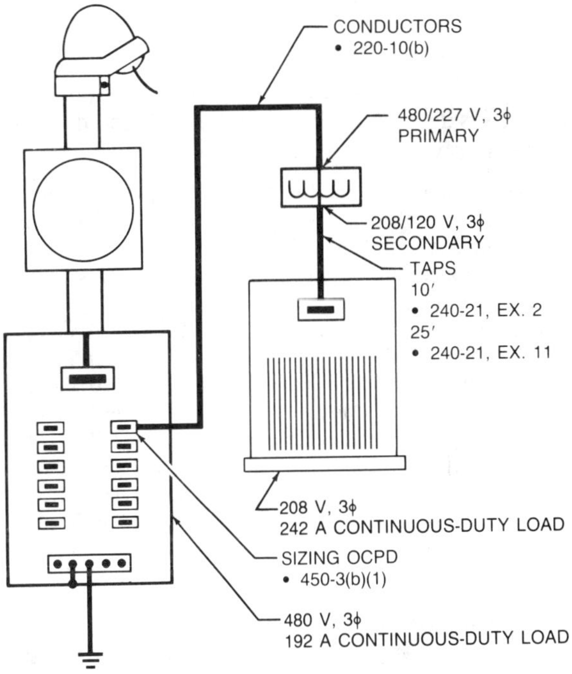

DETERMINING LOAD FOR SERVICE WITH COMBINATION VOLTAGE LOADS

What is the load in VA for the service supplying the connected loads?

Step 1. 208 V load. *220-10(b):*
242 A × 208 V × 1.732 = 87,120 VA

Step 2. 480 V load. *220-10(b):*
192 A × 480 V × 1.732 = 159,552 VA

Step 3. Total loads.
208 V loads = 87,120 VA
480 V loads = 159,552 VA
Total = 246,672 VA

Answer. **Load = 246,672 VA**

Figure 8-14. The load in VA for the service using combination voltage is derived by multiplying the load in amps by the corresponding voltage and adding the VA.

Step 3. Special loads. *220-10(b):*

80 A × 208 V =	16,640 VA
110 A × 831 V =	91,410 VA
Total =	108,050 VA

Step 4. Heating or A/C load. *220-21:*
48 A × 831 V = 39,888 VA

Step 5. Motor loads. *430-25:*

21 A × 831 V =	17,451 VA
27 A × 831 V =	22,437 VA
34 A × 831 V =	28,254 VA
Total =	68,142 VA

Step 6. Largest motor load. *220-14:*
34 A × 831 V × 25% = 7063.5 VA
Total loads = 250,560.5 VA

Step 7. Finding amps. $I = \frac{VA}{V \times \sqrt{3}}$

$$I = \frac{250{,}560.5 \text{ VA}}{480 \text{ V} \times 1.732} = 301.5 \text{ A}$$

Answer. **Load = 301.5 A**

The service load in VA for a service with four voltage levels is determined by adding all VA and dividing by the supply voltage of the service.

What is the service load in VA and amps for a service supplied by a 4160 V, 3ϕ, 4-wire system with the following loads? All loads have been computed at continuous duty or noncontinuous duty with demand factors applied where permitted.

Lighting loads	50 A, 120 V, 1ϕ
Receptacle loads	45 A, 120 V, 1ϕ
Personal computer loads	12 A, 120 V, 1ϕ
Calculator loads	15 A, 120 V, 1ϕ
Lighting loads	42 A, 277 V, 1ϕ
A/C unit	23 A, 480 V, 3ϕ
Heating unit	60 A, 480 V, 3ϕ
Motor	42 A, 230 V, 3ϕ
Motor	54 A, 230 V, 3ϕ
Motor	68 A, 208 V, 3ϕ
Motor	40 A, 208 V, 3ϕ
Special loads	80 A, 208 V, 1ϕ
Special loads	110 A, 208 V, 3ϕ
Motor loads	77 A, 480 V, 3ϕ
Motor loads	52 A, 480 V, 3ϕ
Motor loads	40 A, 480 V, 3ϕ
Processing machine	12 A, 4160 V, 3ϕ
Processing machine	22 A, 4160 V, 3ϕ

Step 1. Lighting loads. *220, Part B:*

50 A × 120 V =	6000 VA
42 A × 277 V =	11,634 VA
Total =	17,634 VA

Step 2. Receptacle loads. *220-3(c)(5):*

45 A × 120 V =	5400 VA
12 A × 120 V =	1440 VA
15 A × 120 V =	1800 VA
Total =	8640 VA

Step 3. Special appliance loads. *220-10(b):*
VA = *A* × *V* × 1.732

80 A × 208 V × 1.732 =	28,800 VA
110 A × 208 V × 1.732 =	39,600 VA
12 A × 4160 V × 1.732 =	86,460 VA
22 A × 4160 V × 1.732 =	158,510 VA
Total =	313,370 VA

Step 4. Heating or A/C loads. *220-21:*
60 A × 480 V × 1.732 = 49,860 VA

Step 5. Motor loads.
430-25: VA = *A* × *V* × 1.732

42 A × 230 V × 1.732 =	16,716 VA
54 A × 230 V × 1.732 =	21,492 VA
68 A × 208 V × 1.732 =	24,480 VA
40 A × 208 V × 1.732 =	14,400 VA
77 A × 480 V × 1.732 =	63,987 VA
52 A × 480 V × 1.732 =	43,212 VA
40 A × 480 V × 1.732 =	33,240 VA
Total =	217,527 VA

Step 6. Largest motor load. *220-14:*
77 A × 831 V × 25% = 15,996.8 VA
Total loads = 623,027.8 VA

Finding amps. $I = \frac{VA}{V \times \sqrt{3}}$

$$I = \frac{623{,}027.8 \text{ VA}}{4160 \text{ V} \times 1.732} = 86.5 \text{ A}$$

Answer. **Load = 86.5 A**

EXAMPLES

The load in VA used to select the elements of the service equipment for industrial and commercial buildings is determined by the method in which the load is used. Loads are computed at either 100% or 125% with demand factors applied. The following examples are procedures for calculating loads based on usage to determine the VA or amp rating required to select elements of the service.

Store Building

The load in VA or amps for a store building is determined by the VA of each load and the amount of time each load takes to perform work. The VA of each load is divided by the applied voltage to obtain the amperage. This amperage is multiplied by a percentage based on the load and the time required to do the work.

120/240 V, 1ϕ, 3-wire Supply. A 120/240 V, 1ϕ, 3-wire service is used for smaller industrial and commercial buildings. A 120/240 V, 1ϕ, 3-wire service has two 120 V phase (hot) conductors; therefore, the neutral is shared by the two phase conductors.

An 80′-0″ × 70′-0″ store building has a 120/240 V, 1ϕ, 3-wire service and the following loads. At least 50% or more of the lighting load is electric discharge. Find the load in VA and amps. Use the form, STANDARD CALCULATION—COMMERCIAL AND INDUSTRIAL. See Figure 8-15.

65 linear feet of show window	120 V, 1ϕ
20′-0″ lighting track	120 V, 1ϕ
Twenty 180 VA ballasts outside lighting (continuous-duty)	120 V, 1ϕ
2400 VA sign lighting (continuous-duty)	120 V, 1ϕ
10 receptacles (continuous-duty)	120 V, 1ϕ

STANDARD CALCULATION—COMMERCIAL AND INDUSTRIAL

1. LIGHTING LOADS—220, Part B

General lighting loads—Table 220-3(b); 220-10(b)
5600 sq. ft. × 3 VA × 125% = 21,000 VA

Show window loads—220-12
65′ × 200 VA = 13,000 VA

Track lighting loads—410-102
20′/2′ × 180 VA = 1800 VA

Outside lighting loads—220-10(b)
180 VA × 20 units × 125% = 4500 VA

Sign lighting loads—220-3, FPN; 600-6(c)
2400 VA × 125% = 3000 VA

		PHASES	NEUTRAL
(total)	43,300 VA	43,300 VA	43,300 VA

2. RECEPTACLE LOADS—220-3(c)(5)

Noncontinuous duty—220-10(b); Table 220-13
— × 180 VA = — VA = — VA

Applying demand factors—Table 220-13
First 10,000 VA × 100% = — VA = — VA
Remaining — VA × 50% = — VA

Continuous duty—220-10(b) # 10 × 180 VA × 125% = 2250 VA

Multioutlet assembly—220-3(c)
Heavy duty = 25′ × 180 VA = 4500 VA
Light duty = 15′/5′ × 180 VA = 540 VA

		PHASES	NEUTRAL
(total)	7290 VA	7290 VA	7290 VA

3. SPECIAL LOADS—220-10(b)

Freezer = 3600 VA × 100% = 3600 VA
Ice box = 2880 VA × 100% = 2880 VA
Water htr. = 8000 VA × 100% = 8000 VA
Cooler = 3840 VA × 100% = 3840 VA
Drinkbox = 1800 VA × 100% = 1800 VA

		PHASES	NEUTRAL
(total)	20,120 VA	20,120 VA	1800 VA

4. HEATING OR A/C LOAD—220-21

		PHASES	NEUTRAL
	35,000 VA	35,000 VA	— VA

5. MOTOR LOADS—430-22(a); 430-24; 430-25

					3ϕ only		
—	=	— A ×	— % ×	— V	$\times \sqrt{3}$	=	— VA
—	=	— A ×	— % ×	— V	$\times \sqrt{3}$	=	— VA
—	=	— A ×	— % ×	— V	$\times \sqrt{3}$	=	— VA
—	=	— A ×	— % ×	— V	$\times \sqrt{3}$	=	— VA

		PHASES	NEUTRAL
(total)	— VA	— VA	— VA

6. LARGEST MOTOR LOAD—220-14

				3ϕ only			PHASES	NEUTRAL
ϕ	16 A ×	100 % ×	240 V	$\times \sqrt{3}$	× 25% =	960 VA	960 VA	— VA
N	15 A ×	100 % ×	120 V	$\times \sqrt{3}$	× 25% =	450 VA	— VA	450 VA

TOTAL = 106,670 VA | 52,840 VA

1ϕ service: $I = \frac{VA}{V}$

PHASES $I = \frac{106{,}670\ VA}{240\ V} = 444.5\ A$

NEUTRAL $I = \frac{52{,}840\ VA}{240\ V} = 220\ A$

3ϕ service: $I = \frac{VA}{V \times \sqrt{3}}$

$I = \frac{VA}{V \times \sqrt{3}} =$ ____ A

$I = \frac{VA}{V \times \sqrt{3}} =$ ____ A

If over 200 A—220-22; First 200 A × 100% = 200 A
Remaining ____ A × 70% = ____ A
____ A

Figure 8-15. The form, STANDARD CALCULATION—COMMERCIAL AND INDUSTRIAL is used to find VA and amps for store buildings with a 120/240 V, 1ϕ, 3-wire supply.

15′-0″ multioutlet assembly
(light-duty) .... 120 V, 1ϕ
25′-0″ multioutlet assembly
(heavy-duty) .... 120 V, 1ϕ
3600 VA freezer .... 240 V, 1ϕ
2880 VA ice cream boxes .... 240 V, 1ϕ
8000 VA water heater .... 240 V, 1ϕ
3840 VA walk-in cooler .... 240 V, 1ϕ
35,000 VA heating unit .... 240 V, 1ϕ
17,280 VA A/C .... 240 V, 1ϕ
1800 VA soft drink box .... 120 V, 1ϕ

(1) What size THHN Cu conductors paralleled two times per phase is required for phases A and B and the neutral in one conduit (overhead service)?

Phases A and B

Step 1. *Note 10, Part C to Tables 310-16–310-19; Table 310-16:* #300 kcmil = 320 A

Step 2. *Note 8 to Tables 310-16–310-19:* 320 A × 80% = 256 A

Answer. **Conductors = #300 kcmil THHN Cu**

Note: #250 kcmil can be used where connected to 90 °C terminals.

Neutral

Step 1. *Text:* $\frac{220\text{ A}}{2} = 110\text{ A}$

Step 2. *310-4:* #1/0 THHN Cu

Answer. **Conductors = #1/0 THHW Cu**

(2) What size main OCPD is required in the service equipment to protect the elements of the service?

Step 1. *240-3, Ex. 4; 240-6:* 256 A × 2 = 512 A

Answer. **OCPD = 600 A**

(3) What size 3-wire panelboard is required?

Step 1. *Chart:* 600 A OCPD requires 600 A panelboard

Answer. **Panelboard = 600 A**

120/208 V, 3ϕ, 4-wire Supply. A 120/208 V, 3ϕ, 4-wire service produces three hot conductors to ground. The three hot conductors can share one neutral that carries the unbalanced current of the three hot conductors. A 120/208 V, 3ϕ, 4-wire system is used where there is a large amount of 120 V lighting and receptacle loads.

An 8100 sq. ft. store building has a 120/208 V, 3ϕ, 4-wire service and the following loads. Find the load in VA and amps. Note that 50% of the lighting load is electric discharge. The service conduit extends 3′-0″ from an underground trench to the service equipment. Use the form, STANDARD CALCULATION—COMMERCIAL AND INDUSTRIAL. See Figure 8-16.

80 linear feet of show window .... 120 V, 1ϕ
120′-0″ lighting track .... 120 V, 1ϕ
Thirty 180 VA ballasts outside lighting
(continuous-duty) .... 120 V, 1ϕ
3600 VA sign lighting
(continuous-duty) .... 120 V, 1ϕ
65 receptacles (noncontinuous-duty) ... 120 V, 1ϕ
28 receptacles (continuous-duty) .... 120 V, 1ϕ
40′-0″ multioutlet assembly
(heavy-duty) .... 120 V, 1ϕ
7380 VA freezer .... 208 V, 3ϕ
5580 VA ice cream boxes .... 208 V, 3ϕ
1296 VA (six .6 A) fans .... 208 V, 3ϕ
10,000 VA water heater .... 208 V, 3ϕ
9540 VA walk-in cooler .... 208 V, 3ϕ
50,000 VA heating unit .... 208 V, 3ϕ
11,520 VA A/C .... 208 V, 3ϕ
3801.6 VA water pump .... 208 V, 3ϕ

(1) Are the service-entrance conductors required to be derated where they extend from the meter to the service?

Step 1. *Notes 8 and 10 to Tables 310-16–310-19:* Yes

Answer. **Yes**

(2) Can the neutral be reduced 70% for all amperage exceeding 200 A?

Step 1. *220-22; Note 10(c) to Tables 310-16–310-19:* No

Answer. **No**

(3) What is the percentage by which the service-entrance conductors can be derated?

Step 1. *Note 8 to Tables 310-16–310-19:* Four conductors = 80% derating factor

Answer. **80%**

Combination Voltage. Combination voltage is used for heavy equipment supplied by the higher voltage and lighter equipment and lighting and receptacle outlets supplied by the lower voltage. Loads on the neutral conductor shall not be calculated on service entrance conductors for a service supplying step-down transformers with step-down voltages.

A 10,000 sq. ft. store building has a 277/480 V, 3ϕ, 4-wire service and the following loads. Find the load in VA and amps. Note that 50% or more of the lighting load is electric discharge. Use the form, STANDARD CALCULATION—COMMERCIAL AND INDUSTRIAL. See Figure 8-17.

General lighting load .... 277 V, 1ϕ
80 linear feet of show window .... 120 V, 1ϕ
100′-0″ lighting track .... 120 V, 1ϕ
Forty 180 VA ballasts outside lighting
(continuous-duty) .... 277 V, 1ϕ
5720 VA sign lighting
(continuous-duty) .... 120 V, 1ϕ
72 receptacles (noncontinuous-duty) ... 120 V, 1ϕ
50 receptacles (continuous-duty) .... 120 V, 1ϕ
15′-0″ multioutlet assembly
(light-duty) .... 120 V, 1ϕ
30′-0″ multioutlet assembly
(heavy-duty) .... 120 V, 1ϕ
8310 VA freezer .... 480 V, 3ϕ
8725.5 VA ice cream boxes .... 480 V, 3ϕ
31,578 VA (five 5 HP) fans .... 480 V, 3ϕ
12,000 VA water heater .... 480 V, 3ϕ
9972 VA walk-in cooler .... 480 V, 3ϕ
70,000 VA heating unit .... 480 V, 3ϕ

STANDARD CALCULATION—COMMERCIAL AND INDUSTRIAL

1. LIGHTING LOADS—220, Part B

General lighting loads—Table 220-3(b); 220-10(b)

8100 sq. ft. × 3 VA × 125 % = 30,375 VA

Show window loads—220-12

80 ′ × 200 VA = 16,000 VA

Track lighting loads—410-102

$\frac{120'}{2'}$ × 180 VA = 10,800 VA

Outside lighting loads—220-10(b)

180 VA × 30 units × 125 % = 6750 VA

Sign lighting loads—220-3, FPN; 600-6(c)

3600 VA × 125 % = 4500 VA

		PHASES	NEUTRAL
(total)	68,425 VA	68,425 VA	68,425 VA

2. RECEPTACLE LOADS—220-3(c)(5)

Noncontinuous duty—220-10(b); Table 220-13

\# 65 × 180 VA = 11,700 VA = — VA

Applying demand factors—Table 220-13

First 10,000 VA × 100% = 10,000 VA = 10,000 VA

Remaining 1700 VA × 50% = 850 VA

Continuous duty—220-10(b) # 28 × 180 VA × 125% = 6300 VA

Multioutlet assembly—220-3(c)

Heavy duty = 40 ′ × 180 VA = 7200 VA

Light duty = $\frac{—'}{5'}$ × 180 VA = — VA

		PHASES	NEUTRAL
(total)	24,350 VA	24,350 VA	24,350 VA

3. SPECIAL LOADS—220-10(b)

Freezer = 7380 VA × 100 % = 7380 VA

Ice box = 5580 VA × 100 % = 5580 VA

Water htr. = 10,000 VA × 100 % = 10,000 VA

Cooler = 9540 VA × 100 % = 9540 VA

— = — VA × — % = — VA

		PHASES	NEUTRAL
(total)	32,500 VA	32,500 VA	— VA

4. HEATING OR A/C LOAD—220-21

		PHASES	NEUTRAL
	50,000 VA	50,000 VA	— VA

5. MOTOR LOADS—430-22(a); 430-24; 430-25

						3ϕ only		
Fans	=	3.6 A ×	100 % ×	208 V		× $\sqrt{3}$	=	1296 VA
Pump	=	9.6 A ×	110 % ×	208 V		× $\sqrt{3}$	=	3801.6 VA
—	=	— A ×	— % ×	— V		× $\sqrt{3}$	=	— VA
—	=	— A ×	— % ×	— V		× $\sqrt{3}$	=	— VA

		PHASES	NEUTRAL
(total)	5097.6 VA	5097.6 VA	— VA

6. LARGEST MOTOR LOAD—220-14

				3ϕ only			PHASES	NEUTRAL
ϕ	26.5 A ×	100 % ×	208 V	× $\sqrt{3}$	× 25% =	2385 VA	2385 VA	— VA
N	— A ×	— % ×	— V	× $\sqrt{3}$	× 25% =	— VA	— VA	— VA
						TOTAL =	182,757.6 VA	92,775 VA

1ϕ service: $I = \frac{VA}{V}$

PHASES I = $\frac{\_\_\_ VA}{V}$ = ___ A

NEUTRAL I = $\frac{\_\_\_ VA}{V}$ = ___ A

3ϕ service: $I = \frac{VA}{V \times \sqrt{3}}$

I = $\frac{182,757.6 \text{ VA}}{208 \text{ V} \times \sqrt{3}}$ = 507.7 A

I = $\frac{92,775 \text{ VA}}{208 \text{ V} \times \sqrt{3}}$ = 257.7 A

If over 200 A—220-22; First 200 A × 100% = 200 A

Remaining ___ A × 70% = ___ A

___ A

Figure 8-16. The form, STANDARD CALCULATION—COMMERCIAL AND INDUSTRIAL is used to find VA and amps for store buildings with a 120/208 V, 3ϕ, 4-wire supply.

STANDARD CALCULATION—COMMERCIAL AND INDUSTRIAL

1. LIGHTING LOADS—220, Part B

General lighting loads—Table 220-3(b); 220-10(b)

10,000 sq. ft. × 3 VA × 125% = 37,500 VA

Show window loads—220-12

80 ′ × 200 VA = 16,000 VA

Track lighting loads—410-102

100 ′ / 2′ × 180 VA = 9000 VA

Outside lighting loads—220-10(b)

40 VA × 180 units × 125% = 9000 VA

Sign lighting loads—220-3, FPN; 600-6(c)

5720 VA × 125% = 7150 VA

		PHASES	NEUTRAL
(total)	78,650 VA	78,650 VA	46,500 VA

2. RECEPTACLE LOADS—220-3(c)(5)

Noncontinuous duty—220-10(b); Table 220-13

72 × 180 VA = 12,960 VA = — VA

Applying demand factors—Table 220-13

First 10,000 VA × 100% = 10,000 VA = 10,000 VA

Remaining 2960 VA × 50% = 1480 VA

Continuous duty—220-10(b) # 50 × 180 VA × 125% = 11,250 VA

Multioutlet assembly—220-3(c)

Heavy duty = 30 ′ × 180 VA = 5400 VA

Light duty = 15 ′ / 5′ × 180 VA = 540 VA

		PHASES	NEUTRAL
(total)	28,670 VA	28,670 VA	— VA

3. SPECIAL LOADS—220-10(b)

Freezer = 8310 VA × 100% = 8310 VA

Ice box = 8725.5 VA × 100% = 8725.5 VA

Water htr. = 12,000 VA × 100% = 12,000 VA

Cooler = 9972 VA × 100% = 9972 VA

Copy mch. = 1560 VA × 100% = 1560 VA

		PHASES	NEUTRAL
(total)	40,567.5 VA	40,567.5 VA	— VA

4. HEATING OR A/C LOAD—220-21

	70,000 VA	70,000 VA	— VA

5. MOTOR LOADS—430-22(a); 430-24; 430-25

				3ϕ only		
Fans	= 38 A	× 100 %	× 480 V	× √3	= 31,578 VA	
Pump	= 11 A	× 100 %	× 480 V	× √3	= 9141 VA	
—	= — A	× — %	× — V	× √3	= — VA	
—	= — A	× — %	× — V	× √3	= — VA	

		PHASES	NEUTRAL
(total)	40,719 VA	40,719 VA	— VA

6. LARGEST MOTOR LOAD—220-14

				3ϕ only			PHASES	NEUTRAL
ϕ	12 A ×	100 % ×	480 V	× √3	× 25% =	2493 VA	2493 VA	— VA
N	— A ×	— % ×	— V	× √3	× 25% =	— VA	— VA	— VA

TOTAL = 261,099.5 VA | 46,500 VA

1ϕ service: I = VA / V

PHASES I = ______ VA / V = ______ A

NEUTRAL I = ______ VA / V = ______ A

3ϕ service: I = VA / (V × √3)

I = 261,099.5 VA / (480 V × √3) = 314 A

I = 46,500 VA / (480 V × √3) = 55.9 A

If over 200 A—220-22; First 200 A × 100% = 200 A

Remaining ______ A × 70% = ______ A

______ A

Figure 8-17. The form, STANDARD CALCULATION—COMMERCIAL AND INDUSTRIAL is used to find VA and amps for store buildings with combination voltages.

30,000 VA A/C . . . . 480 V, 3ϕ
9141 VA (7½ HP) water pump . . . . 480 V, 3ϕ
1560 VA copying machine . . . . 120 V, 1ϕ

(1) What size service-entrance conductors, using THWN Cu, paralleled three times per phase is required?

Step 1. *Note 10(c) to Tables 310-16–310-19; 310-4:*

$$\text{Size load} = \frac{314 \text{ A}}{3} = 104.7 \text{ A}$$

Step 2. *Note 8 to Tables 310-16–310-19; Table 310-16:*
#1/0 THWN Cu = 150 A
150 A × 80% = 120 A

Step 3. Total ampacity. 120 A × 3 = 360 A

Answer. **Conductors = #1/0 THWN Cu**

(2) What size copper grounding electrode conductor is required?

Step 1. *Table 8, Ch. 9:* #1/0 = 105,600 CM
105,600 CM × 3 = 316,800 CM

Step 2. Total kcmil. $kcmil = \frac{CM}{1000}$

$$kcmil = \frac{316{,}800 \text{ CM}}{1000} = 316.8 \text{ kcmil}$$

Step 3. *Table 250-94:* 316.8 kcmil requires #2 Cu

Answer. **Conductor = #2 Cu**

(3) What is the size of the neutral conductors using THWN Cu?

Step 1. *310-4:* $A = \frac{N\ load}{No.\ in\ parallel}$

$$A = \frac{55.9 \text{ A}}{3} = 18.6 \text{ A}$$

Step 2. *310-4:* #1/0 THWN Cu = 150 A
150 A × 3 = 450 A

Answer. **Conductors = #1/0 THWN Cu**

(4) What size OCPD, using the minimum rating, is required for the walk-in cooler?

Step 1. *Text:* $A = \frac{VA}{V \times \sqrt{3}}$

$$A = \frac{9972 \text{ VA}}{480 \text{ V} \times 1.732} = 12 \text{ A}$$

Step 2. *440-22(a):* 12 A × 175% = 21 A

Step 3. *240-3, Ex. 6; 240-6:*
20 A is the next lower standard size

Answer. **OCPD = 20 A**

(5) What size THWN Cu conductors is required for the walk-in cooler?

Step 1. *440-32:* 12 A × 125% = 15 A

Step 2. *Table 310-16:* #14 THWN Cu = 20 A

Answer. **Conductors = #14 THWN Cu**

Office Building

Loads in an office building are usually larger for the 120/240 V, 1ϕ loads. Heavy equipment is supplied by the 3ϕ power. For a 120/240 V, 1ϕ service, 120 V is used for the lighting and receptacle outlets and 240 V for the heavy equipment. Larger office buildings use 277/480 V, 3ϕ, 4-wire systems. The higher 480 V, 3ϕ voltage supplies the heavier equipment and the 277 V, 1ϕ voltage supplies lighting. Transformers step down the 480 V to 120/208 V or 120/240 V to serve the loads with lower voltage.

120/240 V, 1ϕ, 3-wire Supply.

A 70′-0″ × 70′-0″ office building has a 120/240 V, 1ϕ, 3-wire service and the following loads. Find the load in VA and amps. Note that 50% or more of the lighting load is electric discharge. Use the form, STANDARD CALCULATION—COMMERCIAL AND INDUSTRIAL. See Figure 8-18.

30 linear feet show window . . . . 120 V, 1ϕ
22′-0″ lighting track . . . . 120 V, 1ϕ
Twelve 180 VA ballasts outside lighting
(continuous-duty) . . . . 120 V, 1ϕ
4800 VA sign lighting
(continuous-duty) . . . . 120 V, 1ϕ
82 receptacles (noncontinuous-duty) . . . 120 V, 1ϕ
21 receptacles (continuous-duty) . . . . 120 V, 1ϕ
80′-0″ multioutlet assembly
(light-duty) . . . . 120 V, 1ϕ
16′-0″ multioutlet assembly
(heavy-duty) . . . . 120 V, 1ϕ
35 kW heating unit . . . . 240 V, 1ϕ
8640 VA A/C . . . . 240 V, 1ϕ
30 HP elevator
(30 minute intermittent duty) . . . . 240 V, 1ϕ
1450 VA copying machine
(special load) . . . . 120 V, 1ϕ
6500 VA water heater . . . . 240 V, 1ϕ
Eight ⅓ HP fans . . . . 240 V, 1ϕ
Fifteen 225 VA data processors
(special loads) . . . . 120 V, 1ϕ
Ten 175 VA word processors . . . . 120 V, 1ϕ
Three 1200 VA printers . . . . 120 V, 1ϕ

(1) What size THWN Cu conductors, paralleled four times per phase and routed through a constant ambient temperature of 120 °F, is required for phases A and B?

Step 1. *310-4:* Size load $= \frac{666.9 \text{ A}}{4} = 166.7 \text{ A}$

Step 2. *Table 310-16:* #300 kcmil THWN Cu = 285 A

Step 3. *Notes 8 and 10(c) to Tables 310-16–310-19:*
285 A × 80% = 228 A

Step 4. *Table 310-16, Footnotes:* 228 A × 75% = 171 A

Answer. **Conductors = #300 kcmil THWN Cu**

(2) What size OCPD is required for the switchgear?

Step 1. *240-3, Ex. 4:* 171 A × 4 = 684 A

Step 2. *240-6:* 700 A is the next higher standard size

Answer. **OCPD = 700 A**

(3) What size switchgear is required for the service equipment?

Step 1. *Chart:* 700 A OCPD requires 800 A switchgear

Answer. **Switchgear = 800 A**

STANDARD CALCULATION—COMMERCIAL AND INDUSTRIAL

1. LIGHTING LOADS—220, Part B

General lighting loads—Table 220-3(b); 220-10(b)

4900 sq. ft. × 3.5 VA × 125 % = 21,437.5 VA

Show window loads—220-12

30 ′ × 200 VA = 6000 VA

Track lighting loads—410-102

22 ′ / 2′ × 180 VA = 1980 VA

Outside lighting loads—220-10(b)

180 VA × 12 units × 125 % = 2700 VA

Sign lighting loads—220-3, FPN; 600-6(c)

4800 VA × 125 % = 6000 VA

(total) 38,117.5 VA — PHASES 38,117.5 VA — NEUTRAL 38,117.5 VA

2. RECEPTACLE LOADS—220-3(c)(5)

Noncontinuous duty—220-10(b); Table 220-13

82 × 180 VA = 14,760 VA = — VA

Applying demand factors—Table 220-13

First 10,000 VA × 100% = 10,000 VA = 10,000 VA

Remaining 4760 VA × 50% = 2380 VA

Continuous duty—220-10(b) # 21 × 180 VA × 125% = 4725 VA

Multioutlet assembly—220-3(c)

Heavy duty = 16 ′ × 180 VA = 2880 VA

Light duty = 80 ′ / 5′ × 180 VA = 2880 VA

(total) 22,865 VA — PHASES 22,865 VA — NEUTRAL 22,865 VA

3. SPECIAL LOADS—220-10(b)

Copy mch. = 1450 VA × 100 % = 1450 VA

Water htr. = 6500 VA × 100 % = 6500 VA

Data Pr. = 3375 VA × 100 % = 3375 VA

Word Pr. = 1750 VA × 100 % = 1750 VA

Printer = 3600 VA × 100 % = 3600 VA

(total) 16,675 VA — PHASES 16,675 VA — NEUTRAL 10,175 VA

4. HEATING OR A/C LOAD—220-21 35,000 VA — PHASES 35,000 VA — NEUTRAL — VA

5. MOTOR LOADS—430-22(a); 430-24; 430-25

					3ϕ only		
Fans	= 28.8 A	× 100 %	× 240 V		× $\sqrt{3}$	=	6912 VA
Elevator	= 135 A	× 100 %	× 240 V		× $\sqrt{3}$	=	32,400 VA
—	= — A	× — %	× — V		× $\sqrt{3}$	=	— VA
—	= — A	× — %	× — V		× $\sqrt{3}$	=	— VA

(total) 39,312 VA — PHASES 39,312 VA — NEUTRAL — VA

6. LARGEST MOTOR LOAD—220-14

				3ϕ only			PHASES	NEUTRAL
ϕ	150 A ×	90 % ×	240 V	× $\sqrt{3}$	× 25% =	8100 VA	8100 VA	— VA
N	— A ×	— % ×	— V	× $\sqrt{3}$	× 25% =	— VA	— VA	— VA

TOTAL = 160,069.5 VA — 71,157.5 VA

1ϕ service: $I = \frac{VA}{V}$

PHASES $I = \frac{160,069.5\ VA}{240\ V}$ = 666.9 A

NEUTRAL $I = \frac{71,157.5\ VA}{240\ V}$ = 296.5 A

3ϕ service: $I = \frac{VA}{V \times \sqrt{3}}$

$I = \frac{\_\_\_\ VA}{\_\_\_\ V \times \sqrt{3}}$ = ___ A

$I = \frac{\_\_\_\ VA}{\_\_\_\ V \times \sqrt{3}}$ = ___ A

If over 200 A—220-22; First 200 A × 100% = 200 A

Remaining ___ A × 70% = ___ A

___ A

Figure 8-18. The form, STANDARD CALCULATION—COMMERCIAL AND INDUSTRIAL is used to find VA and amps for office buildings with a 120/240 V, 1ϕ, 3-wire supply.

120/208 V, 3ϕ, 4-wire Supply.

An 80′-0″ × 80′-0″ office building has a 120/208 V, 3ϕ, 4-wire service. What is the service load in VA and A for the phases and the neutral? Use the form, STANDARD CALCULATION—COMMERCIAL AND INDUSTRIAL. See Figure 8-19.

12 linear feet show window . . . . 120 V, 1ϕ
24′-0″ lighting track . . . . 120 V, 1ϕ
Ten 225 VA outside lighting (continuous-duty) . . . . 120 V, 1ϕ
1200 VA sign lighting (continuous-duty) . . . . 120 V, 1ϕ
90 receptacles (noncontinuous-duty) . . . 120 V, 1ϕ
35 receptacles (continuous-duty) . . . . 120 V, 1ϕ
40′-0″ multioutlet assembly (light-duty) . . . . 120 V, 1ϕ
12′-0″ multioutlet assembly (heavy-duty) . . . . 120 V, 1ϕ
45 kW heating unit . . . . 208 V, 3ϕ
10,800 VA A/C . . . . 208 V, 3ϕ
75 HP elevator (15 minute intermittent duty) . . . . 208 V, 3ϕ
1200 VA copying machine (special load) . . . . 120 V, 1ϕ
9000 VA water heater . . . . 208 V, 3ϕ
Six ¾ HP fans . . . . 208 V, 3ϕ
4320 VA typewriters (continuous-duty, special loads) . . . . . 120 V, 1ϕ
2160 VA word processors (continuous-duty, special loads) . . . . . 120 V, 1ϕ
1728 VA personal computers (continuous-duty, special loads) . . . . . 120 V, 1ϕ

(1) What size THWN Cu conductors is required to supply the elevator?

Step 1. *430-6(a); Table 430-150, Footnotes:*
75 HP = 192 A × 110% = 211.2 A

Step 2. *Table 430-22(a), Ex.:* 211.2 A × 85% = 179.52 A

Step 3. *Table 310-16:* #3/0 THWN Cu = 200 A

Answer. **Conductors = #3/0 THWN Cu**

(2) What size TDF is required? Use the minimum rating.

Step 1. *430-6(a); Table 430-152:*
211.2 A × 175% = 369.6 A

Step 2. *240-3, Ex. 6; 430-52:*
350 A is the next lower standard size

Answer. **TDF = 350 A**

(3) What size EMT is required?

Step 1. *Table 3B, Ch. 9:* Three #3/0 THWN requires 1½″

Answer. **EMT = 1½″**

Combination Voltage.

A 125′-0″ × 125′-0″ office building has a 277/480 V, 3ϕ, 4-wire service and the following loads. Find the load in VA and amps. Note that 50% or more of the lighting load is electric discharge lighting. Use the form, STANDARD CALCULATION—COMMERCIAL AND INDUSTRIAL. See Figure 8-20.

General lighting load . . . . 277 V, 1ϕ
100 linear feet show window . . . . 120 V, 1ϕ
200′-0″ lighting track . . . . 120 V, 1ϕ
Sixty 180 VA outside lighting (continuous-duty) . . . . 120 V, 1ϕ
Four 1200 VA sign lighting (continuous-duty) . . . . 120 V, 1ϕ
250 receptacles (noncontinuous-duty) . . 120 V, 1ϕ
125 receptacles (continuous-duty) . . . . . . 120 V, 1ϕ
150′-0″ multioutlet assembly (light duty) . . . . 120 V, 1ϕ
60′-0″ multioutlet assembly (heavy load) . . . . 120 V, 1ϕ
Four 30 kW heating units . . . . 480 V, 3ϕ
Four 9972 VA A/C . . . . 480 V, 3ϕ
Five 5 HP water pumps . . . . 480 V, 3ϕ
Four 1200 VA soft drink boxes . . . . 120 V, 1ϕ
Eight 1500 VA copying machines (special loads) . . . . 120 V, 1ϕ
Four 8000 VA water heaters . . . . 480 V, 3ϕ
Sixteen 2 HP fans . . . . 480 V, 3ϕ
Four 12 kW ranges (noncontinuous-duty, no neutral load) . . . . 208 V, 1ϕ
Eight 2400 VA personal computers (special loads) . . . . 208 V, 1ϕ

(1) How many 20 A, 2-wire circuits are required to supply the general lighting load?

Step 1. *220-4(a):* $No. = \dfrac{VA}{20\text{ A OCPD} \times 277\text{ V}}$

$$No. = \frac{68{,}359\text{ VA}}{5540} = 12.34\text{ A}$$

Answer. **2-wire circuits = 13**

(2) How many 20 A, 2-wire circuits are required for the general-purpose receptacles?

Step 1. *220-4(a):* $No. = \dfrac{VA}{20\text{ A OCPD} \times 120\text{ V}}$

$$No. = \frac{45{,}000\text{ VA}}{2400} = 18.75\text{ A}$$

Answer. **2-wire circuits = 19 A**

(3) What size transformer is required for the 120/208 V loads?

Step 1. Finding total VA.

Show window loads. *220-12:* 100′-0″ × 200 VA =	20,000 VA
Lighting track loads. *410-102:* $\dfrac{200'\text{-}0''}{2'} \times 180$ VA =	18,000 VA
Outside lighting loads. *220-10(b):* 180 VA × 60 units × 100% =	10,800 VA
Sign lighting load. *220-3, FPN; 600-6(c):* 4800 VA × 100% =	4800 VA
Total =	53,600 VA

STANDARD CALCULATION—COMMERCIAL AND INDUSTRIAL

1. LIGHTING LOADS—220, Part B

General lighting loads—Table 220-3(b); 220-10(b)

6400 sq. ft. × 3.5 VA × 125 % = 28,000 VA

Show window loads—220-12

12 ′ × 200 VA = 2400 VA

Track lighting loads—410-102

24 ′ / 2′ × 180 VA = 2160 VA

Outside lighting loads—220-10(b)

225 VA × 10 units × 125 % = 2812.5 VA

Sign lighting loads—220-3, FPN; 600-6(c)

1200 VA × 100 % = 1200 VA

		PHASES	NEUTRAL
(total)	36,572.5 VA	36,572.5 VA	36,572.5 VA

2. RECEPTACLE LOADS—220-3(c)(5)

Noncontinuous duty—220-10(b); Table 220-13

\# 90 × 180 VA = 16,200 VA = — VA

Applying demand factors—Table 220-13

First 10,000 VA × 100% = 10,000 VA = 10,000 VA

Remaining 6200 VA × 50% = 3100 VA

Continuous duty—220-10(b) # 35 × 180 VA × 125% = 7875 VA

Multioutlet assembly—220-3(c)

Heavy duty = 12 ′ × 180 VA = 2160 VA

Light duty = 40 ′ / 5′ × 180 VA = 1440 VA

		PHASES	NEUTRAL
(total)	24,575 VA	24,575 VA	24,575 VA

3. SPECIAL LOADS—220-10(b)

Copier	1200 VA ×	100 % =	1200 VA
Water htr.	9000 VA ×	100 % =	9000 VA
Typewrts.	4320 VA ×	125 % =	5400 VA
Word proc.	2160 VA ×	125 % =	2700 VA
PC's	1728 VA ×	125 % =	2160 VA

		PHASES	NEUTRAL
(total)	20,460 VA	20,460 VA	11,460 VA

4. HEATING OR A/C LOAD—220-21 45,000 VA | 45,000 VA | — VA

5. MOTOR LOADS—430-22(a); 430-24; 430-25

				3ϕ only		
Elevator =	211.2 A ×	85 % ×	208 V	× $\sqrt{3}$	=	64,627.2 VA
Fans =	16.8 A ×	110 % ×	208 V	× $\sqrt{3}$	=	6652.8 VA
— =	— A ×	— % ×	— V	× $\sqrt{3}$	=	— VA
— =	— A ×	— % ×	— V	× $\sqrt{3}$	=	— VA

		PHASES	NEUTRAL
(total)	71,280 VA	71,280 VA	— VA

6. LARGEST MOTOR LOAD—220-14

				3ϕ only			PHASES	NEUTRAL
ϕ	211.2 A ×	85 % ×	208 V	× $\sqrt{3}$	× 25% =	16,156.8 VA	16,156.8 VA	— VA
N	— A ×	— % ×	— V	× $\sqrt{3}$	× 25% =	— VA	— VA	— VA

TOTAL = 214,044.3 VA | 72,607.5 VA

1ϕ service: I = VA / V

PHASES I = ____ VA / V = ____ A

NEUTRAL I = ____ VA / V = ____ A

3ϕ service: I = VA / (V × $\sqrt{3}$)

I = 214,044.3 VA / (208 V × $\sqrt{3}$) = 594.6 A

I = 72,607.5 VA / (208 V × $\sqrt{3}$) = 201.7 A

If over 200 A—220-22; First 200 A × 100% = 200 A

Remaining ____ A × 70% = ____ A

____ A

Figure 8-19. The form, STANDARD CALCULATION—COMMERCIAL AND INDUSTRIAL is used to find VA and amps for office buildings with a 120/208 V, 3ϕ, 4-wire supply.

STANDARD CALCULATION—COMMERCIAL AND INDUSTRIAL

1. LIGHTING LOADS—220, Part B

General lighting loads—Table 220-3(b); 220-10(b)

15,625 sq. ft. × 3.5 VA × 125 % = 68,359 VA

Show window loads—220-12

100 ′ × 200 VA = 20,000 VA

Track lighting loads—410-102

200 / 2′ ′ × 180 VA = 18,000 VA

Outside lighting loads—220-10(b)

180 VA × 60 units × 125 % = 13,500 VA

Sign lighting loads—220-3, FPN; 600-6(c)

4800 VA × 125 % = 6000 VA

		PHASES	NEUTRAL
(total)	125,859 VA	125,859 VA	68,359 VA

2. RECEPTACLE LOADS—220-3(c)(5)

Noncontinuous duty—220-10(b); Table 220-13

250 × 180 VA = 45,000 VA = — VA

Applying demand factors—Table 220-13

First 10,000 VA × 100% = 10,000 VA = 10,000 VA

Remaining 35,000 VA × 50% = 17,500 VA

Continuous duty—220-10(b) # 125 × 180 VA × 125% = 28,125 VA

Multioutlet assembly—220-3(c)

Heavy duty = 60 ′ × 180 VA = 10,800 VA

Light duty = 150 / 5′ ′ × 180 VA = 5400 VA

		PHASES	NEUTRAL
(total)	71,825 VA	71,825 VA	— VA

3. SPECIAL LOADS—220-10(b)

Copy mch. = 12,000 VA × 100 % = 12,000 VA

Water htrs. = 32,000 VA × 100 % = 32,000 VA

Ranges = 48,000 VA × 100 % = 48,000 VA

PC's = 19,200 VA × 100 % = 19,200 VA

Drink boxes 4800 VA × 100 % = 4800 VA

		PHASES	NEUTRAL
(total)	116,000 VA	116,000 VA	— VA

4. HEATING OR A/C LOAD—220-21 120,000 VA | 120,000 VA | — VA

5. MOTOR LOADS—430-22(a); 430-24; 430-25

				3ϕ only	
Pumps =	38 A ×	100 % ×	480 V	× $\sqrt{3}$	= 31,578 VA
Fans =	54.4 A ×	100 % ×	480 V	× $\sqrt{3}$	= 45,206.4 VA
— =	— A ×	— % ×	— V	× $\sqrt{3}$	= — VA
— =	— A ×	— % ×	— V	× $\sqrt{3}$	= — VA

		PHASES	NEUTRAL
(total)	76,784.4 VA	76,784.4 VA	— VA

6. LARGEST MOTOR LOAD—220-14

				3ϕ only			PHASES	NEUTRAL
ϕ	7.6 A ×	100 % ×	480 V	× $\sqrt{3}$	× 25% =	1578.9 VA	1578.9 VA	— VA
N	— A ×	— % ×	— V	× $\sqrt{3}$	× 25% =	— VA	— VA	— VA

TOTAL = 512,047.3 VA | 68,359 VA

1ϕ service: $I = \frac{VA}{V}$

PHASES I = ______ VA / V = ______ A

NEUTRAL I = ______ VA / V = ______ A

3ϕ service: $I = \frac{VA}{V \times \sqrt{3}}$

I = 512,047.3 VA / 480 V × $\sqrt{3}$ = 616 A

I = 68,359 VA / 480 V × $\sqrt{3}$ = 82 A

If over 200 A—220-22; First 200 A × 100% = 200 A

Remaining ______ A × 70% = ______ A

______ A

Figure 8-20. The form, STANDARD CALCULATION—COMMERCIAL AND INDUSTRIAL is used to find VA and amps for office buildings with combination voltages.

Step 2. Receptacle loads. *220-3(c)(5):*

Noncontinuous-duty. *220-10(b):*
45,000 VA × 100% = 45,000 VA

Continuous-duty. *220-10(b):*
22,500 VA × 100% = 22,500 VA

Multioutlet assembly. *220-3(c):*
Heavy-duty =
60'-0" × 180 VA = 10,800 VA
Light-duty =

$\frac{150'\text{-}0''}{5'\text{-}0''} \times 180 \text{ VA} =$ 5400 VA

Total = 83,700 VA

Step 3. Special loads. *220-3(c)(1):*

Soft drink boxes =
4800 VA × 100% = 4800 VA

Copying machine =
12,000 VA × 100% = 12,000 VA

Range = 48,000 VA × 100% = 48,000 VA

Computers =
19,200 VA × 100% = 19,200 VA

Total = 84,000 VA

Answer. **Loads = 221,300 VA**

Step 4. Selecting kVA. $kVA = \frac{VA}{1000}$

$$kVA = \frac{221{,}300 \text{ VA}}{1000} = 221.3 \text{ kVA}$$

Step 5. Selecting transformer. *Chart:*
221.3 kVA requires a 225 kVA transformer

Answer. **Transformer = 225 kVA**

(4) What size THWN Cu conductors is required to supply the primary side of the transformer? The load is continuous duty.

Step 1. Finding primary amps. $A = \frac{kVA \times 1000}{V \times \sqrt{3}}$

$$A = \frac{225 \text{ kVA} \times 1000}{480 \text{ V} \times 1.732}$$

$$A = \frac{225{,}000 \text{ kVA}}{831 \text{ V}} = 270.8 \text{ A}$$

Step 2. *220-10(b):* 270.8 A × 125% = 338.5 A

Step 3. *Table 310-16:*
338.5 A requires #500 kcmil THWN Cu

Answer. **Conductors = #500 kcmil THWN Cu**

(5) What size OCPD is required for the primary side of the transformer?

Step 1. *450-3(b)(1):* 270.8 A × 125% = 338.5 A

Step 2. *240-3, Ex. 2; 450-3(b)(1), Ex. 1; 240-6:*
338.5 A requires a 350 A OCPD

Answer. **OCPD = 350 A**

Restaurant

The voltage of the kitchen equipment determines the voltage to be selected for the service. For a large number of pieces of kitchen equipment required to prepare food, elements of the electrical service can be selected using higher voltage. This allows smaller components to be used for the branch circuit supplying the equipment. Step-down transformers are used to lower the voltage for the operation of loads requiring a lower voltage. Supply voltage to a restaurant is usually a 120/208 V or 277/408 V, 3ϕ, 4-wire system. Demand factors can be applied for all equipment that is thermostatically controlled or used at intermittent duty. The percentage of the demand factor is based on the number of units in the kitchen area.

A 65'-0" × 85'-0" restaurant has a 120/208 V, 3ϕ, 4-wire service and the following loads. Find the load in VA and amps. Note that 50% or more of the lighting load is incandescent lighting. Use the form, STANDARD CALCULATION—COMMERCIAL AND INDUSTRIAL. See Figure 8-21.

20'-0" lighting track . . . . 120 V, 1ϕ
Eight 180 VA outside lighting
(continuous-duty) . . . . 120 V, 1ϕ
1200 VA sign lighting
(continuous-duty) . . . . 120 V, 1ϕ
38 receptacles (noncontinuous-duty) . . . 120 V, 1ϕ
18 receptacles (continuous-duty) . . . . 120 V, 1ϕ
20'-0" multioutlet assembly
(heavy-duty) . . . . 120 V, 1ϕ
Two 30 kW heating units . . . . 208 V, 3ϕ
Two 7560 VA A/C units . . . . 208 V, 3ϕ
Two 3 HP water pumps . . . . 208 V, 3ϕ
Six ¾ HP exhaust fans . . . . 208 V, 3ϕ

Kitchen equipment

3600 VA boilers . . . . 208 V, 3ϕ
Two 2880 VA deep fat fryers . . . . 208 V, 3ϕ
Two 1600 VA dishwashers . . . . 120 V, 1ϕ
Four 1200 VA toasters . . . . 120 V, 1ϕ
Two 840 VA mixers . . . . 120 V, 1ϕ
17 A walk-in cooler . . . . 208 V, 3ϕ
4320 VA ice cream box . . . . 208 V, 1ϕ
15 A freezer . . . . 208 V, 1ϕ
Two 1680 VA dishwashers . . . . 120 V, 1ϕ
1200 VA dishwasher booster heater . . . . 120 V, 1ϕ
8000 VA water heater . . . . 208 V, 3ϕ
12,000 VA cooktop . . . . 208 V, 1ϕ
Two 10,000 VA ovens . . . . 208 V, 1ϕ
8 A bun warmer . . . . 120 V, 1ϕ
12,000 VA range . . . . 208 V, 1ϕ
Two 1860 VA coffee makers . . . . 120 V, 1ϕ
1500 VA grinder . . . . 120 V, 1ϕ
15 A refrigerator . . . . 208 V, 1ϕ
14.5 A soft drink box . . . . 120 V, 1ϕ

(1) What is the demand load in VA and amps for the kitchen equipment?

Step 1. Kitchen equipment. *220-20:*
Boilers = 3600 VA
Deep fat fryers = 2880 VA × 2 = 5760 VA
Dishwasher = 1600 VA × 2 = 3200 VA
Toasters = 1200 VA × 4 = 4800 VA
Mixers = 840 VA × 2 = 1680 VA
Walk-in cooler = 17 A × 360 V = 6120 VA
Ice cream box = 4320 VA
Freezer = 15 A × 208 V = 3120 VA
Dishwasher = 1680 VA × 2 = 3360 VA

STANDARD CALCULATION—COMMERCIAL AND INDUSTRIAL

1. LIGHTING LOADS—220, Part B

General lighting loads—Table 220-3(b); 220-10(b)
5525 sq. ft. × 2 VA × 125 % = 13,812.5 VA

Show window loads—220-12
— ′ × 200 VA = — VA

Track lighting loads—410-102
20 ′ × 180 VA = 1800 VA
2′

Outside lighting loads—220-10(b)
180 VA × 8 units × 125 % = 1800 VA

Sign lighting loads—220-3, FPN; 600-6(c)
1200 VA × 100 % = 1200 VA

		PHASES	NEUTRAL
(total)	18,612.5 VA	18,612.5 VA	18,612.5 VA

2. RECEPTACLE LOADS—220-3(c)(5)

Noncontinuous duty—220-10(b); Table 220-13
38 × 180 VA = 6840 VA = 6840 VA

Applying demand factors—Table 220-13
First 10,000 VA × 100% = — VA = — VA
Remaining — VA × 50% = — VA
Continuous duty—220-10(b) # 18 × 180 VA × 125% = 4050 VA

Multioutlet assembly—220-3(c)
Heavy duty = 20 ′ × 180 VA = 3600 VA
Light duty = — ′ × 180 VA = — VA
5′

		PHASES	NEUTRAL
(total)	14,490 VA	14,490 VA	14,490 VA

3. SPECIAL LOADS—220-10(b)

Kit. equip. 100,200 VA × 65 % = 65,130 VA
— = — VA × — % = — VA
— = — VA × — % = — VA
— = — VA × — % = — VA
— = — VA × — % = — VA

		PHASES	NEUTRAL
(total)	65,130 VA	65,130 VA	14,404 VA

4. HEATING OR A/C LOAD—220-21 60,000 VA | 60,000 VA | — VA

5. MOTOR LOADS—430-22(a); 430-24; 430-25

				3ϕ only		
Exh. fans =	16.8 A ×	110 % ×	208 V	$\times \sqrt{3}$	=	6652.8 VA
Pump =	19.2 A ×	110 % ×	208 V	$\times \sqrt{3}$	=	7603.2 VA
— =	— A ×	— % ×	— V	$\times \sqrt{3}$	=	— VA
— =	— A ×	— % ×	— V	$\times \sqrt{3}$	=	— VA

		PHASES	NEUTRAL
(total)	14,256 VA	14,256 VA	— VA

6. LARGEST MOTOR LOAD—220-14

				3ϕ only			PHASES	NEUTRAL
ϕ	20.8 A ×	100 % ×	208 V	$\times \sqrt{3}$	× 25% =	1081.6 VA	1081.6 VA	— VA
N	14.5 A ×	100 % ×	120 V	$\times \sqrt{3}$	× 25% =	435 VA	— VA	435 VA

TOTAL = 173,570.1 VA | 47,941.5 VA

1ϕ service: $I = \frac{VA}{V}$

PHASES I = ______ VA / V = ______ A

NEUTRAL I = ______ VA / V = ______ A

3ϕ service: $I = \frac{VA}{V \times \sqrt{3}}$

I = 173,570.1 VA / 208 V × $\sqrt{3}$ = 482 A

I = 47,941.5 VA / 208 V × $\sqrt{3}$ = 133 A

If over 200 A—220-22; First 200 A × 100% = 200 A

Remaining ______ A × 70% = ______ A

______ A

Figure 8-21. The form, STANDARD CALCULATION—COMMERCIAL AND INDUSTRIAL is used to find VA and amps for restaurants.

Dishwasher booster heater =	1200 VA
Water heater =	8000 VA
Cooktop =	12,000 VA
Ovens = 10,000 VA × 2 =	20,000 VA
Bun warmer = 8 A × 120 V =	960 VA
Range =	12,000 VA
Coffee makers = 1860 VA × 2 =	3720 VA
Grinder =	1500 VA
Refrigerator = 15 A × 208 V =	3120 VA
Soft drink box = 14.5 A × 120 V =	1740 VA
Total =	100,200 VA

Step 2. Applying demand factors. *Table 220-20:* 100,200 VA × 65% = 65,130 VA

Answer. **Load = 65,130 VA**

Step 3. Finding amps. *Text:* $A = \dfrac{VA}{V \times \sqrt{3}}$

$$A = \frac{65{,}130 \text{ VA}}{208 \text{ V} \times 1.732} = 180.9 \text{ A}$$

Answer. **Load = 180.9 A**

(2) What size THWN Cu conductors paralleled three times per phase is required for the service-entrance conductors?

Step 1. *310-4:* Load $= \dfrac{482 \text{ A}}{3} = 160.7$ A

Step 2. *Table 310-16:* #2/0 THWN Cu = 175 A

Answer. **Conductors = #2/0 THWN Cu**

(3) What size THWN Cu conductors paralleled three times is required for the neutral conductors?

Step 1. *310-4:* Load $= \dfrac{133 \text{ A}}{3} = 44$ A

Step 2. *310-4; Table 310-16:*
#1/0 THWN Cu is the smallest size conductor allowed to be connected in parallel

Answer. **Conductors = #1/0 THWN Cu**

(4) What size main OCPD, based on the calculated load for the service panelboard, is required?

Step 1. *384-13:* Load = 482 A

Step 2. *240-6:* 482 A requires 500 A OCPD

Answer. **OCPD = 500 A**

Hospital

Hospitals are supplied by 120/240 V, 1ϕ, 3-wire; 120/208 V, 3ϕ, 4-wire; 277/480 V, 3ϕ, 4-wire; and up to 4160 V, 3ϕ, 4-wire services depending on their size. Transformers are installed for the higher voltages to step down voltages to 120/208 V, 3ϕ, 4-wire and 277/480 V, 3ϕ, 4-wire.

A 200′-0″ × 200′-0″ hospital has a 277/480 V, 3ϕ, 4-wire service and the following loads. Find the load in VA and amps. Note that the 277 V general lighting loads are noncontinuous duty. Use the form, STANDARD CALCULATION—COMMERCIAL AND INDUSTRIAL. See Figure 8-22.

14 linear feet show window .... 120 V, 1ϕ
30′-0″ lighting track .... 120 V, 1ϕ
Fourteen 225 VA outside lighting loads
(continuous-duty) .... 120 V, 1ϕ
Three 1200 VA sign lighting loads
(continuous-duty) .... 120 V, 1ϕ
175 receptacles (noncontinuous-duty) .. 120 V, 1ϕ
90 receptacles (continuous-duty) .... 120 V, 1ϕ
60′-0″ multioutlet assembly
(light-duty) .... 120 V, 1ϕ
35′-0″ multioutlet assembly
(heavy-duty) .... 120 V, 1ϕ
Five 35 kW heating units .... 480 V, 3ϕ
Five 19.5 kW A/C units .... 480 V, 3ϕ
Four 10 HP water pumps .... 480 V, 3ϕ
Six 10 A soft drink boxes .... 120 V, 1ϕ
Five 12.9 A copying machines
(special loads) .... 120 V, 1ϕ
Ten 3 HP exhaust fans .... 480 V, 3ϕ
One 10 HP sprinkler pump .... 480 V, 3ϕ

X-ray equipment

Short-time rating .... two each at ¾ second
Rating = 150 MA
Primary = 208 V, 1ϕ
Secondary = 100,000 V, 1ϕ
Long-time rating .... two each at 15 minutes
Rating = 15 MA
Primary = 208 V, 1ϕ
Secondary = 200,000 V, 1ϕ

Note: See 517-143 and 660-6.

Emergency system

45,000 VA life-support branch load (already computed at continuous-duty)
45,000 VA critical branch load (already computed at continuous-duty)

(1) What is the demand load in VA for the general lighting load?

Step 1. *Table 220-3(b):*
200′-0″ × 200′-0″ × 2 VA × 100% = 80,000 VA

Step 2. Applying demand factors.
Table 220-11:

First 50,000 VA × 40% =	20,000 VA
Next 30,000 VA × 20% =	6000 VA
Total =	26,000 VA

Answer. **Load = 26,000 VA**

(2) What is the load in amps for the branch circuit to the x-ray equipment with the short-time rating?

Step 1. Finding amps. $A = \dfrac{MA}{1000} \times \dfrac{\textit{secondary V}}{\textit{primary V}}$

$$A = \frac{150 \text{ MA}}{1000} \times \frac{100{,}000 \text{ V}}{208 \text{ V}} = 72 \text{ A}$$

Step 2. *660-6(a):* 72 A × 50% = 36 A

Answer. **Load = 36 A**

(3) What size THWN Cu conductors is required for the x-ray equipment with the short-time rating?

Step 1. *Table 310-16:*
36 A requires #8 THWN Cu conductors

Answer. **Conductors = #8 THWN Cu**

Note: The next higher standard size OCPD is 40 A with 75 °C terminals based on the load.

STANDARD CALCULATION—COMMERCIAL AND INDUSTRIAL

1. LIGHTING LOADS—220, Part B

General lighting loads—Table 220-3(b); 220-10(b)

— sq. ft. × 26,000 VA × 100 % = 26,000 VA

Show window loads—220-12

14 ′ × 200 VA = 2800 VA

Track lighting loads—410-102

30 / 2′ × 180 VA = 2700 VA

Outside lighting loads—220-10(b)

225 VA × 14 units × 125 % = 3937.5 VA

Sign lighting loads—220-3, FPN; 600-6(c)

3600 VA × 125 % = 4500 VA

(total) 39937.5 VA — PHASES 39,937.5 VA — NEUTRAL 26,000 VA

2. RECEPTACLE LOADS—220-3(c)(5)

Noncontinuous duty—220-10(b); Table 220-13

175 × 180 VA = 31,500 VA = — VA

Applying demand factors—Table 220-13

First 10,000 VA × 100% = 10,000 VA = 10,000 VA

Remaining 21,500 VA × 50% = 10,750 VA

Continuous duty—220-10(b) # 90 × 180 VA × 125% = 20,250 VA

Multioutlet assembly—220-3(c)

Heavy duty = 35 ′ × 180 VA = 6300 VA

Light duty = 60 / 5′ ′ × 180 VA = 2160 VA

(total) 49,460 VA — PHASES 49,460 VA — NEUTRAL — VA

3. SPECIAL LOADS—220-10(b)

Drink boxes = 7200 VA × 100 % = 7200 VA

Copy mch. = 7740 VA × 100 % = 7740 VA

X-ray = 16,140.8 VA × 100 % = 16,140.8 VA

Life sppt. = 45,000 VA × 100 % = 45,000 VA

Critical = 45,000 VA × 100 % = 45,000 VA

(total) 121,080.8 VA — PHASES 121,080.8 VA — NEUTRAL — VA

4. HEATING OR A/C LOAD—220-21 175,000 VA — PHASES 175,000 VA — NEUTRAL — VA

5. MOTOR LOADS—430-22(a); 430-24; 430-25

				3ϕ only		
Pumps =	56 A ×	100 % ×	480 V	× √3	=	46,536 VA
Exh. fans =	48 A ×	100 % ×	480 V	× √3	=	39,888 VA
Sprk. pump =	14 A ×	100 % ×	480 V	× √3	=	11,634 VA
— =	— A ×	— % ×	— V	× √3	=	— VA

(total) 98,058 VA — PHASES 98,058 VA — NEUTRAL — VA

6. LARGEST MOTOR LOAD—220-14

				3ϕ only			PHASES	NEUTRAL
ϕ	14 A ×	100 % ×	480 V	× √3	× 25% =	2908.5 VA	2908.5 VA	— VA
N	— A ×	— % ×	— V	× √3	× 25% =	— VA	— VA	— VA

TOTAL = 486,444.8 VA — 26,000 VA

1ϕ service: I = VA / V

PHASES I = ______ VA / V = ______ A

NEUTRAL I = ______ VA / V = ______ A

3ϕ service: I = VA / (V × √3)

I = 486,444.8 VA / (480 V × √3) = 585 A

I = 26,000 VA / (480 V × √3) = 31 A

If over 200 A—220-22; First 200 A × 100% = 200 A

Remaining ______ A × 70% = ______ A

______ A

Figure 8-22. The form, STANDARD CALCULATION—COMMERCIAL AND INDUSTRIAL is used to find VA and amps for hospitals.

(4) What is the load in amps for the branch circuit to the x-ray equipment with the long-time rating?

Step 1. Finding amps. $A = \frac{MA}{1000} \times \frac{secondary\ V}{primary\ V}$

$$A = \frac{15\ \text{MA}}{1000} \times \frac{200{,}000\ \text{V}}{208\ \text{V}} = 14\ \text{A}$$

Step 2. *660-6(a):* 14 A × 100% = 14 A

Answer. **Load = 14 A**

(5) What size THWN Cu conductors is required for the branch circuit to the x-ray equipment with the long-time rating?

Step 1. *Table 310-16:*
14 A requires #14 THWN Cu conductors

Answer. **Conductors = #14 THWN Cu**

Note: OCPD is 15 A with 75 °C terminals.

(6) What is the total VA and amp ratings for the feeder and service conductors for the x-ray equipment?

Step 1. Finding amps. Short-time rating.

$$A = \frac{MA}{1000} \times \frac{secondary\ V}{primary\ V}$$

$$A = \frac{150\ \text{MA}}{1000} \times \frac{100{,}000\ \text{V}}{208\ \text{V}} = 72\ \text{A}$$

Step 2. Finding VA. *660-6(b):*
Short-time rating =
72 A × 2 × 50% × 208 V × 100% = 14,976 VA
Long-time rating =
14 A × 2 × 208 V = 5824 VA
5824 VA × 20% = 1164.8 VA
Total VA =
14,976 VA + 5824 VA = 16,140.8 VA

Answer. **VA rating = 16,140.8 VA**

Step 3. Finding amps. $A = \frac{VA}{V}$

$$A = \frac{16{,}140.8\ \text{VA}}{208\ \text{V}} = 77.6\ \text{A}$$

Answer. **A rating = 77.6 A**

(7) What size THWN Cu conductors is required to supply power for the feeder to the x-ray equipment?

Step 1. *Table 310-16:*
77.6 A requires #4 THWN Cu conductors

Answer. **Conductors = #4 THWN Cu**

Note: The next higher standard size OCPD is 80 A with 75 °C terminals based on the load.

Welding Shop

The load in VA or amps for a service supplying power to a welding shop is usually less than other service loads due to duty cycles of the welder used. Welders pull full loads momentarily and the service never carries the full-load current of the welders. Elements of the service equipment are not required to be sized as large for welding shops as other similar sized occupancies.

A welding shop has a 120/240 V, 3ϕ, 4-wire service and the following loads. Find the load in VA and amps. Note that 50% or more of the neutral load is harmonically balanced. Use the form, STANDARD CALCULATION—COMMERCIAL AND INDUSTRIAL. See Figure 8-23.

9000 VA inside lighting loads (continuous-duty) .... 120 V, 1ϕ
Ten 180 VA outside lighting loads (continuous-duty) .... 120 V, 1ϕ
1900 VA sign lighting loads (continuous-duty) .... 120 V, 1ϕ
65 receptacles (noncontinuous-duty) ... 120 V, 1ϕ
30 receptacles (continuous-duty) .... 120 V, 1ϕ
Three 10 kW heating units .... 240 V, 3ϕ
4800 VA A/C units .... 240 V, 3ϕ
1200 VA soft drink box .... 120 V, 1ϕ
1080 VA copying machine .... 120 V, 1ϕ
Two 1.5 HP drill presses .... 240 V, 3ϕ
Two 7.5 HP air compressors .... 240 V, 3ϕ
Four 1 HP grinders .... 240 V, 3ϕ
Two 3 HP lathes .... 240 V, 3ϕ

Welders—Resistance (with 50% duty cycle)

One 14 kW .... 240 V, 1ϕ
One 12 kW .... 240 V, 1ϕ
One 10 kW .... 240 V, 1ϕ
One 7.155 kW .... 240 V, 1ϕ

Motors—Generator arc (with 100% duty cycle)

One 15 kW .... 240 V, 1ϕ
One 13 kW .... 240 V, 1ϕ
One 10 kW .... 240 V, 1ϕ
One 9 kW .... 240 V, 1ϕ

AC transformer and DC rectifier arc (with 70% duty cycle)

One 16 kW .... 240 V, 1ϕ
One 14 kW .... 240 V, 1ϕ
One 12 kW .... 240 V, 1ϕ
One 10 kW .... 240 V, 1ϕ

(1) What is the 1ϕ load in VA and amps for phases A and C?

Step 1. Finding VA.

Lighting loads =	15,875 VA
Receptacle loads =	17,600 VA
Special loads =	107,088 VA
Total =	140,563 VA

Answer. **Load = 140,563 VA**

Step 2. Finding amps. $A = \frac{VA}{V}$

$$A = \frac{140{,}563\ \text{VA}}{240\ \text{V}} = 585.7\ \text{A}$$

Answer. **Load = 585.7 A**

(2) What is the 3ϕ load in VA and amps for phase B?

Step 1. Finding VA.

Heating loads =	30,000 VA
Motor loads =	36,608 VA
Largest motor load =	2288 VA
Total =	68,896 VA

Answer. **Load = 68,896 VA**

STANDARD CALCULATION—COMMERCIAL AND INDUSTRIAL

1. LIGHTING LOADS—220, Part B

General lighting loads—Table 220-3(b); 220-10(b)

— sq. ft. × 9000 VA × 125 % = 11,250 VA

Show window loads—220-12

—′ × 200 VA = — VA

Track lighting loads—410-102

—′ (2′) × 180 VA = — VA

Outside lighting loads—220-10(b)

180 VA × 10 units × 125 % = 2250 VA

Sign lighting loads—220-3, FPN; 600-6(c)

1900 VA × 125 % = 2375 VA

		PHASES	NEUTRAL
(total)	15,875 VA	15,875 VA	15,875 VA

2. RECEPTACLE LOADS—220-3(c)(5)

Noncontinuous duty—220-10(b); Table 220-13

65 × 180 VA = 11,700 VA = — VA

Applying demand factors—Table 220-13

First 10,000 VA × 100% = 10,000 VA = 10,000 VA

Remaining 1700 VA × 50% = 850 VA

Continuous duty—220-10(b) # 30 × 180 VA × 125% = 6750 VA

Multioutlet assembly—220-3(c)

Heavy duty = —′ × 180 VA = — VA

Light duty = —′ (5′) × 180 VA = — VA

		PHASES	NEUTRAL
(total)	17,600 VA	17,600 VA	17,600 VA

3. SPECIAL LOADS—220-10(b)

Drink box = 1200 VA × 100 % = 1200 VA

Copy mch. = 1080 VA × 100 % = 1080 VA

Res. wldrs. = 22,360 VA × 100 % = 22,360 VA

Arc wldrs. = 42,800 VA × 100 % = 42,800 VA

AC/DC wldrs. = 39,648 VA × 100 % = 39,648 VA

		PHASES	NEUTRAL
(total)	107,088 VA	107,088 VA	2280 VA

4. HEATING OR A/C LOAD—220-21 30,000 VA | PHASES 30,000 VA | NEUTRAL — VA

5. MOTOR LOADS—430-22(a); 430-24; 430-25

				3ϕ only		
Drill	= 10.4 A ×	100 % ×	240 V	× $\sqrt{3}$	=	4326.4 VA
Air comp.	= 44 A ×	100 % ×	240 V	× $\sqrt{3}$	=	18,304 VA
Grinders	= 14.4 A ×	100 % ×	240 V	× $\sqrt{3}$	=	5990.4 VA
Lathes	= 19.2 A ×	100 % ×	240 V	× $\sqrt{3}$	=	7987.2 VA

		PHASES	NEUTRAL
(total)	36,608 VA	36,608 VA	— VA

6. LARGEST MOTOR LOAD—220-14

				3ϕ only			PHASES	NEUTRAL
ϕ	22 A ×	100 % ×	240 V	× $\sqrt{3}$	× 25% =	2288 VA	2288 VA	— VA
N	10 A ×	100 % ×	120 V	× $\sqrt{3}$	× 25% =	300 VA	— VA	300 VA

TOTAL = 209,459 VA | 36,055 VA

1ϕ service: $I = \frac{VA}{V}$

PHASES I = 140,563 VA / 240 V = 585.7 A

NEUTRAL I = 36,055 VA / 240 V = 150 A

3ϕ service: $I = \frac{VA}{V \times \sqrt{3}}$

I = 68,896 VA / 240 V × $\sqrt{3}$ = 165.6 A

I = — VA / — V × $\sqrt{3}$ = — A

If over 200 A—220-22; First 200 A × 100% = 200 A

Remaining ______ A × 70% = ______ A

______ A

Figure 8-23. The form, STANDARD CALCULATION—COMMERCIAL AND INDUSTRIAL is used to find VA and amps for welding shops.

Step 2. Finding amps. $A = \dfrac{VA}{V \times \sqrt{3}}$

$$A = \frac{68{,}896 \text{ VA}}{240 \text{ V} \times 1.732} = 165.6 \text{ A}$$

Answer. **Load = 165.6 A**

(3) What is the load in VA and amps for the neutral?

Step 1. Finding VA.

Lighting loads =	15,875 VA
Receptacle loads =	17,600 VA
Special loads =	2280 VA
Motor loads =	0 VA
Heating or A/C load =	0 VA
Largest motor load (soft drink box) =	300 VA
Total =	36,055 VA

Answer. **Load = 36,055 VA**

Step 2. Finding amps. $A = \dfrac{VA}{V}$

$$A = \frac{36{,}055 \text{ VA}}{240 \text{ V}} = 150 \text{ A}$$

Answer. **Load = 150 A**

(4) What is the load in VA and amps for the resistance welders?

Step 1. Finding VA. *630-31(a)(b):*

First largest welder at 71% =	
14,000 × 71% =	9940 VA
All others at 71% × 60% =	
12,000 VA × 71% × 60% =	5112 VA
10,000 VA × 71% × 60% =	4260 VA
7155 VA × 71% × 60% =	3048 VA
Total =	22,360 VA

Answer. **Load = 22,360 VA**

Step 2. Finding amps. $A = \dfrac{VA}{V}$

$$A = \frac{22{,}360 \text{ VA}}{240 \text{ V}} = 93 \text{ A}$$

Answer. **Load = 93 A**

(5) What is the load in VA and amps for the group of motor-generator arc welders?

Step 1. Finding VA. *630-21(a)(b):*

First two largest welders at 100% =	
15,000 VA × 100% =	15,000 VA
13,000 VA × 100% =	13,000 VA
Third largest welder at 100% × 85% =	
10,000 VA × 100% × 85% =	8500 VA
Fourth largest welder at 100% × 70% =	
9000 VA × 100% × 70% =	6300 VA
Total =	42,800 VA

Answer. **Load = 42,800 VA**

Step 2. Finding amps. $A = \dfrac{VA}{V}$

$$A = \frac{42{,}800 \text{ VA}}{240 \text{ V}} = 178 \text{ A}$$

Answer. **Load = 178 A**

(6) What is the load in VA and amps for the group of AC transformer and DC rectifier arc welders?

Step 1. Finding VA. *630-11(a)(b):*

First two largest welders at 84% × 100% =	
16,000 VA × 84% × 100% =	13,440 VA
14,000 VA × 84% × 100% =	11,760 VA
Third largest welder at 84% × 85% =	
12,000 VA × 84% × 85% =	8568 VA
Fourth largest welder at 84% × 70% =	
10,000 VA × 84% × 70% =	5880 VA
Total =	39,648 VA

Answer. **Load = 39,648 VA**

Step 2. Finding amps. $A = \dfrac{VA}{V}$

$$A = \frac{39{,}648 \text{ VA}}{240 \text{ V}} = 165 \text{ A}$$

Answer. **Load = 165 A**

(7) What size THHN Cu conductors, paralleled four times per phase, is required for phases A and C? (All service conductors are in the same conduit and connected to 75 °C terminals.)

Step 1. *310-10:*

$$\textit{Size load per conductor} = \frac{751.3 \text{ A}}{4} = 187.8 \text{ A}$$

Step 2. *Notes 8 and 10(c) to Tables 310-16–310-19; Table 310-16:*
#250 kcmil THHN Cu = 290 A × 80% = 232 A

Step 3. 232 A supplies 188 A at 75 °C rating

Answer. **Conductors = *#250* kcmil THHN Cu**

(8) What size THHN Cu conductor is required for phase B? The conductor is connected to a 75 °C terminal.

Step 1. *Notes 8 and 10(c) to Tables 310-16–310-19; Table 310-16:*
#4/0 THHN Cu = 260 A × 80% = 208 A

Step 2. *310-4; 300-5(i):*
208 A supplies 165.6 A at 75 °C rating

Answer. **Conductors = *#4/0* THHN Cu**

(9) What size THHN Cu conductor is required for the neutral? The conductor is connected to a 75 °C terminal.

Step 1. *Notes 8 and 10(c) to Tables 310-16–310-19; Table 310-16:*
#3/0 THHN Cu = 225 A × 80% = 180 A

Step 2. *310-4; 300-5(i):*
180 A supplies 150 A at 75 °C rating

Answer. **Conductor = *#3/0* THHN Cu**

(10) What size conduit, using EMT, is required for phases A, B, and C and the neutral?

Step 1. *Table 5, Ch. 9:*
#250 kcmil THHN =
.4026 sq. in. × 4 = 1.6104 sq. in.
#4/0 THHN =
.3278 sq. in. × 1 = .3278 sq. in.
#2/0 THHN =
.2265 sq. in. × 1 = .2265 sq. in.
Total = 2.1647 sq. in.

Step 2. *Table 4, Ch. 9:* 2.1647 requires 3″ EMT

Answer. **Conduit = 3″ EMT**

School

The load in VA and amps for a school can be computed by two methods. The first method is the standard calculation per 220-10(b)–220-22. The second method is the optional calculation per 220-20 and Table 220-20.

Standard Calculation. The load in VA or amps for a school is found by dividing the loads into six individual loads. Each load is computed by its condition of use and loads are computed at continuous duty or noncontinuous duty. Certain loads can have demand factors applied based on their operation.

What is the load in VA and amps for a 134′-0″ × 150′-0″ school building supplied by a 277/480 V, 3ϕ, 4-wire service? Note that 50% or more of the lighting load is electric discharge lighting. Use the form, STANDARD CALCULATION—COMMERCIAL AND INDUSTRIAL. See Figure 8-24.

General lighting load . . . . 277 V, 1ϕ
50 linear feet show window . . . . 120 V, 1ϕ
200 VA lighting track . . . . 120 V, 1ϕ
Fifty 225 VA outside lighting loads
(continuous-duty) . . . . 277 V, 1ϕ
1200 VA sign lighting loads
(continuous-duty) . . . . 120 V, 1ϕ
150 receptacles (noncontinuous-duty) . . 120 V, 1ϕ
65 receptacles (continuous-duty) . . . . 120 V, 1ϕ
80′-0″ multioutlet assembly
(heavy-duty) . . . . 120 V, 1ϕ
Sixteen 1½HP exhaust fans . . . . 480 V, 3ϕ
Twenty 10 kW heating units . . . . 480 V, 3ϕ
Twenty 6.48 kVA A/C units . . . . 480 V, 3ϕ
40 HP pump motor . . . . 480 V, 3ϕ
Eight 1500 VA soft drink boxes
(noncontinuous-duty) . . . . 120 V, 1ϕ
Five 1020 VA copying machines . . . . 120 V, 1ϕ

Kitchen equipment
(all units equipped with thermostatic controls)

4644 VA boilers . . . . 208 V, 1ϕ
Three 3780 VA deep fat fryers . . . . 208 V, 1ϕ
Three 2880 VA dishwashers . . . . 120 V, 1ϕ
Six 1200 VA toasters . . . . 120 V, 1ϕ
Two 1440 VA mixers . . . . 120 V, 1ϕ
Two 1560 VA ice cream boxes . . . . 120 V, 1ϕ
Two 1740 VA freezers . . . . 120 V, 1ϕ
Two 7920 VA dishwasher
booster heaters . . . . 208 V, 1ϕ
18 kW cooktops . . . . 480 V, 3ϕ
24 kW oven . . . . 480 V, 3ϕ
Three 1200 VA bun warmers . . . . 120 V, 1ϕ
22 kW range . . . . 208 V, 3ϕ
Two 1920 VA coffee makers . . . . 120 V, 1ϕ
Two 3420 VA grinders . . . . 208 V, 1ϕ
Four 1740 VA refrigerators . . . . 120 V, 1ϕ
20 kW water heater . . . . 480 V, 3ϕ

(1) What is the kitchen equipment load in VA and amps for phases A, B, and C?

Step 1. Finding total load. *220-20:*
4644 VA boilers . . . . . 208 V, 1ϕ = 4644 VA
Three 3780 VA deep fat
fryers . . . . 208 V, 1ϕ = 11,340 VA
Three 2880 VA
dishwashers . . . . 120 V, 1ϕ = 8640 VA
Six 1200 VA
toasters . . . . 120 V, 1ϕ = 7200 VA
Two 1440 VA
mixers . . . . 120 V, 1ϕ = 2880 VA
Two 1560 VA
ice cream boxes . . . 120 V, 1ϕ = 3120 VA
Two 1740 VA
freezers . . . . 120 V, 1ϕ = 3480 VA
Two 7920 VA dishwasher booster
heaters . . . . 208 V, 1ϕ = 15,840 VA
18 kW cooktops . . . 480 V, 3ϕ = 18,000 VA
24 kW oven . . . . 480 V, 3ϕ = 24,000 VA
Three 1200 VA
bun warmers . . . . . . 120 V, 1ϕ = 3600 VA
22 kW range . . . . . . 208 V, 3ϕ = 22,000 VA
Two 1920 VA coffee
makers . . . . 120 V, 1ϕ = 3840 VA
Two 3420 VA
grinders . . . . 208 V, 1ϕ = 6840 VA
Four 1740 VA
refrigerators . . . . 120 V, 1ϕ = 6960 VA
20 kW water
heater . . . . 480 V, 3ϕ = 20,000 VA
Total = 162,384 VA

Step 2. Selecting demand factors. *Table 220-20:*
Over six units = 65%

Step 3. Applying demand factors.
162,384 VA × 65% = 105,549.6 VA

Answer. **Load = 105,549.6 VA**

Step 4. Finding amps. *220-22:* $I = \frac{VA}{V \times \sqrt{3}}$

$$I = \frac{105{,}549.6\text{ VA}}{480\text{ V} \times 1.732} = 127\text{ A}$$

Answer. **Load = 127 A**

STANDARD CALCULATION—COMMERCIAL AND INDUSTRIAL

1. LIGHTING LOADS—220, Part B

General lighting loads—Table 220-3(b); 220-10(b)

20,100 sq. ft. × 3 VA × 125 % = 75,375 VA

Show window loads—220-12

50 ′ × 200 VA = 10,000 VA

Track lighting loads—410-102

200 ′ / 2′ × 180 VA = 18,000 VA

Outside lighting loads—220-10(b)

225 VA × 50 units × 125 % = 14,062.5 VA

Sign lighting loads—220-3, FPN; 600-6(c)

1200 VA × 100 % = 1200 VA

		PHASES	NEUTRAL
(total)	118,637.5 VA	118,637.5 VA	89,437.5 VA

2. RECEPTACLE LOADS—220-3(c)(5)

Noncontinuous duty—220-10(b); Table 220-13

150 × 180 VA = 27,000 VA = — VA

Applying demand factors—Table 220-13

First 10,000 VA × 100% = 10,000 VA = 10,000 VA

Remaining 17,000 VA × 50% = 8500 VA

Continuous duty—220-10(b) # 65 × 180 VA × 125% = 14,625 VA

Multioutlet assembly—220-3(c)

Heavy duty = 80 ′ × 180 VA = 14,400 VA

Light duty = — ′ / 5′ × 180 VA = — VA

(total) 47,525 VA | 47,525 VA | — VA

3. SPECIAL LOADS—220-10(b)

Drink boxes = 12,000 VA × 100 % = 12,000 VA

Copy mch. = 5100 VA × 100 % = 5100 VA

Kit. equip. = 162,384 VA × 65 % = 105,549.6 VA

— = — VA × — % = — VA

— = — VA × — % = — VA

(total) 122,649.6 VA | 122,649.6 VA | — VA

4. HEATING OR A/C LOAD—220-21 200,000 VA | 200,000 VA | — VA

5. MOTOR LOADS—430-22(a); 430-24; 430-25

				3ϕ only	
Exh. fans =	41.6 A ×	100 % ×	480 V	× √3	= 34,569.6 VA
Pumps =	52 A ×	100 % ×	480 V	× √3	= 43,212 VA
— =	— A ×	— % ×	— V	× √3	= — VA
— =	— A ×	— % ×	— V	× √3	= — VA

(total) 77,781.6 VA | 77,781.6 VA | — VA

6. LARGEST MOTOR LOAD—220-14

				3ϕ only				
ϕ	52 A ×	100 % ×	480 V	× √3	× 25% =	10,803 VA	10,803 VA	— VA
N	— A ×	— % ×	— V	× √3	× 25% =	— VA	— VA	— VA

TOTAL = 577,396.7 VA | 89,437.5 VA

1ϕ service: I = VA / V

PHASES I = ____ VA / V = ____ A

NEUTRAL I = ____ VA / V = ____ A

3ϕ service: I = VA / (V × √3)

I = 577,396.7 VA / (480 V × √3) = 694.8 A

I = 89,437.5 VA / (480 V × √3) = 107.6 A

If over 200 A—220-22; First 200 A × 100% = 200 A

Remaining ____ A × 70% = ____ A

____ A

Figure 8-24. The form, STANDARD CALCULATION—COMMERCIAL AND INDUSTRIAL is used to find VA and amps for schools.

(2) What is the load in VA and amps for the neutral?

Step 1. Finding total load. *220-20:*

Three 2880 VA dishwashers....... 120 V, 1ϕ = 8640 VA
Six 1200 VA toasters.......... 120 V, 1ϕ = 7200 VA
Two 1440 VA mixers........... 120 V, 1ϕ = 2880 VA
Two 1560 VA ice cream boxes............ 120 V, 1ϕ = 3120 VA
Two 1740 VA freezers.......... 120 V, 1ϕ = 3480 VA
Three 1200 VA bun warmers.......... 120 V, 1ϕ = 3600 VA
Two 1920 VA coffee makers........... 120 V, 1ϕ = 3840 VA
Four 1740 VA refrigerators...... 120 V, 1ϕ = 6960 VA
Total = 39,720 VA

Step 2. Selecting demand factors. *Table 220-20:*
Over six units = 65%

Step 3. Applying demand factors.
39,720 VA × 65% = 25,818 VA

Answer. **Load = 25,818 VA**

Step 4. Finding amps. *220-22:* $I = \dfrac{VA}{V \times \sqrt{3}}$

$$I = \frac{25{,}818 \text{ VA}}{208 \text{ V} \times 1.732} = 71.7 \text{ A}$$

Answer. **Load = 71.7 A**

Optional Calculation. The load in VA and amps can be calculated by the optional calculation instead of the standard calculation. The optional calculation involves dividing the total connected load in VA by the total square footage to derive the VA per sq. ft. The square footage is multiplied by the demand factors in Table 220-34, which is based on the VA per sq. ft. These totals are added to determine the total VA and amp loads.

What is the load in VA and amps for a school with a 25,000 sq. ft. classroom area, 6000 sq. ft. auditorium area, 3000 sq. ft. cafeteria area, and 950 sq. ft. hall area supplied by a 120/208 V, 3ϕ, 4-wire service? Use the form, OPTIONAL CALCULATION—SCHOOL. See Figure 8-25.

Lighting loads

62,550.08 VA inside lighting loads..... 120 V, 1ϕ
12,000 VA outside lighting loads...... 120 V, 1ϕ
6000 VA stage lighting loads.......... 120 V, 1ϕ

Receptacle loads

33,300 VA receptacles (general-purpose).................. 120 V, 1ϕ
13,500 VA receptacles (continuous-duty)................. 120 V, 1ϕ

Special loads

62,000 VA special loads.............. 208 V, 3ϕ
50 kW heating units.................. 208 V, 3ϕ
Six 5 HP A/C units.................. 208 V, 3ϕ

Motor loads

Three 7.5 HP motors............... 208 V, 3ϕ
Three 15 HP motors................. 208 V, 3ϕ
Four ¾ HP hood fans............... 208 V, 1ϕ
Two ½ HP grill vent fans............ 208 V, 1ϕ

Kitchen equipment (cafeteria)

Three 16 kW ranges................. 208 V, 1ϕ
Two 10 kW ovens................... 208 V, 1ϕ
Four 3.5 kW fryers.................. 208 V, 1ϕ
Two 2.5 kW heaters................. 208 V, 1ϕ
5 kW booster heater................. 208 V, 1ϕ
Two 1.5 kW toasters................ 120 V, 1ϕ
10 kW water heater................. 208 V, 1ϕ
Three 1.9 kW refrigerators........... 120 V, 1ϕ
Two 1.86 kW freezers............... 120 V, 1ϕ

Hotels and Motels

Hotels and motels differ from dwelling units in that the guest rooms are not provided with permanent cooking facilities. Dwelling units shall be equipped with space for eating, living, sleeping, sanitation, and provisions for permanent cooking.

What is the VA and amps for a hotel or motel with 60 units of 300 sq. ft. each? The supply service is 120/208 V, 3ϕ, 4-wire. Note that 50% or more of the neutral load is harmonically balanced. Use the form, STANDARD CALCULATION—HOTELS AND MOTELS. See Figure 8-26.

15 linear feet show window........... 120 V, 1ϕ
12′-0″ lighting track................. 120 V, 1ϕ
Fourteen 180 VA outside lighting loads (continuous-duty)................. 120 V, 1ϕ
Three 1200 VA sign lighting loads (continuous-duty)................. 120 V, 1ϕ
50 receptacles (continuous-duty)....... 120 V, 1ϕ
Two 10 kW heating units (offices and lobby)................ 208 V, 3ϕ
Sixty 5 HP A/C and heat pump units... 208 V, 1ϕ
Six 1 HP soft drink boxes............ 120 V, 1ϕ
Three 1020 VA copying machines..... 120 V, 1ϕ
Six 8 kW water heaters.............. 208 V, 3ϕ
25,200 VA house loads (continuous-duty)................. 208 V, 3ϕ
6000 VA house loads (continuous-duty)................. 120 V, 1ϕ
24,960 VA laundry facilities load (continuous-duty)................. 208 V, 1ϕ
Sixty ⅙ HP exhaust fans (noncontinuous-duty).............. 120 V, 1ϕ

(1) What size OCPD is required for the service supplied by copper busduct?

Step 1. Selecting busduct. *Chart:*
1693.9 A requires 2000 A busduct

Step 2. *240-3; 240-6:*
2000 A busduct requires 2000 A OCPD

Answer. **OCPD = 2000 A**

(2) What size switchgear is required for the service equipment?

OPTIONAL CALCULATION—SCHOOL

1. TOTAL VA—Table 220-34

Total connected load = 438,359.22 VA

2. TOTAL SQUARE FOOTAGE—Table 220-34

Square footage = 34,950 sq. ft.

3. VA PER SQ. FT.—Table 220-34

438,359.22 VA / 34,950 sq. ft. = 12.54246695 VA/sq. ft.

4. APPLYING DEMAND FACTORS—Table 220-34

First 3 VA per sq. ft. × 100%

Next 3 VA to 20 VA per sq. ft. × 75%

Remaining 21 VA per sq. ft × 25%

	PHASES
3 VA/sq. ft. × 34,950 sq. ft. × 100% =	104,850 VA
9.54246695 VA/sq. ft. × 34,950 sq. ft. × 75% =	250,131.9 VA
— VA/sq. ft. × — sq. ft. × 25% =	— VA
TOTAL =	354,981.9 VA

NEUTRAL

(Loads from Standard Calculation)

1. Lighting loads = — VA
2. Receptacle loads = — VA
3. Special loads = — VA
4. Heating or A/C load = — VA
5. Motor loads = — VA
6. Largest motor load = — VA

TOTAL = — VA

— A

1ϕ service: I = VA / V

PHASES I = ______ VA / V = ______ A

NEUTRAL I = ______ VA / V = ______ A

3ϕ service: I = VA / (V × √3)

I = 354,981.9 VA / 208 V × √3 = 986 A

I = ______ VA / V × √3 = ______ A

If over 200 A—220-22; First 200 A × 100% = 200 A

Remaining ______ A × 70% = ______ A

______ A

Figure 8-25. The form, OPTIONAL CALCULATION—SCHOOL is used to find VA and amps for schools.

STANDARD CALCULATION—HOTELS AND MOTELS

1. GENERAL LIGHTING LOADS—Table 220-3(b)

300 sq. ft. × 2 VA × 60 units = 36,000 VA

— sq. ft. × 2 VA × — units = — VA

— sq. ft. × 2 VA × — units = — VA

36,000 VA

Applying demand factors—Table 220-11

First 20,000 VA × 50% = 10,000 VA

Next 16,000 VA × 40% = 6400 VA

Remaining — VA × 30% = — VA

(total) 16,400 VA — 16,400 VA

Show window loads—220-12

15′ × 200 VA = 3000 VA

Track lighting loads—410-102

12′ × 180 VA = 1080 VA

2′

Outside lighting loads—220-10(b)

180 VA × 14 units × 125 % = 3150 VA

Sign lighting loads—220-3, FPN; 600-6(c)

3600 VA × 125 % = 4500 VA

(total) 28,130 VA | PHASES 28,130 VA | NEUTRAL 28,130 VA

2. RECEPTACLE LOADS—220-3(c)(5)

Continuous duty—220-10(b); Table 220-13

9000 VA × 125% = (total) 11,250 VA | PHASES 11,250 VA | NEUTRAL 11,250 VA

3. SPECIAL LOADS—220-10(b)

Load	VA	Units	%	VA
Drink boxes	1920	6	100	11,520
Copy mch.	1020	3	100	3060
Water htr.	8000	6	100	48,000
House	25,200	—	125	31,500
House	6000	—	125	7500
Laundry	24,960	—	125	31,200
Exh. fans	31,680	—	100	31,680

(total) 164,460 VA | PHASES 164,460 VA | NEUTRAL 53,760 VA

4. HEATING OR A/C LOAD—220-21

Heating unit = 10,000 VA × 100% × 2 units = 20,000 VA

A/C unit = — VA × 100% × — units = — VA

Heat pump = 6406.4 VA × 100% × 60 units = 384,384 VA

(largest load) 404,384 VA | PHASES 404,384 VA | NEUTRAL — VA

5. LARGEST MOTOR LOAD—220-14

ϕ 28 A × 110 % × 208 V × 25% = 1601.6 VA | PHASES 1601.6 VA | NEUTRAL — VA

N 16 A × 100 % × 120 V × 25% = 480 VA | PHASES — VA | NEUTRAL 480 VA

TOTAL | PHASES 609,825.6 VA | NEUTRAL 93,620 VA

1ϕ service: $I = \frac{VA}{V}$

PHASES $I = \frac{\_\_\_ VA}{V}$ = ___ A

NEUTRAL $I = \frac{\_\_\_ VA}{V}$ = ___ A

3ϕ service: $I = \frac{VA}{V \times \sqrt{3}}$

$I = \frac{609,825.6 \text{ VA}}{208 \text{ V} \times \sqrt{3}} = 1693.9 \text{ A}$

$I = \frac{93,620 \text{ VA}}{208 \text{ V} \times \sqrt{3}} = 260 \text{ A}$

If over 200 A—220-22; First 200 A × 100% = 200 A

Remaining ___ A × 70% = ___ A

___ A

Figure 8-26. The form, STANDARD CALCULATION—HOTELS AND MOTELS is used to find VA and amps for hotels and motels.

Step 1. Selecting switchgear. *Chart:*
1693.9 A requires 2000 A switchgear

Answer. **Switchgear = 2000 A**

(3) What size copper grounding electrode conductor is required?

Step 1. Finding amps of busduct busbars.
$4'' \times \frac{1}{2}'' \times 1000$ A = 2000 A busbars

Step 2. Finding sq. mils of busbars.
Width = 4″ × 1000 = 4000 mils
Thickness = ½″ × 1000 = 500 mils
Area = *width* × *thickness* (in mils)
Area = 4000 mils × 500 mils = 2,000,000 sq. mils

Step 3. Finding CM of busbars. $CM = \frac{sq.\ mils}{.7854}$

$$CM = \frac{2{,}000{,}000 \text{ sq. mils}}{.7854} = 2{,}546{,}473 \text{ CM}$$

Step 4. Finding kcmil. $kcmil = \frac{CM}{1000}$

$$kcmil = \frac{2{,}546{,}473 \text{ CM}}{1000} = \#2547 \text{ kcmil}$$

Step 5. *Table 250-94:* #2547 kcmil requires #3/0 Cu

Answer. **Conductor = #3/0 Cu**

Bank

The load in VA and amps for a bank is computed using the same method used for office buildings. VA per sq. ft. for the general lighting load is 3.5 VA per Table 220-3(b). If the number of general-purpose receptacles is not known, an extra 1 VA per sq. ft. is computed for the receptacle load.

What is the load in VA and amps for a 4200 sq. ft. bank supplied by a 120/208 V, 3ϕ, 4-wire service? Use the form, STANDARD CALCULATION—COMMERCIAL AND INDUSTRIAL. See Figure 8-27.

30 linear feet show window . . . . 120 V, 1ϕ
40′-0″ lighting track . . . . 120 V, 1ϕ
Fifteen 225 VA outside lighting loads . . 120 V, 1ϕ
1200 VA sign lighting loads . . . . 120 V, 1ϕ
69 receptacles (noncontinuous-duty) . . . 120 V, 1ϕ
24 receptacles (continuous-duty) . . . . 120 V, 1ϕ
10′-0″ multioutlet assembly (heavy-duty) . . . . 120 V, 1ϕ
Twelve ⅙ HP fans . . . . 120 V, 1ϕ
8 kW water heater . . . . 208 V, 3ϕ
35 kW heating unit . . . . 208 V, 3ϕ
7.5 HP A/C . . . . 208 V, 3ϕ
Two 5 HP pump motors . . . . 208 V, 3ϕ
Six 2 HP exhaust fans . . . . 208 V, 3ϕ
4320 VA miscellaneous loads (continuous-duty) . . . . 120 V, 1ϕ
24,300 VA miscellaneous loads (continuous-duty) . . . . 208 V, 1ϕ
Twelve 1200 VA personal computers (continuous-duty) . . . . 120 V, 1ϕ
14 kW range (noncontinuous-duty, no neutral load) . . . . 208 V, 1ϕ

Warehouse and Office

Warehouses are used to store materials or goods. Less than normal VA per sq. ft. is applied for this type of occupancy. Office areas require a greater VA per sq. ft. to provide better lighting for workers. Some areas in an office require more lighting fixtures for desktops where typewriters, word processors, and computers are used.

What is the general lighting load for a 200,000 sq. ft. warehouse and a 50′-0″ × 40′-0″ office? Warehouse lighting is noncontinuous-duty and office lighting is continuous-duty. Lighting loads are supplied by a 120/208 V, 3ϕ feeder circuit.

Office area

Step 1. Lighting loads. *220, Part B:*
General lighting loads. *220-3(b); 220-10(b):*
2000 sq. ft. × 3.5 VA × 125% = 8750 VA

Warehouse

Step 2. Lighting loads. *220, Part B:*
General lighting loads. *220-3(b); 220-10(b):*
200,000 sq. ft. × .25 VA = 50,000 VA

Step 3. Applying demand factors. *Table 220-11:*
First 12,500 VA × 100% = 12,500 VA
Next 37,500 VA × 50% = 18,750 VA
Total = 31,250 VA

Step 3. Total load.
Office area = 8750 VA
Warehouse = 31,250 VA
Total = 40,000 VA

Step 4. $I = \frac{VA}{V \times \sqrt{3}}$

$$I = \frac{40{,}000 \text{ VA}}{208 \text{ V} \times 1.732} = 111 \text{ A}$$

Answer. **Load = 111 A**

(1) What is the load in amps for the office area?

Step 1. $I = \frac{VA}{V \times \sqrt{3}}$

$$I = \frac{8750 \text{ VA}}{208 \text{ V} \times 1.732} = 24.3 \text{ A}$$

Answer. **Load = 24 A**

(2) What is the load in amps for the warehouse area?

Step 1. $I = \frac{VA}{V \times \sqrt{3}}$

$$I = \frac{31{,}250 \text{ VA}}{208 \text{ V} \times 1.732} = 86.8 \text{ A}$$

Answer. **Load = 87 A**

(3) How many 20 A, 2-wire circuits are required for the warehouse?

Step 1. *220-4(a): No.* $= \frac{total\ VA}{OCPD \times 120 \text{ V}}$

$$No. = \frac{50{,}000 \text{ VA}}{20 \text{ A} \times 120 \text{ V}} = 20.8$$

Answer. **Circuits = Twenty-one 20 A, 2-wire**

STANDARD CALCULATION—COMMERCIAL AND INDUSTRIAL

		PHASES	NEUTRAL
1. LIGHTING LOADS—220, Part B			
General lighting loads—Table 220-3(b); 220-10(b)			
4200 sq. ft. × 3.5 VA × 125% =	18,375 VA		
Show window loads—220-12			
30 ′ × 200 VA =	6000 VA		
Track lighting loads—410-102			
40 ′ (2′) × 180 VA =	3600 VA		
Outside lighting loads—220-10(b)			
225 VA × 15 units × 125% =	4218.8 VA		
Sign lighting loads—220-3, FPN; 600-6(c)			
1200 VA × 100% =	1200 VA		
(total)	33,393.8 VA	33,393.8 VA	33,393.8 VA
2. RECEPTACLE LOADS—220-3(c)(5)			
Noncontinuous duty—220-10(b); Table 220-13			
# 69 × 180 VA = 12,420 VA =	— VA		
Applying demand factors—Table 220-13			
First 10,000 VA × 100% = 10,000 VA =	10,000 VA		
Remaining 2420 VA × 50% =	1210 VA		
Continuous duty—220-10(b) # 24 × 180 VA × 125% =	5400 VA		
Multioutlet assembly—220-3(c)			
Heavy duty = 10 ′ × 180 VA =	1800 VA		
Light duty = — ′ (5′) × 180 VA =	— VA		
(total)	18,410 VA	18,410 VA	18,410 VA
3. SPECIAL LOADS—220-10(b)			
Water htr. = 8000 VA × 100% =	8000 VA		
Misc. = 4320 VA × 125% =	5400 VA		
Misc. = 24,300 VA × 125% =	30,375 VA		
PC's = 14,400 VA × 125% =	18,000 VA		
Range = 14,000 VA × 100% =	14,000 VA		
(total)	75,775 VA	75,775 VA	23,400 VA
4. HEATING OR A/C LOAD—220-21	35,000 VA	35,000 VA	— VA
5. MOTOR LOADS—430-22(a); 430-24; 430-25 (3φ only)			
Fans = 52.8 A × 100% × 120 V × $\sqrt{3}$ =	6336 VA		
Pumps = 30.4 A × 110% × 208 V × $\sqrt{3}$ =	12,038.4 VA		
Exh. fans = 40.8 A × 110% × 208 V × $\sqrt{3}$ =	16,156.8 VA		
— = — A × —% × — V × $\sqrt{3}$ =	— VA		
(total)	34,531.2 VA	34,531.2 VA	6336 VA
6. LARGEST MOTOR LOAD—220-14 (3φ only)			
φ 15.2 A × 110% × 208 V × $\sqrt{3}$ × 25% =	1504.8 VA	1504.8 VA	— VA
N 4.4 A × 100% × 120 V × $\sqrt{3}$ × 25% =	132 VA	— VA	132 VA
TOTAL =		198,614.8 VA	81,671.8 VA

1φ service: $I = \frac{VA}{V}$

PHASES I = ______ VA / V = ______ A

NEUTRAL I = ______ VA / V = ______ A

3φ service: $I = \frac{VA}{V \times \sqrt{3}}$

$I = \frac{198{,}614.8 \text{ VA}}{208 \text{ V} \times \sqrt{3}} = 551.7 \text{ A}$

$I = \frac{81{,}671.8 \text{ VA}}{208 \text{ V} \times \sqrt{3}} = 226.9 \text{ A}$

If over 200 A—220-22; First 200 A × 100% = 200 A

Remaining 26.9 A × 70% = 18.8 A

218.8 A

Figure 8-27. The form, STANDARD CALCULATION—COMMERCIAL AND INDUSTRIAL is used to find VA and amps for banks.

(4) How many 20 A, 2-wire circuits are required for the office area?

Step 1. *220-4(a):* $No. = \frac{total\ VA}{OCPD \times 120\ V}$

$$No. = \frac{8750\ VA}{20\ A \times 120\ V} = 3.6\ A$$

Answer. **Circuits = Four 20 A, 2-wire**

Auditorium, Corridor, and Stairway

Auditoriums are sometimes located in buildings to seat people for plays, concerts, etc. Corridors are located in buildings to exit people or allow people to move to various areas of the building. Stairways are used for people to escape from fire or move from floor to floor.

What is the general lighting load in VA and amps for a high-rise building with a 100′-0″ × 350′-0″ auditorium, 3000 sq. ft. of corridors, and 4000 sq. ft. of stairways? The lighting load for the auditorium is noncontinuous duty. The lighting load for the corridors and stairways is continuous duty. The supply voltage is a 480/277 V, 3ϕ, 4-wire system.

Auditorium

Step 1. *Table 220-3(b):*
100′-0″ × 350′-0″ × 1 VA = 35,000 VA

Step 2. *Table 220-11:* No demand factors permitted

Step 3. *220-10(b):* 35,000 VA × 100% = 35,000 VA

Answer. **Load = 35,000 VA**

Corridor

Step 1. *Table 220-3(b):*
3000 sq. ft. × 50 VA = 1500 VA

Step 2. *Table 220-11:* No demand factors permitted

Step 3. *220-10(b):* 1500 VA × 125% = 1875 VA

Answer. **Load = 1875 VA**

Stairway

Step 1. *Table 220-3(b):*
4000 sq. ft. × .50 VA = 2000 VA

Step 2. *Table 220-11:* No demand factors permitted

Step 3. *220-10(b):* 2000 VA × 125% = 2500 VA

Answer. **Load = 2500 VA**

(1) How many 20 A, 2-wire circuits are required for the auditorium using 277 V lighting units?

Step 1. *220-4(a):* $No. = \frac{total\ VA}{OCPD \times 277\ V}$

$$No. = \frac{35{,}000\ VA}{20\ A \times 277\ V} = 6.3$$

Answer. **Circuits = Seven 20 A, 2-wire**

(2) How many 20 A, 2-wire circuits are required for the corridors using 120 V lighting units?

Step 1. *220-4(a):* $No. = \frac{total\ VA}{OCPD \times 120\ V}$

$$No. = \frac{1875\ VA}{20\ A \times 120\ V} = .78$$

Answer. **Circuits = One 20 A, 2-wire**

(3) What is the total VA rating of all lighting loads?

Step 1. Total VA.

Auditorium =	35,000 VA
Corridors =	1875 VA
Stairways =	2500 VA
Total =	39,375 VA

Answer. **VA rating = 39,375 VA**

(4) What is the load in amps for the lighting load using a 480/277 V, 3ϕ, 4-wire system?

Step 1. $I = \frac{VA}{V \times \sqrt{3}}$

$$I = \frac{39{,}375\ VA}{480\ V \times 1.732} = 47\ A$$

Answer. **Load = 47 A**

Dividing VA on Phases and Neutral

Some examination boards require the VA rating of each electrical load to be divided as evenly as possible on each phase. VA of each phase is added together to determine the total VA rating per phase.

Divide the VA per phase as evenly as possible and balance the loads on each phase for a 6000 sq. ft. store building. The supply voltage is 120/208 V, 3ϕ, 4-wire. Lighting loads are less than 50% electric discharge lighting. Use the form, STANDARD CALCULATION—COMMERCIAL AND INDUSTRIAL 3ϕ (CARRYING PHASES AND NEUTRAL IN VA). See Figure 8-28.

60 linear feet show window........... 120 V, 1ϕ
200′-0″ lighting track .... 120 V, 1ϕ
Twenty 180 VA outside lighting loads
(continuous duty) .... 120 V, 1ϕ
2400 VA sign lighting loads (three signs)120 V, 1ϕ
68 receptacles (noncontinuous-duty) ... 120 V, 1ϕ
28 receptacles (continuous-duty)....... 120 V, 1ϕ
60′-0″ multioutlet assembly
(heavy-duty) .... 120 V, 1ϕ
32 A freezer .... 208 V, 3ϕ
Eight 3600 VA ice cream boxes .... 208 V, 3ϕ
Ten 2 HP exhaust fans .... 208 V, 3ϕ
42 A walk-in cooler .... 208 V, 3ϕ
40 kW heating unit.................. 208 V, 3ϕ
38 VA A/C unit .... 208 V, 3ϕ
27 A personal computer.............. 208 V, 1ϕ
35 A miscellaneous loads............. 208 V, 1ϕ

(1) What size THWN Cu conductors, paralleled four times per phase, is required for the service-entrance conductors?

Step 1. *Table 310-16, 75°C Column:*

$$Total\ load = \frac{657\ A}{4} = 164\ A$$

Step 2. *Table 310-16; Note 8 to Tables 310-16–310-19:*
#4/0 = 230 A
230 A × 80% = 184 A
184 A supplies 164 A

Answer. **Conductors = #4/0 THWN Cu**

STANDARD CALCULATION—COMMERCIAL AND INDUSTRIAL (CARRYING PHASES AND NEUTRAL IN VA)

	PHASE	PHASE	PHASE	NEUTRAL
1. LIGHTING LOADS—220, Part B				
General lighting loads—Table 220-3(b); 220-10(b) 6000 sq. ft. × 3 VA × 125 % ÷ 3 =	7500 VA	7500 VA	7500 VA	7500 VA
Show window loads—220-12 60 ′ × 200 VA ÷ 3 =	4000 VA	4000 VA	4000 VA	4000 VA
Track lighting loads—410-102 200 ′ × 180 VA ÷ 3 = (2′)	6000 VA	6000 VA	6000 VA	6000 VA
Outside lighting loads—220-10(b) 180 VA × 20 units × 125 % ÷ 3 =	1500 VA	1500 VA	1500 VA	1500 VA
Sign lighting loads—220-3, FPN; 600-6(c) 2400 VA × 125 % ÷ 3 =	1000 VA	1000 VA	1000 VA	1000 VA
2. RECEPTACLE LOADS—220-3(c)(5)				
Noncontinuous duty—220-10(b) 68 VA × 100% = 12,240 VA ÷ — =	— VA	— VA	— VA	— VA
Applying demand factors—Table 220-13				
First 10,000 VA × 100% ÷ 3 =	3333.3 VA	3333.3 VA	3333.3 VA	3333.3 VA
Remaining 2240 VA × 50% ÷ 3 =	373.3 VA	373.3 VA	373.3 VA	373.3 VA
Continuous duty—220-10(b) 5040 VA × 125% ÷ 3 =	2100 VA	2100 VA	2100 VA	2100 VA
Multioutlet assembly—220-3(c)				
Heavy duty = 60 ′ × 180 VA ÷ 3 =	3600 VA	3600 VA	3600 VA	3600 VA
Light duty = — ′ × 180 VA ÷ — = (5′)	— VA	— VA	— VA	— VA
3. SPECIAL LOADS—220-10(b) (load)				
Freezers = 11,520 VA × 100 % ÷ — =	11,520 VA	11,520 VA	11,520 VA	— VA
IC boxes = 28,800 VA × 100 % ÷ — =	28,800 VA	28,800 VA	28,800 VA	— VA
PC's = 5616 VA × 100 % ÷ — =	5616 VA	5616 VA	— VA	— VA
Cooler = 15,120 VA × 100 % ÷ — =	15,120 VA	15,120 VA	15,120 VA	— VA
Misc. = 7280 VA × 100 % ÷ — =	— VA	7280 VA	7280 VA	— VA
4. HEATING OR A/C LOAD—220-21	40,000 VA	40,000 VA	40,000 VA	— VA
5. MOTOR LOADS—430-22(a); 430-24; 430-25 (load) (3ϕ motors only)				
Exh. fans = 74.8 A × 208 V × $\sqrt{3}$ ÷ — =	26,928 VA	26,928 VA	26,928 VA	— VA
— = — A × — V × $\sqrt{3}$ ÷ — =	— VA	— VA	— VA	— VA
— = — A × — V × $\sqrt{3}$ ÷ — =	— VA	— VA	— VA	— VA
— = — A × — V × $\sqrt{3}$ ÷ — =	— VA	— VA	— VA	— VA
6. LARGEST MOTOR LOAD—220-14				
ϕ 42 A × 360 V × 100 % × 25% =	3780 VA	3780 VA	3780 VA	— VA
N — A × — V × — % × 25% =	— VA	— VA	— VA	— VA
TOTAL =	161,170.6 VA	168,450.6 VA	162,834.6 VA	29,406.6 VA

1ϕ service: $I = \frac{VA}{V}$

PHASES $I = \frac{12,896 \text{ VA}}{208 \text{ V}}$ = 62 A

NEUTRAL $I = \frac{29,406.6 \text{ VA}}{120 \text{ V}}$ = 245 A

3ϕ service: $I = \frac{VA}{V \times \sqrt{3}}$

$I = \frac{126,148 \text{ VA}}{208 \text{ V} \times \sqrt{3}}$ = 350 A

$I = \frac{\text{VA}}{\text{V} \times \sqrt{3}}$ = ____ A

Note: Add all 120 V, 1ϕ loads, all 208 V, 1ϕ loads, all 208 V, 3ϕ loads, and divide each by the proper V to find A.

If over 200 A—220-22; First 200 A × 100% = 200 A

Remaining 45 A × 70% = 31.5 A

231.5 A

Figure 8-28. The form, STANDARD CALCULATION—COMMERCIAL AND INDUSTRIAL (CARRYING PHASES AND NEUTRAL IN VA) is used to find VA of commercial and industrial buildings.

(2) What size Cu grounding electrode conductor is required to ground the service equipment to a metal water pipe?

Step 1. *Table 8, Ch. 8:* 167.8 kcmil × 4 = 671.2 kcmil

Answer. ***Table 250-94:* Conductor = #1/0 Cu**

(3) What size THWN Cu conductor, paralleled four times per phase, is required for the neutral?

Step 1. *Text:* $\frac{231.5 \text{ A}}{4} = 57.9 \text{ A}$

Step 2. *310-4; Table 310-16:* Neutral = four #1/0 Cu

Answer. **Conductor = #1/0 Cu**

(4) What size conduit is required for the service-entrance conductor?

Step 1. *Table 5, Ch. 9:*

#4/0 = .3278 sq. in. × 3 =	.9834 sq. in.
#1/0 = .1893 sq. in. × 1 =	.1893 sq. in.
Total =	1.1727 sq. in.

Step 2. *Table 4, Ch. 9:* 1.34 sq. in. = 2″ conduit

Answer. **Conduit = 2″**

Dividing Amps on Phases and Neutral

Some testing boards require that the amps of each piece of electrical equipment be divided as evenly as possible on each phase. The amps on each phase are added to obtain the total amps per phase.

Divide the amps per phase as evenly as possible and balance the loads on each phase for an 8100 sq. ft. bank with the following loads. The service is supplied by a 120/208 V, 3ϕ, 4-wire supply. Use the form, STANDARD CALCULATION—COMMERCIAL AND INDUSTRIAL 3ϕ (CARRYING PHASES AND NEUTRAL IN AMPS). See Figure 8-29.

30 linear feet show window........... 120 V, 1ϕ
160′-0″ lighting track................ 120 V, 1ϕ
Forty-five 180 VA outside lighting loads..................... 120 V, 1ϕ
3600 VA sign lighting loads (three signs)...................... 120 V, 1ϕ
210 receptacles (noncontinuous-duty) .. 120 V, 1ϕ
42 receptacles (continuous-duty)....... 120 V, 1ϕ
125′-0″ multioutlet assembly (light-duty)....................... 120 V, 1ϕ
75′-0″ multioutlet assembly (heavy-duty)...................... 120 V, 1ϕ
12.6 kW water heater................ 208 V, 3ϕ
50 kW heating unit.................. 208 V, 3ϕ
17.8 kW A/C unit.................. 208 V, 3ϕ
20 HP water pump.................. 208 V, 3ϕ
Twelve 2 HP exhaust fans............ 208 V, 3ϕ
Fifteen 5200 VA personal computers... 208 V, 3ϕ
12 kW range (no neutral load)........ 208 V, 3ϕ
¾ HP soft drink box................ 120 V, 1ϕ
Two ½ HP refrigerators.............. 120 V, 1ϕ

(1) What is the minimum ampacity of each conductor paralleled five times per phase?

Step 1. *310-4:* $A = \frac{\textit{total A}}{\textit{No. of conductors}}$

$$A = \frac{940.04 \text{ A}}{5} = 188 \text{ A}$$

Answer. **Minimum ampacity = 188 A**

(2) What size THWN Cu conductors is required to be paralleled per phase with 50% of the lighting load being electric discharge lighting?

Step 1. *Table 310-16:* #250 kcmil THWN Cu = 255 A

Step 2. *Table 310-16; Note 8 to Tables 310-16–310-19:*
255 A × 80% = 204 A
204 A supplies 188 A

Answer. **Conductors = #250 kcmil THWN Cu**

(3) What size OCPD is required for the service?

Step 1. *240-3, Ex. 4; 240-6:*
204 A OCPD × 5 = 1020 A

Answer. **OCPD = 1000 A**

(4) What size Cu grounding electrode conductor is required to ground the service to building steel?

Step 1. Finding kcmil. *Table 8, Ch. 9:*
#250 kcmil = 250,800 CM
kcmil = 250 kcmil × 5 = #1250 kcmil

Step 2. *250-81(b); Table 250-94:*
#1250 kcmil requires #3/0 Cu

Answer. **Conductor = #3/0 Cu**

Optional Calculation for a Restaurant—220-36; Table 220-36

The calculation for the load to a new restaurant is computed by applying the optional method per 220-36. The service or feeder load for a new restaurant with an all-electric load or not all-electric load is figured using the demand factors of Table 220-36. The demand factor is selected based on the kVA of the new restaurant. A new restaurant that is all-electric with a load of 350 kVA has a demand of 50%. The load for the restaurant is 175 kVA (350 kVA × 50% = 175 kVA). A new restaurant with a load of 350 kVA that is not all-electric has a demand load of 245 kVA (350 kVA × 70% = 245 kVA).

Restaurant

A 65′-0″ × 85′-0″ restaurant has a 120/208 V, 3ϕ, 4-wire service and the following loads. Find the load in VA and amps. Note that 50% or more of the lighting load is incandescent lighting. Use the optional method.

STANDARD CALCULATION—COMMERCIAL AND INDUSTRIAL (CARRYING PHASES AND NEUTRAL IN AMPS)

	PHASE	PHASE	PHASE	NEUTRAL
1. LIGHTING LOADS—220, Part B				
General lighting loads—Table 220-3(b); 220-10(b)				
8100 sq. ft. × 3.5 VA × 125 % ÷ 360 V =	98.4 A	98.4 A	98.4 A	98.4 A
Show window loads—220-12				
30 ′ × 200 VA ÷ 360 V =	16.7 A	16.7 A	16.7 A	16.7 A
Track lighting loads—410-102				
160 ′ × 180 VA ÷ 360 V = (2′)	40 A	40 A	40 A	40 A
Outside lighting loads—220-10(b)				
180 VA × 45 units × 125 % ÷ 360 V =	28 A	28 A	28 A	28 A
Sign lighting loads—220-3, FPN; 600-6(c)				
3600 VA × 125 % ÷ 360 V =	12.5 A	12.5 A	12.5 A	12.5 A
2. RECEPTACLE LOADS—220-3(c)(5)				
Noncontinuous duty—220-10(b)				
37,800 VA × 100% = — VA ÷ — V =	— A	— A	— A	— A
Applying demand factors—Table 220-13				
First 10,000 VA × 100% ÷ 360 V =	27.8 A	27.8 A	27.8 A	27.8 A
Remaining 27,800 VA × 50% ÷ 360 V =	38.6 A	38.6 A	38.6 A	38.6 A
Continuous duty—220-10(b) 7560 VA × 125% ÷ 360 V =	26.3 A	26.3 A	26.3 A	26.3 A
Multioutlet assembly—220-3(c)				
Heavy duty = 75 ′ × 180 VA ÷ 360 V =	37.5 A	37.5 A	37.5 A	37.5 A
Light duty = 125 ′ × 180 VA ÷ 360 V = (5′)	12.5 A	12.5 A	12.5 A	12.5 A
3. SPECIAL LOADS—220-10(b) (load)				
Water htr. = 12,600 VA × 100 % ÷ 360 V =	35 A	35 A	35 A	— A
PC's = 78,000 VA × 100 % ÷ 360 V =	216.7 A	216.7 A	216.7 A	— A
Range = 12,000 VA × 100 % ÷ 360 V =	33.33 A	33.33 A	33.33 A	— A
Drink box = 1656 VA × 100 % ÷ 120 V =	13.8 A	— A	— A	13.8 A
Ref. = 2352 VA × 100 % ÷ 240 V =	— A	9.8 A	9.8 A	9.8 A
4. HEATING OR A/C LOAD—220-21	138.9 A	138.9 A	138.9 A	— A
5. MOTOR LOADS—430-22(a); 430-24; 430-25 (load)				
Pump = 59.4 A × 1 # ÷ — =	59.4 A	59.4 A	59.4 A	— A
Exh. fan = 7.48 A × 12 # ÷ — =	89.76 A	89.76 A	89.76 A	— A
— = — A × — # ÷ — =	— A	— A	— A	— A
— = — A × — # ÷ — =	— A	— A	— A	— A
6. LARGEST MOTOR LOAD—220-14				
ϕ 54 A × — V × 110 % × 25% =	14.85 A	14.85 A	14.85 A	— A
N 13.8 A × — V × — % × 25% =	— A	— A	— A	3.45 A
TOTAL =	940.04 A	936.04 A	936.04 A	365.35 A

If over 200 A—220-22; First 200 A × 100% = 200 A

Remaining ______ A × 70% = ______ A

______ A

Figure 8-29. The form, STANDARD CALCULATION—COMMERCIAL AND INDUSTRIAL (CARRYING PHASES AND NEUTRAL IN AMPS) is used to find amps of commercial and industrial buildings.

20′-0″ lighting track .... 120 V, 1ϕ
Eight 180 VA outside lighting
(continuous-duty) .... 120 V, 1ϕ
1200 VA sign lighting
(continuous-duty) .... 120 V, 1ϕ
38 receptacles (noncontinuous-duty) .... 120 V, 1ϕ
18 receptacles (continuous-duty) .... 120 V, 1ϕ
20′-0″ multioutlet assembly
(heavy-duty) .... 120 V, 1ϕ
Two 30 kW heating units .... 208 V, 3ϕ
Two 7560 VA A/C units .... 208 V, 3ϕ
Two 3 HP water pumps .... 208 V, 3ϕ
Six ¾ HP exhaust fans .... 208 V, 3ϕ

Kitchen equipment

3600 VA boilers .... 208 V, 3ϕ
Two 2880 VA deep fat fryers .... 208 V, 3ϕ
Two 1600 VA dishwashers .... 120 V, 1ϕ
Four 1200 VA toasters .... 120 V, 1ϕ
Two 840 VA mixers .... 120 V, 1ϕ
17 A walk-in cooler .... 208 V, 3ϕ
4320 VA ice cream box .... 208 V, 1ϕ
15 A freezer .... 208 V, 1ϕ
Two 1680 VA dishwashers .... 120 V, 1ϕ
1200 VA dishwasher booster heater .... 120 V, 1ϕ
8000 VA water heater .... 208 V, 3ϕ
12,000 VA cooktop .... 208 V, 1ϕ
Two 10,000 VA ovens .... 208 V, 1ϕ
8 A bun warmer .... 120 V, 1ϕ
12,000 VA range .... 208 V, 1ϕ
Two 1860 VA coffee makers .... 120 V, 1ϕ
1500 VA grinder .... 120 V, 1ϕ
15 A refrigerator .... 208 V, 1ϕ
14.5 A soft drink box .... 120 V, 1ϕ

Step 1. *Table 220-3(b):* General lighting load
5525 sq ft × 2 VA × 125% = 13,812.5 VA

430-102: Track lighting load
$\frac{20'}{2'} \times 180$ VA = 1800 VA

220-10(b): Outside lighting load
8 × 180 VA × 125% = 1800 VA

600-6(c): Sign load
1200 VA × 100% = 1200 VA

220-3(c)(5): Receptacle loads
38 × 180 VA = 6840 VA
18 × 180 VA × 125% = 4050 VA
20 × 180 VA = 3600 VA

220-10(b): Cooking equipment load
Boilers = 3600 VA
Deep fat fryers =
2880 VA × 2 = 5760 VA
Dishwashers = 1600 VA × 2 = 3200 VA
Toasters = 1200 VA × 4 = 4800 VA
Mixers = 840 VA × 2 = 1680 VA
Walk-in cooler =
17 A × 360 V = 6120 VA
Ice cream box = 4320 VA
Freezer = 15 A × 208 V = 3120 VA
Dishwasher = 1680 VA × 2 = 3360 VA
Dishwasher booster heater = 1200 VA
Water heater = 8000 VA
Cooktop = 12,000 VA
Ovens = 10,000 VA × 2 = 20,000 VA
Bun warmer = 8 A × 120 V = 960 VA
Range = 12,000 VA
Coffee makers =
1860 VA × 2 = 3720 VA
Grinder = 1500 VA
Refrigerator = 15 A × 208 V = 3120 VA
Soft drink box =
14.5 A × 120 V = 1740 VA

220-15; 220-21: Heating load
Heating load = 60,000 VA

430-25; 430-24: Motor loads
Exhaust fans = 6652.8 VA
Pump motor = 7603.2 VA
207,558.5 VA

Step 2. *Table 220-36:* Applying demand factors
Load = kVA × demand factor
Load = 207,558.5 VA × 80% = 166.0468 kVA

Step 3. Finding VA
VA = kVA × 1000
VA = 166.0468 kVA × 1000 = 166,046.8 VA

Step 4. Finding A

$$A = \frac{VA}{V \times \sqrt{3}}$$

$$A = \frac{166{,}046.8 \text{ VA}}{208 \text{ V} \times 1.732}$$

$A = 461$

Answer. **VA = 166,046.8 VA**
A = 461 A

Calculations—Commercial and Industrial

Date ____________________

Chapter 8—TEST

Name ______________________________

______ **1.** What is the load in VA for the general lighting load of a 1225 sq. ft. beauty parlor (continuous-duty)?

______ **2.** What is the load in VA for 600 sq. ft. of stairways in a commercial building (noncontinuous-duty)?

______ **3.** What is the demand load in VA for the general lighting load of a 125,000 sq. ft. area used for patient rooms in a hospital (noncontinuous-duty)?

______ **4.** What is the load in VA for a garage with a work space of 1000 sq. ft. and an office area of 576 sq. ft. (continuous-duty)?

______ **5.** What is the demand load in VA for two hundred twenty-five 250 sq. ft. guest rooms in a hotel?

______ **6.** What is the load in VA for a store building with a 40′-0″ × 80′-0″ shopping area and a 250′-0″ × 300′-0″ warehouse space? The store is continuous duty and the warehouse is noncontinuous duty.

______ **7.** What is the general lighting load in VA for ten 1800 sq. ft. assembly halls and a 6400 sq. ft. auditorium? The assembly halls have continuous-duty lighting loads and the auditorium has noncontinuous-duty lighting loads.

______ **8.** What is the lighting load in VA for a 650 sq. ft. corridor leading to a 2400 sq. ft. store, 1200 sq. ft. office, 1800 sq. ft. restaurant, 3600 sq.ft. bank, and a 900 sq. ft. barber shop? The lighting loads are continuous duty.

______ **9.** What is the load in VA for a feeder circuit supplying power to the following pieces of cooking equipment?

12 kW cooktop .... 240 V, 1ϕ
10 kW oven .... 240 V, 1ϕ
8 kW oven .... 240 V, 1ϕ
6 kW steamer .... 240 V, 1ϕ
Three 5 kW fryers .... 240 V, 1ϕ
4 kW kettle .... 240 V, 1ϕ
6.5 kW toaster .... 240 V, 1ϕ
6.5 kW roaster .... 240 V, 1ϕ
1.2 kW deep freezer .... 120 V, 1ϕ
5.5 kW dishwasher .... 240 V, 1ϕ
1.4 kW refrigerator .... 120 V, 1ϕ
1.9 kW coffee pot .... 120 V, 1ϕ

______ **10.** What is the load in VA for a school that has a 220,000 sq. ft. area with a connected load of 2,695,000 VA? (Use optional method.)

______ **11.** What is the maximum size load in A that can be added to an existing 120/208 V, 3ϕ, 4-wire service with #250 kcmil THWN Cu conductors? The utility company listed the peak demand for one year as 62 kVA.

______ **12.** What is the demand load in VA for 350 general-purpose receptacles installed in an office building?

____________ **13.** What is the load in VA for 110 receptacles used at continuous duty?

____________ **14.** What is the load in VA for 200′-0″ of light-duty multioutlet assembly and 150′-0″ of heavy-duty multioutlet assembly?

____________ **15.** What is the lighting load in VA for an unlisted occupancy with the following continuous-duty loads?
75 fluorescent lighting fixtures with 180 VA ballasts
25 recessed lighting fixtures with 75 watt bulbs

____________ **16.** What is the load in VA for an 85′-0″ × 95′-0″ store building supplied by a 120/208 V, 3ϕ, 4-wire service with the following loads? Use the form, STANDARD CALCULATION—COMMERCIAL AND INDUSTRIAL.

Load	Supply
70 linear feet show window	120 V, 1ϕ
80′-0″ lighting track	120 V, 1ϕ
Twenty-two 180 VA outside lighting (continuous-duty)	120 V, 1ϕ
2400 VA sign lighting	120 V, 1ϕ
110 receptacles (noncontinuous-duty)	120 V, 1ϕ
12 receptacles (continuous-duty)	120 V, 1ϕ
60′-0″ multioutlet assembly (light-duty)	120 V, 1ϕ
40′-0″ multioutlet assembly (heavy-duty)	120 V, 1ϕ
Four 3 HP freezers	208 V, 3ϕ
Four 5 HP ice cream boxes	208 V, 3ϕ
Six 1.5 HP exhaust fans	208 V, 3ϕ
8000 VA water heater	208 V, 3ϕ
10,080 VA walk-in cooler	208 V, 3ϕ
40 kW heating unit	208 V, 3ϕ
10,260 VA A/C	208 V, 3ϕ
3 HP water pump	208 V, 3ϕ
2 HP sump pump	208 V, 3ϕ
Twelve 225 VA cash registers (special loads, continuous-duty)	120 V, 1ϕ

____________ **16-1.** How many 20 A, 2-wire circuits are required to supply the general lighting load?

____________ **16-2.** How many 20 A, 2-wire circuits are required to supply the general-purpose receptacle loads?

____________ **16-3.** What is the maximum size OCPD required for the walk-in cooler?

____________ **16-4.** What size CB is required to allow the water pump to start and run? (Use lower rating.)

____________ **16-5.** What size THWN Cu conductors, paralleled three times per phase, is required to supply phases A, B, and C?

____________ **16-6.** What size THWN Cu conductors, paralleled three times per phase, is required for the neutral conductors?

_______ **17.** A 2925 sq. ft. office building is supplied by a 120/240 V, 1ϕ service and has the following loads. Less than half of the load is supplied by electric discharge lighting. Use the form, STANDARD CALCULATION—COMMERCIAL AND INDUSTRIAL.

12 linear feet show window .... 120 V, 1ϕ
30′-0″ lighting track .... 120 V, 1ϕ
Five 225 VA outside lighting (continuous-duty) .... 120 V, 1ϕ
1200 VA sign lighting .... 120 V, 1ϕ
65 receptacles (noncontinuous-duty) .... 120 V, 1ϕ
21 receptacles (continuous-duty) .... 120 V, 1ϕ
50′-0″ multioutlet assembly (light-duty) .... 120 V, 1ϕ
35 kW heating unit .... 240 V, 1ϕ
8400 VA A/C .... 240 V, 1ϕ
3 HP sump pump .... 240 V, 1ϕ
1320 VA soft drink box .... 120 V, 1ϕ
2400 VA copying machine .... 240 V, 1ϕ
5500 VA water heater .... 240 V, 1ϕ
Three ⅙ HP exhaust fans .... 120 V, 1ϕ
4800 VA personal computer (continuous-duty) .... 240 V, 1ϕ
Twenty 180 VA typewriters (special loads, continuous-duty) .... 120 V, 1ϕ

_______ **17-1.** What size THWN Cu conductors is required for the service?

_______ **17-2.** What size main OCPD is required for the service?

_______ **17-3.** What size panelboard is required for the service?

_______ **18.** A 15,625 sq. ft. bank is supplied by a 277/480 V, 3ϕ, 4-wire service and has the following loads. Over 50% of the lighting load is supplied by electric discharge units. Use the form, STANDARD CALCULATION—COMMERCIAL AND INDUSTRIAL.

General lighting load .... 277 V, 1ϕ
60 linear feet show window .... 120 V, 1ϕ
120′-0″ lighting track .... 120 V, 1ϕ
Thirty 180 VA outside lighting .... 277 V, 1ϕ
1900 VA sign lighting .... 120 V, 1ϕ
130 receptacles (noncontinuous-duty) .... 120 V, 1ϕ
23 receptacles (continuous-duty) .... 120 V, 1ϕ
40′-0″ multioutlet assembly (heavy-duty) .... 120 V, 1ϕ
50 kW heating unit .... 480 V, 3ϕ
29 VA A/C .... 480 V, 3ϕ
5 HP water pump .... 460 V, 3ϕ
Ten 1 HP exhaust fans .... 460 V, 3ϕ
6480 VA personal computer (special loads) .... 208 V, 3ϕ
1680 VA coffee maker (special loads) .... 120 V, 1ϕ
10 kW water heater .... 480 V, 3ϕ
1140 VA soft drink box .... 120 V, 1ϕ
10,800 VA cash registers (special loads, continuous-duty) .... 120 V, 1ϕ

_______ **18-1.** How many 20 A, 2-wire circuits are required to supply the outside lighting load?

_______ **18-2.** How many 20 A, 2-wire circuits are required to supply the general-purpose receptacle loads (non-continuous duty)?

_______ **18-3.** What size THWN Cu conductors is required to supply the water heater?

_______ **18-4.** What is the load in amps for the motor loads?

_______ **18-5.** What size THWN Cu conductors routed in the same raceway is required for phases A, B, and C and the neutral?

______ **19.** A 4800 sq. ft. restaurant has a 120/208 V, 3ϕ, 4-wire service and the following loads. Use the form, STANDARD CALCULATION—COMMERCIAL AND INDUSTRIAL.

15 linear feet show window .... 120 V, 1ϕ
100′-0″ lighting track .... 120 V, 1ϕ
Ten 175 VA outside lighting (continuous-duty) .... 120 V, 1ϕ
1800 VA sign lighting .... 120 V, 1ϕ
30 receptacles (noncontinuous-duty) .... 120 V, 1ϕ
15 receptacles (continuous-duty) .... 120 V, 1ϕ
35 kW heating unit .... 208 V, 3ϕ
5.5 kVA A/C .... 208 V, 3ϕ
12 kW water heater .... 208 V, 3ϕ
2692.8 VA blower motor .... 208 V, 3ϕ

Kitchen equipment

8000 VA broiler .... 208 V, 3ϕ
Two 6000 VA deep fryers .... 208 V, 3ϕ
4800 VA dishwasher .... 208 V, 3ϕ
Two 1600 VA toasters .... 120 V, 1ϕ
Three 1800 VA mixers .... 120 V, 1ϕ
4160 VA walk-in cooler .... 208 V, 3ϕ
1800 VA ice cream box .... 120 V, 1ϕ
6656 VA freezer .... 208 V, 3ϕ
6000 VA dishwasher booster heater .... 208 V, 3ϕ
14 kW cooktop .... 208 V, 3ϕ
Two 10 kW ovens .... 208 V, 3ϕ
Three 1400 VA bun warmers .... 120 V, 1ϕ
Two 1920 VA coffee makers .... 120 V, 1ϕ
3328 VA grinder .... 208 V, 3ϕ
Three 1680 VA refrigerators .... 120 V, 1ϕ
1740 VA soft drink box .... 120 V, 1ϕ

______ **19-1.** What is the demand load in VA for the kitchen equipment?

______ **19-2.** What size THWN Cu conductors is required for phases A, B, and C for the feeder circuit to the kitchen?

______ **19-3.** What size OCPD is required for the feeder circuit to the kitchen?

______ **19-4.** What size THWN Cu conductors is required for the heating unit?

______ **19-5.** What size OCPD is required for the heating unit?

______ **19-6.** What is the load in VA for the restaurant using the optional calculation?

______________ **20.** What is the load in VA for phases A, B, and C and the neutral for a 80,000 sq. ft. school building supplied by a 277/480 V, 3ϕ, 4-wire system? The inside load is 853,256.6 VA and the outside load is 222,320 VA. Use the form, OPTIONAL CALCULATION—SCHOOL. Use the form, STANDARD CALCULATION—COMMERCIAL AND INDUSTRIAL for the neutral. The neutral contains over 50% harmonic currents.

General lighting load .... 277 V, 1ϕ
100 linear feet show window .... 277 V, 1ϕ
150′-0″ lighting track .... 120 V, 1ϕ
Thirty 225 VA outside lighting (continuous-duty) .... 277 V, 1ϕ
2400 VA sign lighting (continuous-duty) .... 120 V, 1ϕ
400 receptacles (noncontinuous-duty) .... 120 V, 1ϕ
50 receptacles (continuous-duty) .... 277 V, 1ϕ
220′-0″ multioutlet assembly (light-duty) .... 120 V, 1ϕ
90′-0″ multioutlet assembly (heavy-duty) .... 120 V, 1ϕ
175 kW heating unit .... 480 V, 3ϕ
29 kVA A/C .... 480 V, 3ϕ
Four 5 kW water pumps .... 480 V, 3ϕ
Five 1800 VA copying machines .... 120 V, 1ϕ

Kitchen equipment

Three 10 kW broilers .... 480 V, 3ϕ
Three 12 kW deep fryers .... 480 V, 3ϕ
Four 2400 VA vegetable peelers .... 480 V, 3ϕ
Two 3600 VA mixers .... 480 V, 3ϕ
9000 VA walk-in cooler .... 480 V, 3ϕ
16,620 VA ice cream box .... 480 V, 3ϕ
12,000 VA freezer .... 480 V, 3ϕ
Three 29,085 VA dishwashers .... 480 V, 3ϕ
Three 3000 VA dishwasher booster heaters .... 480 V, 3ϕ
Ten 2 HP exhaust fans (continuous-duty) .... 480 V, 3ϕ
Four 10 kW ovens .... 480 V, 3ϕ
Five 9972 VA bun warmers .... 480 V, 3ϕ
Four 1445 VA coffee makers .... 120 V, 1ϕ
Two 9.6 kW grinders .... 480 V, 3ϕ
Four 1200 VA refrigerators .... 120 V, 1ϕ
Four 12 kW cooktops .... 480 V, 3ϕ

_______________ **21.** What is the load in VA for a complex that has an office and warehouse supplied by a 277/480 V, 3ϕ, 4-wire system with the following loads? Use the form, STANDARD CALCULATION—COMMERCIAL AND INDUSTRIAL to find VA and amps for the office. Use the form, STANDARD CALCULATION—COMMERCIAL AND INDUSTRIAL to find VA and amps for the warehouse.

3500 sq. ft. office

Lighting units .... 277 V, 1ϕ
40′-0″ lighting track .... 120 V, 1ϕ
Ten 180 VA outside lighting (continuous-duty) .... 120 V, 1ϕ
1200 VA sign lighting (continuous-duty) .... 120 V, 1ϕ
42 receptacles (noncontinuous-duty) .... 120 V, 1ϕ
20 receptacles (continuous-duty) .... 120 V, 1ϕ
30′-0″ multioutlet assembly (light-duty) .... 120 V, 1ϕ
15′-0″ multioutlet assembly (heavy-duty) .... 120 V, 1ϕ
Two 7000 VA water heaters .... 480 V, 3ϕ
40 kW heating unit .... 480 V, 3ϕ
20,775 VA heat pump (operates with heating unit) .... 480 V, 3ϕ
10,000 VA special loads (continuous-duty) .... 208 V, 3ϕ
5000 VA special loads (noncontinuous-duty) .... 120 V, 1ϕ

20,000 sq. ft. warehouse

Lighting units (continuous-duty) .... 277 V, 1ϕ
40 receptacles (noncontinuous-duty) .... 120 V, 1ϕ
20 floor receptacles (continuous-duty) .... 120 V, 1ϕ
18 wall receptacles (continuous-duty) .... 120 V, 1ϕ
Sixty-five 3000 VA heating units .... 208 V, 1ϕ
Twenty 2 HP exhaust fans .... 480 V, 3ϕ
Ten 19,944 VA battery chargers (continuous-duty) .... 480 V, 3ϕ

_______________ **21-1.** What size THWN Cu conductors is required to supply phases A, B, and C and the neutral?

_______________ **21-2.** What size 120/208 V, 3ϕ, 4-wire transformer is required to supply power to the lower voltage loads in the office?

_______________ **21-3.** What size OCPD is required in the primary side of the 45 kVA transformer?

_______________ **21-4.** What size 10′ tap is required from the secondary side of the transformer? THWN Cu conductors supply the 120/208 V loads and are sized at continuous duty from the secondary FLC of the transformer.

_______________ **21-5.** What size main OCPD is required for the 120/208 V, 3ϕ, 4-wire panel?

_______________ **21-6.** What size panel is required for the secondary 120/208 V, 3ϕ, 4-wire load?

______ **22.** What is the load in VA and amps for phases A, B, and C for a welding shop with a 277/480 V, 3ϕ, 4-wire service and the following loads? Use the form, STANDARD CALCULATION—COMMERCIAL AND INDUSTRIAL.

Seventy-five 225 VA inside lighting units (continuous-duty) .... 277 V, 1ϕ
Thirty 180 VA outside lighting (continuous-duty) .... 277 V, 1ϕ
1900 VA sign lighting (continuous-duty) .... 120 V, 1ϕ
75 receptacles (noncontinuous-duty) .... 120 V, 1ϕ
80 receptacles (continuous-duty) .... 120 V, 1ϕ
90′-0″ multioutlet assembly (heavy-duty) .... 120 V, 1ϕ
Four 20 kW heating units .... 480 V, 3ϕ
Four 15,789 VA A/C units .... 480 V, 3ϕ
Two 5 HP water pumps .... 480 V, 3ϕ
1800 VA soft drink box .... 120 V, 1ϕ
Three 1200 VA copying machines (special loads) .... 120 V, 1ϕ
Three 3 HP drill presses .... 480 V, 3ϕ
Three 1.5 HP grinders .... 480 V, 3ϕ
Three 7.5 HP lathes .... 480 V, 3ϕ

Resistance welders

29.085 kW, 40% duty cycle .... 480 V, 3ϕ
23.268 kW, 40% duty cycle .... 480 V, 3ϕ
18.282 kW, 40% duty cycle .... 480 V, 3ϕ
15.373 kW, 40% duty cycle .... 480 V, 3ϕ

AC transformers and DC arc welders

35 kW, 80% duty cycle .... 480 V, 3ϕ
32 kW, 80% duty cycle .... 480 V, 3ϕ
28.5 kW, 80% duty cycle .... 480 V, 3ϕ
26.5 kW, 80% duty cycle .... 480 V, 3ϕ

______ **22-1.** What size THWN Cu conductors is required for a 3ϕ feeder circuit to the resistance welders?

______ **22-2.** What size OCPD is required for the feeder circuit to the resistance welders?

______ **22-3.** What size THWN Cu conductors is required to supply the largest resistance welder? The welder is manually operated.

______ **22-4.** What size THWN Cu conductors is required to supply a 3ϕ feeder circuit to the AC transformer arc welders?

______ **22-5.** What size OCPD is required for the feeder circuit to the AC transformer arc welders?

______ **23.** What is the load in VA for a motel complex with the following loads? The service equipment is a 277/480 V, 3ϕ, 4-wire system. Four transformers are installed to step the voltage down from 277/480 V to a 120/208 V, 3ϕ, 4-wire system. Feeders of 120/208 V, 3ϕ are routed to each complex containing 100 rooms. Less than 50% of the lighting load is electric discharge. Use the form, STANDARD CALCULATION—HOTELS AND MOTELS.

One hundred 375 sq. ft. rooms
One hundred 395 sq. ft. rooms
One hundred 400 sq. ft. rooms
12 linear feet show window .... 120 V, 1ϕ
90′-0″ lighting track .... 120 V, 1ϕ
Twenty-five 180 VA outside lighting (continuous-duty) .... 120 V, 1ϕ
28 receptacles (continuous-duty) .... 120 V, 1ϕ
Combination heating and A/C per room .... 208 V, 1ϕ
    8000 VA heating
    4368 VA A/C
Six 1800 VA soft drink boxes .... 120 V, 1ϕ
Two 1200 VA copying machines (special loads) .... 120 V, 1ϕ
Twenty 6500 VA water heaters .... 208 V, 3ϕ
35,000 VA house loads .... 208 V, 3ϕ
22,000 VA house loads .... 120 V, 1ϕ
57,600 VA laundry loads .... 208 V, 3ϕ

______ **23-1.** What size THWN Cu conductors, paralleled eight times per phase, is required for phases A, B, and C?

______ **23-2.** What size 277/480 V, 3ϕ, 4-wire switchgear is required for the service equipment?

______ **23-3.** What size feeder circuit is required to supply the one hundred 400 sq. ft. rooms? Use the form, STANDARD CALCULATION—HOTELS AND MOTELS. The feeder circuit has the following loads:

Combination heating and A/C per room .... 208 V, 1ϕ
    8000 VA heating
    4368 VA A/C
Five 6500 VA water heaters .... 208 V, 3ϕ
Two 1800 VA soft drink boxes .... 120 V, 1ϕ
8750 VA house loads .... 208 V, 3ϕ
5500 VA laundry loads .... 208 V, 3ϕ

______ **23-4.** What size THWN Cu conductors, paralleled seven times per phase, is required for the 120/208 V, 3ϕ, 4-wire feeder circuit?

______ **23-5.** What size OCPD is required for the feeder circuit?

______ **23-6.** What size Cu EGC is required for the feeder circuit?

______ **23-7.** What size conduit, using PVC, is required for the feeder circuit? There are five conduit runs with five conductors in each run.

______ **24.** A 60′-0″ × 60′-0″ store is supplied by a 120/208 V, 3ϕ, 4-wire service with the loads divided as evenly as possible on the phases and neutral. Over 50% of the lighting load is electric discharge. Use the form, STANDARD CALCULATION—COMMERCIAL AND INDUSTRIAL (CARRYING PHASES AND NEUTRAL IN VA).

30 linear feet show window .... 120 V, 1ϕ
150′-0″ lighting track .... 120 V, 1ϕ
Twelve 180 VA outside lighting (continuous-duty) .... 120 V, 1ϕ
1200 VA sign lighting .... 120 V, 1ϕ
75 receptacles (noncontinuous-duty) .... 120 V, 1ϕ
35 receptacles (continuous-duty) .... 120 V, 1ϕ
100′-0″ multioutlet assembly (heavy-duty) .... 120 V, 1ϕ
Four 9000 VA freezers .... 208 V, 3ϕ
Four 7560 VA ice cream boxes .... 208 V, 3ϕ
Two 8000 VA water heaters .... 208 V, 3ϕ
11,520 VA walk-in cooler .... 208 V, 3ϕ
40 kW heating unit .... 208 V, 3ϕ
10,080 VA A/C unit .... 208 V, 3ϕ
3 HP water pump .... 208 V, 3ϕ
1620 soft drink box .... 120 V, 1ϕ
1920 VA exhaust fans .... 120 V, 1ϕ

______ **24-1.** Which phase is used to size the service-entrance conductors based on the amperage?

______ **24-2.** What size Cu conductors is required to parallel phases A, B, and C three times per phase?

______ **24-3.** What size THWN Cu conductors, paralleled three times per phase, is required for the service?

______ **25.** A 4900 sq. ft. bank building has a 120/208 V, 3ϕ, 4-wire system. Over 50% of the lighting loads is electric discharge. The load is divided as evenly as possible on the phases and neutral. Use the form, STANDARD CALCULATION—COMMERCIAL AND INDUSTRIAL (CARRYING PHASES AND NEUTRAL IN AMPS).

30 linear feet show window .... 120 V, 1ϕ
130′-0″ lighting track .... 120 V, 1ϕ
Twenty 225 VA outside lighting (continuous-duty) .... 120 V, 1ϕ
1860 VA sign lighting .... 120 V, 1ϕ
62 receptacles (noncontinuous-duty) .... 120 V, 1ϕ
30 receptacles (continuous-duty) .... 120 V, 1ϕ
60′-0″ multioutlet assembly (light-duty) .... 120 V, 1ϕ
60′-0″ multioutlet assembly (heavy-duty) .... 120 V, 1ϕ
1680 VA soft drink box .... 120 V, 1ϕ
8 kW water heater .... 208 V, 3ϕ
16,200 VA personal computer (continuous-duty) .... 208 V, 3ϕ
40 kW heating unit .... 208 V, 3ϕ
11,520 VA A/C unit .... 208 V, 3ϕ
Two 3 HP water pumps .... 208 V, 3ϕ
Four 2 HP exhaust fans .... 208 V, 3ϕ
12 kW special loads (noncontinuous-duty) .... 208 V, 3ϕ
5760 VA special loads (continuous-duty) .... 120 V per phase

______ **25-1.** Which phase is used to size the service-entrance conductors?

______ **25-2.** What size THWN Cu conductors, paralleled four times per phase, is required for phases A, B, and C?

______ **25-3.** What size Cu grounding electrode conductor is required to ground the service to a metal water pipe?

______ **25-4.** What size OCPD is required for the service main?

______ **25-5.** What size 3ϕ, 4-wire panelboard is required for the service main?

STANDARD CALCULATION—COMMERCIAL AND INDUSTRIAL

Use for Problem 16

1. LIGHTING LOADS—220, Part B

General lighting loads—Table 220-3(b); 220-10(b)

_______ sq. ft. × _______ VA × _______% = _______ VA

Show window loads—220-12

_______' × 200 VA = _______ VA

Track lighting loads—410-102

$\frac{\_\_\_\_\_\_'}{2'}$ × 180 VA = _______ VA

Outside lighting loads—220-10(b)

_______ VA × _______ units × _______% = _______ VA

Sign lighting loads—220-3, FPN; 600-6(c)

_______ VA × _______% = _______ VA

	PHASES	NEUTRAL
(total) _______ VA	VA	VA

2. RECEPTACLE LOADS—220-3(c)(5)

Noncontinuous duty—220-10(b); Table 220-13

#_______ × 180 VA = _______ VA = _______ VA

Applying demand factors—Table 220-13

First 10,000 VA × 100% = _______ VA = _______ VA

Remaining _______ VA × 50% = _______ VA

Continuous duty—220-10(b) #_______ × 180 VA × 125% = _______ VA

Multioutlet assembly—220-3(c)

Heavy duty = _______' × 180 VA = _______ VA

Light duty = $\frac{\_\_\_\_\_\_'}{5'}$ × 180 VA = _______ VA

(total) _______ VA | VA | VA

3. SPECIAL LOADS—220-10(b)

_______ = _______ VA × _______% = _______ VA

_______ = _______ VA × _______% = _______ VA

_______ = _______ VA × _______% = _______ VA

_______ = _______ VA × _______% = _______ VA

_______ = _______ VA × _______% = _______ VA

(total) _______ VA | VA | VA

4. HEATING OR A/C LOAD—220-21 _______ VA | VA | VA

5. MOTOR LOADS—430-22(a); 430-24; 430-25

				3ϕ only		
_______ =	_______ A ×	_______% ×	_______ V	$\times \sqrt{3}$	=	_______ VA
_______ =	_______ A ×	_______% ×	_______ V	$\times \sqrt{3}$	=	_______ VA
_______ =	_______ A ×	_______% ×	_______ V	$\times \sqrt{3}$	=	_______ VA
_______ =	_______ A ×	_______% ×	_______ V	$\times \sqrt{3}$	=	_______ VA

(total) _______ VA | VA | VA

6. LARGEST MOTOR LOAD—220-14

				3ϕ only				
ϕ	_______ A ×	_______% ×	_______ V	$\times \sqrt{3}$	× 25% =	_______ VA	VA	VA
N	_______ A ×	_______% ×	_______ V	$\times \sqrt{3}$	× 25% =	_______ VA	VA	VA

TOTAL = VA | VA

1ϕ service: $I = \frac{VA}{V}$

PHASES $I = \frac{\_\_\_\_\_\_ VA}{V}$ = _______ A

NEUTRAL $I = \frac{\_\_\_\_\_\_ VA}{V}$ = _______ A

3ϕ service: $I = \frac{VA}{V \times \sqrt{3}}$

$I = \frac{\_\_\_\_\_\_ VA}{V \times \sqrt{3}}$ = _______ A

$I = \frac{\_\_\_\_\_\_ VA}{V \times \sqrt{3}}$ = _______ A

If over 200 A—220-22; First 200 A × 100% = 200 A

Remaining _______ A × 70% = _______ A

A

STANDARD CALCULATION—COMMERCIAL AND INDUSTRIAL

Use for Problem 17

1. LIGHTING LOADS—220, Part B

General lighting loads—Table 220-3(b); 220-10(b)

_______ sq. ft. × _______ VA × _______% = _______ VA

Show window loads—220-12

_______′ × 200 VA = _______ VA

Track lighting loads—410-102

$\frac{\_\_\_\_\_\_\_'}{2'}$ × 180 VA = _______ VA

Outside lighting loads—220-10(b)

_______ VA × _______ units × _______% = _______ VA

Sign lighting loads—220-3, FPN; 600-6(c)

_______ VA × _______% = _______ VA

		PHASES	NEUTRAL
(total)	_______ VA	VA	VA

2. RECEPTACLE LOADS—220-3(c)(5)

Noncontinuous duty—220-10(b); Table 220-13

#_______ × 180 VA = _______ VA = _______ VA

Applying demand factors—Table 220-13

First 10,000 VA × 100% = _______ VA = _______ VA

Remaining _______ VA × 50% = _______ VA

Continuous duty—220-10(b) #_______ × 180 VA × 125% = _______ VA

Multioutlet assembly—220-3(c)

Heavy duty = _______′ × 180 VA = _______ VA

Light duty = $\frac{\_\_\_\_\_\_\_'}{5'}$ × 180 VA = _______ VA

(total) _______ VA | VA | VA

3. SPECIAL LOADS—220-10(b)

_______ = _______ VA × _______% = _______ VA

_______ = _______ VA × _______% = _______ VA

_______ = _______ VA × _______% = _______ VA

_______ = _______ VA × _______% = _______ VA

_______ = _______ VA × _______% = _______ VA

(total) _______ VA | VA | VA

4. HEATING OR A/C LOAD—220-21 _______ VA | VA | VA

5. MOTOR LOADS—430-22(a); 430-24; 430-25

_______ = _______ A × _______% × _______ V [3ϕ only: × $\sqrt{3}$] = _______ VA

_______ = _______ A × _______% × _______ V [× $\sqrt{3}$] = _______ VA

_______ = _______ A × _______% × _______ V [× $\sqrt{3}$] = _______ VA

_______ = _______ A × _______% × _______ V [× $\sqrt{3}$] = _______ VA

(total) _______ VA | VA | VA

6. LARGEST MOTOR LOAD—220-14

ϕ _______ A × _______% × _______ V [3ϕ only: × $\sqrt{3}$] × 25% = _______ VA | VA | VA

N _______ A × _______% × _______ V [× $\sqrt{3}$] × 25% = _______ VA | VA | VA

TOTAL = | VA | VA

1ϕ service: $I = \frac{VA}{V}$

PHASES $I = \frac{\_\_\_\_\_\_\_ VA}{V}$ = _______ A

NEUTRAL $I = \frac{\_\_\_\_\_\_\_ VA}{V}$ = _______ A

3ϕ service: $I = \frac{VA}{V \times \sqrt{3}}$

$I = \frac{\_\_\_\_\_\_\_ VA}{V \times \sqrt{3}}$ = _______ A

$I = \frac{\_\_\_\_\_\_\_ VA}{V \times \sqrt{3}}$ = _______ A

If over 200 A—220-22; First 200 A × 100% = 200 A

Remaining _______ A × 70% = _______ A

A

STANDARD CALCULATION—COMMERCIAL AND INDUSTRIAL

Use for Problem 18

1. LIGHTING LOADS—220, Part B

General lighting loads—Table 220-3(b); 220-10(b)

_______ sq. ft. × _______ VA × _______% = _______ VA

Show window loads—220-12

_______′ × 200 VA = _______ VA

Track lighting loads—410-102

_______′ / 2′ × 180 VA = _______ VA

Outside lighting loads—220-10(b)

_______ VA × _______ units × _______% = _______ VA

Sign lighting loads—220-3, FPN; 600-6(c)

_______ VA × _______% = _______ VA

		PHASES	NEUTRAL
(total)	_______ VA	VA	VA

2. RECEPTACLE LOADS—220-3(c)(5)

Noncontinuous duty—220-10(b); Table 220-13

#_______ × 180 VA = _______ VA = _______ VA

Applying demand factors—Table 220-13

First 10,000 VA × 100% = _______ VA = _______ VA

Remaining _______ VA × 50% = _______ VA

Continuous duty—220-10(b) #_______ × 180 VA × 125% = _______ VA

Multioutlet assembly—220-3(c)

Heavy duty = _______′ × 180 VA = _______ VA

Light duty = _______′ / 5′ × 180 VA = _______ VA

		PHASES	NEUTRAL
(total)	_______ VA	VA	VA

3. SPECIAL LOADS—220-10(b)

_______ = _______ VA × _______% = _______ VA

_______ = _______ VA × _______% = _______ VA

_______ = _______ VA × _______% = _______ VA

_______ = _______ VA × _______% = _______ VA

_______ = _______ VA × _______% = _______ VA

		PHASES	NEUTRAL
(total)	_______ VA	VA	VA
4. HEATING OR A/C LOAD—220-21	_______ VA	VA	VA

5. MOTOR LOADS—430-22(a); 430-24; 430-25

_______ = _______ A × _______% × _______ V [3ϕ only: × $\sqrt{3}$] = _______ VA

_______ = _______ A × _______% × _______ V [3ϕ only: × $\sqrt{3}$] = _______ VA

_______ = _______ A × _______% × _______ V [3ϕ only: × $\sqrt{3}$] = _______ VA

_______ = _______ A × _______% × _______ V [3ϕ only: × $\sqrt{3}$] = _______ VA

		PHASES	NEUTRAL
(total)	_______ VA	VA	VA

6. LARGEST MOTOR LOAD—220-14

		PHASES	NEUTRAL
ϕ _______ A × _______% × _______ V [3ϕ only: × $\sqrt{3}$] × 25% =	_______ VA	VA	VA
N _______ A × _______% × _______ V [3ϕ only: × $\sqrt{3}$] × 25% =	_______ VA	VA	VA
TOTAL =		VA	VA

1ϕ service: $I = \frac{VA}{V}$

PHASES $I = \frac{\_\_\_\_\_\_\_ VA}{V}$ = _______ A

NEUTRAL $I = \frac{\_\_\_\_\_\_\_ VA}{V}$ = _______ A

3ϕ service: $I = \frac{VA}{V \times \sqrt{3}}$

$I = \frac{\_\_\_\_\_\_\_ VA}{V \times \sqrt{3}}$ = _______ A

$I = \frac{\_\_\_\_\_\_\_ VA}{V \times \sqrt{3}}$ = _______ A

If over 200 A—220-22; First 200 A × 100% = 200 A

Remaining _______ A × 70% = _______ A

A

STANDARD CALCULATION—COMMERCIAL AND INDUSTRIAL

Use for Problem 19

1. LIGHTING LOADS—220, Part B

General lighting loads—Table 220-3(b); 220-10(b)

________ sq. ft. × ________ VA × ______% = ________ VA

Show window loads—220-12

________′ × 200 VA = ________ VA

Track lighting loads—410-102

$\frac{\_\_\_\_\_\_'}{2'}$ × 180 VA = ________ VA

Outside lighting loads—220-10(b)

________ VA × ________ units × ______% = ________ VA

Sign lighting loads—220-3, FPN; 600-6(c)

________ VA × ________% = ________ VA

		PHASES	NEUTRAL
(total)	________ VA	VA	VA

2. RECEPTACLE LOADS—220-3(c)(5)

Noncontinuous duty—220-10(b); Table 220-13

#________ × 180 VA = ________ VA = ________ VA

Applying demand factors—Table 220-13

First 10,000 VA × 100% = ________ VA = ________ VA

Remaining ________ VA × 50% = ________ VA

Continuous duty—220-10(b) #________ × 180 VA × 125% = ________ VA

Multioutlet assembly—220-3(c)

Heavy duty = ________′ × 180 VA = ________ VA

Light duty = $\frac{\_\_\_\_\_\_'}{5'}$ × 180 VA = ________ VA

(total) ________ VA | VA | VA

3. SPECIAL LOADS—220-10(b)

________ = ________ VA × ______% = ________ VA

________ = ________ VA × ______% = ________ VA

________ = ________ VA × ______% = ________ VA

________ = ________ VA × ______% = ________ VA

________ = ________ VA × ______% = ________ VA

(total) ________ VA | VA | VA

4. HEATING OR A/C LOAD—220-21 ________ VA | VA | VA

5. MOTOR LOADS—430-22(a); 430-24; 430-25

					3ϕ only		
________	= ________ A	× ______%	× ______ V		$\times \sqrt{3}$	=	________ VA
________	= ________ A	× ______%	× ______ V		$\times \sqrt{3}$	=	________ VA
________	= ________ A	× ______%	× ______ V		$\times \sqrt{3}$	=	________ VA
________	= ________ A	× ______%	× ______ V		$\times \sqrt{3}$	=	________ VA

(total) ________ VA | VA | VA

6. LARGEST MOTOR LOAD—220-14

				3ϕ only				
ϕ	________ A ×	______% ×	________ V	$\times \sqrt{3}$	× 25% =	________ VA	VA	VA
N	________ A ×	______% ×	________ V	$\times \sqrt{3}$	× 25% =	________ VA	VA	VA

TOTAL = VA | VA

1ϕ service: $I = \frac{VA}{V}$

PHASES $I = \frac{\_\_\_\_\_\_ VA}{V}$ = ________ A

NEUTRAL $I = \frac{\_\_\_\_\_\_ VA}{V}$ = ________ A

3ϕ service: $I = \frac{VA}{V \times \sqrt{3}}$

$I = \frac{\_\_\_\_\_\_ VA}{V \times \sqrt{3}}$ = ________ A

$I = \frac{\_\_\_\_\_\_ VA}{V \times \sqrt{3}}$ = ________ A

If over 200 A—220-22; First 200 A × 100% = 200 A

Remaining ________ A × 70% = ________ A

A

OPTIONAL CALCULATION—SCHOOL

Use for Problem 20

1. TOTAL VA—Table 220-34

Total connected load = ________ VA

2. TOTAL SQUARE FOOTAGE—Table 220-34

Square footage = ________ sq. ft.

3. VA PER SQ. FT.—Table 220-34

$\frac{\text{\_\_\_\_\_\_\_\_ VA}}{\text{sq. ft.}}$ = ________ VA/sq. ft.

4. APPLYING DEMAND FACTORS—Table 220-34

First 3 VA per sq. ft. × 100%

Next 3 VA to 20 VA per sq. ft. × 75%

Remaining 21 VA per sq. ft × 25%

	PHASES
3 VA/sq. ft. × ________ sq. ft. × 100% =	VA
________ VA/sq. ft. × ________ sq. ft. × 75% =	VA
________ VA/sq. ft. × ________ sq. ft. × 25% =	VA
TOTAL =	VA

NEUTRAL

(Loads from Standard Calculation)

1. Lighting loads = ________ VA
2. Receptacle loads = ________ VA
3. Special loads = ________ VA
4. Heating or A/C load = ________ VA
5. Motor loads = ________ VA
6. Largest motor load = ________ VA

TOTAL = ________ VA

________ A

1ϕ service: $I = \frac{VA}{V}$

PHASES $I = \frac{\text{\_\_\_\_\_\_\_\_ VA}}{V}$ = ________ A

NEUTRAL $I = \frac{\text{\_\_\_\_\_\_\_\_ VA}}{V}$ = ________ A

3ϕ service: $I = \frac{VA}{V \times \sqrt{3}}$

$I = \frac{\text{\_\_\_\_\_\_\_\_ VA}}{V \times \sqrt{3}}$ = ________ A

$I = \frac{\text{\_\_\_\_\_\_\_\_ VA}}{V \times \sqrt{3}}$ = ________ A

If over 200 A—220-22; First 200 A × 100% = 200 A

Remaining ________ A × 70% = ________ A

________ A

Use for Problem 20

STANDARD CALCULATION—COMMERCIAL AND INDUSTRIAL

1. LIGHTING LOADS—220, Part B

General lighting loads—Table 220-3(b); 220-10(b)

________ sq. ft. × ________ VA × ______% = ________ VA

Show window loads—220-12

________′ × 200 VA = ________ VA

Track lighting loads—410-102

$\frac{\_\_\_\_\_\_\_\_'}{2'}$ × 180 VA = ________ VA

Outside lighting loads—220-10(b)

________ VA × ________ units × ______% = ________ VA

Sign lighting loads—220-3, FPN; 600-6(c)

________ VA × ________% = ________ VA

		PHASES	NEUTRAL
(total)	________ VA	VA	VA

2. RECEPTACLE LOADS—220-3(c)(5)

Noncontinuous duty—220-10(b); Table 220-13

#________ × 180 VA = ________ VA = ________ VA

Applying demand factors—Table 220-13

First 10,000 VA × 100% = ________ VA = ________ VA

Remaining ________ VA × 50% = ________ VA

Continuous duty—220-10(b) #________ × 180 VA × 125% = ________ VA

Multioutlet assembly—220-3(c)

Heavy duty = ________′ × 180 VA = ________ VA

Light duty = $\frac{\_\_\_\_\_\_\_\_'}{5'}$ × 180 VA = ________ VA

(total) ________ VA — VA — VA

3. SPECIAL LOADS—220-10(b)

________ = ________ VA × ______% = ________ VA

________ = ________ VA × ______% = ________ VA

________ = ________ VA × ______% = ________ VA

________ = ________ VA × ______% = ________ VA

________ = ________ VA × ______% = ________ VA

(total) ________ VA — VA — VA

4. HEATING OR A/C LOAD—220-21 ________ VA — VA — VA

5. MOTOR LOADS—430-22(a); 430-24; 430-25 (3ϕ only)

________ = ________ A × ______% × ______ V × $\sqrt{3}$ = ________ VA

________ = ________ A × ______% × ______ V × $\sqrt{3}$ = ________ VA

________ = ________ A × ______% × ______ V × $\sqrt{3}$ = ________ VA

________ = ________ A × ______% × ______ V × $\sqrt{3}$ = ________ VA

(total) ________ VA — VA — VA

6. LARGEST MOTOR LOAD—220-14 (3ϕ only)

ϕ ________ A × ______% × ________ V × $\sqrt{3}$ × 25% = ________ VA — VA — VA

N ________ A × ______% × ________ V × $\sqrt{3}$ × 25% = ________ VA — VA — VA

TOTAL = VA — VA

1ϕ service: $I = \frac{VA}{V}$

PHASES $I = \frac{\_\_\_\_\_\_\_\_ VA}{V}$ = ________ A

NEUTRAL $I = \frac{\_\_\_\_\_\_\_\_ VA}{V}$ = ________ A

3ϕ service: $I = \frac{VA}{V \times \sqrt{3}}$

$I = \frac{\_\_\_\_\_\_\_\_ VA}{V \times \sqrt{3}}$ = ________ A

$I = \frac{\_\_\_\_\_\_\_\_ VA}{V \times \sqrt{3}}$ = ________ A

If over 200 A—220-22; First 200 A × 100% = 200 A

Remaining ______ A × 70% = ________ A

A

Use for Problem 21

STANDARD CALCULATION—COMMERCIAL AND INDUSTRIAL

1. LIGHTING LOADS—220, Part B

General lighting loads—Table 220-3(b); 220-10(b)

_______ sq. ft. × _______ VA × _______% = _______ VA

Show window loads—220-12

_______′ × 200 VA = _______ VA

Track lighting loads—410-102

$\frac{\_\_\_\_\_\_\_'}{2'}$ × 180 VA = _______ VA

Outside lighting loads—220-10(b)

_______ VA × _______ units × _______% = _______ VA

Sign lighting loads—220-3, FPN; 600-6(c)

_______ VA × _______% = _______ VA

		PHASES	NEUTRAL
(total)	_______ VA	VA	VA

2. RECEPTACLE LOADS—220-3(c)(5)

Noncontinuous duty—220-10(b); Table 220-13

#_______ × 180 VA = _______ VA = _______ VA

Applying demand factors—Table 220-13

First 10,000 VA × 100% = _______ VA = _______ VA

Remaining _______ VA × 50% = _______ VA

Continuous duty—220-10(b) #_______ × 180 VA × 125% = _______ VA

Multioutlet assembly—220-3(c)

Heavy duty = _______′ × 180 VA = _______ VA

Light duty = $\frac{\_\_\_\_\_\_\_'}{5'}$ × 180 VA = _______ VA

		PHASES	NEUTRAL
(total)	_______ VA	VA	VA

3. SPECIAL LOADS—220-10(b)

_______ = _______ VA × _______% = _______ VA

_______ = _______ VA × _______% = _______ VA

_______ = _______ VA × _______% = _______ VA

_______ = _______ VA × _______% = _______ VA

_______ = _______ VA × _______% = _______ VA

		PHASES	NEUTRAL
(total)	_______ VA	VA	VA

4. HEATING OR A/C LOAD—220-21 _______ VA | VA | VA

5. MOTOR LOADS—430-22(a); 430-24; 430-25

				3ϕ only		
_______ =	_______ A ×	_______% ×	_______ V	$\times \sqrt{3}$	=	_______ VA
_______ =	_______ A ×	_______% ×	_______ V	$\times \sqrt{3}$	=	_______ VA
_______ =	_______ A ×	_______% ×	_______ V	$\times \sqrt{3}$	=	_______ VA
_______ =	_______ A ×	_______% ×	_______ V	$\times \sqrt{3}$	=	_______ VA

		PHASES	NEUTRAL
(total)	_______ VA	VA	VA

6. LARGEST MOTOR LOAD—220-14

				3ϕ only			PHASES	NEUTRAL
ϕ	_______ A ×	_______% ×	_______ V	$\times \sqrt{3}$	× 25% =	_______ VA	VA	VA
N	_______ A ×	_______% ×	_______ V	$\times \sqrt{3}$	× 25% =	_______ VA	VA	VA
						TOTAL =	VA	VA

1ϕ service: $I = \frac{VA}{V}$

PHASES $I = \frac{\_\_\_\_\_\_\_ VA}{V}$ = _______ A

NEUTRAL $I = \frac{\_\_\_\_\_\_\_ VA}{V}$ = _______ A

3ϕ service: $I = \frac{VA}{V \times \sqrt{3}}$

$I = \frac{\_\_\_\_\_\_\_ VA}{V \times \sqrt{3}}$ = _______ A

$I = \frac{\_\_\_\_\_\_\_ VA}{V \times \sqrt{3}}$ = _______ A

If over 200 A—220-22; First 200 A × 100% = 200 A

Remaining _______ A × 70% = _______ A

A

STANDARD CALCULATION—COMMERCIAL AND INDUSTRIAL

Use for Problem 21

1. LIGHTING LOADS—220, Part B

General lighting loads—Table 220-3(b); 220-10(b)

______ sq. ft. × ______ VA × ______% = ______ VA

Show window loads—220-12

______′ × 200 VA = ______ VA

Track lighting loads—410-102

$\frac{\_\_\_\_\_\_'}{2'}$ × 180 VA = ______ VA

Outside lighting loads—220-10(b)

______ VA × ______ units × ______% = ______ VA

Sign lighting loads—220-3, FPN; 600-6(c)

______ VA × ______% = ______ VA

		PHASES	NEUTRAL
(total)	______ VA	VA	VA

2. RECEPTACLE LOADS—220-3(c)(5)

Noncontinuous duty—220-10(b); Table 220-13

#______ × 180 VA = ______ VA = ______ VA

Applying demand factors—Table 220-13

First 10,000 VA × 100% = ______ VA = ______ VA

Remaining ______ VA × 50% = ______ VA

Continuous duty—220-10(b) #______ × 180 VA × 125% = ______ VA

Multioutlet assembly—220-3(c)

Heavy duty = ______′ × 180 VA = ______ VA

Light duty = $\frac{\_\_\_\_\_\_'}{5'}$ × 180 VA = ______ VA

		PHASES	NEUTRAL
(total)	______ VA	VA	VA

3. SPECIAL LOADS—220-10(b)

______ = ______ VA × ______% = ______ VA

______ = ______ VA × ______% = ______ VA

______ = ______ VA × ______% = ______ VA

______ = ______ VA × ______% = ______ VA

______ = ______ VA × ______% = ______ VA

		PHASES	NEUTRAL
(total)	______ VA	VA	VA

4. HEATING OR A/C LOAD—220-21 ______ VA | VA | VA

5. MOTOR LOADS—430-22(a); 430-24; 430-25

				3ϕ only		
______ =	______ A ×	______% ×	______ V	$\times \sqrt{3}$	=	______ VA
______ =	______ A ×	______% ×	______ V	$\times \sqrt{3}$	=	______ VA
______ =	______ A ×	______% ×	______ V	$\times \sqrt{3}$	=	______ VA
______ =	______ A ×	______% ×	______ V	$\times \sqrt{3}$	=	______ VA

		PHASES	NEUTRAL
(total)	______ VA	VA	VA

6. LARGEST MOTOR LOAD—220-14

				3ϕ only			PHASES	NEUTRAL
ϕ	______ A ×	______% ×	______ V	$\times \sqrt{3}$	× 25% =	______ VA	VA	VA
N	______ A ×	______% ×	______ V	$\times \sqrt{3}$	× 25% =	______ VA	VA	VA
						TOTAL =	VA	VA

1ϕ service: $I = \frac{VA}{V}$

PHASES I = $\frac{\_\_\_\_\_\_ VA}{V}$ = ______ A

NEUTRAL I = $\frac{\_\_\_\_\_\_ VA}{V}$ = ______ A

3ϕ service: $I = \frac{VA}{V \times \sqrt{3}}$

I = $\frac{\_\_\_\_\_\_ VA}{V \times \sqrt{3}}$ = ______ A

I = $\frac{\_\_\_\_\_\_ VA}{V \times \sqrt{3}}$ = ______ A

If over 200 A—220-22; First 200 A × 100% = 200 A

Remaining ______ A × 70% = ______ A

A

Use for Problem 22

STANDARD CALCULATION—COMMERCIAL AND INDUSTRIAL

1. LIGHTING LOADS—220, Part B

General lighting loads—Table 220-3(b); 220-10(b)

______ sq. ft. × ______ VA × ______% = ______ VA

Show window loads—220-12

______′ × 200 VA = ______ VA

Track lighting loads—410-102

$\frac{\text{\_\_\_\_\_\_}'}{2'}$ × 180 VA = ______ VA

Outside lighting loads—220-10(b)

______ VA × ______ units × ______% = ______ VA

Sign lighting loads—220-3, FPN; 600-6(c)

______ VA × ______% = ______ VA

		PHASES	NEUTRAL
(total)	______ VA	VA	VA

2. RECEPTACLE LOADS—220-3(c)(5)

Noncontinuous duty—220-10(b); Table 220-13

#______ × 180 VA = ______ VA = ______ VA

Applying demand factors—Table 220-13

First 10,000 VA × 100% = ______ VA = ______ VA

Remaining ______ VA × 50% = ______ VA

Continuous duty—220-10(b) #______ × 180 VA × 125% = ______ VA

Multioutlet assembly—220-3(c)

Heavy duty = ______′ × 180 VA = ______ VA

Light duty = $\frac{\text{\_\_\_\_\_\_}'}{5'}$ × 180 VA = ______ VA

		PHASES	NEUTRAL
(total)	______ VA	VA	VA

3. SPECIAL LOADS—220-10(b)

______ = ______ VA × ______% = ______ VA

______ = ______ VA × ______% = ______ VA

______ = ______ VA × ______% = ______ VA

______ = ______ VA × ______% = ______ VA

______ = ______ VA × ______% = ______ VA

		PHASES	NEUTRAL
(total)	______ VA	VA	VA

4. HEATING OR A/C LOAD—220-21 ______ VA | VA | VA

5. MOTOR LOADS—430-22(a); 430-24; 430-25

				3ϕ only		
______ =	______ A ×	______% ×	______ V	× $\sqrt{3}$	=	______ VA
______ =	______ A ×	______% ×	______ V	× $\sqrt{3}$	=	______ VA
______ =	______ A ×	______% ×	______ V	× $\sqrt{3}$	=	______ VA
______ =	______ A ×	______% ×	______ V	× $\sqrt{3}$	=	______ VA

		PHASES	NEUTRAL
(total)	______ VA	VA	VA

6. LARGEST MOTOR LOAD—220-14

				3ϕ only			PHASES	NEUTRAL
ϕ	______ A ×	______% ×	______ V	× $\sqrt{3}$	× 25% =	______ VA	VA	VA
N	______ A ×	______% ×	______ V	× $\sqrt{3}$	× 25% =	______ VA	VA	VA
						TOTAL =	VA	VA

1ϕ service: $I = \frac{VA}{V}$

PHASES $I = \frac{\text{\_\_\_\_\_\_ VA}}{V}$ = ______ A

NEUTRAL $I = \frac{\text{\_\_\_\_\_\_ VA}}{V}$ = ______ A

3ϕ service: $I = \frac{VA}{V \times \sqrt{3}}$

$I = \frac{\text{\_\_\_\_\_\_ VA}}{V \times \sqrt{3}}$ = ______ A

$I = \frac{\text{\_\_\_\_\_\_ VA}}{V \times \sqrt{3}}$ = ______ A

If over 200 A—220-22; First 200 A × 100% = 200 A

Remaining ______ A × 70% = ______ A

A

STANDARD CALCULATION—HOTELS AND MOTELS

Use for Problem 23

1. GENERAL LIGHTING LOADS—Table 220-3(b)

_____ sq. ft. × 2 VA × _____ units = _____ VA

_____ sq. ft. × 2 VA × _____ units = _____ VA

_____ sq. ft. × 2 VA × _____ units = _____ VA

_____ VA

Applying demand factors—Table 220-11

First 20,000 VA × 50% = 10,000 VA

Next _____ VA × 40% = _____ VA

Remaining _____ VA × 30% = _____ VA

(total) _____ VA _____ VA

Show window loads—220-12

_____′ × 200 VA = _____ VA

Track lighting loads—410-102

$\frac{\text{\_\_\_\_\_}'}{2'}$ × 180 VA = _____ VA

Outside lighting loads—220-10(b)

_____ VA × _____ units × _____% = _____ VA

Sign lighting loads—220-3, FPN; 600-6(c)

_____ VA × _____% = _____ VA

		PHASES	NEUTRAL
(total)	_____ VA	VA	VA

2. RECEPTACLE LOADS—220-3(c)(5)

Continuous duty—220-10(b); Table 220-13

_____ VA × 125% = (total)	_____ VA	VA	VA

3. SPECIAL LOADS—220-10(b)

_____ = _____ VA × _____ units × _____% = _____ VA

_____ = _____ VA × _____ units × _____% = _____ VA

_____ = _____ VA × _____ units × _____% = _____ VA

_____ = _____ VA × _____ units × _____% = _____ VA

_____ = _____ VA × _____ units × _____% = _____ VA

_____ = _____ VA × _____ units × _____% = _____ VA

_____ = _____ VA × _____ units × _____% = _____ VA

(total)	_____ VA	VA	VA

4. HEATING OR A/C LOAD—220-21

Heating unit = _____ VA × 100% × _____ units = _____ VA

A/C unit = _____ VA × 100% × _____ units = _____ VA

Heat pump = _____ VA × 100% × _____ units = _____ VA

(largest load)	_____ VA	VA	VA

5. LARGEST MOTOR LOAD—220-14

ϕ _____ A × _____% × _____ V × 25% =	_____ VA	VA	VA
N _____ A × _____% × _____ V × 25% =	_____ VA	VA	VA
TOTAL =		**VA**	**VA**

1ϕ service: $I = \frac{VA}{V}$

PHASES $I = \frac{\text{\_\_\_\_\_ VA}}{V}$ = _____ A

NEUTRAL $I = \frac{\text{\_\_\_\_\_ VA}}{V}$ = _____ A

3ϕ service: $I = \frac{VA}{V \times \sqrt{3}}$

$I = \frac{\text{\_\_\_\_\_ VA}}{V \times \sqrt{3}}$ = _____ A

$I = \frac{\text{\_\_\_\_\_ VA}}{V \times \sqrt{3}}$ = _____ A

If over 200 A—220-22; First 200 A × 100% = 200 A

Remaining _____ A × 70% = _____ A

A

STANDARD CALCULATION—HOTELS AND MOTELS

Use for Problem 23

1. GENERAL LIGHTING LOADS—Table 220-3(b)

_____ sq. ft. × 2 VA × _____ units = _____ VA

_____ sq. ft. × 2 VA × _____ units = _____ VA

_____ sq. ft. × 2 VA × _____ units = _____ VA

_____ VA

Applying demand factors—Table 220-11

First 20,000 VA × 50% = 10,000 VA

Next _____ VA × 40% = _____ VA

Remaining _____ VA × 30% = _____ VA

(total) _____ VA _____ VA

Show window loads—220-12

_____' × 200 VA = _____ VA

Track lighting loads—410-102

_____' × 180 VA = _____ VA

2'

Outside lighting loads—220-10(b)

_____ VA × _____ units × _____% = _____ VA

Sign lighting loads—220-3, FPN; 600-6(c)

_____ VA × _____% = _____ VA

		PHASES	NEUTRAL
(total)	_____ VA	VA	VA

2. RECEPTACLE LOADS—220-3(c)(5)

Continuous duty—220-10(b); Table 220-13

_____ VA × 125% = (total) _____ VA | VA | VA

3. SPECIAL LOADS—220-10(b)

_____ = _____ VA × _____ units × _____% = _____ VA

_____ = _____ VA × _____ units × _____% = _____ VA

_____ = _____ VA × _____ units × _____% = _____ VA

_____ = _____ VA × _____ units × _____% = _____ VA

_____ = _____ VA × _____ units × _____% = _____ VA

_____ = _____ VA × _____ units × _____% = _____ VA

_____ = _____ VA × _____ units × _____% = _____ VA

(total) _____ VA | VA | VA

4. HEATING OR A/C LOAD—220-21

Heating unit = _____ VA × 100% × _____ units = _____ VA

A/C unit = _____ VA × 100% × _____ units = _____ VA

Heat pump = _____ VA × 100% × _____ units = _____ VA

(largest load) _____ VA | VA | VA

5. LARGEST MOTOR LOAD—220-14

ϕ _____ A × _____% × _____ V × 25% = _____ VA | VA | VA

N _____ A × _____% × _____ V × 25% = _____ VA | VA | VA

TOTAL = VA | VA

1ϕ service: $I = \frac{VA}{V}$

PHASES $I = \frac{\_\_\_\_\_ VA}{V}$ = _____ A

NEUTRAL $I = \frac{\_\_\_\_\_ VA}{V}$ = _____ A

3ϕ service: $I = \frac{VA}{V \times \sqrt{3}}$

$I = \frac{\_\_\_\_\_ VA}{V \times \sqrt{3}}$ = _____ A

$I = \frac{\_\_\_\_\_ VA}{V \times \sqrt{3}}$ = _____ A

If over 200 A—220-22; First 200 A × 100% = 200 A

Remaining _____ A × 70% = _____ A

A

Use for Problem 24

STANDARD CALCULATION—COMMERCIAL AND INDUSTRIAL (CARRYING PHASES AND NEUTRAL IN VA)

	PHASE	PHASE	PHASE	NEUTRAL
1. LIGHTING LOADS—220, Part B				
General lighting loads—Table 220-3(b); 220-10(b)				
______ sq. ft. × ______ VA × ______% ÷ ______ =	______ VA	______ VA	______ VA	______ VA
Show window loads—220-12				
______′ × 200 VA ÷ ______ =	______ VA	______ VA	______ VA	______ VA
Track lighting loads—410-102				
______′ (2′) × 180 VA ÷ ______ =	______ VA	______ VA	______ VA	______ VA
Outside lighting loads—220-10(b)				
______ VA × ______ units × ______% ÷ ______ =	______ VA	______ VA	______ VA	______ VA
Sign lighting loads—220-3, FPN; 600-6(c)				
______ VA × ______% ÷ ______ =	______ VA	______ VA	______ VA	______ VA
2. RECEPTACLE LOADS—220-3(c)(5)				
Noncontinuous duty—220-10(b)				
______ VA × 100% = ______ VA ÷ ______ =	______ VA	______ VA	______ VA	______ VA
Applying demand factors—Table 220-13				
First 10,000 VA × 100% ÷ ______ =	______ VA	______ VA	______ VA	______ VA
Remaining ______ VA × 50% ÷ ______ =	______ VA	______ VA	______ VA	______ VA
Continuous duty—220-10(b) ______ VA × 125% ÷ ______ =	______ VA	______ VA	______ VA	______ VA
Multioutlet assembly—220-3(c)				
Heavy duty = ______′ × 180 VA ÷ ______ =	______ VA	______ VA	______ VA	______ VA
Light duty = ______′ (5′) × 180 VA ÷ ______ =	______ VA	______ VA	______ VA	______ VA
3. SPECIAL LOADS—220-10(b)				
(load)				
______ = ______ VA × ______% ÷ ______ =	______ VA	______ VA	______ VA	______ VA
______ = ______ VA × ______% ÷ ______ =	______ VA	______ VA	______ VA	______ VA
______ = ______ VA × ______% ÷ ______ =	______ VA	______ VA	______ VA	______ VA
______ = ______ VA × ______% ÷ ______ =	______ VA	______ VA	______ VA	______ VA
______ = ______ VA × ______% ÷ ______ =	______ VA	______ VA	______ VA	______ VA
4. HEATING OR A/C LOAD—220-21	______ VA	______ VA	______ VA	______ VA
5. MOTOR LOADS—430-22(a); 430-24; 430-25				
(load) [3ϕ motors only: $\times \sqrt{3}$]				
______ = ______ A × ______ V $\times \sqrt{3}$ ÷ ______ =	______ VA	______ VA	______ VA	______ VA
______ = ______ A × ______ V $\times \sqrt{3}$ ÷ ______ =	______ VA	______ VA	______ VA	______ VA
______ = ______ A × ______ V $\times \sqrt{3}$ ÷ ______ =	______ VA	______ VA	______ VA	______ VA
______ = ______ A × ______ V $\times \sqrt{3}$ ÷ ______ =	______ VA	______ VA	______ VA	______ VA
6. LARGEST MOTOR LOAD—220-14				
ϕ ______ A × ______ V × ______% × 25% =	______ VA	______ VA	______ VA	______ VA
N ______ A × ______ V × ______% × 25% =	______ VA	______ VA	______ VA	______ VA
TOTAL =	**VA**	**VA**	**VA**	**VA**

1ϕ service: $I = \frac{VA}{V}$

PHASES $I = \frac{\_\_\_\_\_\_ VA}{V}$ = ______ A

NEUTRAL $I = \frac{\_\_\_\_\_\_ VA}{V}$ = ______ A

3ϕ service: $I = \frac{VA}{V \times \sqrt{3}}$

$I = \frac{\_\_\_\_\_\_ VA}{V \times \sqrt{3}}$ = ______ A

$I = \frac{\_\_\_\_\_\_ VA}{V \times \sqrt{3}}$ = ______ A

If over 200 A—220-22; First 200 A × 100% =	200 A
Remaining ______ A × 70% =	______ A
	A

Use for Problem 25

STANDARD CALCULATION—COMMERCIAL AND INDUSTRIAL (CARRYING PHASES AND NEUTRAL IN AMPS)

	PHASE	PHASE	PHASE	NEUTRAL
1. LIGHTING LOADS—220, Part B				
General lighting loads—Table 220-3(b); 220-10(b)				
____ sq. ft. × ____ VA × ____% ÷ ____ V =	____ A	____ A	____ A	____ A
Show window loads—220-12				
____′ × 200 VA ÷ ____ V =	____ A	____ A	____ A	____ A
Track lighting loads—410-102				
____′ × 180 VA ÷ ____ V =	____ A	____ A	____ A	____ A
2′				
Outside lighting loads—220-10(b)				
____ VA × ____ units × ____% ÷ ____ V =	____ A	____ A	____ A	____ A
Sign lighting loads—220-3, FPN; 600-6(c)				
____ VA × ____% ÷ ____ V =	____ A	____ A	____ A	____ A
2. RECEPTACLE LOADS—220-3(c)(5)				
Noncontinuous duty—220-10(b)				
____ VA × 100% = ____ VA ÷ ____ V =	____ A	____ A	____ A	____ A
Applying demand factors—Table 220-13				
First 10,000 VA × 100% ÷ ____ V =	____ A	____ A	____ A	____ A
Remaining ____ VA × 50% ÷ ____ V =	____ A	____ A	____ A	____ A
Continuous duty—220-10(b) ____ VA × 125% ÷ ____ V =	____ A	____ A	____ A	____ A
Multioutlet assembly—220-3(c)				
Heavy duty = ____′ × 180 VA ÷ ____ V =	____ A	____ A	____ A	____ A
Light duty = ____′ × 180 VA ÷ ____ V =	____ A	____ A	____ A	____ A
5′				
3. SPECIAL LOADS—220-10(b)				
(load)				
____ = ____ VA × ____% ÷ ____ V =	____ A	____ A	____ A	____ A
____ = ____ VA × ____% ÷ ____ V =	____ A	____ A	____ A	____ A
____ = ____ VA × ____% ÷ ____ V =	____ A	____ A	____ A	____ A
____ = ____ VA × ____% ÷ ____ V =	____ A	____ A	____ A	____ A
____ = ____ VA × ____% ÷ ____ V =	____ A	____ A	____ A	____ A
4. HEATING OR A/C LOAD—220-21	____ A	____ A	____ A	____ A
5. MOTOR LOADS—430-22(a); 430-24; 430-25				
(load)				
____ = ____ A × ____# ÷ ____ =	____ A	____ A	____ A	____ A
____ = ____ A × ____# ÷ ____ =	____ A	____ A	____ A	____ A
____ = ____ A × ____# ÷ ____ =	____ A	____ A	____ A	____ A
____ = ____ A × ____# ÷ ____ =	____ A	____ A	____ A	____ A
6. LARGEST MOTOR LOAD—220-14				
ϕ ____ A × ____ V × ____% × 25% =	____ A	____ A	____ A	____ A
N ____ A × ____ V × ____% × 25% =	____ A	____ A	____ A	____ A
TOTAL =	A	A	A	A

If over 200 A—220-22; First 200 A × 100% = 200 A

Remaining ____ A × 70% = ____ A

A

1000 Questions

9

The 1990 NEC® contains the following:

	Article(s)
Introduction	**90**
Chapter 1. General	**100–110**
Chapter 2. Wiring Design and Protection	**200–280**
Chapter 3. Wiring Methods and Materials	**300–384**
Chapter 4. Equipment for General Use	**400–480**
Chapter 5. Special Occupancies	**500–555**
Chapter 6. Special Equipment	**600–690**
Chapter 7. Special Conditions	**700–780**
Chapter 8. Communications Systems	**800–820**
Chapter 9. Tables and Examples	

Master electrician's examinations include questions based on the introduction and chapters 1–8. Questions on exams may be True-False, Completion, Multiple-Choice, or Short Answer as decided by the examining board.

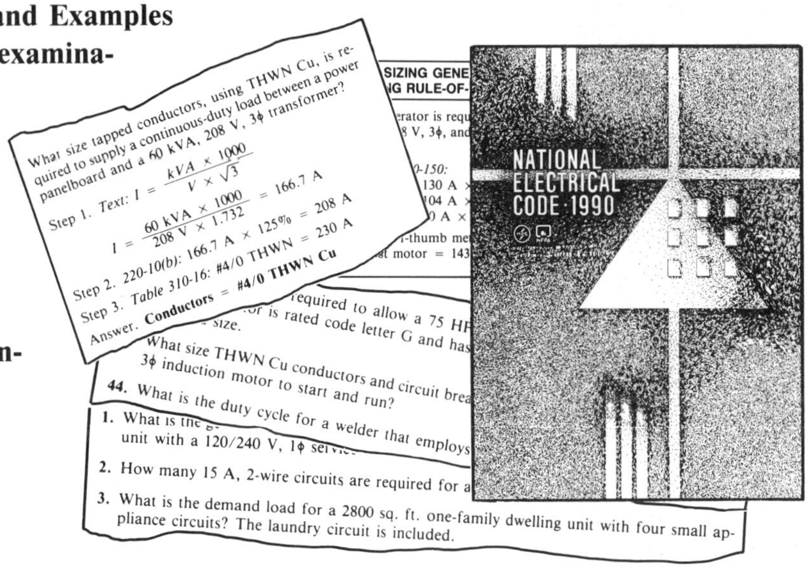

NEC® QUESTIONS

The questions in this chapter are based on the 1990 NEC®. The Code® range, or subject area, is given for each test. Tests included are

- Test 1 . . . . General—100
- Test 2 . . . . Wiring Design and Protection—200-280
- Test 3 . . . . Wiring Design and Protection—200-280
- Test 4 . . . . Wiring Design and Protection—200-280
- Test 5 . . . . Wiring Design and Protection—200-280
- Test 6 . . . . Wiring Methods and Materials—300-384
- Test 7 . . . . Wiring Methods and Materials—300-384
- Test 8 . . . . Wiring Methods and Materials—300-384
- Test 9 . . . . Wiring Methods and Materials—300-384
- Test 10 . . . . Wiring Methods and Materials—300-384
- Test 11 . . . . Equipment for General Use—400-480
- Test 12 . . . . Equipment for General Use—400-480
- Test 13 . . . . Special Occupancies—500-555
- Test 14 . . . . Special Occupancies—530-555

General—100

Date ____________

Chapter 9—TEST 1

Name ____________

Section	Answer	
________	T F	**1.** Electrical equipment under the exclusive control of utility companies for the purpose of generating electricity is exempt from the NEC®.
________	T F	**2.** The authority having jurisdiction can waive certain requirements in the NEC® if no hazard is foreseen.
________	T F	**3.** If persons disagree with a section in the NEC®, they can request a formal interpretation from the NFPA.
________	T F	**4.** A main purpose of the NEC® is for the protection of persons from dangers that can occur from the use of electricity.
________	T F	**5.** Inspectors shall inspect all electrical equipment for factory-wired assemblies at thc timc of installation.
________	T F	**6.** A junction box in an attic with a pull-down stairway is not accessible.
________	T F	**7.** Good electrical designing should include space in the raceway for future use.
________	T F	**8.** Installations per the NEC® always provide capacity for future expansion.
________	T F	**9.** The NEC® covers the installation of wiring methods in underground mines.
________	T F	**10.** Circuit conductors tapped from the secondary side of a transformer and terminated in a panelboard are not considered feeder conductors.
________	T F	**11.** The neutral is a current-carrying conductor for electric discharge lighting supplied by phases A, B, and C.
________	T F	**12.** The highest current at which an overcurrent protection device can be interrupted at rated voltage is the interrupting rating of the device.
________	________	**13.** Bonding is the ________ together of metal parts to provide electrical continuity.
________	________	**14.** Individual branch circuits are used to supply only ________ piece(s) of electrical equipment.
________	________	**15.** A grounded conductor is a circuit conductor in an electrical system that is ________ grounded.
________	________	**16.** Receptacle outlets are outlets where ________ or more receptacles are installed.
________	________	**17.** Service is the conductors and equipment used to deliver energy from the ________ supply to the wiring system on the premise.
________	________	**18.** A(n) ________ device can close or open one or more electrical circuits.
________	________	**19.** ________ housing more than one conductor shall be listed for such use.

__________ __________ **20.** The minimum clearance in front of service equipment is __________′.

A. 2½
B. 3
C. 3½
D. 4

__________ __________ **21.** Work space shall not be less than __________″ in front of electrical equipment.

A. 30
B. 35
C. 40
D. 42

__________ __________ **22.** The minimum headroom clearance above switchgear shall be at least __________′.

A. 5½
B. 6¼
C. 7
D. 7¼

__________ __________ **23.** The minimum clearance for a switchgear rated at 480 V using Condition 3 in Table 110-16(a) is __________′.

A. 3
B. 3½
C. 4
D. 4½

__________ __________ **24.** Service equipment rated over 600 V located on mezzanine floors is required to be accessible by permanent __________.

A. stairways
B. ladders
C. both A and B
D. neither A or B

__________ __________ **25.** At least one entrance __________″ wide and 6½′ high shall be provided at each end of switchgears rated 1200 A or more and over 6′-0″ wide.

A. 24
B. 36
C. 42
D. 48

Wiring Design and Protection—200–280

Date ______________

Chapter 9—TEST 2

Name ______________________

Section	Answer	
________	T F	1. Grounded conductors #6 and smaller can be identified by painting the insulation white or gray (general rule).
________	T F	2. Grounded conductors #4 and larger can be identified by painting white or gray at terminations and junctions.
________	T F	3. The grounded conductor can be identified by notching the insulation.
________	T F	4. Terminals for attachment of grounded conductors are usually identified by a white finish.
________	T F	5. Devices with screw-shells shall have the grounded conductor connected to the screw-shell.
________	T F	6. Grounded conductors connected to leads are not required to be terminated to prevent reverse polarity.
________	T F	7. Grounded conductors at 600 V can be used with ungrounded conductors rated at 1000 V or greater.
________	T F	8. Hallways 10′-0″ or more in length shall have a receptacle outlet installed.
________	________	9. Multiwire circuits feeding ________ loads on the same yoke shall be disconnected simultaneously.
________	________	10. The general rule requires multiwire circuits to supply only phase-to-________ loads.
________	________	11. Voltage between conductors supplying terminals of lighting fixtures in dwelling units shall not exceed ________ V.
________	________	12. At least one 15 or 20 A receptacle shall be installed in the basement of dwelling units and be ________-protected.
________	________	13. Receptacles in hotel and motel rooms can be located for the ________ layout provided the spacing requirements per 210-52(a) are followed.
________	________	14. Dwelling units with detached garages shall have a lighting outlet controlled by a wall switch where ________ is provided.
________	________	15. A bathroom is a(n) ________ or room with a sink, toilet, and tub.
________	________	16. Receptacle outlets installed in the ________ of a dwelling unit may serve as one of the required wall outlets.
________	________	17. At least one receptacle for general use shall be installed in a(n) ________ garage of a dwelling unit.

________ ________ **18.** Receptacle outlets over countertops in kitchens of dwelling units shall be installed to provide an outlet every ________" of wall line.
A. 12
B. 24
C. 48
D. 60

________ ________ **19.** In dwelling units, all 15 or 20 A receptacles located within ________' of the kitchen sink shall be GFCI-protected per 210-8(b).
A. 3
B. 4
C. 5
D. 6

________ ________ **20.** A central A/C unit on a commercial roof shall have at least one 15 or 20 A, 120 V receptacle within ________' of its location.
A. 25
B. 50
C. 75
D. 100

________ ________ **21.** Lighting fixtures exceeding ________ to 600 V between conductors shall be mounted no lower than 22'-0" on a pole.
A. 120
B. 208
C. 277
D. 300

________ ________ **22.** Receptacle outlets in dwelling units shall be installed on any wall space over ________" in length.
A. 12
B. 24
C. 36
D. 42

________ ________ **23.** Receptacle outlets in dwelling units may be spaced ________'-0" apart on walls to ensure 6'-0" spacing of outlets.
A. 12
B. 14
C. 15
D. 16

________ ________ **24.** In dwelling units, any kitchen counter space exceeding ________" in length shall have an additional plug installed.
A. 8
B. 10
C. 12
D. 18

________ ________ **25.** Receptacles located ________' above the floor shall not be counted as required outlets in a dwelling unit.
A. 3½
B. 4
C. 4½
D. 5½

Wiring Design and Protection—200–280

Date ____________

Chapter 9—TEST 3

Name ____________________

Section	Answer		
________	T	F	**1.** Voltage drop not exceeding 3 V is allowed for feeders.
________	T	F	**2.** Feeders with a common neutral shall be allowed to supply two or three sets of 3-wire feeders.
________	T	F	**3.** Feeder circuits supplying 15 and 20 A receptacle branch circuits are not permitted to be protected by a GFCI.
________	T	F	**4.** Feeder circuits shall be rated not less than 40 A to be classified as feeders.
________	T	F	**5.** Note 3 to Notes to Ampacity Tables of 0 to 2000 Volts can be applied to size conductors to a mobile home.
________	T	F	**6.** All feeder circuit conductors, with a common neutral, pulled in metal raceways shall be grouped together.
________	T	F	**7.** Feeder circuit conductors shall be capable of supplying the computed load.
________	________		**8.** Neutral conductors are considered current-carrying conductors where ________ processing equipment are supplied.
________	________		**9.** If the heating load is larger than the A/C load, the A/C load may be ________.
________	________		**10.** Electrical clock receptacles may be installed on either of the small ________ branch circuits in a kitchen, dining room, breakfast room, or pantry.
________	________		**11.** The number of branch circuits required for general lighting loads is computed by ________ the total load by the capacity of the device.
________	________		**12.** The laundry circuit in a dwelling unit shall not serve any outlets except the ________ outlets.
________	________		**13.** Existing demand is calculated at ________% and added to the new load according to the optional method.
________	________		**14.** Before the optional method for dwelling units can be applied, the ampacity of the service-entrance or feeder conductors shall be rated at least ________ A.
________	________		**15.** When the optional calculation is used to determine the rating of the service-entrance conductors to supply a multifamily dwelling complex, the demand factor is based on the ________ of units.
________	________		**16.** Multioutlet assemblies are calculated at ________ VA for each foot installed when serving heavy loads such as appliances.
________	________		**17.** There shall be ________ or more household clothes dryers before a demand factor can be applied.

__________ __________ **18.** The demand load for 225 receptacles in a department store supplying non-continuous loads is __________ VA.

__________ __________ **19.** The total connected load for 32 receptacles serving loads used for eight hours each day in a commercial building is __________ VA.

__________ __________ **20.** The neutral load for 600 A of incandescent lighting is __________ A.

__________ __________ **21.** The optional calculation for dwelling units requires one electric space heating unit to be computed at __________% and added to other loads.

__________ __________ **22.** The load for a household electric clothes dryer in a dwelling is the larger of the nameplate rating or __________ VA.

A. 4000
B. 5000
C. 6000
D. 8000

__________ __________ **23.** When computing the service load for a dwelling unit using the standard method, a 20 kW electric space heating unit shall be computed at __________%.

A. 80
B. 90
C. 100
D. 125

__________ __________ **24.** When computing the service load for a dwelling unit, more than three fixed appliances are computed with a demand factor of __________% of the nameplate rating.

A. 50
B. 60
C. 75
D. 80

__________ __________ **25.** The branch-circuit load for continuous-duty receptacles shall be calculated at __________ VA per receptacle.

A. 150
B. 175
C. 180
D. 225

Wiring Design and Protection—200–280

Date ____________________

Chapter 9—TEST 4

Name ______________________________

Section	Answer	
________	T F	**1.** Lighting fixtures rated at 277 V can be installed outdoors if mounted at least 2′ from windows.
________	T F	**2.** An overhead span of conductors shall not be supported by trees.
________	T F	**3.** The minimum size conductor permitted for festoon lighting, when not supported by a messenger cable, is #12.
________	T F	**4.** Festoon lighting is a string of outdoor lights suspended between two points more than 18′-0″ apart.
________	T F	**5.** Feeder circuits that run outside over sidewalks shall have a 10′-0″ clearance from finished grade to conductors.
________	T F	**6.** Feeder circuits attached on the outside of buildings shall have a point of attachment at least 12′-0″ from the finished grade.
________	T F	**7.** Conductors attached to poles shall have a separation of not less than 2½′ if not placed on racks or brackets.
________	T F	**8.** Festoon messenger wires shall not be supported by a fire escape on a building.
________	T F	**9.** Lighting fixtures rated at 277 V shall not be used for outdoor stairstep illumination of commercial buildings.
________	________	**10.** Each service may have ________ disconnecting means to disconnect the service equipment.
________	________	**11.** Two to six disconnecting means for each service shall be ________ for easy access by the user.
________	________	**12.** The smallest copper conductor permitted for an underground service-entrance conductor is #________ Cu (general rule).
________	________	**13.** The point of attachment to a building shall be a minimum of ________′ from finished grade.
________	________	**14.** A(n) ________ service from the main service may be installed to supply a fire pump.
________	________	**15.** Flexible metal conduit in ________′ or shorter lengths can be used in the service raceway run to the equipment.
________	________	**16.** Each occupant of a multifamily building shall have ________ to the occupancy's service disconnecting means (general rule).
________	________	**17.** Testing of ground-fault protection equipment is performed only by the ________.

__________ __________ **18.** The service point for services over 600 V is the point of connection between the __________ equipment and the property owner's equipment.

__________ __________ **19.** The disconnecting means for a circuit controlling the ground-fault protection of a service is not counted as one of the __________ disconnects.
A. six
B. seven
C. eight
D. none of the above

__________ __________ **20.** Conductors from the service point to the service disconnecting means are considered service __________.
A. subpanels
B. conductors
C. both A and B
D. neither A or B

__________ __________ **21.** A warning sign shall be posted where __________ voltage is available in service equipment.
A. high
B. low
C. both A and B
D. neither A or B

__________ __________ **22.** Control conductors used for load management can be routed with the service-entrance conductors in the same __________.
A. raceway
B. cable
C. either A or B
D. neither A or B

__________ __________ **23.** Service conductors shall not be run in such a manner as to block __________ to buildings.
A. openings
B. driveways
C. both A and B
D. neither A or B

__________ __________ **24.** Voltage drop shall be considered for service laterals run a __________ distance.
A. short
B. long
C. both A and B
D. neither A or B

__________ __________ **25.** Flexible metal conduit in 6′-0″ or shorter lengths can be used between __________ run to the service equipment.
A. gutters
B. wireways
C. raceways
D. all of the above

Wiring Design and Protection—200-280

Date ____________________

Chapter 9—TEST 5

Name ________________________________

Section	Answer	
________	T F	**1.** Overcurrent protection devices may be installed in clothes closets.
________	T F	**2.** All ungrounded conductors supplying a 240 V branch circuit may be opened at one time by a single-pole CB.
________	T F	**3.** A small set of subfeeders not over 25′-0″ in length with a current-carrying capacity of one-third the current-carrying capacity of larger feeders may be tapped to the larger feeders under certain conditions.
________	T F	**4.** For the 100′ tap rule to be applied, the point of tap shall be no more than 24′-0″ from the floor in high-bay manufacturing buildings.
________	T F	**5.** Size #16 fixture wire protected by a 20 A overcurrent protection device may extend no more than 50′-0″.
________	T F	**6.** CBs not marked with an IC rating are considered to have an IC rating of 5000 A.
________	T F	**7.** When applying the 25′ tap rule for transformers, the secondary protection devices shall not exceed the output of the transformer.
________	T F	**8.** When the 25′ tap rule is applied, the ampacity of the tap conductors shall be at least one-third the rating of the feeder conductors or overcurrent protection devices.
________	T F	**9.** Supplementary overcurrent protection devices are not required to be readily accessible.
________	________	**10.** The path-to-ground from circuits, equipment, and conductor enclosures shall be permanent and ________.
________	________	**11.** Class 1 control transformers rated at ________ VA or less are not required to have a separate grounding electrode conductor.
________	________	**12.** A #6 grounding electrode conductor free from exposure to ________ shall be permitted to be run without a raceway.
________	________	**13.** When rigid conduit is threaded ________ tight in threaded bosses on enclosures of electrical equipment, no further bonding is required.
________	________	**14.** The supplemental electrode shall be permitted to be bonded to the grounding electrode conductor, the grounded service-entrance conductor, the grounded service raceway, or the interior metal water piping at any ________ point.
________	________	**15.** When separate conduits are bonded to the grounded busbar in the service switch, the grounded conductor shall not be less than 12½% of the ________ phase conductor.

_______ _______ 16. When only one _______ circuit is supplying another building with no equipment requiring grounding, a separate grounding electrode conductor is not required.

_______ _______ 17. Water piping of a building may be used for the grounding electrode system when _______' or more is in the ground.

_______ _______ 18. Grounding electrode conductors connected to made electrodes shall be no larger than #_______ AWG Cu or #_______ AWG Al.

_______ _______ 19. For high-impedance grounding, the system _______ conductor shall not be connected to the ground except through the grounding impedance.
A. line
B. neutral
C. both A and B
D. neither A or B

_______ _______ 20. _______-insulated office equipment does not require grounding with an equipment grounding conductor.
A. Single
B. Partly
C. Double
D. none of the above

_______ _______ 21. _______ shall not be used as the sole equipment grounding conductor.
A. Earth
B. Conduit
C. Cable
D. none of the above

_______ _______ 22. Standard locknuts and _______ shall not be used as a bond for the service-entrance raceway.
A. connectors
B. bushings
C. within 50′-0″ of service equipment
D. copper only

_______ _______ 23. The points of attachment for the bonding jumper used to bond metal water pipe to the service-equipment enclosure shall be _______.
A. accessible
B. inaccessible
C. both A and B
D. neither A or B

_______ _______ 24. If more than one electrode is used, each system electrode shall be separated at least _______' from the other.
A. 2
B. 3
C. 4
D. 6

_______ _______ 25. A bare #4 conductor may be concrete-encased and serve as the grounding electrode conductor if at least _______' in length.
A. 10
B. 15
C. 20
D. 25

Wiring Methods and Materials—300–384

Date ______________

Chapter 9—TEST 6

Name ____________________________

Section	Answer	
________	T F	**1.** Aluminum conductors carrying over 50 A shall be grouped in AC circuits to prevent inductive heating of metal enclosures and raceways.
________	T F	**2.** Minimum burial depth for underground conduits is measured from the top of the conduit to the finished grade.
________	T F	**3.** Class 1 control circuits may be installed in the same raceway with power conductors when associated with the same motor.
________	T F	**4.** A single enclosure or raceway may be used for AC-DC systems of 600 V or less.
________	T F	**5.** Phase conductors for AC lighting and power circuits may be run in separate metal raceway systems.
________	T F	**6.** Grounding electrode conductors shall be grouped with phase conductors when installed in a raceway.
________	T F	**7.** It is permissible to run 120 V lighting conductors in the same raceway with 2300 V power conductors when all conductors are insulated for the higher voltage.
________	T F	**8.** Raceways installed in notched studs shall be covered by a metal plate.
________	T F	**9.** A Class 1 circuit occupying the same enclosure as the power conductors meets NEC® requirements.
________	T F	**10.** Conductor loads passing through fire-rated partitions with fireproofing applied around the conduit may be derated.
________	________	**11.** Approved handle ties can be used on two single-pole CBs for simultaneous disconnection of ________ circuits used for temporary wiring.
________	________	**12.** ________ wire branch circuits shall not be laid on the floor or ground at construction sites.
________	________	**13.** All lamps used for temporary wiring at construction sites shall be protected by a(n) ________ over the lampholder.
________	________	**14.** Conductors smaller than #________ can be connected in parallel for frequencies less than 360 Hz.
________	________	**15.** THW insulation may be used in ________ or dry locations.
________	________	**16.** Conductors used for direct burial shall be ________ for such use.
________	________	**17.** For multiconductor cables installed in cable trays, the sum of the diameter of the cables shall not exceed the width of the cable tray for conductors #________ and larger.

_____ _____ **18.** The maximum fill area for multiconductor cables in a 24″ cable tray is _____ sq. in.

_____ _____ **19.** The maximum fill area for an 18″ ladder type cable tray holding single conductor cables is _____ sq. in.

_____ _____ **20.** Nails used to mount knobs supporting open wiring on insulators shall not be smaller than _____ d.

A. 6
B. 8
C. 10
D. 12

_____ _____ **21.** Open wiring on insulators shall be permitted for wiring systems of _____ V or less.

A. 150
B. 240
C. 300
D. 600

_____ _____ **22.** Open conductors on insulators routed within _____′ from the floor are exposed to physical damage.

A. 6
B. 7
C. 8
D. 10

_____ _____ **23.** Messenger supported wiring is permitted only in _____ establishments where maintenance personnel service the system.

A. industrial
B. commercial
C. both A and B
D. neither A or B

_____ _____ **24.** The ampacity of conductors supported by a messenger wiring system is determined per Section _____.

A. 310-12
B. 310-13
C. 310-14
D. 310-15

_____ _____ **25.** Splices and _____ in messenger wire shall be made by approved methods.

A. devices
B. taps
C. both A and B
D. neither A or B

Wiring Methods and Materials—300–384

Date ____________

Chapter 9—TEST 7

Name ____________

Section	Answer	
______	T F	1. Concealed knob-and-tube wiring can be installed, with special permission, in hollow spaces of walls and ceilings of dwelling units.
______	T F	2. Knob-and-tube wiring installed in inaccessible attics shall be run through bored holes in the joists, studs, or rafters, or run along the sides.
______	T F	3. Conductors of knob-and-tube wiring can be tied to solid knobs with any type of insulated tie.
______	T F	4. A #250 kcmil, IGS cable has an ampacity of 168 A.
______	T F	5. The maximum size IGS cable is 4750 kcmil.
______	T F	6. IGS cable can be used as service-entrance conductors when routed underground.
______	T F	7. MV cable is available in 2000 V or less.
______	T F	8. The ampacity of medium-voltage cable is determined per 310-15 (general rule).
______	T F	9. Medium voltage cables exposed to sunlight shall be identified for such use.
______	______	10. A flat conductor cable consists of three or more ______ conductors.
______	______	11. General-purpose branch circuits are limited to ______ A or less.
______	______	12. The grounding electrode conductor shall be ______ for 2-wire FCC systems.
______	______	13. Mineral-insulated, metal-sheathed cable (MI) can be used in ______ locations.
______	______	14. MI cable consists of solid copper conductors that correspond to standard ______ sizes.
______	______	15. MI cable shall be supported at intervals not exceeding ______′ (general rule).
______	______	16. Electrical nonmetallic tubing (ENT) is available in sizes ½″ to ______″.
______	______	17. ENT shall be supported at least every ______′.
______	______	18. ENT shall not be buried directly in the ______.
______	______	19. BX shall be supported at intervals not exceeding 4½′ and within ______″ of each outlet, junction box, cabinet, or fitting. A. 10 B. 12 C. 14 D. 15

__________ __________ **20.** AC cable shall have a clearance of at least __________′ from the edge of a scuttle hole.

A. 3
B. 4
C. 5
D. 6

__________ __________ **21.** Exposed AC cable shall closely follow the __________ of the building.

A. finish
B. studs
C. both A and B
D. neither A or B

__________ __________ **22.** Metal-clad cable (MC) can be used in systems of 600 V or __________.

A. less
B. more
C. both A and B
D. neither A or B

__________ __________ **23.** MC cable shall be supported at intervals not exceeding __________′-0″.

A. 3
B. 4
C. 5
D. 6

__________ __________ **24.** MC cable insulation shall have a maximum operating temperature of not less than __________°C (600 V).

A. 60
B. 75
C. 85
D. 90

__________ __________ **25.** The ampacity of conductors in nonmetallic-sheathed cable shall be used at __________°C.

A. 60
B. 75
C. 85
D. 90

Wiring Methods and Materials—300–384

Date ____________________

Chapter 9—TEST 8

Name ________________________________

Section	Answer		
________	T F	1.	Romex shall bc supported at least every 5′-0″ and within 12″ of every outlet and fitting.
________	T F	2.	SE cable shall be surface-marked at intervals not exceeding 36″.
________	T F	3.	SE cable shall have three insulated conductors with an equipment grounding electrode conductor to supply a subfed panelboard.
________	T F	4.	SE cable with a bare conductor between the service equipment and range can serve as the neutral and equipment ground.
________	T F	5.	Underground feeder and branch-circuit cable (UF) is available in sizes #14 Cu through #250 kcmil Cu.
________	T F	6.	UF cable shall not be used when exposed to sunlight unless identified for such use.
________	T F	7.	UF cable can be used for service-entrance.
________	T F	8.	Power and control tray cable (TC) can be used for direct burial in the earth when identified for such use.
________	T F	9.	The ampacity of conductors in TC cable is selected per 400-5 for sizes smaller than #14.
________	________	10.	The maximum size conductor permitted in TC cable is #________ kcmil Cu.
________	________	11.	Splices and taps are not permitted to be made in ________ lengths of nonmetallic extensions.
________	________	12.	Nonmetallic surface extensions shall be secured at intervals not exceeding ________″.
________	________	13.	Nonmetallic extensions shall not be routed through floors or partitions or outside rooms from which they ________.
________	________	14.	Underplaster extensions can be used to extend ________ branch circuits in a building of fire-resistant construction.
________	________	15.	Underplaster extensions shall not extend beyond the floor from which they ________ unless installed in an approved wiring method.
________	________	16.	Underplaster extensions shall be buried in the ________ finish of ceilings or walls.
________	________	17.	IMC with threadless connectors shall be supported every ________′ and within 3′-0″ of each fitting.
________	________	18.	Intermediate metal conduit is permitted for voltages exceeding ________ V.

__________ __________ **19.** Rigid metal conduit entering a box shall be protected using a(n) __________ unless the box or fitting provides equivalent protection.

__________ __________ **20.** The maximum size conductor permitted in NM cable is #__________ Cu.

A. 1
B. 2
C. 3
D. 4

__________ __________ **21.** IMC may be installed in or under cinder fill subject to permanent moisture if protected on all sides by a layer of non-cinder concrete not less than __________" thick.

A. 2
B. 3
C. 4
D. 6

__________ __________ **22.** The NEC® permits __________ 90° bends in a single conduit run.

A. two
B. three
C. four
D. five

__________ __________ **23.** A run of 1½" rigid metal conduit with threaded couplings shall be supported at intervals of __________'.

A. 10
B. 12
C. 14
D. 15

__________ __________ **24.** Conductors without a lead sheath that are pulled through 2" rigid metal conduit shall have a conduit radius bend of __________".

A. 6
B. 8
C. 10
D. 12

__________ __________ **25.** A 1½" PVC conduit shall be supported every __________'.

A. 5
B. 12
C. 14
D. 20

Wiring Methods and Materials—300–384

Date ____________________

Chapter 9—TEST 9

Name ________________________________

Section	Answer	
________	T F	**1.** Rigid nonmetallic conduit shall not be installed where subject to physical damage unless approved for the installation.
________	T F	**2.** PVC shall be supported at least every 8′-0″ for sizes 3½″ to 5″.
________	T F	**3.** Setscrew couplings may be used when EMT is installed in concrete slabs.
________	T F	**4.** EMT shall be supported within 3′-0″ of each outlet box, junction box, cabinet, or fitting.
________	T F	**5.** All cut ends of EMT shall be reamed to remove rough edges.
________	T F	**6.** The maximum number of #12 conductors permitted in ⅜″ flexible metallic tubing with outside fittings is two.
________	T F	**7.** The maximum size flexible metallic tubing is 1″.
________	T F	**8.** Flexible metallic tubing cannot be used in hoistways.
________	T F	**9.** Flexible metal conduit in lengths not exceeding 6′-0″ and in ⅜″ to ½″ diameters may be used when protected by 20 A devices.
________	T F	**10.** Flexible metal conduit in lengths of 6′-0″ or less may be used for grounding where served by a 20 A branch circuit.
________	T F	**11.** Flexible metal conduit, used for fixture whips, shall be supported in 6′-0″ lengths or less.
________	T F	**12.** Motor leads may be enclosed with ⅜″ liquidtight flexible nonmetallic conduit (use minimum size).
________	________	**13.** An equipment grounding electrode conductor can be pulled inside or routed outside of a(n) ________′ piece of liquidtight flexible nonmetallic conduit.
________	________	**14.** Supports are not required when liquidtight flexible metal conduit is ________.
________	________	**15.** Surface metal raceways can be routed through walls and floors without ________ lengths.
________	________	**16.** Splices and taps can be made in surface metal raceways with accessible, removable ________.
________	________	**17.** Surface nonmetal raceways shall not contain conductors with insulation temperatures exceeding ________°C.
________	________	**18.** Multioutlet assemblies can extend through dry partitions where no cap, outlet, or cover is in the ________.
________	________	**19.** Underfloor raceways shall have dead ends ________.

_________ _________ **20.** Underfloor raceways shall have _________ near or at each end to locate the last insertion.

_________ _________ **21.** Cellular metal floor raceways, except by permission, shall not have a conductor larger than #_________ installed.
A. 1/0
B. 2/0
C. 3/0
D. 4/0

_________ _________ **22.** Cellular concrete floor raceways shall not be filled with conductors to more than _________% of the cross-sectional area of the cell.
A. 20
B. 30
C. 40
D. 50

_________ _________ **23.** Wireways shall not contain more than _________ current-carrying conductors without derating.
A. 20
B. 25
C. 30
D. 40

_________ _________ **24.** Conductors installed in wireways shall not contain more than _________% of the interior cross-sectional area.
A. 10
B. 20
C. 30
D. 40

_________ _________ **25.** Conductors with splices shall not fill wireway to more than _________% of its area to that point.
A. 20
B. 40
C. 50
D. 75

Wiring Methods and Materials—300–384

Date ____________

Chapter 9—TEST 10

Name ____________

Section	Answer	
______	T F	1. Cablebus framework can be used as an equipment grounding electrode conductor for feeders when bonded properly.
______	T F	2. Cablebus shall be supported at intervals not exceeding 12½′ (general rule).
______	T F	3. Cablebus shall extend 6′-0″ above the floor that it passes through and be totally enclosed.
______	T F	4. For a U pull, the wall with the most conduits shall be used to size the junction box.
______	T F	5. The addition of extension rings increases the volume of a box.
______	T F	6. Listed outlet boxes can be used to mount ceiling fans.
______	T F	7. Insulated bushings shall be marked with their maximum temperature ratings.
______	______	8. Busways shall be secured at intervals not less than ______′.
______	______	9. Minimum wire-bending space at terminals for two #250 kcmil conductors per terminal when using an S-bend or a Z-bend is ______″.
______	______	10. Minimum wire-bending space at terminals for a #2/0 conductor when using an L-bend is ______″.
______	______	11. Conductors, including splices and taps, shall not fill the gutter to more than ______% of its area.
______	______	12. The cross-sectional area of an auxiliary gutter shall not be filled to more than ______%.
______	______	13. Generally, the maximum number of conductors that may be installed at any one cross-sectional area of an auxiliary gutter is ______.
______	______	14. The maximum height (measured to the operating handle) for mounting disconnecting switches is ______′.
______	______	15. Conductors used for switch loops to control lighting fixtures are not required to be routed with a(n) ______ conductor.
______	______	16. An AC general-use snap switch supplying a motor load can be loaded to ______% of its rating.
______	______	17. An overcurrent protection device approved for continuous duty may be loaded to ______% of its rating.
______	______	18. The next ______ size can be selected when the ampacity of busways does not correspond to a standard OCPD. A. lower B. higher C. both A and B D. neither A or B

__________ __________ **19.** A fusible disconnect can be mounted higher than __________ from finished grade when a hookstick is provided.

A. 6′-0″
B. 6′-6″
C. both A and B
D. neither A or B

__________ __________ **20.** The size OCPD required to protect a continuous-rated load of 80 A is __________ A.

A. 80
B. 90
C. 100
D. 125

__________ __________ **21.** The maximum number of overcurrent protection devices (other than the main) in a lighting and appliance branch-circuit panelboard is __________.

A. 36
B. 40
C. 42
D. 48

__________ __________ **22.** The high leg of a 3ϕ, 4-wire delta system may be marked __________ to identify the higher voltage to ground.

A. orange
B. blue
C. black
D. red

__________ __________ **23.** Each lighting and appliance branch-circuit panelboard shall be protected individually on the supply side by not more than two main CBs or two sets of fuses having a combined rating not __________ than that of the panelboard.

A. less
B. greater
C. both A and B
D. neither A or B

__________ __________ **24.** Conductors may travel horizontally through __________ sections of switchboards if isolated from busbars by a barrier.

A. horizontal
B. vertical
C. both A and B
D. neither A or B

__________ __________ **25.** Continuous-duty loads shall be figured at __________% for branch circuits.

A. 100
B. 110
C. 115
D. 125

Equipment for General Use—400–480

Date ____________

Chapter 9—TEST 11

Name ______________________

Section	Answer		
______	T	F	**1.** Type G portable power cable can be used as a wiring method.
______	T	F	**2.** Type W flexible cable cannot be used as a wiring method.
______	T	F	**3.** Fixture wires shall not be smaller than #14.
______	T	F	**4.** A handhole of not less than 2″ × 4″ shall be provided where metal poles support lighting fixtures (general rule).
______	T	F	**5.** A recessed fixture installed in clothes closets shall have a minimum of 6″ clearance from the storage area.
______	T	F	**6.** Electric discharge lighting fixtures installed over an outlet box shall have an opening providing access to the box.
______	T	F	**7.** Recessed fixtures are not required to be spaced ½″ from wood if suitable for direct contact.
______	______		**8.** A flexible cord with three #8 conductors has an ampacity of ______ A.
______	______		**9.** Fixture wires shall not be used to supply ______ circuits.
______	______		**10.** A #16 conductor used as fixture wire has an ampacity of ______ A.
______	______		**11.** A(n) ______°C rated THW insulated conductor may be run within 3″ of ballast in the fixture compartment.
______	______		**12.** Electric discharge lighting fixtures may be connected by a flexible cord if the cord is ______ for its entire length.
______	______		**13.** Any fixture weighing ______ pounds or less may be supported by an outlet box.
______	______		**14.** Ceiling fans weighing ______ pounds or less can be mounted to ceiling boxes approved for such use.
______	______		**15.** The load for storage water heaters rated at 120 gallons or less is calculated at ______% of the heater's nameplate rating.
______	______		**16.** A single appliance without a motor and rated at 13.3 A or less can be served by a ______ A overcurrent protection device. A. 20 B. 25 C. 30 D. 35
______	______		**17.** SPT-3 cords 36″ to ______″ long may be used to connect dishwashers. A. 42 B. 48 C. 52 D. 60

__________ __________ **18.** Conductors supplying a heating unit shall be calculated at __________% times the heating load plus the blower motor.
A. 110
B. 115
C. 120
D. 125

__________ __________ **19.** Resistance elements in heating units shall be protected at not more than __________ A.
A. 30
B. 40
C. 50
D. 60

__________ __________ **20.** A disconnecting means shall be installed in sight of the heating unit for a self-contained blower motor over __________ HP.
A. 1/8
B. 1/4
C. 1/3
D. 1/2

__________ __________ **21.** Panels or units of embedded de-icing and snow-melting equipment shall not exceed __________ VA per sq. ft. of a heated area.
A. 80
B. 90
C. 120
D. 180

__________ __________ **22.** Conductors supplying outdoor electric de-icing and snow-melting equipment shall be at least __________% of the heating load.
A. 125
B. 150
C. both A and B
D. neither A or B

__________ __________ **23.** Exposed outdoor electric de-icing equipment operating above __________°C shall be installed to protect against personnel contact.
A. 30
B. 40
C. 50
D. 60

__________ __________ **24.** Pipelines with impedance heating shall not operate at greater than __________ V AC.
A. 24
B. 30
C. 50
D. 120

__________ __________ **25.** Secondaries of transformers supplying voltage for impedance heating of vessels are computed at not less than __________% of the heating load.
A. 80
B. 100
C. 125
D. 150

Equipment for General Use—400–480

Date ____________________

Chapter 9—TEST 12

Name ________________________________

Section	Answer	
________	T F	**1.** Table 430-148 is used to select the FLC ratings of 1ϕ motors to size conductors and OCPDs for motors.
________	T F	**2.** The FLC rating on the motor's nameplate is used to size overload protection devices.
________	T F	**3.** Single-phase conductors to a phase converter shall be at least 125% of the motor's FLC (same voltage).
________	T F	**4.** When the minimum percentage does not allow a hermetic motor to start and attain running speed, 225% or less of the hermetic motor's FLC is used to size the OCPD.
________	T F	**5.** Single-phase conductors to phase converters on a feeder are sized 1.73 times the load plus 25% of the largest motor of the group.
________	T F	**6.** A 3ϕ window, console, or in-wall A/C unit may be connected by a cord and plug.
________	T F	**7.** The ampacity of conductors between terminals of a generator and an OCPD shall be computed at least 125% of the generator's FLC.
________	T F	**8.** The grounded neutral conductor between terminals of a generator and equipment can be sized per 220-22 and 250-23(b).
________	T F	**9.** Generators operating above 120 V shall have live parts guarded from accidental contact where accessible to unqualified personnel.
________	________	**10.** A CB for a squirrel-cage induction motor with full-voltage starting and code letter G is selected at ________%.
________	________	**11.** The minimum kVA per HP for code letter B is ________ for a 10 HP, 240 V, 3ϕ motor.
________	________	**12.** Part-winding motors use only ________ of their primary winding when starting.
________	________	**13.** Where an induction motor will not start and run, the next ________ size OCPD can be selected.
________	________	**14.** Minimum running overload protection is selected and based on the temperature rise or service ________ of the motor.
________	________	**15.** Motor control transformers rated less than ________ VA do not require OCPD protection.
________	________	**16.** A nonfused disconnect shall be equal to at least ________% of the motor's FLC.

__________ __________ **17.** For a number of motors with other loads, the OCPD for the service can be selected from the __________ of the circuit conductors.

__________ __________ **18.** For two or more motors with the same FLC, only __________ is selected as the largest motor.

__________ __________ **19.** The maximum percentage times the full current allowed to start and run a motor compressor is __________%.

__________ __________ **20.** There are __________ levels of current in which to select percentages to size OCPDs for the primary side of transformers.

__________ __________ **21.** The OCPD in the primary side of transformers shall not exceed 125% for a primary current of __________ A or more.

__________ __________ **22.** The OCPD in the primary side of transformers shall not exceed 167% for a primary current of __________ A to 9 A or less.

__________ __________ **23.** The ampacity of capacitor circuit conductors shall not be less than __________% of the rated current of the capacitor.

__________ __________ **24.** The ampacity of capacitor circuit conductors shall be at least __________ of the ampacity of the motor's circuit conductors.

__________ __________ **25.** The disconnecting means for a capacitor circuit conductor shall be at least __________% of the capacitor's FLC rating.

Special Occupancies—500–555

Date ____________

Chapter 9—TEST 13

Name ____________

Section	Answer	
________	T F	**1.** The area inside a gasoline pump to a height of 4′-0″ is a Class I, Division 1 location.
________	T F	**2.** Seals shall be provided within 24″ of each arcing device in hazardous locations.
________	T F	**3.** Self-sealed devices in hazardous locations do not require additional seals.
________	T F	**4.** An area 24″ high and extending 20′-0″ in all directions from a gasoline pump is a Class I, Division 2 location.
________	T F	**5.** Portable lighting equipment meeting requirements for fixed lighting equipment is permitted for use in Class I, Division 2 locations.
________	T F	**6.** A totally enclosed pipe-ventilated motor shall not be installed in a Class III, Division 2 location.
________	T F	**7.** Electric cranes with contacts isolated from all other systems shall be equipped with an alarm.
________	T F	**8.** PVC installed beneath a gas island shall be buried at least 24″.
________	T F	**9.** The area beneath a gas island is a Class I, Division 1 location.
________	________	**10.** ________ branch circuits shall supply outlets at a patient bed location.
________	________	**11.** Receptacles installed in psychiatric locations shall be ________-proof.
________	________	**12.** Anesthetizing locations using nonflammable, anesthetizing agents are not classified as ________ locations.
________	________	**13.** If mounted ________′ above the floor, receptacles in flammable, anesthetizing locations need not be totally enclosed.
________	________	**14.** IMC or EMT may be installed ________ a hazardous location.
________	________	**15.** The electrostatic shield of a(n) ________ power transformer is connected to the reference grounding point.
________	________	**16.** Branch circuits supplying listed, fixed equipment are exempt from the ungrounded system requirement in anesthetizing locations where installed ________ the hazardous area.
________	________	**17.** Conductors and overcurrent protection devices shall be rated at least ________% of the current rating of medical, therapy, or X-ray equipment.
________	________	**18.** Equipment grounding terminals in the normal service panelboard shall be bonded equipment grounding terminals in the essential system's panelboard with not less than #________ AWG Cu conductors.

__________ __________ **19.** A place of assembly is any room or space in an occupancy where __________ persons or more assemble.
A. 50
B. 75
C. 100
D. 150

__________ __________ **20.** Nonmetallic raceways installed in places of assembly shall be encased with __________" of concrete.
A. 2
B. 4
C. 5
D. 6

__________ __________ **21.** BX can be used in buildings or portions of buildings (places of assembly) not required to be of __________-rated construction.
A. weatherproof
B. fire
C. both A and B
D. neither A or B

__________ __________ **22.** Temporary wiring to supply power to display booths in exhibition halls shall comply with Article __________.
A. 300
B. 305
C. 310
D. 700

__________ __________ **23.** Dimmers installed in ungrounded conductors shall be protected by OCPDs not exceeding __________% of their rating.
A. 100
B. 115
C. 125
D. 150

__________ __________ **24.** Border lights shall be installed around stages in theaters on circuits rated at __________ A or less.
A. 20
B. 25
C. 30
D. 35

__________ __________ **25.** Receptacles located on stages in theaters shall not exceed __________% of their ratings for continuous-duty loads.
A. 50
B. 60
C. 70
D. 80

Special Occupancies—530–555

Date ____________

Chapter 9—TEST 14

Name ____________

Section	Answer		
________	T	F	**1.** A set in a motion picture studio can be wired with PVC schedule 40 conduit.
________	T	F	**2.** Receptacles wired with DC power and located in plugging boxes of television studios shall not exceed 20 A.
________	T	F	**3.** Plugging boxes located on stages in motion picture studios shall not be supplied by AC systems.
________	T	F	**4.** DC pendant and portable lamps used to illuminate stages in motion picture studios are not required to be grounded at 120 V to ground.
________	T	F	**5.** Flexible cords supplying stages in motion picture studios shall not be spliced on circuits rated at 20 A or less.
________	T	F	**6.** Professional projectors use 35mm film having a minimum width of $1\frac{3}{8}''$.
________	T	F	**7.** AC plugs and receptacles located in television studios shall be grounded and polarized.
________	T	F	**8.** Portable electric equipment can be located in storage vaults containing cellulose nitrate film.
________	T	F	**9.** Manufactured buildings shall have their service-entrance conductors installed after they have been erected at the building site.
________	T	F	**10.** Electrical conductors and equipment for manufactured buildings shall be protected where subjected to physical damage while in transit.
________	________		**11.** The service equipment to a mobile home shall not be ________ on the mobile home.
________	________		**12.** A mobile home supplied by gas for heating and cooking can be served by a(n) ________ A listed cord.
________	________		**13.** Cables installed in agricultural buildings shall be secured within ________″ of boxes.
________	________		**14.** Equipment grounding electrode conductors routed underground to agricultural buildings shall be ________.
________	________		**15.** Lighting fixtures shall be installed to minimize the entrance of ________, foreign matter, and dust.
________	________		**16.** The size bonding jumper required to bond the elements of an equipotential plane in agricultural buildings is #________ Cu.
________	________		**17.** Flexible connections for equipment in agricultural buildings can be made with ________ flexible conduit.

A. liquidtight
B. Greenfield
C. both A and B
D. neither A or B

__________ __________ **18.** Lighting fixtures exposed to cleansing water in agricultural buildings shall be __________.
A. drip-proof
B. watertight
C. both A and B
D. neither A or B

__________ __________ **19.** The feeder assembly shall enter the mobile home at a point in the exterior of the __________.
A. wall
B. floor
C. roof
D. all of the above

__________ __________ **20.** There shall be at least __________ small appliance circuits provided in mobile homes.
A. two
B. three
C. four
D. none of the above

__________ __________ **21.** Receptacle outlets shall be installed in a recreational vehicle so that no point along the wall is more than __________'-0" from an outlet (general rule).
A. 3
B. 5
C. 6
D. 12

__________ __________ **22.** The demand load for a 9 kW range in a recreational vehicle is __________ VA.
A. 6500
B. 7200
C. 8000
D. 8500

__________ __________ **23.** One set of service-entrance conductors shall be permitted to supply more than __________ set(s) of service equipment.
A. one
B. two
C. three
D. none of the above

__________ __________ **24.** The demand load for thirty-five 20 A receptacles on a loading dock in a marina is __________ A.
A. 400
B. 420
C. 520
D. 610

__________ __________ **25.** Floating buildings shall be supplied by __________ set(s) of feeder conductors from their service equipment (general rule).
A. one
B. two
C. both A and B
D. neither A or B

Special Equipment—600–690

Date ______________

Chapter 9—TEST 15

Name ______________________________

Section	Answer	
______	T F	**1.** Wiring methods used to wire signs and outline lighting are required to terminate in the enclosure of the sign or transformer (general rule).
______	T F	**2.** The sign enclosure can be used to route a branch circuit to supply floodlights to illuminate parking lots.
______	T F	**3.** A disconnecting means is required in sight from an outdoor sign and shall be located not more than 50′-0″ from the sign (general rule).
______	T F	**4.** Illuminated, portable outdoor signs can be provided with a GFCI in the attachment cap of the supply cord.
______	T F	**5.** A 15 A branch circuit shall be brought to the front of each commercial store building to serve a sign with grade level access.
______	T F	**6.** A 20 A branch circuit for a sign can feed combination loads such as ballasts, lamps, and transformers.
______	T F	**7.** Locking receptacles and connectors shall be used in manufactured wiring systems.
______	T F	**8.** Component parts used in a manufactured wiring system shall be listed as an assembly for that system.
______	T F	**9.** Receptacle outlets for office furnishings are permitted to be installed in lighting accessories.
______	T F	**10.** Individual partitions for office furnishings are permitted to have multiwired circuits.
______	T F	**11.** Wiring on cranes shall be installed in raceways (general rule).
______	______	**12.** Wiring on electrical cranes can be routed in MC ______ (general rule).
______	______	**13.** Conductors external to motors and controls on cranes shall be at least #______ AWG (general rule).
______	______	**14.** A hoist motor with a 30-minute short-time rating and #6 THHN 90°C conductors has an ampacity of ______ A.
______	______	**15.** Secondary conductors for resistors separated from controllers having a time of 5 seconds on and 75 seconds off have a demand of ______% for cranes.
______	______	**16.** Supports along runways for main contacts of conductors to cranes shall not exceed ______′.
______	______	**17.** The minimum size conductors required for traveling cables of elevators is #______ AWG.

__________ __________ **18.** Feeder conductors without __________ can be installed in a hoistway of an existing elevator.

__________ __________ **19.** Auxiliary gutters used in a dumbwaiter installation are not subjected to the __________ of conductors.

__________ __________ **20.** The disconnecting means for escalators and moving sidewalks shall be located so that it is readily accessible to __________ persons.

__________ __________ **21.** Motors used to drive elevators are not required to be protected from __________.

A. grounds
B. overloads
C. both A and B
D. neither A or B

__________ __________ **22.** A spot welder with a 70% duty cycle has a(n) __________% demand factor.

A. 45
B. 55
C. 78
D. 84

__________ __________ **23.** The OCPD for arc welders with transformers shall not exceed __________% of the primary FLC.

A. 200
B. 250
C. both A and B
D. neither A or B

__________ __________ **24.** Conductors supplying a group of motor-generator arc welders are sized at __________% of the third largest welder plus the percentages of the other welders.

A. 65
B. 70
C. 75
D. 85

__________ __________ **25.** The OCPD for resistance welders shall not exceed __________% of the conductor's ampacity supplying the circuit.

A. 200
B. 250
C. 300
D. 400

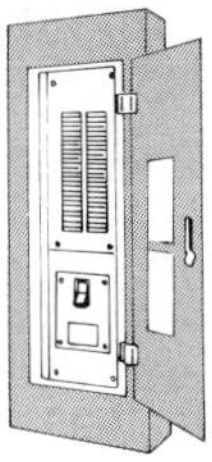

Special Equipment—600–690

Date ____________

Chapter 9—TEST 16

Name ____________

Section	Answer	
______	______	**1.** The generator output circuit shall be ______ from ground when supplying components of induction heating equipment.
______	______	**2.** The supply circuit disconnecting means can serve as the ______ for an individual piece of dielectric heating equipment.
______	______	**3.** Circuit conductors supplying power to equipment other than a motor/generator for induction heating shall be sized at not less than ______% of the nameplate rating.
______	______	**4.** Cell line conductors for electrolytic cells can be either covered, bare, or ______.
______	______	**5.** Removable links or removable conductors can serve as the ______ means for electrolytic cells.
______	______	**6.** Circuit conductors supplying components of electroplating equipment shall be sized at ______% of the load.
______	______	**7.** The ampacity of copper busbars used for electroplating equipment shall not exceed ______ A per sq. in.
______	______	**8.** Bare copper conductors can be used to connect conversion equipment to an electrolytic tank if supported on ______.
______	______	**9.** Taps from ______ circuits can be made to supply power to industrial machinery.
______	______	**10.** All machines in industrial machinery systems shall have individual ______ means.
______	______	**11.** Tap conductors to motors for irrigation machines shall not be less than #______ AWG.
______ ______	______ ______	**12.** Branch-circuit conductors and OCPDs shall be selected at ______% of the largest motor plus ______% of the remaining motors of the irrigation system.
______ ______	______ ______	**13.** The LRC of irrigation machinery shall equal ______ times the LRC of the largest motor plus ______% of the FLC rating of the remaining motors.
______	______	**14.** Irrigation cable shall be supported at intervals not exceeding ______′.
______	______	**15.** Wireways with circuits supplying power or signals to sound-recording equipment shall fill not more than ______% of its depth.
______	______	**16.** Circuits to the heater of an electronic tub shall have overcurrent protection rated at ______ A or less when supplied by storage batteries of over 20 A-hour capacity.

_________ _________ **17.** Branch conductors to data processing equipment shall be calculated at _________% of the connected load.
A. 100
B. 115
C. 125
D. 150

_________ _________ **18.** Flexible cords or data processing cables used to connect computer units shall be _________ as part of the system.
A. isolated
B. approved
C. both A and B
D. neither A or B

_________ _________ **19.** Branch circuits to receptacles under raised floors in computer rooms shall be wired with _________.
A. EMT
B. IMC
C. AC cable
D. all of the above

_________ _________ **20.** Self-excited generators supplying power to organs shall have a potential of not more than _________ V.
A. 15
B. 24
C. 30
D. 60

_________ _________ **21.** Circuit conductors to organs shall be protected with OCPDs rated at _________ A or less.
A. 10
B. 15
C. 20
D. 25

_________ _________ **22.** The smallest conductor used for organ circuits is #_________ AWG.
A. 14
B. 16
C. 18
D. 26

_________ _________ **23.** The momentary rating used to size branch circuit conductors to x-ray equipment is _________%.
A. 25
B. 50
C. 75
D. 80

_________ _________ **24.** The long-time rating used to select OCPDs to protect circuits to x-ray equipment is _________%.
A. 25
B. 50
C. 75
D. 100

_________ _________ **25.** The long-time rating for x-ray equipment is based on an operating time of _________ minutes or longer.
A. five
B. seven
C. eight
D. nine

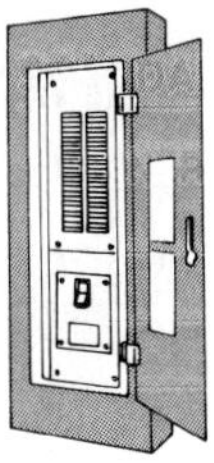

Special Equipment—600-690

Date ____________

Chapter 9—TEST 17

Name ____________

Section	Answer	
________	T F	1. Deck boxes shall be located at least 3′-0″ from the inside walls of the pool.
________	T F	2. Underground wiring shall be located at least 4′-0″ from the inside walls of the pool (general rule).
________	T F	3. Lighting fixtures installed in the walls of pools shall have the top of the lens at least 18″ below the normal water level (general rule).
________	T F	4. Overhead utility conductors shall have a clearance of 18′-0″ above the maximum water level when supplied by 120/240 V, 1ϕ service.
________	T F	5. Switching devices shall be located at least 5′-0″ from the inside walls of the pool.
________	T F	6. A #10 grounding electrode conductor shall be pulled for a wet-niche fixture connected with PVC.
________	T F	7. Junction boxes shall be installed at least 8″ above the deck level of the pool.
________	T F	8. All 120 V, 15 or 20 A receptacles located within 20′-0″ of the inside walls of the pool shall be GFCI-protected.
________	T F	9. Existing lighting fixtures located less than 5′-0″ horizontally and 5′-0″ vertically from the maximum water level of the pool are permitted.
________	T F	10. Circuits supplying 120 V lighting fixtures in pool walls shall be protected by a GFCI circuit.
________	T F	11. Receptacles located behind walls having hinged or sliding doors can be located within 10′-0″ of the pool.
________	T F	12. Lighting fixtures located at least 12′-0″ above the maximum water level of an outdoor pool are permitted.
________	T F	13. Receptacles used to supply power for a recirculating pump motor may be installed within 5′-0″ of the inside walls of the pool.
________	T F	14. Flexible cords supplying fixed swimming pool equipment shall not exceed 3′-0″ in length.
________	T F	15. The minimum mounting height for lighting fixtures over hot tubs is 8′-0″ when installed indoors.
________	T F	16. GFCI protection shall be provided for all existing receptacles within 5′-0″ of a hydromassage tub installed in an existing bathroom.
________	T F	17. Motor circuits for electrical pool covers shall be GFCI-protected.
________	T F	18. Flexible cords supplying 240 V, 1ϕ power to storable swimming pools are not required to be protected by a GFCI circuit.

__________ T F **19.** An orderly shutdown is required to minimize damage to equipment of integrated electrical systems.

__________ __________ **20.** A separately derived system of _________ V or less to ground can be ungrounded for continuity of an integrated electrical system.
A. 50
B. 125
C. 150
D. 277

__________ __________ **21.** A photovoltaic power source having one conductor of a 2-wire system over _________ V shall be solidly grounded.
A. 50
B. 125
C. 150
D. 175

__________ __________ **22.** Storage batteries in solar photovoltaic systems for dwellings shall have cells operating at less than _________ V.
A. 12
B. 20
C. 30
D. 50

__________ __________ **23.** Solar photovoltaic systems in dwelling units shall have OCPDs protecting circuits to busbars sized at _________% less of the busbars.
A. 100
B. 110
C. 120
D. 125

__________ __________ **24.** A _________ phase power conditioning unit shall automatically disconnect all ungrounded conductors of a solar photovoltaic system.
A. 1
B. 2
C. both A and B
D. neither A or B

__________ __________ **25.** Solar photovoltaic systems in a one-family dwelling unit with circuits rated over _________ V to ground while energized shall not be accessible to other than qualified persons.
A. 50
B. 150
C. 175
D. 200

Special Conditions—700–780

Date ____________________

Chapter 9—TEST 18

Name ____________________________________

Section	Answer	
________	T F	**1.** Signs indicating type and location of emergency power sources shall be placed on service-entrance equipment.
________	T F	**2.** Emergency equipment shall be rated to operate all loads simultaneously.
________	T F	**3.** Ground-fault indicating means are not required on emergency systems.
________	T F	**4.** Emergency systems shall be tested periodically to verify proper operation.
________	T F	**5.** Inadvertent paralleling is permitted for isolation switches used to isolate transfer switch equipment.
________	T F	**6.** The general rule requires all emergency wiring to be routed independently of the normal wiring.
________	T F	**7.** A separate service shall not be used for emergency power.
________	T F	**8.** A tap ahcad of the main to serve an emergency system shall not be made in the main service disconnecting means.
________	T F	**9.** Under certain conditions, a generator can take more than 10 seconds to develop power to serve the load.
________	________	**10.** Emergency wiring and normal wiring can enter the same ________ to supply power to transfer equipment.
________	________	**11.** An emergency generator can be used for peak load ________.
________	________	**12.** Unit equipment can provide emergency power to ________ offices should the normal power fail.
________	________	**13.** Switches used to control emergency circuits shall be located so only ________ persons can have control.
________	________	**14.** Three- and ________-way switches shall not be used to control emergency lighting circuits.
________	________	**15.** OCPDs for branch circuits supplying ________ power equipment shall be accessible to authorized personnel only.
________	________	**16.** Battery pack units may supply emergency power if connected ________. A. on the lighting circuit of the area B. on any receptacle circuit C. on any branch circuit D. ahead of the main
________	________	**17.** An emergency system can be used for ________ load shaving in an industrial complex. A. peak B. optional C. both A and B D. neither A or B

_____ _____ **18.** All _____ containing emergency circuits shall be marked so they are readily identifiable from other circuits.

A. boxes
B. enclosures
C. both A and B
D. neither A or B

_____ _____ **19.** A separate branch circuit for unit equipment with a lock-on device can be routed to a(n) _____ area with three additional lighting circuits.

A. uninterrupted
B. hazardous
C. both A and B
D. neither A or B

_____ _____ **20.** Power of the emergency system shall be maintained to ensure the _____ of high-intensity discharge lighting until normal power is restored.

A. operation
B. restriking
C. both A and B
D. neither A or B

_____ _____ **21.** Emergency power panel conductors supplying a building are tapped on _____.

A. the line side of the service
B. any subfed panel
C. any circuit breaker main
D. any feeder circuit

_____ _____ **22.** A storage battery used to supply emergency power shall be capable of serving the load for at least _____ hours.

A. 1½
B. 2
C. 2½
D. 3

_____ _____ **23.** Signs indicating type and location of emergency power sources shall be placed on _____ equipment.

A. controlling
B. service-entrance
C. both A and B
D. neither A or B

_____ _____ **24.** Overcurrent protection devices in emergency systems shall _____.

A. be coordinated
B. clear in steps
C. not trip the main device
D. all of the above

_____ _____ **25.** Generators may be sized to carry only _____- selected loads for optional standby systems.

A. user
B. nonuser
C. both A and B
D. neither A or B

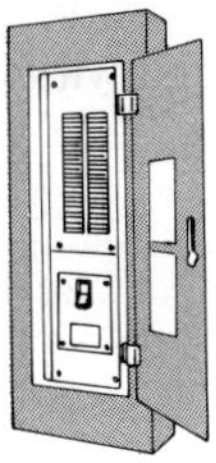

Special Conditions—700–780

Date ____________

Chapter 9—TEST 19

Name ____________

Section	Answer	
______	T F	**1.** Legally required standby systems are required by all governmental agencies having jurisdiction.
______	T F	**2.** The authority having jurisdiction is not required to witness a test on the complete installation of a legally required standby system.
______	T F	**3.** A written record shall be kept on test results of legally required standby systems.
______	T F	**4.** The wiring of legally required standby systems can occupy the same raceways as general wiring.
______	T F	**5.** Prime movers shall be solely dependent upon public utility gas companies for fuel supply for legally required standby systems.
______	T F	**6.** A tap ahead of the main supplying a legally required standby system can be made if the service main has sufficient separation.
______	T F	**7.** Generators supplying power to legally required standby systems shall have an on-site fuel supply that provides fuel for at least 1½ hours.
______	T F	**8.** Legally required standby systems shall be located to minimize hazards, floods, fire, etc.
______	T F	**9.** The wiring to optional standby systems cannot occupy the same junction boxes as the normal wiring.
______	T F	**10.** Optional standby systems are not mandatory.
______	______	**11.** The general rule requires interconnected electric power production sources to be connected and ______ at one location.
______	______	**12.** It may not be feasible to match the primary source wave shape to a(n) ______ power production source.
______	______	**13.** An electric power production source shall be disconnected if the ______ power source is lost.
______	______	**14.** Underground conductors rated at 23 kV and routed in PVC shall be buried at least ______″ deep.
______	______	**15.** It is permissible to splice high-voltage direct ______ cables without a junction box.
______	______	**16.** High-voltage non-shielded underground conductors routed in PVC conduit shall be encased in at least ______″ of concrete (general rule).
______	______	**17.** Switches with fused interrupters shall have the terminals for supply conductors located at the ______ of the enclosure.

__________ __________ **18.** Circuits of 50 V or less serving one or more appliances shall be wired with at least #__________ Cu.

__________ __________ **19.** Class __________ circuits shall not occupy the same raceway as communication cables.

__________ __________ **20.** Class 1 control circuits shall not exceed __________ V.

A. 120
B. 208
C. 277
D. 600

__________ __________ **21.** OCPDs for Class 1 #18 circuit conductors shall not exceed __________ A.

A. 5
B. 7
C. 10
D. 15

__________ __________ **22.** Class 1, Class 2, and Class 3 conductors are __________ permitted to be routed in the same raceway.

A. not
B. always
C. both A and B
D. neither A or B

__________ __________ **23.** Class 3 single circuit conductors shall not be less than #__________ AWG.

A. 12
B. 14
C. 16
D. 18

__________ __________ **24.** Fire protective signaling circuits shall have conductors that are bunch-tinned or __________ copper.

A. stranded
B. solid
C. both A and B
D. neither A or B

__________ __________ **25.** Stranded copper conductors used for fire-protective signaling circuits shall have __________ strands for size #16 or #18 AWG.

A. five
B. six
C. seven
D. none of the above

Communications Systems—800–820

Date ____________

Chapter 9—TEST 20

Name ______________________

Section	Answer	
________	T F	**1.** Communications circuit conductors shall be routed independently of other circuit wiring (general rule).
________	T F	**2.** The general rule requires communications circuits routed in shafts to be separated at least 1½″ from other power circuits.
________	T F	**3.** Communications circuits installed in buildings shall be Type CM cables.
________	T F	**4.** Communications cables routed vertically in a shaft shall be CMR cables.
________	T F	**5.** Approved listed communications cables can be placed under carpets.
________	T F	**6.** General communications cables not exceeding 50′-0″ in length can enter buildings.
________	T F	**7.** Communications conductors passing over buildings shall have a clearance of 10′-0″ if the roof can be easily walked upon.
________	T F	**8.** Open communications conductors shall be separated from open power conductors on buildings by at least 4″.
________	T F	**9.** Communications conductors run overhead shall be at least 6′-0″ from lightning conductors.
________	T F	**10.** A grounding electrode conductor used to ground communications equipment shall not be smaller than #16 AWG.
________	T F	**11.** The grounding electrode conductor shall be routed in a straight line to the grounding electrode if possible.
________	________	**12.** The bonding jumper between the communications grounding electrode and service equipment electrode shall be #________ AWG Cu.
________	________	**13.** Television antennas and indoor lead-ins shall not be routed nearer than ________″ from the general wiring.
________	________	**14.** Television antennas and indoor lead-ins can occupy the same box with other general wiring provided there is a(n) ________ between the wiring.
________	________	**15.** The minimum size copper grounding electrode conductor for grounding radio and television equipment is #________ AWG.
________	________	**16.** All inside conductors for transmitting stations shall be separated at least ________″ from electric power circuits (general rule).
________	________	**17.** The operating grounding electrode conductor for transmitting stations shall not be smaller than #________ AWG Cu.

________ ________ **18.** All external metal handles used by personnel to control a transmitter shall be ________ effectively.

________ ________ **19.** Coaxial cable installed in buildings shall be Type ________, listed as resistant to the spread of fire.
- A. CATV
- B. COTV
- C. both A and B
- D. neither A or B

________ ________ **20.** Coaxial cable run in raceways can be installed in buildings to connect ________ to television equipment.
- A. lighting
- B. antennas
- C. both A and B
- D. neither A or B

________ ________ **21.** Service equipment grounding electrodes shall be bonded to the antenna system electrode with a #________ Cu conductor.
- A. 6
- B. 8
- C. 12
- D. 14

________ ________ **22.** Underground coaxial cable shall be separated by a barrier when entering ________.
- A. ducts
- B. pedestals
- C. handholes
- D. all of the above

________ ________ **23.** For open spans of hard-drawn copper antenna conductors less than ________'-0" used for transmitting and receiving stations, the minimum size is #14 AWG.
- A. 150
- B. 175
- C. both A and B
- D. neither A or B

________ ________ **24.** Antenna conductors for television equipment shall be installed so they will not ________ under open electric light or power conductors where possible.
- A. parallel
- B. cross
- C. both A and B
- D. neither A or B

________ ________ **25.** Lead-in antenna conductors for television or radio equipment shall be securely ________ to antennas.
- A. attached
- B. fastened
- C. both A and B
- D. neither A or B

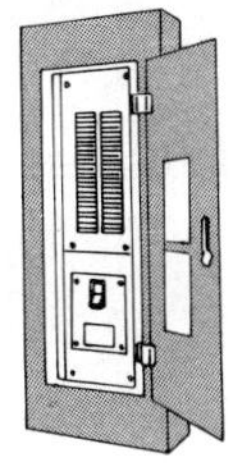

Transformer Elements—450

Date ________

Chapter 9—TEST 21

Name ________

Section	Answer

1. When the primary current of a transformer rated 600 V or less is ________ A or more and a 125% factor does not correspond to a standard size device, the next higher standard rating may be used.

2. When the primary current of a transformer rated 600 V or less is less than 9 A but more than 2 A, an overcurrent protection device shall not be sized at more than ________% of the primary current.

3. When the primary current of a transformer rated 600 V or less is less than ________ A, the overcurrent protection device shall not exceed 300% of the primary current.

4. When the primary side of a transformer rated over 600 V is individually protected by a circuit breaker, the setting shall not exceed ________% of the primary current.

5. When overcurrent protection is provided on the primary and secondary sides of transformers rated over 600 V, the percentage is determined by the ________ of the transformer (unsupervised locations).

6. Transformers rated 600 V or less and ________ kVA or less may be installed in the fire-resistant hollow spaces of buildings.

7. A 2-wire primary shall not protect the ________ of a transformer having a 3-wire or 4-wire hookup.

8. The primary side of a 2-wire transformer rated 9 A or more shall be protected at ________% of the transformer's FLC.

9. Conductors supplying a transformer are calculated at ________% for continuous loads and ________% for noncontinuous loads.

10. The size of a transformer is determined by the ________ served.

11. Transformers shall be installed in vaults when rated over ________ V.

12. Transformers rated 600 V or less shall be installed in a room of fire-resistant construction when rated more than ________ kVA (general rule).

13. Transformers rated 600 V or less installed indoors and transformers rated 112½ kVA or less shall be separated at least ________" from combustible materials.

14. If the total capacity of oil-insulated transformers does not exceed 112½ kVA, the vault may be constructed of reinforced concrete not less than ________" thick.

15. A door leading into a transformer vault shall have a minimum fire rating of ________ hour(s).

16. When transformer vaults are protected with an automatic sprinkler system, the door leading to a vault may have a minimum fire rating of ________ hour(s).

__________ __________ **17.** Transformer vaults shall not be used to store __________.

__________ __________ **18.** Transformer vaults shall not have pipe or ducts passing through vaults that are __________.

A. part of the system
B. provided for fire protection
C. foreign to the system
D. used for transformer cooling

__________ __________ **19.** Dampers used to close ventilation openings in vaults and protect against fire shall be rated not less than __________ hour(s).

A. 1
B. 1½
C. 2
D. 3

__________ __________ **20.** A door sill or curb in a transformer vault shall have a height of at least __________″.

A. 3
B. 4
C. 5
D. 6

__________ __________ **21.** Doors leading into a transformer vault shall be kept locked and access allowed only to __________.

A. the owner
B. any person working in the building
C. qualified persons
D. general public

__________ __________ **22.** Completely enclosed, ventilated transformers equipped with an 80 °C rise insulation may be installed in a room __________.

A. with concrete walls
B. designed as a vault
C. of fire-resistant construction
D. built with tile blocks

__________ __________ **23.** The primary overcurrent protection device shall protect the primary and secondary of the transformer in a __________ hookup.

A. 2-wire to 2-wire
B. 2-wire to 3-wire
C. 3-wire to 3-wire
D. 3-wire to 4-wire

__________ __________ **24.** Transformers that are installed in unsupervised locations and rated over 600 V are protected by fuses on the primary side and cannot exceed __________% of the primary FLC.

A. 150
B. 175
C. 250
D. 300

__________ __________ **25.** Transformers rated over 600 V and installed in supervised locations are protected by circuit breakers on the primary side and cannot exceed __________% of the primary FLC.

A. 250
B. 300
C. 500
D. 600

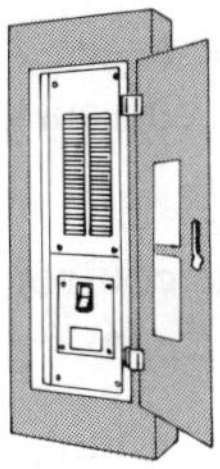

Motor Circuits and Components--430

Date ____________

Chapter 9–TEST 22

Name ____________

Section	Answer	
________	T F	**1.** Conductors supplying a single motor shall be computed at 135% of the motor's FLC.
________	T F	**2.** Branch-circuit conductors supplying multispeed motors shall be sized and selected from the largest amperage on the motor's nameplate based on the motor's RPM.
________	T F	**3.** The windings of motors with a duty cycle have a chance to rest during the OFF cycle.
________	T F	**4.** Conductors supplying individual motors with a duty cycle are sized from the starting and stopping application.
________	T F	**5.** If a motor is used for 15 minutes to drive a load, the motor shall rest for 35 minutes.
________	T F	**6.** Conductors supplying two or more motors shall be sized by multiplying all the motors in the group by 125%.
________	T F	**7.** Feeder conductors supplying two or more motors can be routed to a gutter and spliced to the conductors feeding each motor.
________	T F	**8.** FLC ratings selected to size conductors for motor circuits are taken from the motor's nameplate.
________	T F	**9.** The ampacity of conductors supplying motor circuits is selected from Table 310-16.
________	T F	**10.** The conductors supplying power to a wound-rotor motor are sized at 110% of the full-load current rating.
________	T F	**11.** Resistor leads are the conductors between the resistor bank and drum controller.
________	T F	**12.** Conductors supplying capacitors used for power factor correction are sized at 125% of the full-load current rating of the capacitor.
________	T F	**13.** Where a capacitor is connected on the load side of the magnetic starter, additional overload protection is not required.
________	T F	**14.** An additional disconnecting means is not required for capacitors installed on the load side of the magnetic starter.
________	T F	**15.** The rating of nonfusible disconnecting means shall be sized at least 125% of the motor's full-load current rating.
________	________	**16.** Circuit conductors supplying a single motor shall have a current-carrying capacity of ________% of the motor's FLC.

_________ _________ **17.** Conductors supplying the feeder for two or more motors shall be at least _________% of the largest motor plus the sum of the full-load currents of the remaining motors of the group.

_________ _________ **18.** Control-circuit conductors #18 and #16, which extend beyond the control equipment enclosure, are not protected when fused at _________ A.

_________ _________ **19.** Motors marked with a temperature rise not over 40 °C shall have running protection set at not more than _________% of the nameplate's full-load current rating.

_________ _________ **20.** For motors marked with a service factor of not less than 1.15, _________% of the full-load current of the motor shall be used to size the overload protection device.

_________ _________ **21.** For all motors not marked with a temperature rise or service factor, _________% shall be used to size the overload protection device.

_________ _________ **22.** Control conductors routed to remote start-stop stations may be protected by a device set at _________% of the conductor's ampacity.

_________ _________ **23.** Control conductors inside a starter enclosure that run to the start-stop station on the cover of the enclosure can be protected at _________ times the control conductor's ampacity.

_________ _________ **24.** List in any order four methods of starting a motor.

_________ _________

_________ _________

_________ _________

25. List the percentage of the device used to protect a 25 HP, 3ϕ squirrel-cage motor with full-voltage starting and no code letter using the following:

_________ _________ Nontime-delay fuse

_________ _________ Time-delay fuse

_________ _________ ITCB

_________ _________ Instantaneous trip CB

Dwelling Units—200-280

Date ____________

Chapter 9—TEST 23

Name ____________

Section	Answer	
______	______	**1.** The load for a household electric clothes dryer in a dwelling unit is computed at ______ VA or the nameplate rating, whichever is greater.
______	______	**2.** The general lighting and receptacle load for a dwelling unit is computed at ______ VA times the square footage.
______	______	**3.** Small appliance circuits are computed at ______ VA each.
______	______	**4.** Four or more fixed appliances installed in a dwelling unit may have a demand factor of ______% applied.
______	______	**5.** The general lighting load of a dwelling unit may have a demand factor of ______% applied to all loads exceeding 3000 VA but not more than 120,000 VA.
______	______	**6.** Ranges rated over 8 3/4 kW may have a demand factor applied according to the number listed in Column ______ of Table 220-19.
______	______	**7.** There shall be ______ or more household clothes dryers before a demand factor can be applied.
______	______	**8.** An individual motor shall have conductors with an ampacity of at least ______% of the motor's FLC.
______	______	**9.** The total load of a dwelling unit after demand factors are applied shall have ______% of the largest motor's FLC added to determine the service load.
______	______	**10.** When using the optional calculation to determine the service load, the first 10,000 VA of other loads is computed at 100% and the remainder at ______%.
______	______	**11.** Before the optional method can be applied, the ampacity of the service-entrance or feeder conductors shall be rated at least ______ A or larger.
______	______	**12.** The optional calculation requires one electric space heating unit to be computed at ______% and added to the other loads.
______	______	**13.** The other loads in the optional calculation include all loads except the ______ load and the electric space heating load.
______	______	**14.** Phase-to-______ loads are not considered in determining the rating of the neutral load.
______	______	**15.** A demand factor of ______% may be applied to that portion of the neutral load exceeding 200 A for resistive loads.
______	______	**16.** When the optional calculation is used to determine the rating of service-entrance conductors for a multifamily dwelling, the demand factor is based on the ______ of units.

__________ __________ **17.** To apply the optional calculation, each dwelling unit shall be equipped with electric __________ equipment, A/C, or an electric space heating unit.

__________ __________ **18.** All electric discharge lighting shall be computed at __________% when determining the neutral load.

__________ __________ **19.** House loads for multifamily dwellings are computed by the __________ calculation and added to the total load computed by using the optional calculation.

__________ __________ **20.** A demand factor of __________% may be applied to the nameplate rating of a range rated between 3½ kW and 8¾ kW.

__________ __________ **21.** The load for four 8 kW ranges is computed with a demand factor of __________%.

A. 30
B. 50
C. 65
D. 80

__________ __________ **22.** When computing the service load with the standard method, a 20 kW electric space heating unit is computed at __________%.

A. 80
B. 90
C. 100
D. 125

__________ __________ **23.** When computing the service load with the standard method, more than three fixed appliances are computed with a demand factor of __________% of the nameplate rating.

A. 50
B. 60
C. 75
D. 80

__________ __________ **24.** The continuous load supplied by a branch circuit shall not exceed the branch-circuit rating by more than __________%.

A. 50
B. 60
C. 80
D. 90

__________ __________ **25.** All 125 V, 1ϕ, 15 and 20 A receptacles installed above the countertop and within __________′ of the kitchen shall be GFCI-protected.

Commercial and Industrial Locations—200-280

Date ______________

Chapter 9—TEST 24

Name ______________

Section	Answer	
______	______	1. A branch circuit supplying a continuous load shall not exceed ______% of the OCPD rating.
______	______	2. An OCPD approved for continuous duty may be loaded to ______% of its rating.
______	______	3. Receptacle outlets supplying loads used for three hours or more are computed at ______ VA.
______	______	4. Multioutlet assemblies supplying light loads may be calculated at 180 VA for each ______′ installed.
______	______	5. General lighting loads supplying bank buildings may have a demand factor of ______% applied to the total lighting load.
______	______	6. The lighting load for a courtroom is calculated at ______ VA per sq. ft.
______	______	7. The feeder circuit to serve continuous loads is computed at ______% of the total connected load.
______	______	8. Medium-base lighting outlets in a show window (when listed) are calculated at ______ VA each.
______	______	9. Multioutlet assemblies are calculated at ______ A for each foot installed when serving heavy loads such as appliances.
______	______	10. The capacity of a 20 A CB is ______ VA when computing the number of general lighting circuits.
______	______	11. The existing demand for commercial buildings is calculated at ______% and added to the new load per the optional method.
______	______	12. Three electric cooking units in a restaurant may have the total load reduced by a(n) ______% multiplier.
______	______	13. One resistance welder with a 50% duty cycle rating has a(n) ______% multiplier.
______	______	14. The demand factor for 12,000 VA of lighting for a warehouse is ______%.
______	______	15. A 20′-0″ × 20′-0″ lodge room is calculated at ______ VA per sq. ft.
______	______	16. Each foot of a multioutlet assembly shall be computed at ______ VA.
______	______	17. Heavy-duty lampholders are calculated at ______ VA.
______	______	18. Conductors supplying a resistance welder shall be protected by an OCPD set at ______% or less.

__________ __________ **19.** A motor-generator arc welder shall have overcurrent protection set at not more than __________% of its rated primary current.

__________ __________ **20.** The largest motor of a system with several motors is computed at __________%.

__________ __________ **21.** The first __________ kW or less of receptacle outlets used at noncontinuous duty is computed at 100%.

__________ __________ **22.** The demand factor that may be applied to __________ or more pieces of commercial cooking equipment is 65%.

__________ __________ **23.** When the heating load is larger than the A/C load, the __________ load may be disregarded.

__________ __________ **24.** The total connected load, up to 3 VA per sq. ft., may be computed at __________% when applying the optional method for school buildings.

__________ __________ **25.** Maximum demand data using the optional method shall be available for at least __________ year(s) when adding new loads to an existing service.

Grounding—250

Date ____________

Chapter 9—TEST 25

Name ____________

Section	Answer	
______	______	1. A single electrode consisting of a rod, pipe, or plate shall have a resistance to ground of ______ Ω or less. A. 3 B. 15 C. 20 D. 25
______	______	2. Multiple rods, pipes, or plate electrodes shall be installed at least ______′ apart. A. 3 B. 6 C. 10 D. 12
______	______	3. A driven rod or pipe electrode shall be at least ______′ in length. A. 6 B. 7 C. 8 D. 12
______	______	4. When bonding circuits over ______ V to ground, two locknuts and a bushing attached to a box with cleanly punched holes will comply. A. 125 B. 250 C. 300 D. 480
______	______	5. When equipment bonding jumpers are installed on the outside of the raceway or enclosure, the bonding jumper shall not exceed ______′. A. 6 B. 8 C. 10 D. 12
______	______	6. If service-entrance conductors exceed 1100 kcmil for copper, the bonding jumper shall be at least ______% of the largest phase conductor. A. 8 B. 10 C. 12½ D. 15
______	______	7. The grounded conductor used to ground the frames of ranges shall not be smaller than #______ Cu. A. 6 B. 8 C. 10 D. 12

_______ _______ **8.** The insulation of equipment grounding conductors #6 or smaller shall be green, green with yellow stripes, or _______.

A. black
B. bare
C. white
D. red

_______ _______ **9.** Electrical equipment shall be spaced at least _______′ away from lightning rods.

A. 5
B. 6
C. 8
D. 10

_______ _______ **10.** If two or more buildings are supplied by a grounded system from one main service, each building shall have separate _______.

A. phases
B. neutrals
C. grounding electrode systems
D. service drops

_______ _______ **11.** Transformers used to step-down voltage for general use are classified as _______ systems.

A. separately derived
B. classified
C. direct
D. emergency

_______ _______ **12.** The grounded conductor shall be equal to the _______ conductor.

A. largest phase
B. largest bonding
C. ungrounded service
D. largest equipment

_______ _______ **13.** Circuits for electric cranes operating in Class III locations over combustible fibers shall not be _______.

A. grounded
B. pulled in raceways
C. spliced
D. THWN conductors

_______ _______ **14.** The neutral conductor in a 120 V, 1ϕ, 2-wire grounded system shall be _______.

A. black
B. green
C. yellow
D. white or natural gray

_______ _______ **15.** Flexible metal conduit used as a fixture whip to bond the fixture enclosure to the circuit junction box shall not exceed _______′.

A. 4
B. 6
C. 8
D. 10

16. The first choice for the grounding electrode of a separately derived system is a ________.
   A. ground ring
   B. water pipe
   C. building steel
   D. driven rod

17. Connection devices or fittings shall not connect grounding electrode conductors to equipment by means of ________.
   A. solder
   B. lugs
   C. pressure connectors
   D. approved clamps

18. Grounding electrode conductor size is based on the size of service-entrance conductors listed in ________.
   A. Table 250-94
   B. Table 250-95
   C. 250-42
   D. 250-57

19. Lighting fixtures fastened in place or connected by a permanent wiring method shall be grounded per ________.
   A. 250-43
   B. 250-50
   C. 250-94
   D. 250-95

20. Bonding provides electrical continuity and safely conducts any ________.
   A. voltage on the system
   B. unbalanced current
   C. fault current that may occur
   D. load of the system

21. A bare #4 conductor may be concrete-encased and serve as the grounding electrode if at least ________′ in length.
   A. 10
   B. 15
   C. 20
   D. 25

22. The interior metal water piping system shall be ________ to the service equipment enclosure, the grounded conductor at the service, and the grounding electrode conductor.
   A. bonded
   B. taped
   C. both A and B
   D. neither A or B

23. Equipment grounding conductors larger than #________ can be taped green, colored green, green striped, or bare.
   A. 6
   B. 8
   C. 10
   D. neither A or B

__________ __________ **24.** Equipment grounding conductors smaller than #__________ shall be green, green with yellow stripes, or stripped bare.

A. 6
B. 8
C. 10
D. 12

__________ __________ **25.** Equipment bonding jumpers may be installed inside or outside of a raceway or enclosure in __________′ lengths or less.

A. 2
B. 3
C. 4
D. 6

Services—230

Date ______________

Chapter 9—TEST 26

Name ______________________________

Section	Answer	
________	T F	**1.** A 600 A, 3-pole main circuit breaker supplies a 600 A, 3ϕ, 4-wire panel in building #2. A main in the panel of building #2 or the main in building #1 can serve as the disconnect.
________	T F	**2.** There are seven exceptions to the rule that a building shall have only one service drop or lateral.
________	T F	**3.** A service for a building shall pass through the inside of one building to supply another.
________	T F	**4.** A service drop is an overhead supply and a lateral is an underground supply.
________	T F	**5.** A panel with seven 2-wire, 20 A circuits requires a 100 A service.
________	T F	**6.** Service-entrance conductors can be spliced before termination in a switchboard.
________	T F	**7.** Service-entrance conductors in a raceway under at least 2″ of concrete are considered outside the building.
________	T F	**8.** The service-entrance conduit, meter base housing, and service disconnect enclosure shall be grounded.
________	T F	**9.** A service with six disconnecting means shall be grouped.
________	T F	**10.** A 277/480 V supply service of more than 150 V to ground with 1000 A OCPDs shall have ground-fault protection.
________	T F	**11.** Service equipment is required to be built to withstand available short circuit current without damage to the equipment.
________	T F	**12.** A CB or fuse used as the service disconnect shall be able to interrupt the available short circuit current at its terminals.
________	T F	**13.** Disconnects for fire pumps or emergency service shall be located at a remote point from the disconnect for the normal service.
________	________	**14.** The point of attachment for a service drop to a building is a minimum of ________′ above minimum grade.
________	________	**15.** A service drop clearance of at least ________′ is required from windows, porches, etc.
________	________	**16.** The minimum clearance for a service drop over a flat roof is ________′.
________	________	**17.** The minimum clearance for a service drop over a street is ________′.
________	________	**18.** The minimum clearance for a service drop over a commercial parking lot not subject to truck traffic is ________′.

_______ _______ **19.** A minimum clearance of 10′-0″ is required for voltage drops of _______ V or less to ground.

_______ _______ **20.** The Code® permits _______ disconnects to disconnect the service equipment.

_______ _______ **21.** In general, for service equipment of 150 V or less to ground, the minimum clearance between the service disconnect and grounded walls shall be at least _______′ on each side of the service equipment.

_______ _______ **22.** The service disconnecting means for any premise shall be installed in a readily accessible location and shall be as close as possible to the point where the _______ enters the building.

_______ _______ **23.** The _______ is considered the service conductor at 600 V or less with a transformer located outside the building and the secondaries entering the building.

_______ _______ **24.** List four approved wiring methods that are used as service-entrance conductors.

_______ _______

_______ _______

_______ _______

_______ _______ **25.** Each occupant of a multiple-occupancy building shall have _______ to the OCPD (general rule).

Swimming Pools—680

Date ____________________

Chapter 9—TEST 27

Name ______________________________________

Section	Answer	
__________	T F	**1.** All 120 V, 15 or 20 A receptacles located within 20′-0″ of the inside walls of a pool shall be GFCI-protected.
__________	T F	**2.** Equipment grounding electrode conductors for pool-associated motors shall be #12 AWG Cu or larger.
__________	T F	**3.** A switching device shall be located over 8′-0″ from the water's edge of a swimming pool.
__________	T F	**4.** Lighting fixtures located at least 12′-0″ above the maximum water level of a swimming pool are permitted.
__________	T F	**5.** Lighting fixtures are permitted within 5′-0″ horizontally or 8′-0″ vertically from the inside walls of a pool.
__________	T F	**6.** Lighting fixtures installed over indoor pools are permitted if located at least 8′-0″ above the maximum water level of the pool.
__________	T F	**7.** Receptacles rated 125 V and 15 or 20 A for an existing dwelling shall be installed at least 15′-0″ from the water's edge of a pool.
__________	T F	**8.** Pool steel shall be used as a common grid to bond metal parts of the pool and pool equipment.
__________	T F	**9.** Circuits supplying lighting fixtures in the walls of a pool shall be protected by a GFCI circuit.
__________	T F	**10.** When rigid metal conduit is used, a #8 AWG solid insulated copper conductor shall be installed between the swimming pool to form a shell for a wet-niche lighting fixture and the junction box.
__________	T F	**11.** Overhead utility conductors shall have a clearance of 18′-0″ above the maximum water level when supplied by a 120/240 V, 1ϕ service.
__________	T F	**12.** A flush deck may be used on systems of 24 V or less if the box is at least 4′-0″ from the water's edge and filled with compound.
__________	T F	**13.** Deck boxes shall be located at least 3′-0″ from the inside walls of the pool.
__________	__________	**14.** Lighting fixtures identified for such use can be installed in a swimming pool at a minimum depth of _________″.
__________	__________	**15.** Under certain conditions, lighting fixtures can be installed below _________′ over a spa or hot tub.
__________	__________	**16.** Conductors run on the outside of dwelling units can be installed in _________ and routed to the dry-niche fixture in PVC to swimming pools.

__________ __________ **17.** Heating elements of electric water heaters for spas and hot tubs shall be subdivided into loads of __________ A or less.

__________ __________ **18.** The maximum voltage permitted between conductors for underwater pool lighting is __________ V.

A. 12
B. 50
C. 115
D. 150

__________ __________ **19.** A permanently installed tub equipped with a recirculating piping system is classified as a __________.

A. spa
B. hot tub
C. hydromassage bathtub
D. all of the above

__________ __________ **20.** Conductors supplying water heaters for pools shall be rated at least __________% of the heating element load.

A. 110
B. 115
C. 125
D. 150

__________ __________ **21.** Cord- and plug-connected water pump motors shall be at least __________′ from a pool's water edge.

A. 2
B. 3
C. 4
D. 5

__________ __________ **22.** Stranded or solid #__________ insulated copper conductors shall be installed in rigid nonmetallic conduit when the deck box is connected to the wet-niche fixture of a pool.

A. 6
B. 8
C. 10
D. 12

__________ __________ **23.** Interior wiring between a dwelling and pool motor shall be wired with __________.

A. Romex
B. BX
C. EMT
D. all of the above

__________ __________ **24.** Motor circuits for electrical pool covers shall be __________-protected.

A. GFCI
B. GFI
C. either A or B
D. neither A or B

__________ __________ **25.** A deck box cannot be located less than __________′ from the water's edge.

A. 1
B. 2
C. 3
D. 4

Hazardous Locations—500–516

Date ____________________

Chapter 9—TEST 28

Name ______________________________

Section	Answer	
________	T F	**1.** A 4′-0″ run of conduit between two enclosures (one with a switching device) located in a Class II, Division 1 area shall have a seal.
________	T F	**2.** In hazardous locations, seals shall be provided within 24″ of each arcing device.
________	T F	**3.** Self-sealed devices in hazardous locations do not require additional seals.
________	T F	**4.** Squirrel-cage induction motors without arcing devices shall be explosion-proof when installed in Class I, Division 2 locations.
________	T F	**5.** Portable lighting equipment is permitted for use in Class I, Division 2 locations if it meets the requirements for fixed lighting equipment.
________	T F	**6.** A totally enclosed and pipe-ventilated motor shall not be installed in a Class III, Division 2 location.
________	T F	**7.** PVC installed beneath a gas island shall be buried at least 24″.
________	T F	**8.** The area beneath a gas island is a Class I, Division 1 location.
________	T F	**9.** An area 24″ high and extending 20′-0″ in all directions from a gasoline pump is a Class I, Division 2 location.
________	T F	**10.** The area inside a gasoline pump, to a height of 4′-0″, is a Class I, Division 1 location.
________	________	**11.** Exposed surfaces of space heaters used to prevent moisture condensation during shut-down periods shall not exceed ________% of the ignition temperature or gas involved.
________	________	**12.** Where pyrophoric materials are the only materials used or handled, the location does not have to be ________.
________	________	**13.** Cables without gastight or vaportight sheaths shall be ________ at the boundary of Division 2 and nonhazardous locations.
________	________	**14.** Approved cartridge fuses installed in lighting fixtures are permitted in Class ________, Division 2 locations.
________	________	**15.** Wiring that does not release enough energy to create a hazard under normal operating conditions can be used in Class ________, Division 1 locations.
________	________	**16.** In Class II, Division 2 locations, flexible metal conduit can be used in 6′-0″ or shorter lengths without a bonding jumper for loads of ________ A or less.
________	________	**17.** A bonding jumper in parallel with each ________ shall be installed for parallel runs of flexible conduit in Class III locations.

__________ __________ **18.** __________ protection shall be provided for electrical automotive diagnostic equipment that is cord- and plug-connected to a 120 V, 15 or 20 A receptacle.

__________ __________ **19.** Underground wiring emerging from a Class I, Division 1 location is Class I, Division 2 wiring to __________" above grade in a garage.

__________ __________ **20.** Rigid metal or __________ conduit can be used in Class II, Division 1 locations.

__________ __________ **21.** A flexible __________ or liquidtight flexible metal conduit can be used in Class II, Division 1 locations.

__________ __________ **22.** Totally enclosed, pipe-__________ motors shall be permitted to be installed in Class II, Division 1 locations.

__________ __________ **23.** EMT conduit can be used for general wiring in Class __________, Division 1 locations.

__________ __________ **24.** Class III locations are areas that contain ignitible __________ or flyings.

__________ __________ **25.** Seals are not required in conduit runs in Class __________, Divisions 1 and 2 locations.

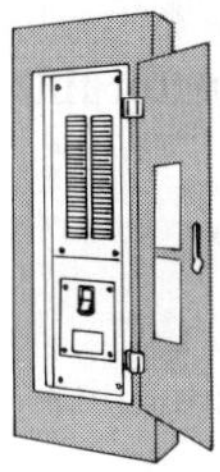

Wiring Methods—300–370

Date ____________________

Chapter 9—TEST 29

Name ______________________________

Section	Answer	
________	________	1. What size device box is required for a receptacle and three #12 THW Cu conductors?
________	________	2. What size octagonal box is required for three #12 THWN Cu conductors, one fixture stud with a hickey, and two #16 fixture wires?
________	________	3. What size square box is required for four #10 THWN Al conductors with two #10 equipment grounding conductors?
________	________	4. What size device box is required for a switch with two #14 conductors and an equipment grounding conductor with two #14 conductors passing through?
________	________	5. What size square box is required for one receptacle and two #12/2 Romex with grounds? (The box contains two cable clamps with the conductors of one Romex passing through.)
________	________	6. What size octagonal box is required for hanging a fixture having one fixture stud, one hickey, two #12 conductors, and two #18 fixture wires?
________	________	7. How many conductors are counted for four equipment grounding conductors in a box?
________	________	8. How many conductors are counted for one hickey and one fixture stud in a box?
________	________	9. How many conductors are counted for two cable clamps in a box?
________	________	10. How many conductors are counted for a total of four spliced conductors in a box?
________	________	11. How many conductors are counted for a total of eight spliced conductors and two conductors passing through a box?
________	________	12. How many conductors are counted for two equipment grounds, two pigtails, two branch circuit conductors, and one receptacle in a box?
________	________	13. How many conductors are counted for one receptacle, one switch, one pigtail, two spliced conductors, and four supply conductors in one box?
________	________	14. How many conductors are counted for three equipment grounds, one hickey, one fixture stud, four supply conductors, one lighting fixture, and two pigtails in a box?
________	________	15. How many #12 conductors can be installed in a $4'' \times 2\frac{1}{8}'' \times 1\frac{1}{2}''$ device box?
________	________	16. How many #14 conductors can be installed in a $4'' \times 1\frac{1}{2}''$ square box?
________	________	17. How many #10 conductors can be installed in a $4\frac{11}{16}'' \times 2\frac{1}{8}''$ square box?

_______ _______ **18.** How many cubic inches does a 4″ × 2⅛″ square box have for conductor fill?

_______ _______ **19.** How many cubic inches are in a 4¹¹⁄₁₆″ × 2⅛″ square box?

_______ _______ **20.** How many cubic inches are in a 3″ × 2″ × 2″ device box?

_______ _______ **21.** What size junction box is required for a straight pull in a 4″ conduit?

_______ _______ **22.** What is the minimum length for each side of a U pull with one 2½″, one 2″, and one 1″ conduit?

_______ _______ **23.** What number is used to multiply the largest conduit to size the minimum length for a junction box used for a straight pull?

_______ _______ **24.** What is the cubic inch rating for a #14 conductor?

_______ _______ **25.** What is the cubic inch rating for a #8 conductor?

Wiring Methods—300–370

Date ____________________

Chapter 9—TEST 30

Name ____

Section	Answer		
________	T	F	**1.** Junction boxes containing #14 conductors or larger shall have bushings on the conduit to protect conductors from abrasives.
________	T	F	**2.** Round boxes should not be used when connecting conduit.
________	T	F	**3.** A box with a minimum length of 28″ can be used for a straight pull in 3″ conduit.
________	T	F	**4.** Pull boxes and junction boxes shall be provided with covers.
________	T	F	**5.** Two or more extension rings can be added to find conductor fill for a box.
________	T	F	**6.** AC cable is a fabricated assembly of insulated conductors in a flexible metallic enclosure.
________	T	F	**7.** A 2½″ piece of PVC shall be supported every 6′-0″.
________	T	F	**8.** EMT shall be supported within 12″ of a junction box.
________	T	F	**9.** Fittings shall protect wires from abrasion at all points where AC cables terminate in a box.
________	T	F	**10.** A junction box is not required for splicing conductors where AC cable changes from one wiring method to another.
________	T	F	**11.** For flexibility, AC cable is permitted in lengths of not more than 24″ at terminals.
________	T	F	**12.** NM or NMC cable shall not be used as service-entrance cable.
________	T	F	**13.** For NM or NMC cable run at angles with joists in unfinished basements, cables not smaller than two #6 or three #8 conductors can be secured directly to the lower edges of the joists.
________	T	F	**14.** Metallic and nonmetallic boxes shall not be used as junction boxes for NM or NMC cable.
________	T	F	**15.** The ampacity of Romex shall be used at 60 °C.
________	T	F	**16.** SE cable supplying a subpanel shall have an EGC.
________	T	F	**17.** A 2″ rigid metal conduit screwed together with threaded couplings shall be supported every 18′-0″.
________	T	F	**18.** Threadless couplings and connectors used with IMC and buried in masonry or concrete shall be concrete-tight.
________	T	F	**19.** Cut ends of rigid metal conduit shall be reamed or otherwise finished to remove rough edges for protection of conductors from abrasion.
________	T	F	**20.** A bushing shall be provided to protect the wire from abrasion where IMC enters a box or fitting.

__________ T F **21.** Running threads are permitted to be used on rigid metal conduit for connections at couplings.

__________ T F **22.** The NEC® permits five 90° bends in a single rigid metal conduit run.

__________ T F **23.** A #12 green insulated conductor can be used in flexible metal conduit as a grounding means if protected by a 20 A or less OCPD.

__________ T F **24.** Table 3A of Ch. 9 lists the number of same-size conductors permitted in EMT.

__________ T F **25.** Romex shall be supported every 5′-0″ and within 12″ of every outlet and fitting.

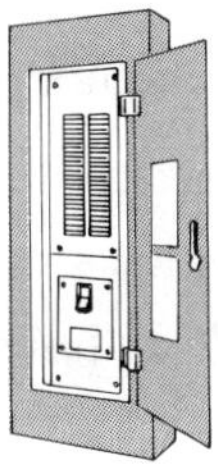

Overcurrent Protection—240

Date ____________

Chapter 9—TEST 31

Name ____________

Section	Answer	
________	T F	**1.** An OCPD can be sized above the conductor rating to allow a motor to start.
________	T F	**2.** A 25 A CB is not a standard device.
________	T F	**3.** A 3 A fuse is a standard rating and can be used for motor control circuits.
________	T F	**4.** Each occupant of a dwelling shall have access to overcurrent protection devices (general rule).
________	________	**5.** When the ampacity of a conductor does not correspond with a standard size OCPD, the next higher size up to ________ A can be used.
________	________	**6.** The primary protection on a transformer shall not be used to protect the secondary conductors of a 2-wire to ________-wire transformer.
________	________	**7.** A(n) ________′ tap can be taken from larger feeder conductors if sized to feed the load or device served.
________	________	**8.** Standard OCPDs start at odd sizes up to ________ A and even sizes up to 100 A.
________	________	**9.** CBs not marked with an IC rating can have an IC rating of ________ A.
________	________	**10.** OCPDs are not required to be ________ to all occupancies in a multiple-occupancy building under certain conditions.
________	________	**11.** When the ________′ tap rule is applied, the ampacity of the tap conductors shall be at least one-third the rating of the feeder conductors or OCPD.
________	________	**12.** Supplementary OCPDs are not required to be ________ accessible.
________	________	**13.** The general rule does not permit an OCPD to be placed in ________ with a grounded conductor.
________	________	**14.** OCPDs shall not be ________ in clothes closets.
________	________	**15.** A 10′ tap shall be made without overcurrent protection at the ________ of tap.
________	________	**16.** All ungrounded conductors, including the ________, shall be disconnected from the circuit when routed through or to a gasoline dispenser pump.
________	________	**17.** A(n) ________ A CB is required for a 25 HP, 230 V, 3ϕ induction motor with no code letter. Use next size.
________	________	**18.** A(n) ________ A NTDF is required for a 62 A, 3ϕ motor. Use minimum.
________	________	**19.** A(n) ________ A TDF is required for a feeder with one 10 HP, one 15 HP, and one 50 HP induction motor supplied by a 208 V, 3ϕ circuit. (The motors are full-voltage starting with code letter F.)

__________ __________ **20.** A(n) __________ A CB is required for an A/C rated at 52 A (minimum).

__________ __________ **21.** When using the maximum rating for an A/C, __________ % is used to size the OCPD.

__________ __________ **22.** The maximum percentage listed for a motor circuit using time delay fuses is __________%.

__________ __________ **23.** A(n) __________ A CB is required to protect a load of 100 A used for six hours.

__________ __________ **24.** A(n) __________ A fuse and #__________ conductors are required for a 25′ tap from a 600 A feeder circuit. (The load served by the tap is 200 A, continuous duty.)

__________ __________ **25.** A(n) __________ A fuse is required to supply a noncontinuous load of 50 A.

Lighting Fixtures—410

Date ____________

Chapter 9—TEST 32

Name ____________

Section	Answer	
______	______	**1.** Tap conductors shall be at least ______′ to ______′ in length from the tap to the point of connection to the fixture. A. 2, 8 B. 3, 7 C. 4, 6 D. 5 10
______	______	**2.** The frames of recessed fixtures shall have a clearance of at least ______″ from combustible material. A. ¼ B. ½ C. ¾ D. 1
______	______	**3.** Receptacles used to attach portable cords to serve lighting units shall be rated at no less than ______ A for 125 V. A. 10 B. 12 C. 15 D. 20
______	______	**4.** Conductors used for fixture wiring shall not be smaller than #______ AWG. A. 12 B. 14 C. 16 D. 18
______	______	**5.** Receptacles rated at ______ A or less and connected to aluminum conductors shall be marked for copper and aluminum. A. 20 B. 25 C. either A or B D. neither A or B
______	______	**6.** Fluorescent lighting fixtures installed in ______ ceilings shall be attached to the framing of the ceiling by clips, bolts, screws, or rivets. A. rocked B. suspended C. both A and B D. neither A or B

_________ _________ **7.** Fixtures with exposed metal parts shall be grounded if supplied by a _________ circuit, or be doubled-insulated or have insulating material.
A. grounding
B. neutral
C. both A and B
D. neither A or B

_________ _________ **8.** A 75 °C rated THW insulated conductor can be run within _________" of ballast in the fixture compartment.
A. 1
B. 2
C. 2½
D. 3

_________ _________ **9.** Fixture chains shall be wired with _________ wire from the outlet box connections to the sockets of the fixture.
A. solid
B. stranded
C. either A and B
D. neither A or B

_________ _________ **10.** Electric discharge lighting fixtures shall be connected by flexible cord if the cord is visible for _________% of its entire length.
A. 25
B. 50
C. 75
D. 100

_________ _________ **11.** A fixture frame is considered grounded if connected by _________ in a continuous run.
A. EMT
B. IMC
C. either A and B
D. neither A or B

_________ _________ **12.** Lighting fixtures approved for damp locations shall be installed only in _________ locations.
A. damp
B. wet
C. flooded
D. none of the above

_________ _________ **13.** Electric discharge lighting fixtures installed _________ an outlet box shall have an opening with access to the box.
A. under
B. over
C. on either side
D. none of the above

_________ _________ **14.** Fixtures and their lampholders installed over combustible material shall be at least _________' above the floor (general rule).
A. 6½
B. 7
C. 7½
D. 8

_______ _______ **15.** Any fixture weighing _______ pounds or less may be supported by an outlet box.
A. 35
B. 40
C. 50
D. 60

_______ _______ **16.** Externally wired fixtures are permitted to be installed in _______.
A. show windows
B. clothes closets
C. both A and B
D. neither A nor B

_______ _______ **17.** The screw-shell of a lampholder can support a fixture weighing _______ pounds and not exceeding 16″ in any dimension.
A. 6
B. 8
C. 10
D. 12

_______ _______ **18.** Outlet boxes are not required to have blank covers to prevent the escape of _______ and _______.
A. heat, temperature
B. arcs, sparks
C. dust, odors
D. none of the above

_______ _______ **19.** The general lighting and receptacle loads for a dwelling unit shall be computed at _______ VA times the square footage of the unit.
A. 2
B. 3
C. 3.5
D. none of the above

_______ _______ **20.** Fixture studs in octagonal boxes used to mount lighting fixtures are computed on the _______ conductor entering the box.
A. largest
B. smallest
C. longest
D. shortest

_______ _______ **21.** Rosettes used for lighting shall be rated at _______ VA.
A. 100
B. 150
C. 300
D. 660

_______ _______ **22.** Incandescent recessed fixtures shall have thermal protection and be _______ as thermally protected.
A. identified
B. listed
C. both A and B
D. neither A or B

__________ __________ **23.** Heavy-duty lighting tracks shall be identified to exceed __________ A in rating.

A. 20
B. 25
C. 30
D. none of the above

__________ __________ **24.** Screw-shell type lampholders shall have the __________ conductor connected to the screw-shell.

A. hot
B. grounded
C. either A and B
D. neither A or B

__________ __________ **25.** Candelabra sockets can be wired with #__________ AWG rubber-covered conductors.

A. 12
B. 14
C. 16
D. 18

Combination Questions

Date ____________

Chapter 9–TEST 33

Name ______________________

Section	Answer	
________	T F	**1.** Any suitable wiring method listed in the NEC® is permitted to be installed in any type of occupancy unless otherwise provided.
________	T F	**2.** GFCIs are required in bathrooms of residential occupancies.
________	T F	**3.** The approved sizes of liquidtight flexible metal conduit are ¼″ to 6″ (general rule).
________	T F	**4.** Rigid metal galvanized conduit can be installed in any location.
________	________	**5.** Raceways extending from an area of one temperature into an area of a widely different temperature shall be ________.
________	________	**6.** Three types of made electrodes are driven rods, pipes, and ________.
________	T F	**7.** Control, relay, or ammeter conductors used with a motor or starter can be installed in the same enclosure as the motor circuit conductors.
________	________	**8.** A driven rod shall be installed so that a minimum of ________′ is in the ground.
________	T F	**9.** If the secondary voltage on electric discharge lamps is 1000 V or less, wiring may occupy the same fixture enclosures as the branch-circuit conductors.
________	T F	**10.** Conductors of 600 V or less of different systems shall not occupy the same enclosures.
________	T F	**11.** Grounding electrode conductors can be spliced (general rule).
________	________	**12.** Grounding requirements for lightning rods and conductors require a minimum spacing of ________′.
________	T F	**13.** Solder may be used to attach connections to the grounding electrode conductor.
________	T F	**14.** Running threads are permitted on rigid metal galvanized iron conduit systems.
________	________	**15.** Flexible metal conduit is allowed as a grounding means if the total length in any ground return does not exceed ________′ and meets other requirements.
________	________	**16.** The smallest size grounding electrode conductor permissible is #________.
________	________	**17.** ________ can be grounded to the neutral conductor in a dwelling unit. A. Ranges B. Cooktops and ovens C. Dryers D. all of the above
________	T F	**18.** Fuses shall not be installed in parallel (general rule).

__________ __________ **19.** If electrode resistance is less than __________ Ω for a driven rod, an additional electrode shall be installed.

__________ T F **20.** Conductors, other than service-entrance conductors, are permitted in the same raceway or cable (general rule).

__________ __________ **21.** The minimum size service-entrance conductors allowable is #__________ copper (general rule).

__________ T F **22.** The largest size conductor permitted in a wireway cannot exceed design specifications for the wireway.

__________ __________ **23.** The minimum clearance of service drops over alleys and public roads is __________′.

__________ __________ **24.** Enclosure for overload devices shall be spaced __________″ from the mounting surface of wet locations.

__________ __________ **25.** Underground service conductors carried up a pole shall be given mechanical protection to a height of __________′.

__________ __________ **26.** The minimum clearance of service drops over residential driveways is __________′.

__________ __________ **27.** The minimum size service for a one-family dwelling with an initial load of 10 kW or more is __________ A.

__________ __________ **28.** The smallest service-entrance conductors normally allowed for limited loads of a single branch circuit are #__________.

__________ __________ **29.** When figuring the neutral load to electric ranges, __________% is the maximum unbalanced load.

__________ __________ **30.** The minimum point-of-attachment height of service-drop conductors rated at 600 V or less is __________′.

__________ __________ **31.** Receptacles in dwellings shall be no more than __________′ apart, starting no less than 6′-0″ from the door opening.

__________ __________ **32.** The minimum clearance of service drops over sidewalks used only for pedestrian traffic is __________′-0″.

A. 8
B. 10
C. 12
D. 15

__________ __________ **33.** The minimum clearance of service-drop conductors over residential property is __________′-0″ (general rule).

A. 8
B. 10
C. 12
D. 15

__________ __________ **34.** The formula, __________ VA × the square footage, is used to find the general lighting load for dwelling units.

A. 1
B. 3
C. 3.5
D. none of the above

______ ______ **35.** The minimum clearance of feeder conductors over commercial areas, parking lots, and other areas subject to truck traffic is ______'-0".
A. 8
B. 12
C. 15
D. 18

______ ______ **36.** Conductors used only for grounding shall be ______.
A. green, green with yellow stripes, or green and yellow
B. green, yellow with green stripes, or yellow
C. green, green with yellow stripes, or bare
D. green, yellow, or bare

______ ______ **37.** The load figured for show-window lighting is ______ VA per sq. ft.
A. 120
B. 180
C. 200
D. 240

______ T F **38.** Service conductors can be soldered.

______ ______ **39.** A minimum of ______" of conductor shall be allowed at outlets and switch points for connections and splicing.

______ ______ **40.** The smallest conductor permitted for a feeder circuit is #______.

______ ______ **41.** The maximum rating permitted for a 15 or 20 A branch circuit supplying cord- and plug-connected appliances is ______%.

______ ______ **42.** The maximum rating permitted for fixed appliances on a 15 or 20 A branch circuit with a lighting fixture is ______%.

______ ______ **43.** Receptacle outlets shall be provided in dwellings for all dining room walls exceeding ______'.

______ ______ **44.** The high-leg conductor of a 4-wire delta system is identified by the color ______.
A. red
B. orange
C. yellow
D. none of the above

______ ______ **45.** The connection for an equipment grounding conductor of a separately derived, ungrounded system shall be made by bonding the ______ to the grounding electrode conductor.
A. OCPD
B. neutral
C. service conductors
D. none of the above

______ ______ **46.** The minimum size service required for a dwelling with over six 2-wire branch circuits is ______ A.

______ T F **47.** Wooden plugs can be used for mounting equipment in masonry, concrete, or plaster.

______ ______ **48.** The maximum rating of a plug fuse between conductors is ______ V (general rule).

______ ______ **49.** Insulated service-entrance conductors larger than #6 are colored ______ to indicate the grounded conductor.

_______ T F **50.** For #6 or smaller conductors, the EGC can be stripped bare.

_______ T F **51.** For 120/240 V multiwire branch circuits feeding through an outlet box, the neutral can be broken and connected by means of the screws on the receptacle.

_______ _______ **52.** Cord- and plug-connected loads rated at 120 V maximum between conductors shall be rated at less than _______ VA.

_______ T F **53.** Grounded conductors shall not be attached to terminals so as to reverse polarity.

_______ T F **54.** A service switch can be used as the disconnecting means for a single motor if located in sight from the motor.

_______ T F **55.** A cable with a white conductor can be used for single-pole or 3- and 4-way switchloops.

_______ T F **56.** One white and one black conductor can be used for a switchloop in a conduit system.

_______ _______ **57.** What minimum size square box is required for a junction box with cable clamps and four #12-2 w/ground Romex conductors terminating in the box?

____

_______ _______ **58.** What is the minimum size square box required for eight #10 circuit conductors terminating in a box with clamps?

_______ _______ **59.** A junction box with cable clamps and a fixture stud has three #14/3 conductor cables with grounding wire and two #16 fixture wires that meet in the box. What is the minimum size 4″ square box required?

_______ _______ **60.** What is the minimum size square box required for six #12 and three #10 conductors that terminate in the box?

_______ _______ **61.** What is the minimum length box required for a straight run of two 3″, two 2″, and one 1″ pieces of conduit?

_______ _______ **62.** What size box is required for a right angle pull of four 3″, two 2″, and three 1½″ conduits?

_______ _______ **63.** What size raceway is required for four #1 THW, three #4 THW, and six #12 THW conductors?

____

_______ T F **64.** Made plate electrodes shall be embedded below the permanent moisture level.

_______ T F **65.** Romex can be installed perpendicular to joists or studs in joist or stud spaces used for environmental air in dwelling units.

_______ T F **66.** Receptacles can be installed on the same circuits as lighting for temporary wiring on construction sites.

_______ T F **67.** Underground wiring for gasoline dispensing equipment can be installed in rigid nonmetallic conduit when buried at least 18″ deep.

_______ T F **68.** An outlet for a sign shall be located outside the occupancy and supplied by an individual 20 A branch circuit for each commercial building with ground-floor footage accessible to pedestrians.

_______ T F **69.** A stranded, bare #6 conductor can be used as a concrete-encased electrode system.

________ ________ **70.** The size THWN Cu conductors required to serve a 15 HP, 230 V, 3ϕ squirrel-cage motor with code letter F, 40 °C rise, and operating at continuous-duty is #________.

____

________ ________ **71.** What instrument is used to measure voltage?

________ T F **72.** Voltage is the effective difference of potential between any two conductors in a circuit.

________ T F **73.** A volt amp is a unit used to measure the amount of electricity.

________ ________ **74.** Voltage is stepped ________ to transmit electricity long distances.

________ T F **75.** All windings are not required to be connected with the same polarity to ensure the operation of a subtractive-connected bank of transformers.

________ T F **76.** A delta-connected secondary is used primarily to supply 3ϕ loads.

________ ________ **77.** A wye-connected secondary is generally used to supply ________ single and 3-phase loads.

________ ________ **78.** The square root of 3 is ________.

________ T F **79.** High head pressure on a central A/C unit can cause the overload protection device to trip open.

________ ________ **80.** A fluctuating voltage to ground is normally caused by losing the system ________.

________ ________ **81.** Resistance in a circuit is measured in ________.

________ ________ **82.** Two 3-way and ________ 4-way switches can be used in a circuit to control the circuit from five different locations.

________ ________ **83.** Conductors can be pulled through a conduit more easily when a pulling ________ is used.

________ T F **84.** Romex shall normally be buried 24″ in the earth to protect the cables from damage by future digging (general rule).

________ ________ **85.** When capacitors are installed to correct power factors, the ratings of overload protection devices shall be adjusted to protect the motor ________ (general rule).

________ ________ **86.** Nails and copper conductors shall not be used to bridge the handles of CBs protecting ________ V, 1ϕ equipment.

________ T F **87.** Stop buttons shall be connected in series to stop a motor controlled by a magnetic starter.

________ ________ **88.** Start buttons shall be connected in ________ to energize the coil of a magnetic starter.

________ T F **89.** A #14 conductor for use as a switch leg can be tapped off a #12 conductor used to control a lighting fixture.

________ ________ **90.** Insulation failure can occur when conductors have a large amount of ________ drop.

________ ________ **91.** A(n) ________ can be used to fish Romex from a ceiling to a receptacle outlet located in a wall.

________ T F **92.** Switch boxes are not designed to support lighting fixtures.

________ T F **93.** Regular fluorescent lighting fixtures can be installed in the bay of a carwash.

__________ T F **94.** A megger can be used to detect faulty insulation on a copper conductor.

__________ __________ **95.** The polarity of connections for receptacle outlets installed in dwelling units should be checked for the __________ of the user.

__________ __________ **96.** An induction motor with six poles can be connected to provide __________ different speeds.

__________ __________ **97.** Single-phase motors shall have an additional winding to __________.

__________ __________ **98.** A 3ϕ motor can be reversed by reversing any two of the motor __________.

__________ T F **99.** The main winding of an induction motor has a higher resistance than the starting winding.

__________ __________ **100.** The starting winding of an induction motor is wound on __________ of the main winding.

Combination Questions

Date ____________________

Chapter 9—TEST 34

Name ______________________________

Section	Answer	
________	T F	1. Type SE cable can be used in interior wiring systems where circuit conductors are rubber-covered or thermoplastic.
________	T F	2. A motor having a controller adjacent with a lockable disconnecting means installed over 50′-0″ away and out of sight shall have an additional disconnecting means adjacent to the motor.
________	T F	3. A chain-supported fluorescent lighting fixture can be fed from its outlet box with Romex.
________	________	4. A power cord to a mobile home shall contain ________ conductors.
________	T F	5. Class 1 circuits can be run in the same raceway with communication circuits.
________	________	6. A fluorescent lighting fixture in a clothes closet shall be at least ________″ from a storage area or combustible material.
________	________	7. The minimum rating for power-supply cords to mobile homes is ________ A.
________	T F	8. Control conductors inside the starter enclosure that run to the start-stop station on the enclosure cover are protected at 400% of the control conductor's ampacity.
________ ________	T F	9. Class 1 control circuits may be installed in the same raceway with power conductors as long as they are associated with the same motor.
________	________	10. The minimum size for a fixture wire is #________ AWG.
________	T F	11. Separate running overload protection devices are based and sized on the motor's nameplate current rating.
________	________	12. The largest stranded conductor that can be connected to a panel with screws and terminal plates is #________.
________ ________	________	13. ________ disconnecting means are required for a rooftop A/C unit with controllers supplied by three different sources of current.
________	T F	14. PVC is permitted to be installed above a drop-in ceiling.
________	________	15. Conductors in sizes #________ and larger can be run in parallel or multiple.
________	T F	16. A 3ϕ room A/C unit, installed in a wall or window, can be cord- and plug-connected.
________	T F	17. The LRC of a motor remains the same for starting, regardless of the code letter rating.

______ ______ **18.** Edison-base plug fuses are recognized for ______ only.

______ T F **19.** A standard isolating transformer can be used on pool lighting circuits.

______ ______ **20.** Lights used in the walls of swimming pools shall operate at not more than ______ V.

______ T F **21.** CBs shall open all ungrounded conductors simultaneously.

______ T F **22.** Branch-circuit conductors on the line side of the controller for multispeed motors shall be sized at not less than 125% of the largest FLC on the motor's nameplate.

______ ______ **23.** Conductors approved for damp and ______ locations only, such as Type RH and THHN, shall not be installed in locations that extend to and include wet areas.

______ ______ **24.** Nonmetallic-sheathed cable shall be protected for a minimum of ______′ when installed in an attic without permanent access.

______

______ T F **25.** A permanent barrier installed in a junction box classifies each section as a separate box.

______ T F **26.** The NEC® gives no special provisions for metallic raceways installed in storage battery rooms.

______ ______ **27.** Two types of surface raceways are ______ and ______.

______

______ ______ **28.** A nipple is ______″ or less in length.

A. 6
B. 12
C. 18
D. 24

______ ______ **29.** Romex shall be protected at least ______″ above the finished floor.

A. 6
B. 12
C. 18
D. 24

______ T F **30.** Carpet squares installed over flat conductor cable shall not exceed 36″ × 36″.

______ T F **31.** Raised covers connected to the surface of metal boxes with #6-32 screws are considered grounded.

______ T F **32.** All equipment grounding conductors entering a metal box shall be spliced together with an approved connector and grounded to the box.

______ T F **33.** A #4 bare copper conductor used as a concrete-encased electrode shall be supplied by a #4 or larger copper grounding electrode conductor.

______ T F **34.** Conductors supplying 240 V loads and protected by CBs shall be protected by a 2-pole CB.

______ T F **35.** A #18 fixture wire supplying a fixture can be run in lengths of ______′ when protected by a 15 A or smaller device.

A. 50
B. 100
C. 200
D. 500

__________ T F **36.** When using the standard calculation, the dryer load on the neutral is reduced 60%.

__________ T F **37.** The disconnecting means for any sign (other than portable) shall be within sight or capable of being locked open by the controller.

__________ __________ **38.** Conductors supplying power to electrical equipment shall be sized and selected so as not to exceed the temperature of the conductor's __________.

__________ __________ **39.** Vertical clearance for service-drop conductors is based on the use of __________°C conductors.

__________ T F **40.** Separate grounding electrode conductors are permitted for connecting the grounded service conductor and the service equipment to the grounding electrode at the service.

__________ T F **41.** Shielding tape and other metallic shielding components shall be grounded on shielded conductors exceeding 600 V (general rule).

__________ __________ **42.** Conductors to each unit of a moving walk shall have a(n) __________ means (general rule).

__________ T F **43.** GFCI protection for personnel can be provided in the feeder circuit instead of the 20 A branch circuit.

__________ T F **44.** A minimum of two small appliance circuits is required to supply countertop receptacles in the kitchen.

__________ T F **45.** Receptacles in bathrooms of hotels and motels shall be protected by GFCI receptacles or GFCI circuits.

__________ T F **46.** Cable trays shall be identified for grounding before use as an equipment grounding conductor.

__________ T F **47.** Service-entrance cable can be used for interior wiring of a 2-wire, 240 V circuit with the bare conductor serving as an equipment grounding conductor.

__________ T F **48.** Busway cannot be used as a branch circuit.

__________ __________ **49.** A(n) __________" pull box is required for a straight pull in 4" conduit.

__________ T F **50.** A $\frac{3}{8}$" flexible metallic tubing can be used for a 6′-0" fixture whip.

__________ T F **51.** The NEC® now requires a minimum nominal voltage of 115/230 V, unless otherwise listed.

__________ T F **52.** Electrical continuity cannot be assured by standard threadless couplings and connectors for bonding service equipment (bonded type).

__________ __________ **53.** CBs can be installed in parallel only __________.

A. on the job site
B. if specified by the architect
C. as an approved factory assembly
D. all of the above

__________ T F **54.** A connector or plug and receptacle combination in the supply line to a wall-mounted oven can serve as the disconnecting means.

__________ T F **55.** Multioutlet assemblies shall not be run through an interior partition.

__________ T F **56.** Receptacles located within 20′-0" of a swimming pool shall be protected by a GFCI circuit.

__________ T F **57.** The wiring of a legally required standby system can occupy the same raceways as general wiring.

__________ T F **58.** Switchboards and panelboards rated at 600 V shall be located in rooms or spaces dedicated exclusively to such equipment.

__________ __________ **59.** When the number of OCPDs in a panelboard is counted, a 3-pole breaker is counted as __________ OCPD(s).

__________ __________ **60.** Any load that operates three hours or more is considered a continuous load and shall be figured at __________% for branch circuits.

__________ __________ **61.** Ground-fault protection is not required for the alternate __________ of power of emergency systems.

__________ __________ **62.** Continuous loads supplying inductive lighting shall not exceed __________% of the branch-circuit rating.

__________ __________ **63.** Rosettes shall be rated at 250 V, 660 W with a maximum operating current of __________ A.

__________ T F **64.** Branch-circuit conductors supplying a single hermetic refrigerant compressor motor shall have an ampacity of not less than 125% of either the motor's compressor-rated load current or the branch-circuit selection current, whichever is greater.

__________ T F **65.** Short circuits or ground faults are considered overloads for circuits supplying motors.

__________ T F **66.** IMC is available in lengths shorter than 10′-0″ for special applications.

__________ T F **67.** Cord pendants can be used as extensions from wireways.

__________ T F **68.** Switched receptacles are permitted instead of switched lighting outlets in dwelling kitchens.

__________ T F **69.** Shielded nonmetallic cable shall be properly marked on its exterior surface and each conductor within the cable shall be numbered for identification.

__________ T F **70.** A continuous white- or gray-striped covering on a conductor shall be used only for the grounded conductor.

__________ __________ **71.** A three-conductor flexible cord shall be permitted to supply a residential garbage disposal unit if not less than __________″ nor more than 36″ in length.

__________ __________ **72.** The minimum permissible diameter of a copper grounding electrode (driven rod) is __________″.

__________ T F **73.** The terminals of fixed appliances shall be marked for identification of the grounded conductor.

__________ T F **74.** Heavy-duty admedium lampholders shall be rated at 600 VA and above.

__________ __________ **75.** Tubing with cut threads used for arms and stems of lighting fixtures shall not be less than __________″ in thickness (pressed threads).

A. .020
B. .025
C. .030
D. none of the above

__________ __________ **76.** Communication conductors shall not be placed in raceways with __________ conductors (general rule).

__________ T F **77.** The bonding together of all separate grounding electrodes is permitted for communication equipment.

__________ __________ **78.** The grounding electrode conductor for CATV cable shall have a(n) __________ or other suitable insulation.

__________ __________ **79.** Overhead communication circuit conductors shall be at least __________′ above any roof that may be readily walked upon.

__________ T F **80.** Generally, coaxial cable used for CATV distribution shall not be placed in the same junction box with light and power circuits.

__________ __________ **81.** Where practical, the minimum separation between coaxial cable and lightning conductors shall be __________′.

__________ __________ **82.** The ampacity of capacitor circuit conductors shall not be less than __________% of the capacitor's rated current.

__________ T F **83.** Capacitors with 10 gallons of flammable liquid shall be installed in vaults.

__________ T F **84.** Capacitors shall have a means of discharging when disconnected from the line.

__________ T F **85.** Capacitors connected directly to a motor winding shall have other means of discharging.

__________ __________ **86.** Capacitors over 600 V shall reduce residual voltage to 50 V or less within __________ minutes.

A. two
B. three
C. four
D. five

__________ T F **87.** Running overload protection should be changed when installing capacitors on the line side of a controller to correct power factor.

__________ T F **88.** In general, capacitor cases should be grounded.

__________ __________ **89.** The minimum rating of grouped capacitor switches for circuits exceeding 600 V is __________%.

A. 80
B. 100
C. 125
D. 135

__________ T F **90.** A motor located 75′-0″ from its controller with no obstructions to prevent vision is considered in sight from the controller.

__________ __________ **91.** Circuit conductors supplying a single motor shall have a current-carrying capacity of __________% of the motor's FLC (general rule).

__________ T F **92.** Motor overload includes protection against ground faults.

__________ __________ **93.** Motors marked with a temperature rise not over 40 °C shall have running protection set at not more than __________% of the nameplate FLC.

A. 80
B. 100
C. 125
D. 140

_________ _________ **94.** If a motor is not in sight from the controller, a _________.

A. journeyman shall be present to oversee maintenance
B. disconnect by the controller shall be capable of being locked in the open position
C. disconnect shall be installed halfway between the motor and the disconnect
D. none of the above

_________ T F **95.** The disconnecting means between a motor and its controller shall be installed with one disconnect switch.

_________ _________ **96.** Switches used as disconnecting means for motors shall be rated in _________ (general rule).

_________ _________ **97.** The disconnecting means for a motor circuit of 600 V or less shall have a rating of not less than _________% of the motor's FLC.

A. 115
B. 125
C. 140
D. 150

_________ _________ **98.** Frames of portable motors operating at more than 150 V to ground shall be guarded or _________.

_________ T F **99.** A switch can serve as a disconnecting means and a controller under certain conditions.

_________ T F **100.** Conductors supplying power to motors rated over 600 V shall have an ampacity of at least the trip-open setting of the overload protection device.

Calculation Problems

Chapter 10

Standard and optional calculations are used to find electrical loads to size the service and elements of the service for all types of dwellings and commercial and industrial locations. The standard calculation, which yields larger values than the optional calculation, was the first method used to calculate loads. The optional calculation is relatively easy to use compared to the standard calculation.

Examination boards may use either or both calculation methods on their tests. Some boards require that VA and amps be carried with the phases and neutral. Amperage is often required to be divided evenly on the phases and neutral. Other examination boards require that the total VA be divided by supply voltage to find amps.

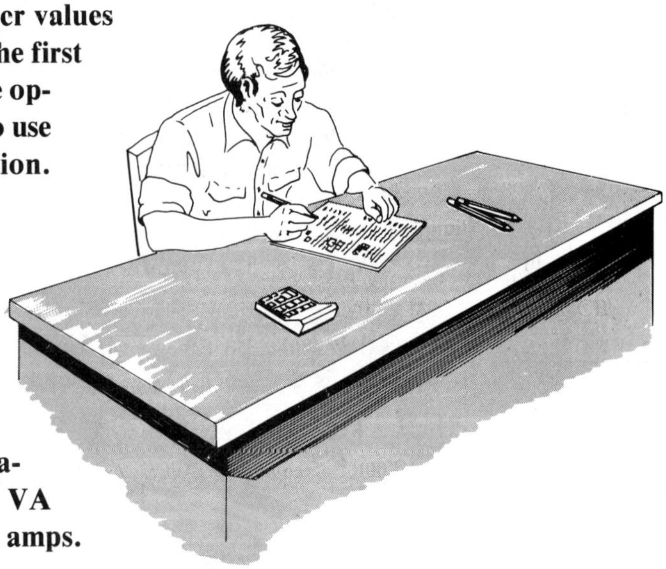

STANDARD AND OPTIONAL CALCULATIONS

Calculation problems for residential, commercial, and industrial locations are solved using the standard or optional method. The problems in this chapter give the method to be used. When specified in the problem, use the appropriate standard or optional form found at the end of this chapter. Each of the forms is cross-referenced to the number of the problem.

Forms used to solve these calculation problems include:

- Standard Calculation—One-Family Dwelling
- Optional Calculation—One-Family Dwelling (New or Existing)
- Optional Calculation—One-Family Dwelling (Existing Only)
- Standard Calculation—One-Family Dwelling 1ϕ (Carrying Phases and Neutral in Amps)
- Standard Calculation—One-Family Dwelling 1ϕ (Carrying Phase and Neutral in VA)
- Standard Calculation—Commercial and Industrial
- Optional Calculation—School
- Standard Calculation—Hotels and Motels
- Standard Calculation—Commercial and Industrial 3ϕ (Carrying Phases and Neutral in VA)

1. A 55′-0″ × 55′-0″ one-family dwelling is supplied by a 120/240 V, 1ϕ service and has the following loads. Use the standard calculation.

Three small appliance circuits
One laundry circuit
25 kW heating unit .... 240 V, 1ϕ
1080 VA blower motor .... 240 V, 1ϕ
5040 VA A/C unit .... 240 V, 1ϕ
12 kW cooktop .... 240 V, 1ϕ
10 kW oven .... 240 V, 1ϕ
9 kW oven .... 240 V, 1ϕ
5 kW dryer .... 240 V, 1ϕ
6 kW water heater .... 240 V, 1ϕ
780 VA attic fan .... 120 V, 1ϕ
1656 VA sump pump .... 240 V, 1ϕ
1920 VA water pump .... 240 V, 1ϕ
1080 VA pool circulating pump .... 240 V, 1ϕ
620 VA stereo intercom .... 120 V, 1ϕ
880 VA overhead garage door motor .... 120 V, 1ϕ
1250 VA microwave .... 120 V, 1ϕ
1600 VA dishwasher .... 120 V, 1ϕ
1015 VA disposal .... 120 V, 1ϕ
950 VA compactor .... 120 V, 1ϕ

__________ **1-1.** What is the general lighting load in VA and amps?

____

__________ **1-2.** What is the cooking equipment load in VA and amps?

____

__________ **1-3.** What is the dryer equipment load in VA and amps?

____

__________ **1-4.** What is the fixed appliance load in VA and amps?

____

__________ **1-5.** What is the service load for the heating or A/C unit load in VA and amps?

____

__________ **1-6.** What is the largest motor load in VA and amps?

____

2. A 45′-0″ × 60′-0″ one-family dwelling has a 120/240 V, 1ϕ service and the following loads. Use the form, STANDARD CALCULATION—ONE-FAMILY DWELLING to find the VA and amps of the six loads.

Four small appliance circuits
One laundry circuit
30 kW heating unit . . . . 240 V, 1ϕ
6000 VA A/C unit . . . . 240 V, 1ϕ
10 kW cooktop . . . . 240 V, 1ϕ
12 kW oven . . . . 240 V, 1ϕ
9 kW oven . . . . 240 V, 1ϕ
5 kW dryer . . . . 240 V, 1ϕ
1176 VA blower motor . . . . 240 V, 1ϕ
5500 VA water heater . . . . 240 V, 1ϕ
484 VA overhead garage door motor . . . . 120 V, 1ϕ
1200 VA microwave . . . . 120 V, 1ϕ
1500 VA dishwasher . . . . 120 V, 1ϕ
1092 VA disposal . . . . 120 V, 1ϕ
880 VA compactor . . . . 120 V, 1ϕ
983 VA pool pump motor . . . . 240 V, 1ϕ

3. A 2250 sq. ft. one-family dwelling has a 115/230 V, 1ϕ service and the following loads. Use the form, STANDARD CALCULATION—ONE-FAMILY DWELLING.

Three small appliance circuits
One laundry circuit
20 kW heating unit . . . . 230 V, 1ϕ
667 VA blower motor . . . . 230 V, 1ϕ
4370 VA A/C unit . . . . 230 V, 1ϕ
8.5 kW cooktop . . . . 230 V, 1ϕ
6.5 kW oven . . . . 230 V, 1ϕ
6 kW oven . . . . 230 V, 1ϕ
5 kW dryer . . . . 230 V, 1ϕ
5.5 kW water heater . . . . 230 V, 1ϕ
759 VA attic fan . . . . 115 V, 1ϕ
390 VA stereo intercom . . . . 115 V, 1ϕ
667 VA overhead garage door motor . . . . 115 V, 1ϕ
1440 VA dishwasher . . . . 115 V, 1ϕ
980 VA disposal . . . . 115 V, 1ϕ
1020 VA compactor . . . . 115 V, 1ϕ

__________ **3-1.** What size THWN Cu conductors is required for phases A and B?

__________ **3-2.** What size THWN Cu conductor is required for the neutral?

__________ **3-3.** How many 15 A, 2-wire circuits are required for the general lighting load?

__________ **3-4.** What size branch circuit conductors is required to supply the 8.5 kW cooktop, 6.5 kW oven, and 6 kW oven? (Use nonmetallic-sheathed cable with THHN Cu conductors.)

__________ **3-5.** What size THWN Al conductors is required to supply the heating unit?

__________ **3-6.** What is the maximum size OCPD required for the A/C unit?

4. A 3300 sq. ft. one-family dwelling has a 120/240 V, 1ϕ service and the following loads. Use the form, STANDARD CALCULATION—ONE-FAMILY DWELLING.

Two small appliance circuits
One laundry circuit
25 kW heating unit .... 240 V, 1ϕ
936 VA blower motor .... 240 V, 1ϕ
Four 12.5 A A/C window units .... 120 V, 1ϕ
4440 VA A/C unit .... 240 V, 1ϕ
8 kW cooktop .... 240 V, 1ϕ
6 kW oven .... 240 V, 1ϕ
5 kW oven .... 240 V, 1ϕ
5.5 kW dryer .... 240 V, 1ϕ
6.5 kW water heater .... 240 V, 1ϕ
420 VA attic fan .... 120 V, 1ϕ
840 VA water pump .... 240 V, 1ϕ
1320 VA pool pump motor .... 240 V, 1ϕ
1550 VA dishwasher .... 120 V, 1ϕ
960 VA disposal .... 120 V, 1ϕ
840 VA compactor .... 120 V, 1ϕ

__________ **4-1.** What maximum size THWN Al conductors is required for phases A and B and the neutral?

____

__________ **4-2.** What size panelboard is required for the service-entrance conductors?

__________ **4-3.** What size main OCPD is required for the panelboard to protect the service-entrance conductors?

__________ **4-4.** What size aluminum grounding electrode conductor is required to ground the service panelboard?

__________ **4-5.** How many 20 A, 2-wire circuits are required to supply the general lighting and receptacle load and the small appliance load?

__________ **4-6.** What size Romex Cu conductors is required to supply the water heater?

5. A 2116 sq. ft. one-family dwelling has a 120/240 V, 1ϕ service and the following loads. Use the form, STANDARD CALCULATION—ONE-FAMILY DWELLING.

Two small appliance circuits
One laundry circuit
15 kW heating unit .... 240 V, 1ϕ
840 VA blower motor .... 240 V, 1ϕ
3960 VA heat pump (operates with heating unit) .... 240 V, 1ϕ
5000 VA heat strips for heat pump .... 240 V, 1ϕ
560 VA A/C condenser motor .... 240 V, 1ϕ
20 kW range .... 240 V, 1ϕ
4.5 kW dryer .... 240 V, 1ϕ
6 kW water heater .... 240 V, 1ϕ
120 VA stereo intercom .... 120 V, 1ϕ
1350 VA microwave .... 120 V, 1ϕ
1600 VA dishwasher .... 120 V, 1ϕ
980 VA disposal .... 120 V, 1ϕ
880 VA compactor .... 120 V, 1ϕ

5-1. What maximum size THHN Cu conductors is required for phases A and B and the neutral using 75 °C lugs?

5-2. What size conduit is required for the service-entrance conductors?

5-3. What size OCPD is required for the service equipment based on the ampacity of the service conductors?

5-4. What size branch-circuit conductors and maximum size OCPDs is required for the heat pump and heat strips? (Use Romex.)

6. A 50′-0″ × 58′-0″ one-family dwelling has a 110/220 V, 1ϕ service and the following loads. Use the form, OPTIONAL CALCULATION—ONE-FAMILY DWELLING (NEW OR EXISTING).

Two small appliance circuits
One laundry circuit

25 kW heating unit	220 V, 1ϕ
1078 VA blower motor	220 V, 1ϕ
4840 VA A/C unit	220 V, 1ϕ
330 VA condenser motor	220 V, 1ϕ
5.5 kW dryer	220 V, 1ϕ
6 kW water heater	220 V, 1ϕ
770 VA attic fan	110 V, 1ϕ
979 VA water pump	110 V, 1ϕ
2398 VA pool pump motor	220 V, 1ϕ
220 VA overhead garage door motor	110 V, 1ϕ
1500 VA dishwasher	110 V, 1ϕ
605 VA disposal	110 V, 1ϕ
715 VA compactor	110 V, 1ϕ

6-1. What is the maximum and minimum size THWN Cu conductors required for phases A and B?

6-2. What is the maximum size THWN Cu conductor required for the neutral?

6-3. What size THWN Cu conductors is required to supply the water heater? (Use Romex.)

6-4. What size copper grounding electrode conductor is required to ground the maximum size service to an underground metal water pipe?

6-5. What size copper grounding electrode conductor is required to ground the minimum size service to an underground metal water pipe?

6-6. What size nonmetallic-sheathed cable (Romex) is required to supply the branch circuit to the A/C unit?

7. A 60′-0″ × 70′-0″ one-family dwelling has a 120/240 V, 1ϕ service and the following loads. Use the form, OPTIONAL CALCULATION—ONE-FAMILY DWELLING (NEW OR EXISTING).

Five small appliance circuits
One laundry circuit
Four 10 kW heating units .... 240 V, 1ϕ
Four 864 VA blower motors .... 240 V, 1ϕ
Four 4680 VA A/C units .... 240 V, 1ϕ
10 kW cooktop .... 240 V, 1ϕ
Two 9 kW ovens .... 240 V, 1ϕ
4.5 kW dryer .... 240 V, 1ϕ
6.5 kW water heater .... 240 V, 1ϕ
235 VA overhead garage door motor .... 120 V, 1ϕ
1150 VA microwave .... 120 V, 1ϕ
1600 VA dishwasher .... 120 V, 1ϕ
980 VA disposal .... 120 V, 1ϕ
960 VA compactor .... 120 V, 1ϕ

__________ **7-1.** What is the maximum ampacity rating of phases A and B for the service load?

__________ **7-2.** What is the maximum ampacity rating of the neutral for the service load?

__________ **7-3.** What maximum size TW Cu conductors is required for the service-entrance conductors for phases A and B?

__________ **7-4.** What maximum size TW Cu conductor is required for the neutral?

__________ **7-5.** What size copper grounding electrode conductor is required to ground the service to an underground water pipe?

__________ **7-6.** What size copper grounding electrode conductor is required for a supplementary ground to a driven rod?

__________ **7-7.** What size OCPD is required based on the service load?

__________ **7-8.** What size OCPD is required based on the ampacity of the conductors?

8. A 3575 sq. ft. one-family dwelling has a 115/230 V, 1ϕ service and the following loads. Use the form, OPTIONAL CALCULATION—ONE-FAMILY DWELLING (NEW OR EXISTING).

Four small appliance circuits
One laundry circuit
Three 15 kW heating units .... 230 V, 1ϕ
Three 828 VA blower motors .... 230 V, 1ϕ
Three 4830 VA A/C units .... 230 V, 1ϕ
8 kW range .... 230 V, 1ϕ
6 kW dryer .... 230 V, 1ϕ
6 kW water heater .... 230 V, 1ϕ
1840 VA water pump .... 230 V, 1ϕ
1587 VA pool pump .... 230 V, 1ϕ
1788 VA microwave .... 115 V, 1ϕ
1782.5 VA dishwasher .... 115 V, 1ϕ
1028 VA disposal .... 115 V, 1ϕ
1127 VA compactor .... 115 V, 1ϕ

______________ **8-1.** What maximum size THWN Cu conductors is required for phases A and B?

____

______________ **8-2.** What maximum size THWN Cu conductor is required for the neutral?

______________ **8-3.** What size panelboard is required based on the conductors supplying the service equipment?

______________ **8-4.** What size EMT is required for the service-entrance conductors?

______________ **8-5.** What size copper grounding electrode conductor is required to ground the service to a concrete-encased electrode?

______________ **8-6.** What minimum and maximum size CB is required for phases A and B?

____

9. A 3500 sq. ft. one-family dwelling has a 120/240 V, 1ϕ service and the following loads. Use the form, OPTIONAL CALCULATION—ONE FAMILY DWELLING (EXISTING ONLY) to find the ampacity of the neutral. Can two 25 A, 240 V, 1ϕ A/C window units be added to the existing dwelling with #2 THWN Cu conductors supplying the service equipment?

Two small appliance circuits
One laundry circuit
5 kW dryer .... 240 V, 1ϕ
780 VA attic fan .... 120 V, 1ϕ
1200 VA microwave .... 120 V, 1ϕ
1350 VA dishwasher .... 120 V, 1ϕ
695 VA disposal .... 120 V, 1ϕ
720 VA compactor .... 120 V, 1ϕ
120 VA floor furnace motor .... 120 V, 1ϕ

10. A 1600 sq. ft. one-family dwelling has a 120/240 V, 1ϕ service and the following loads. Use the form, STANDARD CALCULATION—ONE-FAMILY DWELLING 1ϕ (CARRYING PHASES AND NEUTRAL IN VA).

Two small appliance circuits
One laundry circuit
15 kW heating unit .... 240 V, 1ϕ
840 VA blower motor .... 240 V, 1ϕ
4440 VA A/C unit .... 240 V, 1ϕ
12 kW cooktop .... 240 V, 1ϕ
Two 10 kW ovens .... 240 V, 1ϕ
5 kW dryer .... 240 V, 1ϕ
5.5 kW water heater .... 240 V, 1ϕ
880 VA attic fan .... 120 V, 1ϕ
910 VA water pump .... 120 V, 1ϕ
1728 VA pool pump .... 240 V, 1ϕ
1800 VA dishwasher .... 120 V, 1ϕ
980 VA disposal .... 120 V, 1ϕ
890 VA compactor .... 120 V, 1ϕ

______________ **10-1.** What is the amps for phases A and B?

____

______________ **10-2.** What size THW Al conductors is required for phases A and B?

______________ **10-3.** What size THW Al conductor is required for the neutral?

11. A 3600 sq. ft. one-family dwelling has a 120/240 V, 1ϕ service and the following loads. Use the form, STANDARD CALCULATION—ONE-FAMILY DWELLING 1ϕ (CARRYING PHASES AND NEUTRAL IN AMPS).

Four small appliance circuits
One laundry circuit
30 kW heating unit .... 240 V, 1ϕ
1080 VA blower motor .... 240 V, 1ϕ
6000 VA A/C unit .... 240 V, 1ϕ
8.5 kW cooktop .... 240 V, 1ϕ
8 kW oven .... 240 V, 1ϕ
6 kW oven .... 240 V, 1ϕ
5 kW dryer .... 240 V, 1ϕ
6.5 kW water heater .... 240 V, 1ϕ
300 VA stereo intercom .... 120 V, 1ϕ
264 VA overhead garage door motor .... 120 V, 1ϕ
1500 VA microwave .... 120 V, 1ϕ
1800 VA dishwasher .... 120 V, 1ϕ
864 VA disposal .... 120 V, 1ϕ
1176 VA compactor .... 120 V, 1ϕ

11-1. Find the amps for phases A and B using the form, STANDARD CALCULATION—ONE-FAMILY DWELLING, and compare to the amps for phases A and B from the form, STANDARD CALCULATION—ONE-FAMILY DWELLING 1ϕ (CARRYING PHASES AND NEUTRAL IN AMPS).

11-2. Find the amps for the neutral using the form, STANDARD CALCULATION—ONE-FAMILY DWELLING, and compare to the amps for the neutral from the form, STANDARD CALCULATION—ONE-FAMILY DWELLING 1ϕ (CARRYING PHASES AND NEUTRAL IN AMPS).

____________ **12.** A commercial building has the following occupancies. All loads are continuous duty except for the closet area and storage space. Use the standard method to find the general lighting load in VA. Do not include the receptacle load.

1600 sq. ft. office space
1800 sq. ft. bank space
800 sq. ft. halls and corridors
4600 sq. ft. warehouse
700 sq. ft. stairways
200 sq. ft. closet area
400 sq. ft. storage space

13. A welding shop is supplied by a 277/480 V, 3ϕ, 4-wire service. A step-down transformer provides a secondary voltage of 120/208 V, 3ϕ, 4-wire. The welding shop has the following loads.

78,800 VA (continuous duty) .... 480 V, 3ϕ
42,000 VA (continuous duty) .... 277 V, 1ϕ

Step-down transformer loads

58,600 VA (continuous duty) .... 208 V, 3ϕ
32,500 VA (continuous duty) .... 120 V, 1ϕ

____________ **13-1.** What is the load in VA and amps for the 277/480 V loads?

____

____________ **13-2.** What is the load in VA and amps for the step-down transformer?

____

13-3. What is the load in amps for the service-entrance conductors?

13-4. What size THWN Cu conductors is required for phases A, B, and C and the neutral of the service equipment?

13-5. What size THWN Cu conductors is required for the primary and secondary side of the step-down transformer?

13-6. What size OCPD is required for the primary and secondary side of the step-down transformer?

14. A welding shop has a 277/480 V, 3ϕ, 4-wire service and the following loads.

- Forty 500 VA mercury fixtures .... 277 V, 1ϕ
- Ten 500 VA mercury floodlights .... 277 V, 1ϕ
- 20 kW resistance welder (50% duty cycle) .... 480 V, 3ϕ
- 19 kW resistance welder (50% duty cycle) .... 480 V, 3ϕ
- 18 kW resistance welder (50% duty cycle) .... 480 V, 3ϕ
- 17 kW resistance welder (50% duty cycle) .... 480 V, 3ϕ
- 25 kW transformer arc welder (60% duty cycle) .... 480 V, 3ϕ
- 24 kW transformer arc welder (60% duty cycle) .... 480 V, 3ϕ
- 23 kW transformer arc welder (60% duty cycle) .... 480 V, 3ϕ
- 22 kW transformer arc welder (60% duty cycle) .... 480 V, 3ϕ
- 21 kW transformer arc welder (60% duty cycle) .... 480 V, 3ϕ
- 18.6 kW motor-generator arc welder (40% duty cycle) .... 480 V, 3ϕ
- 17.8 kW motor-generator arc welder (40% duty cycle) .... 480 V, 3ϕ
- 16.9 kW motor-generator arc welder (40% duty cycle) .... 480 V, 3ϕ
- 15.7 kW motor-generator arc welder (40% duty cycle) .... 480 V, 3ϕ
- 14.5 kW motor-generator arc welder (40% duty cycle) .... 480 V, 3ϕ
- 13.4 kW motor-generator arc welder (40% duty cycle) .... 480 V, 3ϕ
- 5 HP drill press .... 480 V, 3ϕ
- 3 HP punch press .... 480 V, 3ϕ
- Ten 2 HP exhaust fans .... 480 V, 3ϕ
- Four 7½ HP lathes .... 480 V, 3ϕ

14-1. What is the total VA and amps for the welders and motors?

14-2. What size THWN Al conductors is required for the feeder circuit to the group of transformer arc welders?

14-3. What size OCPD is required for the feeder circuit to the group of transformer arc welders?

14-4. What size THWN Cu conductors is required to supply power to the 25 kW, 480 V, 3ϕ transformer arc welder?

14-5. What size OCPD is required for the 25 kW, 480 V, 3ϕ transformer arc welder?

15. A hospital has a 277/480 V, 3ϕ, 4-wire service and the following loads. Use the form, STANDARD CALCULATION—COMMERCIAL AMD INDUSTRIAL.

65,000 sq. ft. area (continuous duty) .... 277 V, 1ϕ

Emergency system loads supplied by generator

50,000 VA life safety branch .... 277/480 V, 3ϕ
60,000 VA critical branch .... 277/480 V, 3ϕ

650 duplex receptacles (noncontinuous duty) .... 120 V, 1ϕ
35,000 VA patient care areas (noncontinuous duty) .... 277 V, 1ϕ
48,000 VA general care areas (noncontinuous duty) .... 277 V, 1ϕ
120,000 VA heating units .... 480 V, 3ϕ
42,000 VA A/C units .... 480 V, 3ϕ
Five 5 HP motors .... 480 V, 3ϕ
Three 10 HP motors .... 480 V, 3ϕ
Six 1½ HP motors .... 480 V, 3ϕ
Four 2 HP motors .... 480 V, 3ϕ
Four diagnostic units (x-ray) .... 208 V, 1ϕ
150 MA momentary rating at 100 kV
Four treatment units (x-ray) .... 208 V, 1ϕ
15 MA long-time rating at 200 kV
200′ lighting track .... 120 V, 1ϕ
Thirty 450 VA outside lighting fixtures .... 277 V, 1ϕ

________ **15-1.** What is the service load in VA and amps?

____

________ **15-2.** What is the load in amps for one diagnostic and one treatment unit?

____

________ **15-3.** What is the total VA and amps for the x-ray equipment?

____

16. A 65′-0″ × 80′-0″ commercial store has a 120/208 V, 3ϕ service and the following loads. Use the form, STANDARD CALCULATION—COMMERCIAL AND INDUSTRIAL to find the service load in VA and amps.

Fifty linear ft. show window lighting .... 120 V, 1ϕ
70′ lighting track .... 120 V, 1ϕ
Thirty-two 156 VA outside lighting units .... 208 V, 1ϕ
2200 VA sign lighting .... 120 V, 1ϕ
68 general-purpose receptacles (noncontinuous duty) .... 120 V, 1ϕ
36 general-purpose receptacles (continuous duty) .... 120 V, 1ϕ
40′ multioutlet assembly (light duty) .... 120 V, 1ϕ
25 HP walk-in cooler .... 208 V, 3ϕ
30 HP refrigeration box .... 208 V, 3ϕ
20 HP produce rack .... 208 V, 3ϕ
9000 VA mainframe computer (continuous duty) .... 208 V, 3ϕ
12,000 VA miscellaneous (continuous duty) .... 208 V, 3ϕ
Six 1 HP motors .... 208 V, 3ϕ
Four 2 HP motors .... 208 V, 3ϕ
Three 5 HP motors .... 208 V, 3ϕ
Two 7½ HP motors .... 208 V, 3ϕ
40 kW heating unit .... 208 V, 3ϕ
42 A A/C unit .... 208 V, 3ϕ

17. A 65′-0″ × 65′-0″ office building is lighted by 277 V, 1ϕ lighting fixtures and has a 277/480 V, 3ϕ, 4-wire service with the following loads. Use the form, STANDARD CALCULATION—COMMERCIAL AND INDUSTRIAL to find the service load in VA and amps.

80′ lighting track .... 120 V, 1ϕ
Twenty-six 225 VA outside lighting units .... 120 V, 1ϕ
1200 VA sign circuit .... 120 V, 1ϕ
225 general-purpose receptacles .... 120 V, 1ϕ
60′ multioutlet assembly (heavy duty) .... 120 V, 1ϕ
35 kW heating unit .... 480 V, 3ϕ
25 A A/C unit .... 480 V, 3ϕ
22 A computer (continuous duty) .... 208 V, 3ϕ
14 A data processor (continuous duty) .... 120 V, 1ϕ
30 word processors (continuous duty) .... 120 V, 1ϕ
18 A copier (continuous duty) .... 208 V, 1ϕ
25 A printer (continuous duty) .... 208 V, 3ϕ
15 HP motor .... 208 V, 3ϕ
20 HP motor .... 480 V, 3ϕ
25 HP motor .... 480 V, 3ϕ
2 HP motor .... 120 V, 1ϕ

18. A 50,000 sq. ft. school building has a 277/480 V, 3ϕ, 4-wire service and the following loads. Use the form, OPTIONAL CALCULATION—SCHOOL to find the total VA and amps.

PHASES

Classrooms .... 60,650 VA
Classroom A/C load .... 45,360 VA
Classroom heating load .... 63,200 VA
Auditorium .... 18,980 VA
Kitchen equipment (12 units) .... 85,500 VA
Cafeteria .... 6850 VA
Office space .... 60,850 VA
Office heating load .... 120,204 VA
Office A/C load .... 18,120 VA

Neutral (Loads from standard calculation)

Lighting loads .... 60,825 VA
Receptacle loads .... 20,600 VA
Special loads .... 133,005 VA
Motor loads .... 5600 VA
Largest motor load .... 360 VA

19. A 70′-0″ × 80′-0″ restaurant has a 120/240 V, 1ϕ, 3-wire service with 90% of the neutral load harmonically balanced and the following loads. Use the form, STANDARD CALCULATION—COMMERCIAL AND INDUSTRIAL.

20 linear ft. show window lighting ..... 120 V, 1ϕ
30′ lighting track ..... 120 V, 1ϕ
Ten 450 VA outside mercury floodlights ..... 120 V, 1ϕ
1200 VA sign circuit ..... 120 V, 1ϕ
30 general-purpose receptacles ..... 120 V, 1ϕ
25 general-purpose receptacles (continuous duty) ..... 120 V, 1ϕ
20′ multioutlet assembly (heavy duty) ..... 120 V, 1ϕ

Kitchen equipment

Two 1½ HP mixers ..... 240 V, 1ϕ
3 HP grinder ..... 240 V, 1ϕ
2 HP dishwasher ..... 240 V, 1ϕ
5 HP walk-in cooler ..... 240 V, 1ϕ
Three 12 kW deep fryers ..... 240 V, 1ϕ
Two 6 kW refrigerators ..... 240 V, 1ϕ
Two 5.28 kW freezers ..... 240 V, 1ϕ
Two 5.5 kW water heaters ..... 240 V, 1ϕ
5.16 kW ice cream box ..... 240 V, 1ϕ
Two 2.5 kW coffee pots ..... 240 V, 1ϕ
Two 18.5 kW toasters ..... 240 V, 1ϕ
Two 10 kW ovens ..... 240 V, 1ϕ
Three 8 kW grills ..... 240 V, 1ϕ

Two 25 kW heating units ..... 240 V, 1ϕ
Two 5.5 kW A/C units ..... 240 V, 1ϕ
5 kW special loads (continuous duty) ..... 240 V, 1ϕ
2 kW special loads (continuous duty) ..... 120 V, 1ϕ
Two 7½ HP motors ..... 240 V, 1ϕ
Four ¾ HP motors ..... 120 V, 1ϕ

______ **19-1.** What is the demand load in VA and amps for the kitchen equipment?

______

20. A 300-unit motel has a 120/208 V, 3ϕ, 4-wire service and the following loads. Use the form, STANDARD CALCULATION—HOTELS AND MOTELS to find the service load in VA and amps.

100 units at 650 sq. ft. each ..... 120 V, 1ϕ
100 units at 600 sq. ft. each ..... 120 V, 1ϕ
100 units at 550 sq. ft. each ..... 120 V, 1ϕ
Three hundred ⅙ HP exhaust fans ..... 120 V, 1ϕ
Three hundred 300 VA heat lamps ..... 120 V, 1ϕ
1200 general-purpose receptacles (four per unit) ..... 120 V, 1ϕ
68,200 VA kitchen equipment (10 units) ..... 208 V, 1ϕ
58 duplex receptacles (continuous duty) ..... 120 V, 1ϕ
30′ lighting track ..... 120 V, 1ϕ
Twenty-five 180 VA outside lighting fixtures ..... 120 V, 1ϕ
1900 VA sign circuit ..... 120 V, 1ϕ
Three hundred 3324 VA A/C units ..... 208 V, 1ϕ
Three hundred 5000 VA heating units ..... 208 V, 1ϕ

______ **20-1.** If the service is 277/480 V, 3ϕ, 4-wire, what is the service load for the phases?

______ **20-2.** If the service is 2400/4160 V, 3ϕ, 4-wire, what is the service load for the phases?

______ **20-3.** What size busduct is required for the service to the 300 units using the ampacities based on the three voltages?

______

______

21. A church has a 120/240 V, 1ϕ service and the following loads. Use the form, STANDARD CALCULATION—COMMERCIAL AND INDUSTRIAL to find the total VA and amps for the service.

650 sq. ft. nursery (continuous duty) .... 120 V, 1ϕ
4000 sq. ft. church area (continuous duty) .... 120 V, 1ϕ
80′ lighting track .... 120 V, 1ϕ
Six 450 VA outside lighting loads (continuous duty) .... 120 V, 1ϕ
68 duplex receptacles (noncontinuous duty) .... 120 V, 1ϕ
20 duplex receptacles (continuous duty) .... 120 V, 1ϕ

Special loads

4800 sq. ft. auditorium (continuous duty) .... 120 V, 1ϕ
800 sq. ft. halls and stairways (continuous duty) .... 120 V, 1ϕ
600 sq. ft. storage space (noncontinuous duty) .... 120 V, 1ϕ
10,000 VA office equipment (continuous duty) .... 240 V, 1ϕ
6400 VA miscellaneous office loads (continuous duty) .... 120 V, 1ϕ

50,000 VA heating units .... 240 V, 1ϕ
28,000 VA A/C units .... 240 V, 1ϕ
Three 1 HP motors .... 240 V, 1ϕ
Three 2 HP motors .... 240 V, 1ϕ
Four 3 HP motors .... 240 V, 1ϕ
Six ¼ HP motors .... 120 V, 1ϕ

______ **21-1.** What is the lighting load in VA for the auditorium, halls and stairways, and storage space?

______ **21-2.** What size THWN Cu conductors, paralleled three times per phase, is required for the service equipment?

______ **21-3.** What size switchboard is required for the service equipment based on the service load?

______ **21-4.** What size THWN Cu conductors, paralleled three times, is required for the neutral?

______ **21-5.** What size EMT is required for each run to the service?

22. A 2500 sq. ft. beauty parlor has a 120/208 V, 1ϕ service and the the following loads. The neutral contains over 80% harmonic currents. Use the form, STANDARD CALCULATION—COMMERCIAL AND INDUSTRIAL.

10′ show window . . . . 120 V, 1ϕ
20′ lighting track . . . . 120 V, 1ϕ
Four 225 VA floodlights . . . . 120 V, 1ϕ
1200 VA sign circuit . . . . 120 V, 1ϕ
80 duplex receptacles (noncontinuous duty) . . . . 120 V, 1ϕ
40′ multioutlet assembly (heavy duty) . . . . 120 V, 1ϕ

Special loads

Ten 1800 VA fixed hair dryers (noncontinuous duty) . . . . 120 V, 1ϕ
15,000 VA tanning bed (noncontinuous duty) . . . . 208 V, 1ϕ
5000 VA sauna (continuous duty) . . . . 208 V, 1ϕ
10,000 VA spa (continuous duty) . . . . 208 V, 1ϕ

25 kW heating unit . . . . 208 V, 1ϕ
4.5 kW A/C unit . . . . 208 V, 1ϕ
Two ⅙ HP bathroom fans . . . . 120 V, 1ϕ
⅓ HP blower motor . . . . 208 V, 1ϕ

________ **22-1.** What size lighting and appliance panelboard is required based on the load?

________ **22-2.** What size main CB is required?

________ **22-3.** What size PVC is required for three runs of THWN Cu conductors, paralleled three times per phase, to the service?

23. A 5000 sq. ft. bank has a 120/240 V, 1ϕ service and the following loads. The neutral has 150 A of harmonically related loads. Use the form, STANDARD CALCULATION—COMMERCIAL AND INDUSTRIAL.

General lighting load . . . . 120 V, 1ϕ
100′ lighting track . . . . 120 V, 1ϕ
Twelve 500 VA outside lighting fixtures . . . . 120 V, 1ϕ
2800 VA sign circuit . . . . 120 V, 1ϕ
190 receptacles (noncontinuous duty) . . . . 120 V, 1ϕ
60 receptacles (continuous duty) . . . . 120 V, 1ϕ
10,000 VA miscellaneous loads (noncontinuous duty) . . . . 120 V, 1ϕ
15,000 VA miscellaneous loads (continuous duty) . . . . 240 V, 1ϕ
12,000 VA computers (continuous duty) . . . . 240 V, 1ϕ
8000 VA office equipment (noncontinuous duty) . . . . 120 V, 1ϕ
9000 VA office equipment (continuous duty) . . . . 240 V, 1ϕ
40,000 VA heating units . . . . 240 V, 1ϕ
8400 VA A/C units . . . . 240 V, 1ϕ
Two 5 HP motors . . . . 240 V, 1ϕ
Two 7½ HP motors . . . . 240 V, 1ϕ
Two 2 HP motors . . . . 120 V, 1ϕ

________ **23-1.** What size THW Cu conductors is required to supply power to a UPS for the computers?

________ **23-2.** What size OCPD is required for the feeder circuit to the UPS?

________ **23-3.** What size copper EGC is required for the feeder circuit to the UPS?

________ **23-4.** What size EMT is required for the feeder circuit to the UPS?

24. A small industrial building with a 2400 sq. ft. office (120 V, 1ϕ) has a 120/240 V, 3ϕ, 4-wire service and the following loads. Use the form, STANDARD CALCULATION—COMMERCIAL AND INDUSTRIAL.

Thirty-five 180 VA outside lighting fixtures . . . . 120 V, 1ϕ
2000 VA sign circuit . . . . 120 V, 1ϕ
40 receptacles (noncontinuous duty) . . . . 120 V, 1ϕ
20 receptacles (continuous duty) . . . . 120 V, 1ϕ

Special loads

20,000 sq. ft. warehouse (continuous duty) . . . . 120 V, 1ϕ
12,000 VA equipment (continuous duty) . . . . 120 V, 1ϕ
43,000 VA equipment (continuous duty) . . . . 240 V, 3ϕ
20,000 VA industrial plant lighting (continuous duty) . . . . 120 V, 1ϕ
25,000 VA miscellaneous loads (continuous duty) . . . . 240 V, 3ϕ

Motors

10 HP motor . . . . 230 V, 3ϕ
15 HP motor . . . . 230 V, 3ϕ
20 HP motor . . . . 230 V, 3ϕ
25 HP motor . . . . 230 V, 3ϕ

60 kW heating load . . . . 240 V, 3ϕ
15 kW A/C load . . . . 230 V, 3ϕ

________ **24-1.** What size THWN Cu conductors, paralleled four times per phase, is required for phases A and C?

________ **24-2.** What size THWN Cu conductors, paralleled four times per phase, is required for phase B?

________ **24-3.** What size THWN Cu conductors, paralleled four times per phase, is required for the neutral?

25. A 10,000 sq. ft. store building has a 120/208 V, 3ϕ, 4-wire service and the following loads. Use the form, STANDARD CALCULATION—COMMERCIAL AND INDUSTRIAL (CARRYING PHASES AND NEUTRAL IN VA) to divide the loads as evenly as possible on each phase and neutral.

60′ show window lighting . . . . 120 V, 1ϕ
60′ lighting track . . . . 120 V, 1ϕ
Twelve 500 VA outside lighting fixtures (continuous duty) . . . . 120 V, 1ϕ
1200 VA sign circuit . . . . 120 V, 1ϕ
75 receptacles (noncontinuous duty) . . . . 120 V, 1ϕ
56 receptacles (continuous duty) . . . . 120 V, 1ϕ
40′ multioutlet assembly (heavy duty) . . . . 120 V, 1ϕ
40′ multioutlet assembly (light duty) . . . . 120 V, 1ϕ

Special loads

15,000 VA equipment (continuous duty) . . . . 120 V, 1ϕ
12,000 VA equipment (continuous duty) . . . . 208 V, 3ϕ
9540 VA walk-in cooler . . . . 208 V, 3ϕ
8820 VA refrigeration . . . . 208 V, 3ϕ
9180 VA produce rack . . . . 208 V, 3ϕ

120 kW heating unit . . . . 208 V, 3ϕ
60 kW A/C unit . . . . 208 V, 3ϕ

Motors

Six ½ HP motors . . . . 120 V, 1ϕ
Six ¾ HP motors . . . . 208 V, 1ϕ
Six 1 HP motors . . . . 208 V, 3ϕ
Three 2 HP motors . . . . 208 V, 3ϕ

26. A 14,400 sq. ft. store building has a 120/208 V, 3ϕ, 4-wire service and the following loads. Use the form, STANDARD CALCULATION—COMMERCIAL AND INDUSTRIAL (CARRYING PHASES AND NEUTRAL IN AMPS) to divide the loads as evenly as possible on each phase and neutral. All lighting is electric discharge.

90′ show window lighting . . . . 120 V, 1ϕ
90′ lighting track . . . . 120 V, 1ϕ
Eighteen 500 VA outside lighting fixtures (continuous duty) . . . . 120 V, 1ϕ
3600 VA sign (supplied by three circuits) . . . . 208 V, 3ϕ, 4-wire
150 receptacles (noncontinuous duty) . . . . 120 V, 1ϕ
60 receptacles (continuous duty) . . . . 120 V, 1ϕ
120′ multioutlet assembly (heavy duty) . . . . 120 V, 1ϕ
120′ multioutlet assembly (light duty) . . . . 120 V, 1ϕ

Special loads

18,000 VA equipment (continuous duty) . . . . 120 V, 1ϕ
15,000 VA equipment (continuous duty) . . . . 208 V, 3ϕ
9000 VA walk-in cooler . . . . 208 V, 3ϕ
9180 VA refrigeration unit . . . . 208 V, 3ϕ
10,080 VA produce rack . . . . 208 V, 3ϕ

150 kW heating unit . . . . 208 V, 3ϕ
70 kVA A/C unit . . . . 208 V, 3ϕ

Motors

Six 1 HP motors . . . . 120 V, 1ϕ
Six 1½ HP motors . . . . 208 V, 1ϕ
Six 3 HP motors . . . . 208 V, 3ϕ
Three 5 HP motors . . . . 208 V, 3ϕ

__________ **26-1.** What phase is used to size a busduct to supply the service equipment?

__________ **26-2.** What size OCPD is required for the service equipment?

__________ **26-3.** What size switchgear is required for the service equipment?

Use for Problem 2

STANDARD CALCULATION—ONE-FAMILY DWELLING

1. GENERAL LIGHTING LOADS—Table 220-3(b)

______ sq. ft. × 3 VA = ______ VA

Small appliance loads—220-16(a)

______ VA × ______ circuits = ______ VA

Laundry load—220-16(b)

______ VA × 1 = ______ VA

______ VA

Applying demand factors—Table 220-11

First 3000 VA × 100% = 3000 VA

Next ______ VA × 35% = ______ VA

Remaining ______ VA × 25% = ______ VA

(total) ______ VA

2. COOKING EQUIPMENT LOADS—Table 220-19; Notes

Col. A ______ VA demand = ______ VA

Col. B ______ VA × ______% = ______ VA

Col. C ______ VA × ______% = ______ VA

(total) ______ VA

3. DRYER LOAD—220-18; Table 220-18

______ VA × ______% = ______ VA

4. FIXED APPLIANCE LOADS—220-17

Dishwasher = ______ VA

Disposal = ______ VA

Compactor = ______ VA

Water heater = ______ VA

______ = ______ VA

______ = ______ VA

______ = ______ VA

______ = ______ VA

(total) ______ VA × 75% = ______ VA

5. HEATING OR A/C LOAD—220-21

Heating unit = ______ VA × 100% = ______ VA

A/C unit = ______ VA × 100% = ______ VA

Heat pump = ______ VA × 100% = ______ VA

(largest load) ______ VA

6. LARGEST MOTOR LOAD—220-14

ϕ ______ VA × 25% = ______ VA

N ______ VA × 25% = ______ VA

Item	PHASES		NEUTRAL
1. General lighting (total)	VA		VA
2. Cooking equipment (total)	VA	× 70% =	VA
3. Dryer load	VA	× 70% =	VA
4. Fixed appliance loads	VA		(120 V Loads × 75%) VA
5. Heating or A/C (largest load)	VA		VA
6. Largest motor load ϕ	VA		
6. Largest motor load N			VA
TOTAL =	VA		VA

1ϕ service: $I = \frac{VA}{V}$

PHASES I = ______ $\frac{VA}{V}$ = ______ A

NEUTRAL I = ______ $\frac{VA}{V}$ = ______ A

If over 200 A—220-22; First 200 A × 100% = 200 A

Remaining ______ A × 70% = ______ A

______ A

STANDARD CALCULATION—ONE-FAMILY DWELLING

Use for Problem 3

1. GENERAL LIGHTING LOADS—Table 220-3(b)

_____ sq. ft. × 3 VA = _____ VA

Small appliance loads—220-16(a)

_____ VA × _____ circuits = _____ VA

Laundry load—220-16(b)

_____ VA × 1 = _____ VA

_____ VA

Applying demand factors—Table 220-11

First 3000 VA × 100% = 3 0 0 0 VA

Next _____ VA × 35% = _____ VA

Remaining _____ VA × 25% = _____ VA

(total) _____ VA | **PHASES** VA | **NEUTRAL** VA

2. COOKING EQUIPMENT LOADS—Table 220-19; Notes

Col. A _____ VA demand = _____ VA

Col. B _____ VA × _____% = _____ VA

Col. C _____ VA × _____% = _____ VA

(total) _____ VA | VA × 70% = VA

3. DRYER LOAD—220-18; Table 220-18

_____ VA × _____% = _____ VA | VA × 70% = VA

4. FIXED APPLIANCE LOADS—220-17

Dishwasher = _____ VA

Disposal = _____ VA

Compactor = _____ VA

Water heater = _____ VA

_____ = _____ VA

_____ = _____ VA

_____ = _____ VA

_____ = _____ VA

(total) _____ VA × 75% = _____ VA | VA | (120 V Loads × 75%) VA

5. HEATING OR A/C LOAD—220-21

Heating unit = _____ VA × 100% = _____ VA

A/C unit = _____ VA × 100% = _____ VA

Heat pump = _____ VA × 100% = _____ VA

(largest load) _____ VA | VA | VA

6. LARGEST MOTOR LOAD—220-14

ϕ _____ VA × 25% = _____ VA | VA

N _____ VA × 25% = _____ VA | VA

TOTAL = VA | VA

1ϕ service: I = $\frac{VA}{V}$

PHASES I = $\frac{\text{\_\_\_\_\_ VA}}{V}$ = _____ A

NEUTRAL I = $\frac{\text{\_\_\_\_\_ VA}}{V}$ = _____ A

If over 200 A—220-22; First 200 A × 100% = 2 0 0 A

Remaining _____ A × 70% = _____ A

A

Use for Problem 4

STANDARD CALCULATION—ONE-FAMILY DWELLING

1. GENERAL LIGHTING LOADS—Table 220-3(b)

______ sq. ft. × 3 VA = ______ VA

Small appliance loads—220-16(a)

______ VA × ______ circuits = ______ VA

Laundry load—220-16(b)

______ VA × 1 = ______ VA

______ VA

Applying demand factors—Table 220-11

First 3000 VA × 100% = 3000 VA

Next ______ VA × 35% = ______ VA

Remaining ______ VA × 25% = ______ VA

(total) ______ VA | PHASES: VA | NEUTRAL: VA

2. COOKING EQUIPMENT LOADS—Table 220-19; Notes

Col. A ______ VA demand = ______ VA

Col. B ______ VA × ______% = ______ VA

Col. C ______ VA × ______% = ______ VA

(total) ______ VA | VA × 70% = VA

3. DRYER LOAD—220-18; Table 220-18

______ VA × ______% = ______ VA | VA × 70% = VA

4. FIXED APPLIANCE LOADS—220-17

Dishwasher = ______ VA

Disposal = ______ VA

Compactor = ______ VA

Water heater = ______ VA

______ = ______ VA

______ = ______ VA

______ = ______ VA

______ = ______ VA

(total) ______ VA × 75% = ______ VA | VA | (120 V Loads × 75%) VA

5. HEATING OR A/C LOAD—220-21

Heating unit = ______ VA × 100% = ______ VA

A/C unit = ______ VA × 100% = ______ VA

Heat pump = ______ VA × 100% = ______ VA

(largest load) ______ VA | VA | VA

6. LARGEST MOTOR LOAD—220-14

ϕ ______ VA × 25% = ______ VA | VA

N ______ VA × 25% = ______ VA | VA

TOTAL = VA | VA

1ϕ service: I = $\frac{\textbf{VA}}{\textbf{V}}$

PHASES I = $\frac{\text{______ VA}}{\text{V}}$ = ______ A

NEUTRAL I = $\frac{\text{______ VA}}{\text{V}}$ = ______ A

If over 200 A—220-22; First 200 A × 100% = 200 A

Remaining ______ A × 70% = ______ A

A

STANDARD CALCULATION—ONE-FAMILY DWELLING

Use for Problem 5

1. GENERAL LIGHTING LOADS—Table 220-3(b)

______ sq. ft. × 3 VA = ______ VA

Small appliance loads—220-16(a)

______ VA × ______ circuits = ______ VA

Laundry load—220-16(b)

______ VA × 1 = ______ VA

______ VA

Applying demand factors—Table 220-11

First 3000 VA × 100% = 3000 VA

Next ______ VA × 35% = ______ VA

Remaining ______ VA × 25% = ______ VA

		PHASES		NEUTRAL
(total) ______ VA		VA		VA

2. COOKING EQUIPMENT LOADS—Table 220-19; Notes

Col. A ______ VA demand = ______ VA

Col. B ______ VA × ______% = ______ VA

Col. C ______ VA × ______% = ______ VA

(total) ______ VA | VA × 70% = VA

3. DRYER LOAD—220-18; Table 220-18

______ VA × ______% = ______ VA | VA × 70% = VA

4. FIXED APPLIANCE LOADS—220-17

Dishwasher = ______ VA

Disposal = ______ VA

Compactor = ______ VA

Water heater = ______ VA

______ = ______ VA

______ = ______ VA

______ = ______ VA

______ = ______ VA

(120 V Loads × 75%)

(total) ______ VA × 75% = ______ VA | VA | VA

5. HEATING OR A/C LOAD—220-21

Heating unit = ______ VA × 100% = ______ VA

A/C unit = ______ VA × 100% = ______ VA

Heat pump = ______ VA × 100% = ______ VA

(largest load) ______ VA | VA | VA

6. LARGEST MOTOR LOAD—220-14

ϕ ______ VA × 25% = ______ VA | VA

N ______ VA × 25% = ______ VA | ______ | ______ VA

TOTAL = VA | VA

1ϕ service: I = $\frac{VA}{V}$

PHASES I = $\frac{\text{______ VA}}{V}$ = ______ A

NEUTRAL I = $\frac{\text{______ VA}}{V}$ = ______ A

If over 200 A—220-22; First 200 A × 100% = 200 A

Remaining ______ A × 70% = ______ A

A

Use for Problem 6

OPTIONAL CALCULATION—ONE-FAMILY DWELLING (NEW OR EXISTING)

1. OTHER LOADS—220-30(b)

General lighting loads—220-30(b)(2)

________ sq. ft. × 3 VA = ________ VA

Small appliance and laundry loads—220-30(b)(1)

________ VA × ________ circuits = ________ VA

Special loads—220-30(b)(3)(4)

Dishwasher = ________ VA

Disposal = ________ VA

Compactor = ________ VA

Water heater = ________ VA

________________ = ________ VA

________________ = ________ VA

________________ = ________ VA

________________ = ________ VA

________________ = ________ VA

________________ = ________ VA

(total) ________ VA

Applying demand factors—Table 220-30(4)

First 10,000 VA × 100% = 10,000 VA

Remaining ________ VA × 40% = ________ VA

(total) ________ VA

PHASES

VA

2. HEATING OR A/C LOAD—Table 220-30(1)(2)(3)

Heating units (3 or less) = ________ VA × 65% = ________ VA

Heating units (4 or more) = ________ VA × 40% = ________ VA

A/C unit = ________ VA × 100% = ________ VA

Heat pump = ________ VA × 100% = ________ VA

(largest load) ________ VA VA

TOTAL = VA

A

NEUTRAL

(Loads from Standard Calculation)

1. General lighting = ________ VA
2. Cooking equipment = ________ VA
3. Dryer = ________ VA
4. Fixed appliance = ________ VA
5. Heating or A/C = ________ VA
6. Largest motor = ________ VA

TOTAL = VA

A

1ϕ service: $I = \frac{VA}{V}$

PHASES $I = \frac{\_\_\_\_\_\_ VA}{V}$ = ________ A

NEUTRAL $I = \frac{\_\_\_\_\_\_ VA}{V}$ = ________ A

OPTIONAL CALCULATION—ONE-FAMILY DWELLING (NEW OR EXISTING)

Use for Problem 7

1. OTHER LOADS—220-30(b)

General lighting loads—220-30(b)(2)

_______ sq. ft. × 3 VA = _______ VA

Small appliance and laundry loads—220-30(b)(1)

_______ VA × _______ circuits = _______ VA

Special loads—220-30(b)(3)(4)

Dishwasher = _______ VA

Disposal = _______ VA

Compactor = _______ VA

Water heater = _______ VA

_______ = _______ VA

_______ = _______ VA

_______ = _______ VA

_______ = _______ VA

_______ = _______ VA

_______ = _______ VA

(total) _______ VA

Applying demand factors—Table 220-30(4)

First 10,000 VA × 100% = 10,000 VA

Remaining _______ VA × 40% = _______ VA

(total) _______ VA

PHASES VA

2. HEATING OR A/C LOAD—Table 220-30(1)(2)(3)

Heating units (3 or less) = _______ VA × 65% = _______ VA

Heating units (4 or more) = _______ VA × 40% = _______ VA

A/C unit = _______ VA × 100% = _______ VA

Heat pump = _______ VA × 100% = _______ VA

(largest load) _______ VA — VA

TOTAL = VA

A

NEUTRAL

(Loads from Standard Calculation)

1. General lighting = _______ VA
2. Cooking equipment = _______ VA
3. Dryer = _______ VA
4. Fixed appliance = _______ VA
5. Heating or A/C = _______ VA
6. Largest motor = _______ VA

TOTAL = VA

A

1ϕ service: $I = \frac{VA}{V}$

PHASES $I = \frac{\text{\_\_\_\_\_\_\_ VA}}{V}$ = _______ A

NEUTRAL $I = \frac{\text{\_\_\_\_\_\_\_ VA}}{V}$ = _______ A

Use for Problem 8

OPTIONAL CALCULATION—ONE-FAMILY DWELLING (NEW OR EXISTING)

1. OTHER LOADS—220-30(b)

General lighting loads—220-30(b)(2)

_______ sq. ft. × 3 VA = _______ VA

Small appliance and laundry loads—220-30(b)(1)

_______ VA × _______ circuits = _______ VA

Special loads—220-30(b)(3)(4)

Dishwasher = _______ VA

Disposal = _______ VA

Compactor = _______ VA

Water heater = _______ VA

_______ = _______ VA

_______ = _______ VA

_______ = _______ VA

_______ = _______ VA

_______ = _______ VA

_______ = _______ VA

(total) _______ VA

Applying demand factors—Table 220-30(4)

First 10,000 VA × 100% = 10,000 VA

Remaining _______ VA × 40% = _______ VA

(total) _______ VA

PHASES VA

2. HEATING OR A/C LOAD—Table 220-30(1)(2)(3)

Heating units (3 or less) = _______ VA × 65% = _______ VA

Heating units (4 or more) = _______ VA × 40% = _______ VA

A/C unit = _______ VA × 100% = _______ VA

Heat pump = _______ VA × 100% = _______ VA

(largest load) _______ VA — _______ VA

TOTAL = _______ VA

_______ A

NEUTRAL

(Loads from Standard Calculation)

1. General lighting = _______ VA
2. Cooking equipment = _______ VA
3. Dryer = _______ VA
4. Fixed appliance = _______ VA
5. Heating or A/C = _______ VA
6. Largest motor = _______ VA

TOTAL = _______ VA

_______ A

1ϕ service: $I = \frac{VA}{V}$

PHASES $I = \frac{\_\_\_\_\ VA}{V}$ = _______ A

NEUTRAL $I = \frac{\_\_\_\_\ VA}{V}$ = _______ A

OPTIONAL CALCULATION—ONE-FAMILY DWELLING (EXISTING ONLY)

Use for Problem 9

1. OTHER LOADS—220-31

General lighting loads—220-31

________ sq. ft. × 3 VA = ________ VA

Small appliance and laundry loads—220-31

________ VA × ______ circuits = ________ VA

Special loads—220-31

Dishwasher = ________ VA

Disposal = ________ VA

Compactor = ________ VA

Water heater = ________ VA

________________ = ________ VA

________________ = ________ VA

________________ = ________ VA

________________ = ________ VA

________________ = ________ VA

________________ = ________ VA

(total) ________ VA

Applying demand factors—220-31

First 8000 VA × 100% = 8000 VA

Remaining ________ VA × 40% = ________ VA

Added appliance = ________ VA × 100% = ________ VA

(total) ________ VA

PHASES

________ VA

________ A

1φ service: I = $\frac{VA}{V}$

I = $\frac{\_\_\_\_\_\_\_\_ VA}{V}$ = ________ A

(*Note:* Total ampacity of service shall not exceed service conductor ampacity. Existing service conductor ampacity = ________ A.)

NEUTRAL

(Loads from Standard Calculation)

1. General lighting = ________ VA
2. Cooking equipment = ________ VA
3. Dryer = ________ VA
4. Fixed appliance = ________ VA
5. Heating or A/C = ________ VA
6. Largest motor = ________ VA

TOTAL = ________ VA

________ A

STANDARD CALCULATION—ONE-FAMILY DWELLING 1ϕ (CARRYING PHASES AND NEUTRAL IN VA)

Use for Problem 10

1. GENERAL LIGHTING LOADS—Table 220-3(b)

________ sq. ft. × 3 VA = ________ VA

Small appliance loads—220-16(a)

________ VA × ________ circuits = ________ VA

Laundry load—220-16(b)

________ VA × 1 = ________ VA

________ VA

Applying demand factors—Table 220-11

First 3000 VA × 100% = 3000 VA

Next ________ VA × 35% = ________ VA

Remaining ________ VA × 25% = ________ VA

	PHASE A	PHASE B	NEUTRAL
(total) ________ VA ÷ 2 =	________ VA	________ VA	________ VA
2. COOKING EQUIPMENT LOADS—Table 220-19, Notes			
Col. A ________ units = ________ VA demand =	________ VA	________ VA × 70% =	________ VA
Col. B ________ VA × ________ units × ________% =	________ VA	________ VA × 70% =	________ VA
Col. C ________ VA × ________ units × ________% =	________ VA	________ VA × 70% =	________ VA
3. DRYER LOAD—220-18; Table 220-18			
________ VA × ________ units × ________% =	________ VA	________ VA × 70% =	________ VA
4. FIXED APPLIANCE LOADS—220-17			
Dishwasher = ________ VA × 75% ÷ ________ =	________ VA	________ VA	________ VA
Disposal = ________ VA × 75% ÷ ________ =	________ VA	________ VA	________ VA
Compactor = ________ VA × 75% ÷ ________ =	________ VA	________ VA	________ VA
Water heater = ________ VA × 75% ÷ ________ =	________ VA	________ VA	________ VA
________ = ________ VA × 75% ÷ ________ =	________ VA	________ VA	________ VA
________ = ________ VA × 75% ÷ ________ =	________ VA	________ VA	________ VA
________ = ________ VA × 75% ÷ ________ =	________ VA	________ VA	________ VA
________ = ________ VA × 75% ÷ ________ =	________ VA	________ VA	________ VA
________ = ________ VA × 75% ÷ ________ =	________ VA	________ VA	________ VA
5. HEATING OR A/C LOAD—220-21			
Heating unit = ________ VA × 100% × ________ =	________ VA	________ VA	________ VA
A/C unit = ________ VA × 100% × ________ =	________ VA	________ VA	________ VA
Heat pump = ________ VA × 100% × ________ =	________ VA	________ VA	________ VA
6. LARGEST MOTOR LOAD—220-14			
ϕ ________ VA × 25% =	________ VA	________ VA	
N ________ VA × 25% =	________ VA	________ VA	________ VA
TOTAL =	VA	VA	VA

1ϕ service: I = $\frac{VA}{V}$

PHASES I = $\frac{\_\_\_\_\_\_\_\_ \text{ VA}}{\text{V}}$ = ________ A

NEUTRAL I = $\frac{\_\_\_\_\_\_\_\_ \text{ VA}}{\text{V}}$ = ________ A

If over 200 A—220-22; First 200 A × 100% = 200 A

Remaining ________ A × 70% = ________ A

A

STANDARD CALCULATION—ONE-FAMILY DWELLING 1ϕ (CARRYING PHASES AND NEUTRAL IN AMPS)

Use for Problem 11

1. GENERAL LIGHTING LOADS—Table 220-3(b)

________ sq. ft. × 3 VA = ________ VA

Small appliance loads—220-16(a)

________ VA × ________ circuits = ________ VA

Laundry load—220-16(b)

________ VA × 1 = ________ VA

________ VA

Applying demand factors—Table 220-11

First 3000 VA × 100% = 3000 VA

Next ________ VA × 35% = ________ VA

Remaining ________ VA × 25% = ________ VA

	PHASE A	PHASE B	NEUTRAL
(total) ________ VA ÷ ________ V =	________ A	________ A	________ A
2. COOKING EQUIPMENT LOADS—Table 220-19, Notes			
Col. A ________ VA demand ÷ ________ V =	________ A	________ A × 70% =	________ A
Col. B ________ VA × ________% ÷ ________ V =	________ A	________ A × 70% =	________ A
Col. C ________ VA × ________% ÷ ________ V =	________ A	________ A × 70% =	________ A
3. DRYER LOAD—220-18; Table 220-18			
________ VA × ________% ÷ ________ V =	________ A	________ A × 70% =	________ A
4. FIXED APPLIANCE LOADS—220-17			
Dishwasher = ________ VA × 75% ÷ ________ V =	________ A	________ A	________ A
Disposal = ________ VA × 75% ÷ ________ V =	________ A	________ A	________ A
Compactor = ________ VA × 75% ÷ ________ V =	________ A	________ A	________ A
Water heater = ________ VA × 75% ÷ ________ V =	________ A	________ A	________ A
________ = ________ VA × 75% ÷ ________ V =	________ A	________ A	________ A
________ = ________ VA × 75% ÷ ________ V =	________ A	________ A	________ A
________ = ________ VA × 75% ÷ ________ V =	________ A	________ A	________ A
________ = ________ VA × 75% ÷ ________ V =	________ A	________ A	________ A
________ = ________ VA × 75% ÷ ________ V =	________ A	________ A	________ A
5. HEATING OR A/C LOAD—220-21			
Heating unit = ________ VA × 100% ÷ ________ V =	________ A	________ A	________ A
A/C unit = ________ VA × 100% ÷ ________ V =	________ A	________ A	________ A
Heat pump = ________ VA × 100% ÷ ________ V =	________ A	________ A	________ A
6. LARGEST MOTOR LOAD—220-14			
ϕ ________ VA × 25% ÷ ________ V =	________ A	________ A	
N ________ VA × 25% ÷ ________ V =			________ A
TOTAL =	A	A	A

If over 200 A—220-22; First 200 A × 100% = 200 A

Remaining ________ A × 70% = ________ A

A

Use for Problems 11-1 and 11-2

STANDARD CALCULATION—ONE-FAMILY DWELLING

1. GENERAL LIGHTING LOADS—Table 220-3(b)

_______ sq. ft. × 3 VA = _______ VA

Small appliance loads—220-16(a)

_______ VA × _______ circuits = _______ VA

Laundry load—220-16(b)

_______ VA × 1 = _______ VA

_______ VA

Applying demand factors—Table 220-11

First 3000 VA × 100% = 3000 VA

Next _______ VA × 35% = _______ VA

Remaining _______ VA × 25% = _______ VA

		PHASES		NEUTRAL
(total)	_______ VA	VA		VA

2. COOKING EQUIPMENT LOADS—Table 220-19; Notes

Col. A _______ VA demand = _______ VA

Col. B _______ VA × _______% = _______ VA

Col. C _______ VA × _______% = _______ VA

(total) _______ VA | VA × 70% = | VA

3. DRYER LOAD—220-18; Table 220-18

_______ VA × _______% = _______ VA | VA × 70% = | VA

4. FIXED APPLIANCE LOADS—220-17

Dishwasher = _______ VA

Disposal = _______ VA

Compactor = _______ VA

Water heater = _______ VA

_______ = _______ VA

_______ = _______ VA

_______ = _______ VA

_______ = _______ VA

(total) _______ VA × 75% = _______ VA | VA | (120 V Loads × 75%) VA

5. HEATING OR A/C LOAD—220-21

Heating unit = _______ VA × 100% = _______ VA

A/C unit = _______ VA × 100% = _______ VA

Heat pump = _______ VA × 100% = _______ VA

(largest load) _______ VA | VA | VA

6. LARGEST MOTOR LOAD—220-14

ϕ _______ VA × 25% = _______ VA | VA

N _______ VA × 25% = _______ VA | | VA

TOTAL = VA | VA

1ϕ service: I = $\frac{VA}{V}$

PHASES I = $\frac{\_\_\_\_\_\_\_ VA}{V}$ = _______ A

NEUTRAL I = $\frac{\_\_\_\_\_\_\_ VA}{V}$ = _______ A

If over 200 A—220-22; First 200 A × 100% = 200 A

Remaining _______ A × 70% = _______ A

A

Use for Problem 15

STANDARD CALCULATION—COMMERCIAL AND INDUSTRIAL

1. LIGHTING LOADS—220, Part B

General lighting loads—Table 220-3(b); 220-10(b)

______ sq. ft. × ______ VA × _____% = ______ VA

Show window loads—220-12

______' × 200 VA = ______ VA

Track lighting loads—410-102

______' / 2' × 180 VA = ______ VA

Outside lighting loads—220-10(b)

______ VA × ______ units × _____% = ______ VA

Sign lighting loads—220-3, FPN; 600-6(c)

______ VA × ______% = ______ VA

		PHASES	NEUTRAL
(total)	______ VA	VA	VA

2. RECEPTACLE LOADS—220-3(c)(5)

Noncontinuous duty—220-10(b); Table 220-13

#______ × 180 VA = ______ VA = ______ VA

Applying demand factors—Table 220-13

First 10,000 VA × 100% = ______ VA = ______ VA

Remaining ______ VA × 50% = ______ VA

Continuous duty—220-10(b) #______ × 180 VA × 125% = ______ VA

Multioutlet assembly—220-3(c)

Heavy duty = ______' × 180 VA = ______ VA

Light duty = ______' / 5' × 180 VA = ______ VA

		PHASES	NEUTRAL
(total)	______ VA	VA	VA

3. SPECIAL LOADS—220-10(b)

______ = ______ VA × _____% = ______ VA

______ = ______ VA × _____% = ______ VA

______ = ______ VA × _____% = ______ VA

______ = ______ VA × _____% = ______ VA

______ = ______ VA × _____% = ______ VA

		PHASES	NEUTRAL
(total)	______ VA	VA	VA

4. HEATING OR A/C LOAD—220-21 ______ VA | VA | VA

5. MOTOR LOADS—430-22(a); 430-24; 430-25

				3ϕ only		
______ =	______ A ×	_____% ×	______ V	$\times \sqrt{3}$	=	______ VA
______ =	______ A ×	_____% ×	______ V	$\times \sqrt{3}$	=	______ VA
______ =	______ A ×	_____% ×	______ V	$\times \sqrt{3}$	=	______ VA
______ =	______ A ×	_____% ×	______ V	$\times \sqrt{3}$	=	______ VA

		PHASES	NEUTRAL
(total)	______ VA	VA	VA

6. LARGEST MOTOR LOAD—220-14

				3ϕ only			PHASES	NEUTRAL
ϕ	______ A ×	_____% ×	______ V	$\times \sqrt{3}$	× 25% =	______ VA	VA	VA
N	______ A ×	_____% ×	______ V	$\times \sqrt{3}$	× 25% =	______ VA	VA	VA
						TOTAL =	VA	VA

1ϕ service: $I = \frac{VA}{V}$

PHASES $I = \frac{\_\_\_\_\_\_ VA}{V}$ = ______ A

NEUTRAL $I = \frac{\_\_\_\_\_\_ VA}{V}$ = ______ A

3ϕ service: $I = \frac{VA}{V \times \sqrt{3}}$

$I = \frac{\_\_\_\_\_\_ VA}{V \times \sqrt{3}}$ = ______ A

$I = \frac{\_\_\_\_\_\_ VA}{V \times \sqrt{3}}$ = ______ A

If over 200 A—220-22; First 200 A × 100% = 200 A

Remaining ______ A × 70% = ______ A

A

Use for Problem 16

STANDARD CALCULATION—COMMERCIAL AND INDUSTRIAL

1. LIGHTING LOADS—220, Part B

General lighting loads—Table 220-3(b); 220-10(b)

_____ sq. ft. × _____ VA × _____% = _____ VA

Show window loads—220-12

_____' × 200 VA = _____ VA

Track lighting loads—410-102

_____' (÷ 2') × 180 VA = _____ VA

Outside lighting loads—220-10(b)

_____ VA × _____ units × _____% = _____ VA

Sign lighting loads—220-3, FPN; 600-6(c)

_____ VA × _____% = _____ VA

		PHASES	NEUTRAL
(total)	_____ VA	VA	VA

2. RECEPTACLE LOADS—220-3(c)(5)

Noncontinuous duty—220-10(b); Table 220-13

#_____ × 180 VA = _____ VA = _____ VA

Applying demand factors—Table 220-13

First 10,000 VA × 100% = _____ VA = _____ VA

Remaining _____ VA × 50% = _____ VA

Continuous duty—220-10(b) #_____ × 180 VA × 125% = _____ VA

Multioutlet assembly—220-3(c)

Heavy duty = _____' × 180 VA = _____ VA

Light duty = _____' (÷ 5') × 180 VA = _____ VA

(total) _____ VA | VA | VA

3. SPECIAL LOADS—220-10(b)

_____ = _____ VA × _____% = _____ VA

_____ = _____ VA × _____% = _____ VA

_____ = _____ VA × _____% = _____ VA

_____ = _____ VA × _____% = _____ VA

_____ = _____ VA × _____% = _____ VA

(total) _____ VA | VA | VA

4. HEATING OR A/C LOAD—220-21 _____ VA | VA | VA

5. MOTOR LOADS—430-22(a); 430-24; 430-25

				3ϕ only		
_____ =	_____ A ×	_____% ×	_____ V	$\times \sqrt{3}$	=	_____ VA
_____ =	_____ A ×	_____% ×	_____ V	$\times \sqrt{3}$	=	_____ VA
_____ =	_____ A ×	_____% ×	_____ V	$\times \sqrt{3}$	=	_____ VA
_____ =	_____ A ×	_____% ×	_____ V	$\times \sqrt{3}$	=	_____ VA

(total) _____ VA | VA | VA

6. LARGEST MOTOR LOAD—220-14

				3ϕ only			PHASES	NEUTRAL
ϕ	_____ A ×	_____% ×	_____ V	$\times \sqrt{3}$	× 25% =	_____ VA	VA	VA
N	_____ A ×	_____% ×	_____ V	$\times \sqrt{3}$	× 25% =	_____ VA	VA	VA
						TOTAL =	VA	VA

1ϕ service: $I = \frac{VA}{V}$

PHASES $I = \frac{\_\_\_\_\_ VA}{V}$ = _____ A

NEUTRAL $I = \frac{\_\_\_\_\_ VA}{V}$ = _____ A

3ϕ service: $I = \frac{VA}{V \times \sqrt{3}}$

$I = \frac{\_\_\_\_\_ VA}{V \times \sqrt{3}}$ = _____ A

$I = \frac{\_\_\_\_\_ VA}{V \times \sqrt{3}}$ = _____ A

If over 200 A—220-22; First 200 A × 100% = 200 A

Remaining _____ A × 70% = _____ A

A

STANDARD CALCULATION—COMMERCIAL AND INDUSTRIAL

Use for Problem 17

1. LIGHTING LOADS—220, Part B

General lighting loads—Table 220-3(b); 220-10(b)

_____ sq. ft. × _____ VA × _____% = _____ VA

Show window loads—220-12

_____′ × 200 VA = _____ VA

Track lighting loads—410-102

_____′ / 2′ × 180 VA = _____ VA

Outside lighting loads—220-10(b)

_____ VA × _____ units × _____% = _____ VA

Sign lighting loads—220-3, FPN; 600-6(c)

_____ VA × _____% = _____ VA

		PHASES	NEUTRAL
(total)	_____ VA	VA	VA

2. RECEPTACLE LOADS—220-3(c)(5)

Noncontinuous duty—220-10(b); Table 220-13

#_____ × 180 VA = _____ VA = _____ VA

Applying demand factors—Table 220-13

First 10,000 VA × 100% = _____ VA = _____ VA

Remaining _____ VA × 50% = _____ VA

Continuous duty—220-10(b) #_____ × 180 VA × 125% = _____ VA

Multioutlet assembly—220-3(c)

Heavy duty = _____′ × 180 VA = _____ VA

Light duty = _____′ / 5′ × 180 VA = _____ VA

(total) _____ VA | VA | VA

3. SPECIAL LOADS—220-10(b)

_____ = _____ VA × _____% = _____ VA

_____ = _____ VA × _____% = _____ VA

_____ = _____ VA × _____% = _____ VA

_____ = _____ VA × _____% = _____ VA

_____ = _____ VA × _____% = _____ VA

(total) _____ VA | VA | VA

4. HEATING OR A/C LOAD—220-21 _____ VA | VA | VA

5. MOTOR LOADS—430-22(a); 430-24; 430-25

				3ϕ only		
_____ =	_____ A ×	_____% ×	_____ V	× $\sqrt{3}$	=	_____ VA
_____ =	_____ A ×	_____% ×	_____ V	× $\sqrt{3}$	=	_____ VA
_____ =	_____ A ×	_____% ×	_____ V	× $\sqrt{3}$	=	_____ VA
_____ =	_____ A ×	_____% ×	_____ V	× $\sqrt{3}$	=	_____ VA

(total) _____ VA | VA | VA

6. LARGEST MOTOR LOAD—220-14

				3ϕ only				
ϕ _____ A ×	_____% ×	_____ V		× $\sqrt{3}$	× 25% =	_____ VA	VA	VA
N _____ A ×	_____% ×	_____ V		× $\sqrt{3}$	× 25% =	_____ VA	VA	VA

TOTAL = VA | VA

1ϕ service: $I = \frac{VA}{V}$

PHASES $I = \frac{\_\_\_\_\_ VA}{V}$ = _____ A

NEUTRAL $I = \frac{\_\_\_\_\_ VA}{V}$ = _____ A

3ϕ service: $I = \frac{VA}{V \times \sqrt{3}}$

$I = \frac{\_\_\_\_\_ VA}{V \times \sqrt{3}}$ = _____ A

$I = \frac{\_\_\_\_\_ VA}{V \times \sqrt{3}}$ = _____ A

If over 200 A—220-22; First 200 A × 100% = 200 A

Remaining _____ A × 70% = _____ A

A

Use for Problem 18

OPTIONAL CALCULATION—SCHOOL

1. **TOTAL VA—Table 220-34**

   Total connected load = ________ VA

2. **TOTAL SQUARE FOOTAGE—Table 220-34**

   Square footage = ________ sq. ft.

3. **VA PER SQ. FT.—Table 220-34**

   $\frac{\_\_\_\_\_\_\_\_ \text{ VA}}{\text{sq. ft.}}$ = ________ VA/sq. ft.

4. **APPLYING DEMAND FACTORS—Table 220-34**

   First 3 VA per sq. ft. × 100%

   Next 3 VA to 20 VA per sq. ft. × 75%

   Remaining 21 VA per sq. ft × 25%

	PHASES
3 VA/sq. ft. × ________ sq. ft. × 100% =	VA
______ VA/sq. ft. × ________ sq. ft. × 75% =	VA
______ VA/sq. ft. × ________ sq. ft. × 25% =	VA
TOTAL =	VA

NEUTRAL

(Loads from Standard Calculation)

1. Lighting loads = ________ VA
2. Receptacle loads = ________ VA
3. Special loads = ________ VA
4. Heating or A/C load = ________ VA
5. Motor loads = ________ VA
6. Largest motor load = ________ VA

TOTAL = ________ VA

________ A

1ϕ service: $I = \frac{VA}{V}$

PHASES $I = \frac{\_\_\_\_\_\_\_\_ \text{ VA}}{\text{V}}$ = ________ A

NEUTRAL $I = \frac{\_\_\_\_\_\_\_\_ \text{ VA}}{\text{V}}$ = ________ A

3ϕ service: $I = \frac{VA}{V \times \sqrt{3}}$

$I = \frac{\_\_\_\_\_\_\_\_ \text{ VA}}{V \times \sqrt{3}}$ = ________ A

$I = \frac{\_\_\_\_\_\_\_\_ \text{ VA}}{V \times \sqrt{3}}$ = ________ A

If over 200 A—220-22; First 200 A × 100% = 200 A

Remaining ______ A × 70% = ________ A

________ A

STANDARD CALCULATION—COMMERCIAL AND INDUSTRIAL

Use for Problem 19

1. **LIGHTING LOADS—220, Part B**
   General lighting loads—Table 220-3(b); 220-10(b)
   ______ sq. ft. × ______ VA × ______% = ______ VA
   Show window loads—220-12
   ______′ × 200 VA = ______ VA
   Track lighting loads—410-102
   $\frac{\_\_\_\_\_\_'}{2'}$ × 180 VA = ______ VA
   Outside lighting loads—220-10(b)
   ______ VA × ______ units × ______% = ______ VA
   Sign lighting loads—220-3, FPN; 600-6(c)
   ______ VA × ______% = ______ VA

		PHASES	NEUTRAL
(total)	______ VA	VA	VA

2. **RECEPTACLE LOADS—220-3(c)(5)**
   Noncontinuous duty—220-10(b); Table 220-13
   #______ × 180 VA = ______ VA = ______ VA
   Applying demand factors—Table 220-13
   First 10,000 VA × 100% = ______ VA = ______ VA
   Remaining ______ VA × 50% = ______ VA
   Continuous duty—220-10(b) #______ × 180 VA × 125% = ______ VA
   Multioutlet assembly—220-3(c)
   Heavy duty = ______′ × 180 VA = ______ VA
   Light duty = $\frac{\_\_\_\_\_\_'}{5'}$ × 180 VA = ______ VA
   (total) ______ VA | VA | VA

3. **SPECIAL LOADS—220-10(b)**
   ______ = ______ VA × ______% = ______ VA
   ______ = ______ VA × ______% = ______ VA
   ______ = ______ VA × ______% = ______ VA
   ______ = ______ VA × ______% = ______ VA
   ______ = ______ VA × ______% = ______ VA
   (total) ______ VA | VA | VA

4. **HEATING OR A/C LOAD—220-21** ______ VA | VA | VA

5. **MOTOR LOADS—430-22(a); 430-24; 430-25**
   ______ = ______ A × ______% × ______ V [3ϕ only: × $\sqrt{3}$] = ______ VA
   ______ = ______ A × ______% × ______ V [× $\sqrt{3}$] = ______ VA
   ______ = ______ A × ______% × ______ V [× $\sqrt{3}$] = ______ VA
   ______ = ______ A × ______% × ______ V [× $\sqrt{3}$] = ______ VA
   (total) ______ VA | VA | VA

6. **LARGEST MOTOR LOAD—220-14**
   ϕ ______ A × ______% × ______ V [3ϕ only: × $\sqrt{3}$] × 25% = ______ VA | VA | VA
   N ______ A × ______% × ______ V [× $\sqrt{3}$] × 25% = ______ VA | VA | VA

TOTAL = | VA | VA

1ϕ service: $I = \frac{VA}{V}$

PHASES $I = \frac{\_\_\_\_\_\_ VA}{V}$ = ______ A

NEUTRAL $I = \frac{\_\_\_\_\_\_ VA}{V}$ = ______ A

3ϕ service: $I = \frac{VA}{V \times \sqrt{3}}$

$I = \frac{\_\_\_\_\_\_ VA}{V \times \sqrt{3}}$ = ______ A

$I = \frac{\_\_\_\_\_\_ VA}{V \times \sqrt{3}}$ = ______ A

If over 200 A—220-22; First 200 A × 100% = 200 A

Remaining ______ A × 70% = ______ A

A

Use for Problem 20

STANDARD CALCULATION—HOTELS AND MOTELS

1. GENERAL LIGHTING LOADS—Table 220-3(b)

_______ sq. ft. × 2 VA × _____ units = _______ VA
_______ sq. ft. × 2 VA × _____ units = _______ VA
_______ sq. ft. × 2 VA × _____ units = _______ VA
_______ VA

Applying demand factors—Table 220-11
First 20,000 VA × 50% = 10,000 VA
Next _______ VA × 40% = _______ VA
Remaining _______ VA × 30% = _______ VA
(total) _______ VA _______ VA

Show window loads—220-12
_______′ × 200 VA = _______ VA

Track lighting loads—410-102
$\frac{\_\_\_\_\_\_\_'}{2'}$ × 180 VA = _______ VA

Outside lighting loads—220-10(b)
_______ VA × _____ units × _____% = _______ VA

Sign lighting loads—220-3, FPN; 600-6(c)
_______ VA × _______% = _______ VA

(total) _______ VA | PHASES VA | NEUTRAL VA

2. RECEPTACLE LOADS—220-3(c)(5)
Continuous duty—220-10(b); Table 220-13
_______ VA × 125% = (total) _______ VA | VA | VA

3. SPECIAL LOADS—220-10(b)

_______ = _______ VA × _____ units × _____% = _______ VA
_______ = _______ VA × _____ units × _____% = _______ VA
_______ = _______ VA × _____ units × _____% = _______ VA
_______ = _______ VA × _____ units × _____% = _______ VA
_______ = _______ VA × _____ units × _____% = _______ VA
_______ = _______ VA × _____ units × _____% = _______ VA
_______ = _______ VA × _____ units × _____% = _______ VA
(total) _______ VA | VA | VA

4. HEATING OR A/C LOAD—220-21

Heating unit = _______ VA × 100% × _____ units = _______ VA
A/C unit = _______ VA × 100% × _____ units = _______ VA
Heat pump = _______ VA × 100% × _____ units = _______ VA
(largest load) _______ VA | VA | VA

5. LARGEST MOTOR LOAD—220-14

ϕ _______ A × _____% × _______ V × 25% = _______ VA | VA | VA
N _______ A × _____% × _______ V × 25% = _______ VA | VA | VA

TOTAL = | VA | VA

1ϕ service: $I = \frac{VA}{V}$

PHASES $I = \frac{\_\_\_\_\_\_\_ VA}{V}$ = _______ A

NEUTRAL $I = \frac{\_\_\_\_\_\_\_ VA}{V}$ = _______ A

3ϕ service: $I = \frac{VA}{V \times \sqrt{3}}$

$I = \frac{\_\_\_\_\_\_\_ VA}{V \times \sqrt{3}}$ = _______ A

$I = \frac{\_\_\_\_\_\_\_ VA}{V \times \sqrt{3}}$ = _______ A

If over 200 A—220-22; First 200 A × 100% = 200 A
Remaining _______ A × 70% = _______ A
A

Use for Problem 21

STANDARD CALCULATION—COMMERCIAL AND INDUSTRIAL

1. LIGHTING LOADS—220, Part B

General lighting loads—Table 220-3(b); 220-10(b)

__________ sq. ft. × __________ VA × _______% = __________ VA

Show window loads—220-12

__________′ × 200 VA = __________ VA

Track lighting loads—410-102

__________′ / 2′ × 180 VA = __________ VA

Outside lighting loads—220-10(b)

__________ VA × _________ units × _______% = __________ VA

Sign lighting loads—220-3, FPN; 600-6(c)

__________ VA × __________% = __________ VA

		PHASES	NEUTRAL
(total)	__________ VA	VA	VA

2. RECEPTACLE LOADS—220-3(c)(5)

Noncontinuous duty—220-10(b); Table 220-13

#__________ × 180 VA = __________ VA = __________ VA

Applying demand factors—Table 220-13

First 10,000 VA × 100% = __________ VA = __________ VA

Remaining __________ VA × 50% = __________ VA

Continuous duty—220-10(b) #__________ × 180 VA × 125% = __________ VA

Multioutlet assembly—220-3(c)

Heavy duty = __________′ × 180 VA = __________ VA

Light duty = __________′ / 5′ × 180 VA = __________ VA

		PHASES	NEUTRAL
(total)	__________ VA	VA	VA

3. SPECIAL LOADS—220-10(b)

__________ = __________ VA × _______% = __________ VA

__________ = __________ VA × _______% = __________ VA

__________ = __________ VA × _______% = __________ VA

__________ = __________ VA × _______% = __________ VA

__________ = __________ VA × _______% = __________ VA

		PHASES	NEUTRAL
(total)	__________ VA	VA	VA

4. HEATING OR A/C LOAD—220-21 __________ VA | VA | VA

5. MOTOR LOADS—430-22(a); 430-24; 430-25

				3ϕ only		
__________ =	__________ A ×	_______% ×	__________ V	× $\sqrt{3}$	=	__________ VA
__________ =	__________ A ×	_______% ×	__________ V	× $\sqrt{3}$	=	__________ VA
__________ =	__________ A ×	_______% ×	__________ V	× $\sqrt{3}$	=	__________ VA
__________ =	__________ A ×	_______% ×	__________ V	× $\sqrt{3}$	=	__________ VA

		PHASES	NEUTRAL
(total)	__________ VA	VA	VA

6. LARGEST MOTOR LOAD—220-14

				3ϕ only			PHASES	NEUTRAL
ϕ	__________ A ×	_______% ×	__________ V	× $\sqrt{3}$	× 25% =	__________ VA	VA	VA
N	__________ A ×	_______% ×	__________ V	× $\sqrt{3}$	× 25% =	__________ VA	VA	VA
						TOTAL =	VA	VA

1ϕ service: $I = \frac{VA}{V}$

PHASES $I = \frac{\_\_\_\_\_\_ VA}{V}$ = __________ A

NEUTRAL $I = \frac{\_\_\_\_\_\_ VA}{V}$ = __________ A

3ϕ service: $I = \frac{VA}{V \times \sqrt{3}}$

$I = \frac{\_\_\_\_\_\_ VA}{V \times \sqrt{3}}$ = __________ A

$I = \frac{\_\_\_\_\_\_ VA}{V \times \sqrt{3}}$ = __________ A

If over 200 A—220-22; First 200 A × 100% = 200 A

Remaining _______ A × 70% = __________ A

A

Use for Problem 22

STANDARD CALCULATION—COMMERCIAL AND INDUSTRIAL

1. LIGHTING LOADS—220, Part B

General lighting loads—Table 220-3(b); 220-10(b)

______ sq. ft. × ______ VA × ______% = ______ VA

Show window loads—220-12

______′ × 200 VA = ______ VA

Track lighting loads—410-102

$\frac{\text{______}'}{2'}$ × 180 VA = ______ VA

Outside lighting loads—220-10(b)

______ VA × ______ units × ______% = ______ VA

Sign lighting loads—220-3, FPN; 600-6(c)

______ VA × ______% = ______ VA

		PHASES	NEUTRAL
(total)	______ VA	VA	VA

2. RECEPTACLE LOADS—220-3(c)(5)

Noncontinuous duty—220-10(b); Table 220-13

#______ × 180 VA = ______ VA = ______ VA

Applying demand factors—Table 220-13

First 10,000 VA × 100% = ______ VA = ______ VA

Remaining ______ VA × 50% = ______ VA

Continuous duty—220-10(b) #______ × 180 VA × 125% = ______ VA

Multioutlet assembly—220-3(c)

Heavy duty = ______′ × 180 VA = ______ VA

Light duty = $\frac{\text{______}'}{5'}$ × 180 VA = ______ VA

(total) ______ VA | VA | VA

3. SPECIAL LOADS—220-10(b)

______ = ______ VA × ______% = ______ VA

______ = ______ VA × ______% = ______ VA

______ = ______ VA × ______% = ______ VA

______ = ______ VA × ______% = ______ VA

______ = ______ VA × ______% = ______ VA

(total) ______ VA | VA | VA

4. HEATING OR A/C LOAD—220-21 ______ VA | VA | VA

5. MOTOR LOADS—430-22(a); 430-24; 430-25

				3ϕ only		
______ =	______ A ×	______% ×	______ V	× $\sqrt{3}$	=	______ VA
______ =	______ A ×	______% ×	______ V	× $\sqrt{3}$	=	______ VA
______ =	______ A ×	______% ×	______ V	× $\sqrt{3}$	=	______ VA
______ =	______ A ×	______% ×	______ V	× $\sqrt{3}$	=	______ VA

(total) ______ VA | VA | VA

6. LARGEST MOTOR LOAD—220-14

				3ϕ only				
ϕ	______ A ×	______% ×	______ V	× $\sqrt{3}$	× 25% =	______ VA	VA	VA
N	______ A ×	______% ×	______ V	× $\sqrt{3}$	× 25% =	______ VA	VA	VA

TOTAL = VA | VA

1ϕ service: $I = \frac{VA}{V}$

PHASES $I = \frac{\text{______ VA}}{V}$ = ______ A

NEUTRAL $I = \frac{\text{______ VA}}{V}$ = ______ A

3ϕ service: $I = \frac{VA}{V \times \sqrt{3}}$

$I = \frac{\text{______ VA}}{V \times \sqrt{3}}$ = ______ A

$I = \frac{\text{______ VA}}{V \times \sqrt{3}}$ = ______ A

If over 200 A—220-22; First 200 A × 100% = 200 A

Remaining ______ A × 70% = ______ A

A

Use for Problem 23

STANDARD CALCULATION—COMMERCIAL AND INDUSTRIAL

1. LIGHTING LOADS—220, Part B

General lighting loads—Table 220-3(b); 220-10(b)

_____ sq. ft. × _____ VA × _____% = _____ VA

Show window loads—220-12

_____′ × 200 VA = _____ VA

Track lighting loads—410-102

$\frac{\_\_\_\_\_'}{2'}$ × 180 VA = _____ VA

Outside lighting loads—220-10(b)

_____ VA × _____ units × _____% = _____ VA

Sign lighting loads—220-3, FPN; 600-6(c)

_____ VA × _____% = _____ VA

		PHASES	NEUTRAL
(total)	_____ VA	VA	VA

2. RECEPTACLE LOADS—220-3(c)(5)

Noncontinuous duty—220-10(b); Table 220-13

#_____ × 180 VA = _____ VA = _____ VA

Applying demand factors—Table 220-13

First 10,000 VA × 100% = _____ VA = _____ VA

Remaining _____ VA × 50% = _____ VA

Continuous duty—220-10(b) #_____ × 180 VA × 125% = _____ VA

Multioutlet assembly—220-3(c)

Heavy duty = _____′ × 180 VA = _____ VA

Light duty = $\frac{\_\_\_\_\_'}{5'}$ × 180 VA = _____ VA

(total) _____ VA | VA | VA

3. SPECIAL LOADS—220-10(b)

_____ = _____ VA × _____% = _____ VA

_____ = _____ VA × _____% = _____ VA

_____ = _____ VA × _____% = _____ VA

_____ = _____ VA × _____% = _____ VA

_____ = _____ VA × _____% = _____ VA

(total) _____ VA | VA | VA

4. HEATING OR A/C LOAD—220-21 _____ VA | VA | VA

5. MOTOR LOADS—430-22(a); 430-24; 430-25

				3ϕ only		
_____ =	_____ A ×	_____% ×	_____ V	× $\sqrt{3}$	=	_____ VA
_____ =	_____ A ×	_____% ×	_____ V	× $\sqrt{3}$	=	_____ VA
_____ =	_____ A ×	_____% ×	_____ V	× $\sqrt{3}$	=	_____ VA
_____ =	_____ A ×	_____% ×	_____ V	× $\sqrt{3}$	=	_____ VA

(total) _____ VA | VA | VA

6. LARGEST MOTOR LOAD—220-14

				3ϕ only				
ϕ	_____ A ×	_____% ×	_____ V	× $\sqrt{3}$	× 25% =	_____ VA	VA	VA
N	_____ A ×	_____% ×	_____ V	× $\sqrt{3}$	× 25% =	_____ VA	VA	VA

TOTAL = VA | VA

1ϕ service: I = $\frac{VA}{V}$

PHASES I = $\frac{\_\_\_\_\_ VA}{V}$ = _____ A

NEUTRAL I = $\frac{\_\_\_\_\_ VA}{V}$ = _____ A

3ϕ service: I = $\frac{VA}{V \times \sqrt{3}}$

I = $\frac{\_\_\_\_\_ VA}{V \times \sqrt{3}}$ = _____ A

I = $\frac{\_\_\_\_\_ VA}{V \times \sqrt{3}}$ = _____ A

If over 200 A—220-22; First 200 A × 100% = 200 A

Remaining _____ A × 70% = _____ A

A

Use for Problem 24

STANDARD CALCULATION—COMMERCIAL AND INDUSTRIAL

1. LIGHTING LOADS—220, Part B

General lighting loads—Table 220-3(b); 220-10(b)

________ sq. ft. × ________ VA × ______% = ________ VA

Show window loads—220-12

________′ × 200 VA = ________ VA

Track lighting loads—410-102

$\frac{\_\_\_\_\_\_\_\_'}{2'}$ × 180 VA = ________ VA

Outside lighting loads—220-10(b)

________ VA × ________ units × ______% = ________ VA

Sign lighting loads—220-3, FPN; 600-6(c)

________ VA × ________% = ________ VA

		PHASES	NEUTRAL
(total)	________ VA	VA	VA

2. RECEPTACLE LOADS—220-3(c)(5)

Noncontinuous duty—220-10(b); Table 220-13

#________ × 180 VA = ________ VA = ________ VA

Applying demand factors—Table 220-13

First 10,000 VA × 100% = ________ VA = ________ VA

Remaining ________ VA × 50% = ________ VA

Continuous duty—220-10(b) #________ × 180 VA × 125% = ________ VA

Multioutlet assembly—220-3(c)

Heavy duty = ________′ × 180 VA = ________ VA

Light duty = $\frac{\_\_\_\_\_\_\_\_'}{5'}$ × 180 VA = ________ VA

(total) ________ VA | VA | VA

3. SPECIAL LOADS—220-10(b)

________ = ________ VA × ______% = ________ VA

________ = ________ VA × ______% = ________ VA

________ = ________ VA × ______% = ________ VA

________ = ________ VA × ______% = ________ VA

________ = ________ VA × ______% = ________ VA

(total) ________ VA | VA | VA

4. HEATING OR A/C LOAD—220-21 ________ VA | VA | VA

5. MOTOR LOADS—430-22(a); 430-24; 430-25

				3ϕ only		
________ =	________ A ×	______% ×	______ V	$\times \sqrt{3}$	=	________ VA
________ =	________ A ×	______% ×	______ V	$\times \sqrt{3}$	=	________ VA
________ =	________ A ×	______% ×	______ V	$\times \sqrt{3}$	=	________ VA
________ =	________ A ×	______% ×	______ V	$\times \sqrt{3}$	=	________ VA

(total) ________ VA | VA | VA

6. LARGEST MOTOR LOAD—220-14

				3ϕ only			PHASES	NEUTRAL
ϕ	________ A ×	______% ×	______ V	$\times \sqrt{3}$	× 25% =	________ VA	VA	VA
N	________ A ×	______% ×	______ V	$\times \sqrt{3}$	× 25% =	________ VA	VA	VA
						TOTAL =	VA	VA

1ϕ service: $I = \frac{VA}{V}$

PHASES I = $\frac{\_\_\_\_\_\_ VA}{V}$ = ________ A

NEUTRAL I = $\frac{\_\_\_\_\_\_ VA}{V}$ = ________ A

3ϕ service: $I = \frac{VA}{V \times \sqrt{3}}$

I = $\frac{\_\_\_\_\_\_ VA}{V \times \sqrt{3}}$ = ________ A

I = $\frac{\_\_\_\_\_\_ VA}{V \times \sqrt{3}}$ = ________ A

If over 200 A—220-22; First 200 A × 100% = 200 A

Remaining ______ A × 70% = ________ A

A

Use for Problem 25

STANDARD CALCULATION—COMMERCIAL AND INDUSTRIAL (CARRYING PHASES AND NEUTRAL IN VA)

	PHASE	PHASE	PHASE	NEUTRAL
1. LIGHTING LOADS—220, Part B				
General lighting loads—Table 220-3(b); 220-10(b)				
________ sq. ft. × ________ VA × ______% ÷ ______ =	________ VA	________ VA	________ VA	________ VA
Show window loads—220-12				
________′ × 200 VA ÷ ______ =	________ VA	________ VA	________ VA	________ VA
Track lighting loads—410-102				
________′ (2′) × 180 VA ÷ ______ =	________ VA	________ VA	________ VA	________ VA
Outside lighting loads—220-10(b)				
________ VA × ______ units × ______% ÷ ______ =	________ VA	________ VA	________ VA	________ VA
Sign lighting loads—220-3, FPN; 600-6(c)				
________ VA × ________% ÷ ______ =	________ VA	________ VA	________ VA	________ VA
2. RECEPTACLE LOADS—220-3(c)(5)				
Noncontinuous duty—220-10(b)				
________ VA × 100% = ________ VA ÷ ______ =	________ VA	________ VA	________ VA	________ VA
Applying demand factors—Table 220-13				
First 10,000 VA × 100% ÷ ______ =	________ VA	________ VA	________ VA	________ VA
Remaining ________ VA × 50% ÷ ______ =	________ VA	________ VA	________ VA	________ VA
Continuous duty—220-10(b) ________ VA × 125% ÷ ______ =	________ VA	________ VA	________ VA	________ VA
Multioutlet assembly—220-3(c)				
Heavy duty = ________′ × 180 VA ÷ ______ =	________ VA	________ VA	________ VA	________ VA
Light duty = ________′ (5′) × 180 VA ÷ ______ =	________ VA	________ VA	________ VA	________ VA
3. SPECIAL LOADS—220-10(b) (load)				
________ = ________ VA × ______% ÷ ______ =	________ VA	________ VA	________ VA	________ VA
________ = ________ VA × ______% ÷ ______ =	________ VA	________ VA	________ VA	________ VA
________ = ________ VA × ______% ÷ ______ =	________ VA	________ VA	________ VA	________ VA
________ = ________ VA × ______% ÷ ______ =	________ VA	________ VA	________ VA	________ VA
________ = ________ VA × ______% ÷ ______ =	________ VA	________ VA	________ VA	________ VA
4. HEATING OR A/C LOAD—220-21	________ VA	________ VA	________ VA	________ VA
5. MOTOR LOADS—430-22(a); 430-24; 430-25 (load) [3ϕ motors only: × $\sqrt{3}$]				
________ = ________ A × ______ V × $\sqrt{3}$ ÷ ______ =	________ VA	________ VA	________ VA	________ VA
________ = ________ A × ______ V × $\sqrt{3}$ ÷ ______ =	________ VA	________ VA	________ VA	________ VA
________ = ________ A × ______ V × $\sqrt{3}$ ÷ ______ =	________ VA	________ VA	________ VA	________ VA
________ = ________ A × ______ V × $\sqrt{3}$ ÷ ______ =	________ VA	________ VA	________ VA	________ VA
6. LARGEST MOTOR LOAD—220-14				
ϕ ________ A × ________ V × ________% × 25% =	________ VA	________ VA	________ VA	________ VA
N ________ A × ________ V × ________% × 25% =	________ VA	________ VA	________ VA	________ VA
TOTAL =	VA	VA	VA	VA

1ϕ service: $I = \frac{VA}{V}$

PHASES $I = \frac{\_\_\_\_\_\_ VA}{V}$ = ________ A

NEUTRAL $I = \frac{\_\_\_\_\_\_ VA}{V}$ = ________ A

3ϕ service: $I = \frac{VA}{V \times \sqrt{3}}$

$I = \frac{\_\_\_\_\_\_ VA}{V \times \sqrt{3}}$ = ________ A

$I = \frac{\_\_\_\_\_\_ VA}{V \times \sqrt{3}}$ = ________ A

If over 200 A—220-22; First 200 A × 100% = 200 A

Remaining ______ A × 70% = ________ A

A

Use for Problem 26

STANDARD CALCULATION—COMMERCIAL AND INDUSTRIAL (CARRYING PHASES AND NEUTRAL IN AMPS)

	PHASE	PHASE	PHASE	NEUTRAL
1. LIGHTING LOADS—220, Part B				
General lighting loads—Table 220-3(b); 220-10(b)				
______ sq. ft. × ______ VA × ______% ÷ ______ V =	______ A	______ A	______ A	______ A
Show window loads—220-12				
______′ × 200 VA ÷ ______ V =	______ A	______ A	______ A	______ A
Track lighting loads—410-102				
______′ (2′) × 180 VA ÷ ______ V =	______ A	______ A	______ A	______ A
Outside lighting loads—220-10(b)				
______ VA × ______ units × ______% ÷ ______ V =	______ A	______ A	______ A	______ A
Sign lighting loads—220-3, FPN; 600-6(c)				
______ VA × ______% ÷ ______ V =	______ A	______ A	______ A	______ A
2. RECEPTACLE LOADS—220-3(c)(5)				
Noncontinuous duty—220-10(b)				
______ VA × 100% = ______ VA ÷ ______ V =	______ A	______ A	______ A	______ A
Applying demand factors—Table 220-13				
First 10,000 VA × 100% ÷ ______ V =	______ A	______ A	______ A	______ A
Remaining ______ VA × 50% ÷ ______ V =	______ A	______ A	______ A	______ A
Continuous duty—220-10(b) ______ VA × 125% ÷ ______ V =	______ A	______ A	______ A	______ A
Multioutlet assembly—220-3(c)				
Heavy duty = ______′ × 180 VA ÷ ______ V =	______ A	______ A	______ A	______ A
Light duty = ______′ (5′) × 180 VA ÷ ______ V =	______ A	______ A	______ A	______ A
3. SPECIAL LOADS—220-10(b)				
(load)				
______ = ______ VA × ______% ÷ ______ V =	______ A	______ A	______ A	______ A
______ = ______ VA × ______% ÷ ______ V =	______ A	______ A	______ A	______ A
______ = ______ VA × ______% ÷ ______ V =	______ A	______ A	______ A	______ A
______ = ______ VA × ______% ÷ ______ V =	______ A	______ A	______ A	______ A
______ = ______ VA × ______% ÷ ______ V =	______ A	______ A	______ A	______ A
4. HEATING OR A/C LOAD—220-21	______ A	______ A	______ A	______ A
5. MOTOR LOADS—430-22(a); 430-24; 430-25				
(load)				
______ = ______ A × ______# ÷ ______ =	______ A	______ A	______ A	______ A
______ = ______ A × ______# ÷ ______ =	______ A	______ A	______ A	______ A
______ = ______ A × ______# ÷ ______ =	______ A	______ A	______ A	______ A
______ = ______ A × ______# ÷ ______ =	______ A	______ A	______ A	______ A
6. LARGEST MOTOR LOAD—220-14				
ϕ ______ A × ______ V × ______% × 25% =	______ A	______ A	______ A	______ A
N ______ A × ______ V × ______% × 25% =	______ A	______ A	______ A	______ A
TOTAL =	A	A	A	A

If over 200 A—220-22; First 200 A × 100% = 200 A

Remaining ______ A × 70% = ______ A

A

Calculation Forms

Calculation forms provide standardized formats that assure all loads are included and demand factors applied as required when sizing the service for dwellings and commercial and industrial locations. Other elements of the service are sized based upon the completed forms.

The following forms are copyrighted by American Technical Publishers, Inc. These forms may be reproduced for instructional use only. They shall not be reproduced and sold.

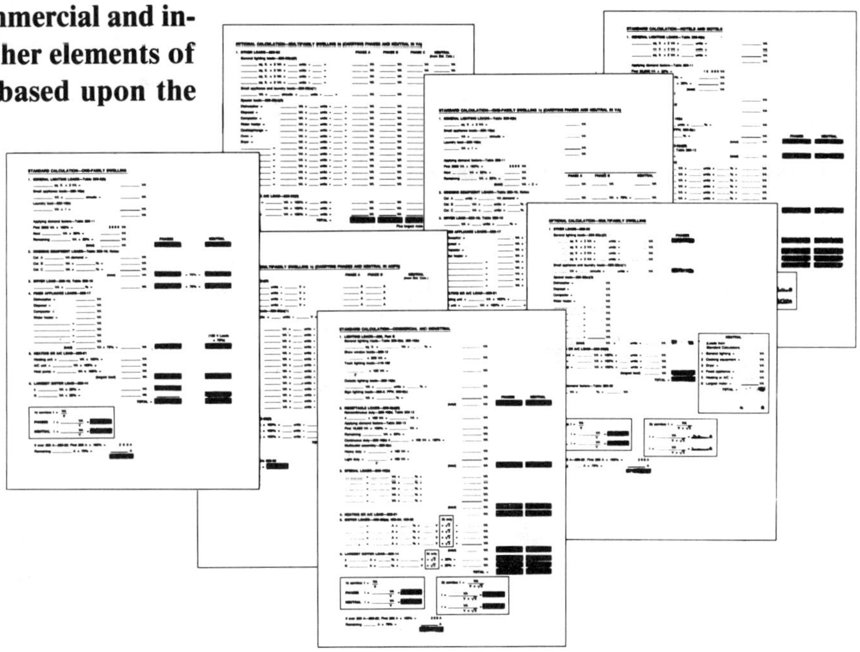

Standard Calculation—One-Family Dwelling
Standard Calculation—One-Family Dwelling 1ϕ (Carrying Phases and Neutral in VA)
Standard Calculation—One-Family Dwelling 1ϕ (Carrying Phases and Neutral in Amps)
Optional Calculation—One-Family Dwelling (Existing Only)
Optional Calculation—One-Family Dwelling (New or Existing)
Standard and Optional Calculations—One-Family Dwelling
Standard Calculation—Multifamily Dwelling
Standard Calculation—Multifamily Dwelling 1ϕ (Carrying Phases and Neutral in VA)
Standard Calculation—Multifamily Dwelling 1ϕ (Carrying Phases and Neutral in Amps)
Standard Calculation—Multifamily Dwelling 3ϕ (Carrying Phases and Neutral in VA)
Standard Calculation—Multifamily Dwelling 3ϕ (Carrying Phases and Neutral in Amps)
Optional Calculation—Multifamily Dwelling
Optional Calculation—Multifamily Dwelling 1ϕ (Carrying Phases and Neutral in VA)
Optional Calculation—Multifamily Dwelling 1ϕ (Carrying Phases and Neutral in Amps)
Optional Calculation—Multifamily Dwelling 3ϕ (Carrying Phases and Neutral in VA)
Standard Calculation—Commercial and Industrial
Standard Calculation—Commercial and Industrial (Carrying Phases and Neutral in VA)
Standard Calculation—Commercial and Industrial (Carrying Phases and Neutral in Amps)
Standard Calculation—Mobile Home
Standard Calculation—Hotels and Motels
Optional Calculation—School

STANDARD CALCULATION—ONE-FAMILY DWELLING

1. GENERAL LIGHTING LOADS—Table 220-3(b)

_______ sq. ft. × 3 VA = _______ VA

Small appliance loads—220-16(a)

_______ VA × _______ circuits = _______ VA

Laundry load—220-16(b)

_______ VA × 1 = _______ VA

_______ VA

Applying demand factors—Table 220-11

First 3000 VA × 100% = 3000 VA

Next _______ VA × 35% = _______ VA

Remaining _______ VA × 25% = _______ VA

	PHASES		NEUTRAL
(total) _______ VA	VA		VA

2. COOKING EQUIPMENT LOADS—Table 220-19; Notes

Col. A _______ VA demand = _______ VA

Col. B _______ VA × _______ % = _______ VA

Col. C _______ VA × _______ % = _______ VA

(total) _______ VA | VA × 70% = VA

3. DRYER LOAD—220-18; Table 220-18

_______ VA × _______ % = _______ VA | VA × 70% = VA

4. FIXED APPLIANCE LOADS—220-17

Dishwasher = _______ VA

Disposal = _______ VA

Compactor = _______ VA

Water heater = _______ VA

_______ = _______ VA

_______ = _______ VA

_______ = _______ VA

_______ = _______ VA

(total) _______ VA × 75% = _______ VA | VA | (120 V Loads × 75%) VA

5. HEATING OR A/C LOAD—220-21

Heating unit = _______ VA × 100% = _______ VA

A/C unit = _______ VA × 100% = _______ VA

Heat pump = _______ VA × 100% = _______ VA

(largest load) _______ VA | VA | VA

6. LARGEST MOTOR LOAD—220-14

ϕ _______ VA × 25% = _______ VA | VA

N _______ VA × 25% = _______ VA | | VA

TOTAL = VA | VA

1ϕ **service:** $I = \frac{VA}{V}$

PHASES $I = \frac{\_\_\_\_\_\_ VA}{V}$ = _______ A

NEUTRAL $I = \frac{\_\_\_\_\_\_ VA}{V}$ = _______ A

If over 200 A—220-22; First 200 A × 100% = 200 A

Remaining _______ A × 70% = _______ A

A

STANDARD CALCULATION—ONE-FAMILY DWELLING 1ϕ (CARRYING PHASES AND NEUTRAL IN VA)

1. GENERAL LIGHTING LOADS—Table 220-3(b)

_____ sq. ft. × 3 VA = _____ VA

Small appliance loads—220-16(a)

_____ VA × _____ circuits = _____ VA

Laundry load—220-16(b)

_____ VA × 1 = _____ VA

_____ VA

Applying demand factors—Table 220-11

First 3000 VA × 100% = 3000 VA

Next _____ VA × 35% = _____ VA

Remaining _____ VA × 25% = _____ VA

(total) _____ VA ÷ 2 =

	PHASE A	PHASE B	NEUTRAL
(total) _____ VA ÷ 2 =	_____ VA	_____ VA	_____ VA

2. COOKING EQUIPMENT LOADS—Table 220-19, Notes

	PHASE A	PHASE B	NEUTRAL
Col. A _____ units = _____ VA demand =	_____ VA	_____ VA × 70% =	_____ VA
Col. B _____ VA × _____ units × _____% =	_____ VA	_____ VA × 70% =	_____ VA
Col. C _____ VA × _____ units × _____% =	_____ VA	_____ VA × 70% =	_____ VA

3. DRYER LOAD—220-18; Table 220-18

	PHASE A	PHASE B	NEUTRAL
_____ VA × _____ units × _____% =	_____ VA	_____ VA × 70% =	_____ VA

4. FIXED APPLIANCE LOADS—220-17

		PHASE A	PHASE B	NEUTRAL
Dishwasher =	_____ VA × 75% ÷ _____ =	_____ VA	_____ VA	_____ VA
Disposal =	_____ VA × 75% ÷ _____ =	_____ VA	_____ VA	_____ VA
Compactor =	_____ VA × 75% ÷ _____ =	_____ VA	_____ VA	_____ VA
Water heater =	_____ VA × 75% ÷ _____ =	_____ VA	_____ VA	_____ VA
_____ =	_____ VA × 75% ÷ _____ =	_____ VA	_____ VA	_____ VA
_____ =	_____ VA × 75% ÷ _____ =	_____ VA	_____ VA	_____ VA
_____ =	_____ VA × 75% ÷ _____ =	_____ VA	_____ VA	_____ VA
_____ =	_____ VA × 75% ÷ _____ =	_____ VA	_____ VA	_____ VA
_____ =	_____ VA × 75% ÷ _____ =	_____ VA	_____ VA	_____ VA

5. HEATING OR A/C LOAD—220-21

	PHASE A	PHASE B	NEUTRAL
Heating unit = _____ VA × 100% × _____ =	_____ VA	_____ VA	_____ VA
A/C unit = _____ VA × 100% × _____ =	_____ VA	_____ VA	_____ VA
Heat pump = _____ VA × 100% × _____ =	_____ VA	_____ VA	_____ VA

6. LARGEST MOTOR LOAD—220-14

	PHASE A	PHASE B	NEUTRAL
ϕ _____ VA × 25% =	_____ VA	_____ VA	
N _____ VA × 25% =	_____ VA	_____ VA	_____ VA
TOTAL =	VA	VA	VA

1ϕ service: I = $\frac{VA}{V}$

PHASES I = $\frac{\_\_\_\_\_ VA}{V}$ = _____ A

NEUTRAL I = $\frac{\_\_\_\_\_ VA}{V}$ = _____ A

If over 200 A—220-22; First 200 A × 100% = 200 A

Remaining _____ A × 70% = _____ A

_____ A

STANDARD CALCULATION—ONE-FAMILY DWELLING 1ϕ (CARRYING PHASES AND NEUTRAL IN AMPS)

1. GENERAL LIGHTING LOADS—Table 220-3(b)

Calculation			
______ sq. ft. × 3 VA =	______ VA		
Small appliance loads—220-16(a)			
______ VA × ______ circuits =	______ VA		
Laundry load—220-16(b)			
______ VA × 1 =	______ VA		
	______ VA		
Applying demand factors—Table 220-11			
First 3000 VA × 100% = 3000 VA			
Next ______ VA × 35% = ______ VA			
Remaining ______ VA × 25% = ______ VA	**PHASE A**	**PHASE B**	**NEUTRAL**
(total) ______ VA ÷ ______ V =	______ A	______ A	______ A
2. COOKING EQUIPMENT LOADS—Table 220-19, Notes			
Col. A ______ VA demand ÷ ______ V =	______ A	______ A × 70% =	______ A
Col. B ______ VA × ______% ÷ ______ V =	______ A	______ A × 70% =	______ A
Col. C ______ VA × ______% ÷ ______ V =	______ A	______ A × 70% =	______ A
3. DRYER LOAD—220-18; Table 220-18			
______ VA × ______% ÷ ______ V =	______ A	______ A × 70% =	______ A
4. FIXED APPLIANCE LOADS—220-17			
Dishwasher = ______ VA × 75% ÷ ______ V =	______ A	______ A	______ A
Disposal = ______ VA × 75% ÷ ______ V =	______ A	______ A	______ A
Compactor = ______ VA × 75% ÷ ______ V =	______ A	______ A	______ A
Water heater = ______ VA × 75% ÷ ______ V =	______ A	______ A	______ A
______ = ______ VA × 75% ÷ ______ V =	______ A	______ A	______ A
______ = ______ VA × 75% ÷ ______ V =	______ A	______ A	______ A
______ = ______ VA × 75% ÷ ______ V =	______ A	______ A	______ A
______ = ______ VA × 75% ÷ ______ V =	______ A	______ A	______ A
______ = ______ VA × 75% ÷ ______ V =	______ A	______ A	______ A
5. HEATING OR A/C LOAD—220-21			
Heating unit = ______ VA × 100% ÷ ______ V =	______ A	______ A	______ A
A/C unit = ______ VA × 100% ÷ ______ V =	______ A	______ A	______ A
Heat pump = ______ VA × 100% ÷ ______ V =	______ A	______ A	______ A
6. LARGEST MOTOR LOAD—220-14			
ϕ ______ VA × 25% ÷ ______ V =	______ A	______ A	
N ______ VA × 25% ÷ ______ V =			______ A
TOTAL =	A	A	A

If over 200 A—220-22; First 200 A × 100% = 200 A

Remaining ______ A × 70% = ______ A

A

STANDARD AND OPTIONAL CALCULATIONS—ONE-FAMILY DWELLING

	STANDARD PHASES	STANDARD NEUTRAL	OPTIONAL PHASES
1. GENERAL LIGHTING LOADS—Table 220-3(b)			
______ sq. ft. × 3 VA =	______ VA	______ VA	______ VA
Small appliance loads—220-16(a)			
______ VA × ______ circuits =	______ VA	______ VA	______ VA
Laundry load—220-16(b)			
______ VA × 1 =	______ VA	______ VA	______ VA
	______ VA	______ VA	
Applying demand factors—Table 220-11			
First 3000 VA × 100% =	______ VA	______ VA	
Next ______ VA × 35% =	______ VA	______ VA	
Remaining ______ VA × 25% =	______ VA	______ VA	
2. COOKING EQUIPMENT LOADS—Table 220-19; Notes			
Col. A ______ VA demand =	______ VA	× 70% = ______ VA	______ VA
Col. B ______ VA × ______% =	______ VA	× 70% = ______ VA	______ VA
Col. C ______ VA × ______% =	______ VA	× 70% = ______ VA	______ VA
3. DRYER LOAD—220-18; Table 220-18			
______ VA × ______% =	______ VA	× 70% = ______ VA	______ VA
4. FIXED APPLIANCE LOADS—220-17			
Dishwasher = ______ VA × 75% =	______ VA	______ VA	______ VA
Disposal = ______ VA × 75% =	______ VA	______ VA	______ VA
Compactor = ______ VA × 75% =	______ VA	______ VA	______ VA
Water heater = ______ VA × 75% =	______ VA	______ VA	______ VA
______ = ______ VA × 75% =	______ VA	______ VA	______ VA
______ = ______ VA × 75% =	______ VA	______ VA	______ VA
______ = ______ VA × 75% =	______ VA	______ VA	______ VA
______ = ______ VA × 75% =	______ VA	______ VA	______ **VA**
		(total) =	______ VA
5. HEATING OR A/C LOAD—220-21			
Heating unit = ______ VA × 100% =	______ VA	______ VA × ______% =	______ VA
A/C unit = ______ VA × 100% =	______ VA	______ VA	______ VA
Heat pump = ______ VA × 100% =	______ VA	______ VA	______ VA
6. LARGEST MOTOR LOAD—220-14			
ϕ ______ VA × 25% =	______ VA		
N ______ VA × 25% =		______ VA	
TOTAL =	**VA**	**VA**	

1ϕ service: $I = \frac{VA}{V}$

STANDARD PHASES	I = ______	$\frac{VA}{V}$ =	______ **A**
OPTIONAL PHASES	I = ______	$\frac{VA}{V}$ =	______ **A**
NEUTRAL	I = ______	$\frac{VA}{V}$ =	______ **A**

Applying demand factors—Table 220-30(4)

First 10,000 VA × 100% =	10,000 VA
Remaining ______ VA × 40% =	______ VA
Heating load =	______ VA
TOTAL =	______ **VA**

If over 200 A—220-22; First 200 A × 100% =	200 A
Remaining ______ A × 70% =	______ A
	A

OPTIONAL CALCULATION—ONE-FAMILY DWELLING (NEW OR EXISTING)

1. OTHER LOADS—220-30(b)

General lighting loads—220-30(b)(2)

__________ sq. ft. × 3 VA = __________ VA

Small appliance and laundry loads—220-30(b)(1)

__________ VA × __________ circuits = __________ VA

Special loads—220-30(b)(3)(4)

Dishwasher = __________ VA

Disposal = __________ VA

Compactor = __________ VA

Water heater = __________ VA

____________________ = __________ VA

____________________ = __________ VA

____________________ = __________ VA

____________________ = __________ VA

____________________ = __________ VA

____________________ = __________ VA

(total) __________ VA

Applying demand factors—Table 220-30(4)

First 10,000 VA × 100% = 10,000 VA

Remaining __________ VA × 40% = __________ VA

(total) __________ VA

PHASES __________ **VA**

2. HEATING OR A/C LOAD—Table 220-30(1)(2)(3)

Heating units (3 or less) = __________ VA × 65% = __________ VA

Heating units (4 or more) = __________ VA × 40% = __________ VA

A/C unit = __________ VA × 100% = __________ VA

Heat pump = __________ VA × 100% = __________ VA

(largest load) __________ VA __________ **VA**

TOTAL = __________ **VA**

__________ **A**

NEUTRAL

(Loads from Standard Calculation)

1. General lighting = __________ VA
2. Cooking equipment = __________ VA
3. Dryer = __________ VA
4. Fixed appliance = __________ VA
5. Heating or A/C = __________ VA
6. Largest motor = __________ VA

TOTAL = __________ **VA**

__________ **A**

1ϕ service: $I = \frac{VA}{V}$

PHASES $I = \frac{\_\_\_\_\_\_ VA}{V}$ = __________ **A**

NEUTRAL $I = \frac{\_\_\_\_\_\_ VA}{V}$ = __________ **A**

OPTIONAL CALCULATION—ONE-FAMILY DWELLING (EXISTING ONLY)

1. OTHER LOADS—220-31

General lighting loads—220-31

________ sq. ft. × 3 VA = ________ VA

Small appliance and laundry loads—220-31

________ VA × ______ circuits = ________ VA

Special loads—220-31

Dishwasher = ________ VA

Disposal = ________ VA

Compactor = ________ VA

Water heater = ________ VA

______________ = ________ VA

______________ = ________ VA

______________ = ________ VA

______________ = ________ VA

______________ = ________ VA

______________ = ________ VA

(total) ________ VA

Applying demand factors—220-31

First 8000 VA × 100% = 8000 VA

Remaining ________ VA × 40% = ________ VA

Added appliance = ________ VA × 100% = ________ VA

(total) ________ VA

PHASES

________ **VA**

________ **A**

1φ service: $I = \frac{VA}{V}$

$I = \frac{\_\_\_\_\_\_ VA}{V}$ = ________ A

(*Note:* Total ampacity of service shall not exceed service conductor ampacity. Existing service conductor ampacity = ________ A.)

NEUTRAL

(Loads from Standard Calculation)

1. General lighting = ________ VA
2. Cooking equipment = ________ VA
3. Dryer = ________ VA
4. Fixed appliance = ________ VA
5. Heating or A/C = ________ VA
6. Largest motor = ________ VA

TOTAL = ________ **VA**

________ **A**

STANDARD CALCULATION—MULTIFAMILY DWELLING

1. GENERAL LIGHTING LOADS—Table 220-3(b)

_______ sq. ft. × 3 VA × _______ units = _______ VA

_______ sq. ft. × 3 VA × _______ units = _______ VA

_______ sq. ft. × 3 VA × _______ units = _______ VA

_______ VA

Small appliance loads—220-16(a)

_______ VA × _______ circuits × _______ units = _______ VA

Laundry load—220-16(b)

_______ VA × 1 × _______ units = _______ VA

_______ VA

Applying demand factors—Table 220-11

First 3000 VA × 100% = 3000 VA

Next _______ VA × 35% = _______ VA

Remaining _______ VA × 25% = _______ VA

(total) _______ VA | **PHASES** _______ VA | **NEUTRAL** _______ VA

2. COOKING EQUIPMENT LOADS—Table 220-19, Notes

Col. A _______ units = _______ VA demand = _______ VA

Col. B _______ VA × _______ units × _______% = _______ VA

Col. C _______ VA × _______ units × _______% = _______ VA

(total) _______ VA | _______ VA × 70% = _______ VA

3. DRYER LOAD—220-18; Table 220-18

_______ VA × _______ units × _______% = _______ VA | _______ VA × 70% = _______ VA

4. FIXED APPLIANCE LOADS—220-17

Dishwasher = _______ VA

Disposal = _______ VA

Compactor = _______ VA

Water heater = _______ VA

_______ = _______ VA

_______ = _______ VA

(total) _______ VA × _______ units × 75% = _______ VA | _______ VA | (120 V Loads × units × 75%) _______ VA

5. HEATING OR A/C LOAD—220-21

Heating unit = _______ VA × 100% × _______ units = _______ VA

A/C unit = _______ VA × 100% × _______ units = _______ VA

Heat pump = _______ VA × 100% × _______ units = _______ VA

(largest load) _______ VA | _______ VA | _______ VA

6. LARGEST MOTOR LOAD—220-14

ϕ _______ VA × 25% = _______ VA | _______ VA

N _______ VA × 25% = _______ VA | _______ VA

TOTAL = _______ VA | _______ VA

1ϕ service: $I = \frac{VA}{V}$

PHASES $I = \frac{\_\_\_\_\_\_ VA}{V}$ = _______ A

NEUTRAL $I = \frac{\_\_\_\_\_\_ VA}{V}$ = _______ A

3ϕ service: $I = \frac{VA}{V \times \sqrt{3}}$

$I = \frac{\_\_\_\_\_\_ VA}{V \times \sqrt{3}}$ = _______ A

$I = \frac{\_\_\_\_\_\_ VA}{V \times \sqrt{3}}$ = _______ A

If over 200 A—220-22; First 200 A × 100% = 200 A

Remaining _______ A × 70% = _______ A

_______ A

STANDARD CALCULATION—MULTIFAMILY DWELLING 1ϕ (CARRYING PHASES AND NEUTRAL IN VA)

1. GENERAL LIGHTING LOADS—Table 220-3(b)

____ sq. ft. × 3 VA × ____ units = ____ VA

____ sq. ft. × 3 VA × ____ units = ____ VA

____ sq. ft. × 3 VA × ____ units = ____ VA

____ VA

Small appliance loads—220-16(a)

____ VA × ____ circuits × ____ units = ____ VA

Laundry load—220-16(b)

____ VA × 1 × ____ units = ____ VA

____ VA

Applying demand factors—Table 220-11

First 3000 VA × 100% = 3000 VA

Next ____ VA × 35% = ____ VA

Remaining ____ VA × 25% = ____ VA

	PHASE A	PHASE B	NEUTRAL
(total) ____ VA ÷ 2 =	____ VA	____ VA	____ VA
2. COOKING EQUIPMENT LOADS—Table 220-19, Notes			
Col. A ____ units = ____ VA demand =	____ VA	____ VA × 70% =	____ VA
Col. B ____ VA × ____ units × ____% =	____ VA	____ VA × 70% =	____ VA
Col. C ____ VA × ____ units × ____% =	____ VA	____ VA × 70% =	____ VA
3. DRYER LOAD—220-18; Table 220-18			
____ VA × ____ units × ____% =	____ VA	____ VA × 70% =	____ VA
4. FIXED APPLIANCE LOADS—220-17			
Dishwasher = ____ VA × ____ units × 75% ÷ ____ =	____ VA	____ VA	____ VA
Disposal = ____ VA × ____ units × 75% ÷ ____ =	____ VA	____ VA	____ VA
Compactor = ____ VA × ____ units × 75% ÷ ____ =	____ VA	____ VA	____ VA
Water heater = ____ VA × ____ units × 75% ÷ ____ =	____ VA	____ VA	____ VA
____ = ____ VA × ____ units × 75% ÷ ____ =	____ VA	____ VA	____ VA
____ = ____ VA × ____ units × 75% ÷ ____ =	____ VA	____ VA	____ VA
____ = ____ VA × ____ units × 75% ÷ ____ =	____ VA	____ VA	____ VA
____ = ____ VA × ____ units × 75% ÷ ____ =	____ VA	____ VA	____ VA
____ = ____ VA × ____ units × 75% ÷ ____ =	____ VA	____ VA	____ VA
5. HEATING OR A/C LOAD—220-21			
Heating unit = ____ VA × 100% × ____ units =	____ VA	____ VA	____ VA
A/C unit = ____ VA × 100% × ____ units =	____ VA	____ VA	____ VA
Heat pump = ____ VA × 100% × ____ units =	____ VA	____ VA	____ VA
6. LARGEST MOTOR LOAD—220-14			
ϕ ____ VA × 25% =	____ VA	____ VA	
N ____ VA × 25% =			____ VA
TOTAL =	**VA**	**VA**	**VA**

1ϕ service: I = $\frac{VA}{V}$

PHASES I = $\frac{\_\_\_\_ VA}{V}$ = ____ **A**

NEUTRAL I = $\frac{\_\_\_\_ VA}{V}$ = ____ **A**

If over 200 A—220-22; First 200 A × 100% = 200 A

Remaining ____ A × 70% = ____ A

A

STANDARD CALCULATION—MULTIFAMILY DWELLING 1ϕ (CARRYING PHASES AND NEUTRAL IN AMPS)

1. GENERAL LIGHTING LOADS—Table 220-3(b)

_______ sq. ft. × 3 VA × _____ units = _______ VA

_______ sq. ft. × 3 VA × _____ units = _______ VA

_______ sq. ft. × 3 VA × _____ units = _______ VA

_______ VA

Small appliance loads—220-16(a)

_______ VA × _______ circuits × _____ units = _______ VA

Laundry load—220-16(b)

_______ VA × 1 × _____ units = _______ VA

_______ VA

Applying demand factors—Table 220-11

First 3000 VA × 100% = 3000 VA

Next _______ VA × 35% = _______ VA

Remaining _______ VA × 25% = _______ VA

	PHASE A	PHASE B	NEUTRAL
(total) _______ VA ÷ _______ V =	_______ A	_______ A	_______ A
2. COOKING EQUIPMENT LOADS—Table 220-19, Notes			
Col. A _____ units = _______ VA demand = _______ VA ÷ _______ V =	_______ A	_______ A × 70% =	_______ A
Col. B _______ VA × _____ units × _____% = _______ VA ÷ _______ V =	_______ A	_______ A × 70% =	_______ A
Col. C _______ VA × _____ units × _____% = _______ VA ÷ _______ V =	_______ A	_______ A × 70% =	_______ A
3. DRYER LOAD—220-18; Table 220-18			
_______ VA × _____ units × _____% ÷ _______ V =	_______ A	_______ A × 70% =	_______ A
4. FIXED APPLIANCE LOADS—220-17			
Dishwasher = _______ VA × _____ units × 75% ÷ _______ V =	_______ A	_______ A	_______ A
Disposal = _______ VA × _____ units × 75% ÷ _______ V =	_______ A	_______ A	_______ A
Compactor = _______ VA × _____ units × 75% ÷ _______ V =	_______ A	_______ A	_______ A
Water heater = _______ VA × _____ units × 75% ÷ _______ V =	_______ A	_______ A	_______ A
_________ = _______ VA × _____ units × 75% ÷ _______ V =	_______ A	_______ A	_______ A
_________ = _______ VA × _____ units × 75% ÷ _______ V =	_______ A	_______ A	_______ A
_________ = _______ VA × _____ units × 75% ÷ _______ V =	_______ A	_______ A	_______ A
_________ = _______ VA × _____ units × 75% ÷ _______ V =	_______ A	_______ A	_______ A
_________ = _______ VA × _____ units × 75% ÷ _______ V =	_______ A	_______ A	_______ A
5. HEATING OR A/C LOAD—220-21			
Heating unit = _______ VA × 100% × _____ units ÷ _______ V =	_______ A	_______ A	_______ A
A/C unit = _______ VA × 100% × _____ units ÷ _______ V =	_______ A	_______ A	_______ A
Heat pump = _______ VA × 100% × _____ units ÷ _______ V =	_______ A	_______ A	_______ A
6. LARGEST MOTOR LOAD—220-14			
ϕ _______ VA × 25% ÷ _______ V =	_______ A	_______ A	_______ A
N _______ VA × 25% ÷ 120 V =	_______ A	_______ A	_______ A
TOTAL =	A	A	A

If over 200 A—220-22; First 200 A × 100% = 200 A

Remaining _______ A × 70% = _______ A

A

STANDARD CALCULATION—MULTIFAMILY DWELLING 3ϕ (CARRYING PHASES AND NEUTRAL IN VA)

1. GENERAL LIGHTING LOADS—Table 220-3(b)

_____ sq. ft. × 3 VA × _____ units = _____ VA

_____ sq. ft. × 3 VA × _____ units = _____ VA

_____ sq. ft. × 3 VA × _____ units = _____ VA

_____ VA

Small appliance loads—220-16(a)

_____ VA × _____ circuits × _____ units = _____ VA

Laundry load—220-16(b)

_____ VA × 1 × _____ units = _____ VA

_____ VA

Applying demand factors—Table 220-11

First 3000 VA × 100% = 3000 VA

Next _____ VA × 35% = _____ VA

Remaining _____ VA × 25% = _____ VA

	PHASE A	PHASE B	PHASE C		NEUTRAL
(total) _____ VA ÷ 3 =	_____ VA	_____ VA	_____ VA		_____ VA
2. COOKING EQUIPMENT LOADS—Table 220-19, Notes					
Col. A _____ units = _____ VA demand = _____ VA ÷ 3 =	_____ VA	_____ VA	_____ VA	× 70% =	_____ VA
Col. B _____ VA × _____ units × _____% = _____ VA ÷ 3 =	_____ VA	_____ VA	_____ VA	× 70% =	_____ VA
Col. C _____ VA × _____ units × _____% = _____ VA ÷ 3 =	_____ VA	_____ VA	_____ VA	× 70% =	_____ VA
3. DRYER LOAD—220-18; Table 220-18					
_____ VA × _____ units × _____% ÷ 3 =	_____ VA	_____ VA	_____ VA	× 70% =	_____ VA
4. FIXED APPLIANCE LOADS—220-17					
Dishwasher = _____ VA × _____ units × 75% ÷ 3 =	_____ VA	_____ VA	_____ VA		_____ VA
Disposal = _____ VA × _____ units × 75% ÷ 3 =	_____ VA	_____ VA	_____ VA		_____ VA
Compactor = _____ VA × _____ units × 75% ÷ 3 =	_____ VA	_____ VA	_____ VA		_____ VA
Water heater = _____ VA × _____ units × 75% ÷ 3 =	_____ VA	_____ VA	_____ VA		_____ VA
_____ = _____ VA × _____ units × 75% ÷ 3 =	_____ VA	_____ VA	_____ VA		_____ VA
_____ = _____ VA × _____ units × 75% ÷ 3 =	_____ VA	_____ VA	_____ VA		_____ VA
_____ = _____ VA × _____ units × 75% ÷ 3 =	_____ VA	_____ VA	_____ VA		_____ VA
_____ = _____ VA × _____ units × 75% ÷ 3 =	_____ VA	_____ VA	_____ VA		_____ VA
_____ = _____ VA × _____ units × 75% ÷ 3 =	_____ VA	_____ VA	_____ VA		_____ VA
5. HEATING OR A/C LOAD—220-21					
Heating unit = _____ VA × 100% × _____ units ÷ 3 =	_____ VA	_____ VA	_____ VA		_____ VA
A/C unit = _____ VA × 100% × _____ units ÷ 3 =	_____ VA	_____ VA	_____ VA		_____ VA
Heat pump = _____ VA × 100% × _____ units ÷ 3 =	_____ VA	_____ VA	_____ VA		_____ VA
6. LARGEST MOTOR LOAD—220-14					
ϕ _____ VA × 25% =	_____ VA	_____ VA	_____ VA		
N _____ VA × 25% =					_____ VA
TOTAL =	**VA**	**VA**	**VA**		**VA**

3ϕ service: $I = \frac{VA}{V \times \sqrt{3}}$

$I = \frac{VA}{V \times \sqrt{3}} =$ _____ **A**

$I = \frac{VA}{V \times \sqrt{3}} =$ _____ **A**

If over 200 A—220-22; First 200 A × 100% = 200 A

Remaining _____ A × 70% = _____ A

A

STANDARD CALCULATION—MULTIFAMILY DWELLING 3ϕ (CARRYING PHASES AND NEUTRAL IN AMPS)

1. GENERAL LIGHTING LOADS—Table 220-3(b)

_________ sq. ft. × 3 VA × _____ units = _________ VA

_________ sq. ft. × 3 VA × _____ units = _________ VA

_________ sq. ft. × 3 VA × _____ units = _________ VA

_________ VA

Small appliance loads—220-16(a)

_________ VA × _________ circuits × _____ units = _________ VA

Laundry load—220-16(b)

_________ VA × 1 × _____ units = _________ VA

_________ VA

Applying demand factors—Table 220-11

First 3000 VA × 100% = 3000 VA

Next _________ VA × 35% = _________ VA

Remaining _________ VA × 25% = _________ VA

	PHASE A	PHASE B	PHASE C		NEUTRAL
(total) _________ VA ÷ _________ V ÷ 3 =	_________ A	_________ A	_________ A		_________ A
2. COOKING EQUIPMENT LOADS—Table 220-19, Notes					
Col. A _____ units = _______ VA demand = _______ VA ÷ _______ V ÷ 3 =	_________ A	_________ A	_________ A	× 70% =	_________ A
Col. B _______ VA × _____ units × _____% = _______ VA ÷ _______ V ÷ 3 =	_________ A	_________ A	_________ A	× 70% =	_________ A
Col. C _______ VA × _____ units × _____% = _______ VA ÷ _______ V ÷ 3 =	_________ A	_________ A	_________ A	× 70% =	_________ A
3. DRYER LOAD—220-18; Table 220-18					
_________ VA × _____ units × _____% ÷ _________ V ÷ 3 =	_________ A	_________ A	_________ A	× 70% =	_________ A
4. FIXED APPLIANCE LOADS—220-17					
Dishwasher = _________ VA × _____ units × 75% ÷ _________ V ÷ 3 =	_________ A	_________ A	_________ A		_________ A
Disposal = _________ VA × _____ units × 75% ÷ _________ V ÷ 3 =	_________ A	_________ A	_________ A		_________ A
Compactor = _________ VA × _____ units × 75% ÷ _________ V ÷ 3 =	_________ A	_________ A	_________ A		_________ A
Water heater = _________ VA × _____ units × 75% ÷ _________ V ÷ 3 =	_________ A	_________ A	_________ A		_________ A
_________ = _________ VA × _____ units × 75% ÷ _________ V ÷ 3 =	_________ A	_________ A	_________ A		_________ A
_________ = _________ VA × _____ units × 75% ÷ _________ V ÷ 3 =	_________ A	_________ A	_________ A		_________ A
_________ = _________ VA × _____ units × 75% ÷ _________ V ÷ 3 =	_________ A	_________ A	_________ A		_________ A
_________ = _________ VA × _____ units × 75% ÷ _________ V ÷ 3 =	_________ A	_________ A	_________ A		_________ A
_________ = _________ VA × _____ units × 75% ÷ _________ V ÷ 3 =	_________ A	_________ A	_________ A		_________ A
5. HEATING OR A/C LOAD—220-21					
Heating unit = _________ VA × 100% × _____ units ÷ _________ V ÷ 3 =	_________ A	_________ A	_________ A		_________ A
A/C unit = _________ VA × 100% × _____ units ÷ _________ V ÷ 3 =	_________ A	_________ A	_________ A		_________ A
Heat pump = _________ VA × 100% × _____ units ÷ _________ V ÷ 3 =	_________ A	_________ A	_________ A		_________ A
6. LARGEST MOTOR LOAD—220-14					
ϕ _________ VA × 25% ÷ _________ V =	_________ A	_________ A	_________ A		
N _________ VA × 25% ÷ 120 V =					_________ A
TOTAL =	A	A	A		A

If over 200 A—220-22; First 200 A × 100% = 200 A

Remaining _________ A × 70% = _________ A

A

OPTIONAL CALCULATION—MULTIFAMILY DWELLING

1. OTHER LOADS—220-32

General lighting loads—220-32(c)(2) **PHASES**

_____ sq. ft. × 3 VA × _____ units = _____ VA **VA**

_____ sq. ft. × 3 VA × _____ units = _____ VA

_____ sq. ft. × 3 VA × _____ units = _____ VA

_____ sq. ft. × 3 VA × _____ units = _____ VA

Small appliance and laundry loads—220-32(c)(1)

_____ VA × _____ circuits × _____ units = _____ VA **VA**

Special loads—220-32(c)(3)

Dishwasher = _____ VA

Disposal = _____ VA

Compactor = _____ VA

Water heater = _____ VA

_____ = _____ VA

_____ = _____ VA

_____ = _____ VA

_____ = _____ VA

_____ = _____ VA

_____ = _____ VA

(total) _____ VA × _____ units = _____ VA **VA**

2. HEATING OR A/C LOAD—220-32(c)(5)

Heating unit = _____ VA × 100% × _____ units = _____ VA

A/C unit = _____ VA × 100% × _____ units = _____ VA

Heat pump = _____ VA × 100% × _____ units = _____ VA

(largest load) _____ VA **VA**

TOTAL = **VA**

Applying demand factors—Table 220-32

_____ VA × _____% = _____ VA

NEUTRAL

(Loads from Standard Calculation)

1. General lighting = _____ VA
2. Cooking equipment = _____ VA
3. Dryer = _____ VA
4. Fixed appliance = _____ VA
5. Heating or A/C = _____ VA
6. Largest motor = _____ VA

TOTAL = **VA**

A

1ϕ service: $I = \frac{VA}{V}$

PHASES $I = \frac{\_\_\_\_\_ VA}{V}$ = _____ **A**

NEUTRAL $I = \frac{\_\_\_\_\_ VA}{V}$ = _____ **A**

3ϕ service: $I = \frac{VA}{V \times \sqrt{3}}$

$I = \frac{\_\_\_\_\_ VA}{V \times \sqrt{3}}$ = _____ **A**

$I = \frac{\_\_\_\_\_ VA}{V \times \sqrt{3}}$ = _____ **A**

If over 200 A—220-22; First 200 A × 100% = 200 A

Remaining _____ A × 70% = _____ A

A

OPTIONAL CALCULATION—MULTIFAMILY DWELLING 1ϕ (CARRYING PHASES AND NEUTRAL IN VA)

1. OTHER LOADS—220-32	PHASE A	PHASE B	NEUTRAL (from Std. Calc.)
General lighting loads—220-32(c)(2)			
________ sq. ft. × 3 VA × _____ units ÷ _____ =	________ VA	________ VA	________ VA
________ sq. ft. × 3 VA × _____ units ÷ _____ =	________ VA	________ VA	________ VA
________ sq. ft. × 3 VA × _____ units ÷ _____ =	________ VA	________ VA	________ VA
________ sq. ft. × 3 VA × _____ units ÷ _____ =	________ VA	________ VA	________ VA
Small appliance and laundry loads—220-32(c)(1)			
________ VA × _____ circuits × _____ units ÷ _____ =	________ VA	________ VA	________ VA
Special loads—220-32(c)(3)			
Dishwasher = ________ VA × _____ units ÷ ________ =	________ VA	________ VA × 75% =	________ VA
Disposal = ________ VA × _____ units ÷ ________ =	________ VA	________ VA × 75% =	________ VA
Compactor = ________ VA × _____ units ÷ ________ =	________ VA	________ VA × 75% =	________ VA
Water heater = ________ VA × _____ units ÷ ________ =	________ VA	________ VA × 75% =	________ VA
Cooktop/range = ________ VA × _____ units ÷ ________ =	________ VA	________ VA × 70% =	________ VA
Oven = ________ VA × _____ units ÷ ________ =	________ VA	________ VA × 70% =	________ VA
Dryer = ________ VA × _____ units ÷ ________ =	________ VA	________ VA × 70% =	________ VA
________________ = ________ VA × _____ units ÷ ________ =	________ VA	________ VA × 75% =	________ VA
________________ = ________ VA × _____ units ÷ ________ =	________ VA	________ VA × 75% =	________ VA
________________ = ________ VA × _____ units ÷ ________ =	________ VA	________ VA × 75% =	________ VA
________________ = ________ VA × _____ units ÷ ________ =	________ VA	________ VA × 75% =	________ VA
________________ = ________ VA × _____ units ÷ ________ =	________ VA	________ VA × 75% =	________ VA
________________ = ________ VA × _____ units ÷ ________ =	________ VA	________ VA × 75% =	________ VA
________________ = ________ VA × _____ units ÷ ________ =	________ VA	________ VA × 75% =	________ VA
2. HEATING OR A/C LOAD—220-32(c)(5)			
Heating unit = ________ VA × 100% × _____ units =	________ VA	________ VA	________ VA
A/C unit = ________ VA × 100% × _____ units =	________ VA	________ VA	________ VA
Heat pump = ________ VA × 100% × _____ units =	________ VA	________ VA	________ VA
TOTAL =	**VA**	**VA**	
		Plus largest motor =	________ VA
			VA

Applying demand factors—Table 220-32

________ VA × ________% = ________ VA

1ϕ service: I = $\frac{\textbf{VA}}{\textbf{V}}$

PHASES I = $\frac{\text{VA}}{\text{V}}$ = ________ A

NEUTRAL I = $\frac{\text{VA}}{\text{V}}$ = ________ A

If over 200 A—220-22; First 200 A × 100% = 200 A

Remaining ______ A × 70% = ________ A

A

OPTIONAL CALCULATION—MULTIFAMILY DWELLING 1ϕ (CARRYING PHASES AND NEUTRAL IN AMPS)

1. OTHER LOADS—220-32	PHASE A	PHASE B		NEUTRAL (from Std. Calc.)
General lighting loads—220-32(c)(2)				
____ sq. ft. × 3 VA × ____ units ÷ ____ V =	____ A	____ A		____ A
____ sq. ft. × 3 VA × ____ units ÷ ____ V =	____ A	____ A		____ A
____ sq. ft. × 3 VA × ____ units ÷ ____ V =	____ A	____ A		____ A
____ sq. ft. × 3 VA × ____ units ÷ ____ V =	____ A	____ A		____ A
Small appliance and laundry loads—220-32(c)(1)				
____ VA × ____ circuits × ____ units ÷ ____ V =	____ A	____ A		____ A
Special loads—220-32(c)(3)				
Dishwasher = ____ VA × ____ units ÷ ____ V =	____ A	____ A	× 75% =	____ A
Disposal = ____ VA × ____ units ÷ ____ V =	____ A	____ A	× 75% =	____ A
Compactor = ____ VA × ____ units ÷ ____ V =	____ A	____ A	× 75% =	____ A
Water heater = ____ VA × ____ units ÷ ____ V =	____ A	____ A	× 75% =	____ A
Cooktop/range = ____ VA × ____ units ÷ ____ V =	____ A	____ A	× 70% =	____ A
Oven = ____ VA × ____ units ÷ ____ V =	____ A	____ A	× 70% =	____ A
Dryer = ____ VA × ____ units ÷ ____ V =	____ A	____ A	× 70% =	____ A
____ = ____ VA × ____ units ÷ ____ V =	____ A	____ A	× 75% =	____ A
____ = ____ VA × ____ units ÷ ____ V =	____ A	____ A	× 75% =	____ A
____ = ____ VA × ____ units ÷ ____ V =	____ A	____ A	× 75% =	____ A
____ = ____ VA × ____ units ÷ ____ V =	____ A	____ A	× 75% =	____ A
____ = ____ VA × ____ units ÷ ____ V =	____ A	____ A	× 75% =	____ A
____ = ____ VA × ____ units ÷ ____ V =	____ A	____ A	× 75% =	____ A
____ = ____ VA × ____ units ÷ ____ V =	____ A	____ A	× 75% =	____ A
2. HEATING OR A/C LOAD—220-32(c)(5)				
Heating unit = ____ VA × 100% × ____ units ÷ ____ V =	____ A	____ A		____ A
A/C unit = ____ VA × 100% × ____ units ÷ ____ V =	____ A	____ A		____ A
Heat pump = ____ VA × 100% × ____ units ÷ ____ V =	____ A	____ A		____ A
TOTAL =	**A**	**A**		
			Plus largest motor =	____ A
				A

Applying demand factors—Table 220-32

____ A × ____ % = **A**

If over 200 A—220-22; First 200 A × 100% = 200 A

Remaining ____ A × 70% = ____ A

A

OPTIONAL CALCULATION—MULTIFAMILY DWELLING 3ϕ (CARRYING PHASES AND NEUTRAL IN VA)

1. **OTHER LOADS—220-32**	**PHASE A**	**PHASE B**	**PHASE C**	**NEUTRAL** (from Std. Calc.)
General lighting loads—220-32(c)(2)				
______ sq. ft. × 3 VA × ____ units ÷ ____ =	______ VA	______ VA	______ VA	______ VA
______ sq. ft. × 3 VA × ____ units ÷ ____ =	______ VA	______ VA	______ VA	______ VA
______ sq. ft. × 3 VA × ____ units ÷ ____ =	______ VA	______ VA	______ VA	______ VA
______ sq. ft. × 3 VA × ____ units ÷ ____ =	______ VA	______ VA	______ VA	______ VA
Small appliance and laundry loads—220-32(c)(1)				
______ VA × ____ circuits × ____ units ÷ ____ =	______ VA	______ VA	______ VA	______ VA
Special loads—220-32(c)(3)				
Dishwasher = ______ VA × ____ units ÷ ____ =	______ VA	______ VA	______ VA	______ VA
Disposal = ______ VA × ____ units ÷ ____ =	______ VA	______ VA	______ VA	______ VA
Compactor = ______ VA × ____ units ÷ ____ =	______ VA	______ VA	______ VA	______ VA
Water heater = ______ VA × ____ units ÷ ____ =	______ VA	______ VA	______ VA	______ VA
Cooktop/range = ______ VA × ____ units ÷ ____ =	______ VA	______ VA	______ VA	______ VA
Oven = ______ VA × ____ units ÷ ____ =	______ VA	______ VA	______ VA	______ VA
Dryer = ______ VA × ____ units ÷ ____ =	______ VA	______ VA	______ VA	______ VA
________ = ______ VA × ____ units ÷ ____ =	______ VA	______ VA	______ VA	______ VA
________ = ______ VA × ____ units ÷ ____ =	______ VA	______ VA	______ VA	______ VA
________ = ______ VA × ____ units ÷ ____ =	______ VA	______ VA	______ VA	______ VA
________ = ______ VA × ____ units ÷ ____ =	______ VA	______ VA	______ VA	______ VA
________ = ______ VA × ____ units ÷ ____ =	______ VA	______ VA	______ VA	______ VA
________ = ______ VA × ____ units ÷ ____ =	______ VA	______ VA	______ VA	______ VA
________ = ______ VA × ____ units ÷ ____ =	______ VA	______ VA	______ VA	______ VA
2. **HEATING OR A/C LOAD—220-32(c)(5)**				
Heating unit = ______ VA × 100% × ____ units =	______ VA	______ VA	______ VA	______ VA
A/C unit = ______ VA × 100% × ____ units =	______ VA	______ VA	______ VA	______ VA
Heat pump = ______ VA × 100% × ____ units =	______ VA	______ VA	______ VA	______ VA
TOTAL =	**VA**	**VA**	**VA**	
			Plus largest motor =	______ VA
				VA

Applying demand factors—Table 220-32

______ VA × ______% = ______ VA

3ϕ service: $I = \dfrac{VA}{V \times \sqrt{3}}$

$I = \dfrac{\text{\_\_\_\_\_\_ VA}}{\text{\_\_\_ V} \times \sqrt{3}} =$ ______ **A**

$I = \dfrac{\text{\_\_\_\_\_\_ VA}}{\text{\_\_\_ V} \times \sqrt{3}} =$ ______ **A**

If over 200 A—220-22; First 200 A × 100% = 200 A

Remaining ______ A × 70% = ______ A

A

STANDARD CALCULATION—COMMERCIAL AND INDUSTRIAL

1. LIGHTING LOADS—220, Part B

General lighting loads—Table 220-3(b); 220-10(b)

_______ sq. ft. × _______ VA × _______% = _______ VA

Show window loads—220-12

_______′ × 200 VA = _______ VA

Track lighting loads—410-102

$\frac{\_\_\_\_\_\_\_'}{2'}$ × 180 VA = _______ VA

Outside lighting loads—220-10(b)

_______ VA × _______ units × _______% = _______ VA

Sign lighting loads—220-3, FPN; 600-6(c)

_______ VA × _______% = _______ VA

		PHASES	NEUTRAL
(total)	_______ VA	VA	VA

2. RECEPTACLE LOADS—220-3(c)(5)

Noncontinuous duty—220-10(b); Table 220-13

#_______ × 180 VA = _______ VA = _______ VA

Applying demand factors—Table 220-13

First 10,000 VA × 100% = _______ VA = _______ VA

Remaining _______ VA × 50% = _______ VA

Continuous duty—220-10(b) #_______ × 180 VA × 125% = _______ VA

Multioutlet assembly—220-3(c)

Heavy duty = _______′ × 180 VA = _______ VA

Light duty = $\frac{\_\_\_\_\_\_\_'}{5'}$ × 180 VA = _______ VA

(total) _______ VA | VA | VA

3. SPECIAL LOADS—220-10(b)

_______ = _______ VA × _______% = _______ VA

_______ = _______ VA × _______% = _______ VA

_______ = _______ VA × _______% = _______ VA

_______ = _______ VA × _______% = _______ VA

_______ = _______ VA × _______% = _______ VA

(total) _______ VA | VA | VA

4. HEATING OR A/C LOAD—220-21 _______ VA | VA | VA

5. MOTOR LOADS—430-22(a); 430-24; 430-25

				3ϕ only		
_______ =	_______ A ×	_______% ×	_______ V	× $\sqrt{3}$	=	_______ VA
_______ =	_______ A ×	_______% ×	_______ V	× $\sqrt{3}$	=	_______ VA
_______ =	_______ A ×	_______% ×	_______ V	× $\sqrt{3}$	=	_______ VA
_______ =	_______ A ×	_______% ×	_______ V	× $\sqrt{3}$	=	_______ VA

(total) _______ VA | VA | VA

6. LARGEST MOTOR LOAD—220-14

				3ϕ only				
ϕ	_______ A ×	_______% ×	_______ V	× $\sqrt{3}$	× 25% =	_______ VA	VA	VA
N	_______ A ×	_______% ×	_______ V	× $\sqrt{3}$	× 25% =	_______ VA	VA	VA

TOTAL = VA | VA

1ϕ service: $I = \frac{VA}{V}$

PHASES I = $\frac{\_\_\_\_\_\_\_ VA}{V}$ = _______ A

NEUTRAL I = $\frac{\_\_\_\_\_\_\_ VA}{V}$ = _______ A

3ϕ service: $I = \frac{VA}{V \times \sqrt{3}}$

I = $\frac{\_\_\_\_\_\_\_ VA}{V \times \sqrt{3}}$ = _______ A

I = $\frac{\_\_\_\_\_\_\_ VA}{V \times \sqrt{3}}$ = _______ A

If over 200 A—220-22; First 200 A × 100% = 200 A

Remaining _______ A × 70% = _______ A

A

STANDARD CALCULATION—COMMERCIAL AND INDUSTRIAL (CARRYING PHASES AND NEUTRAL IN VA)

	PHASE	PHASE	PHASE	NEUTRAL
1. LIGHTING LOADS—220, Part B				
General lighting loads—Table 220-3(b); 220-10(b)				
____ sq. ft. × ____ VA × ____% ÷ ____ =	____ VA	____ VA	____ VA	____ VA
Show window loads—220-12				
____′ × 200 VA ÷ ____ =	____ VA	____ VA	____ VA	____ VA
Track lighting loads—410-102				
____′ × 180 VA ÷ ____ = (2′)	____ VA	____ VA	____ VA	____ VA
Outside lighting loads—220-10(b)				
____ VA × ____ units × ____% ÷ ____ =	____ VA	____ VA	____ VA	____ VA
Sign lighting loads—220-3, FPN; 600-6(c)				
____ VA × ____% ÷ ____ =	____ VA	____ VA	____ VA	____ VA
2. RECEPTACLE LOADS—220-3(c)(5)				
Noncontinuous duty—220-10(b)				
____ VA × 100% = ____ VA ÷ ____ =	____ VA	____ VA	____ VA	____ VA
Applying demand factors—Table 220-13				
First 10,000 VA × 100% ÷ ____ =	____ VA	____ VA	____ VA	____ VA
Remaining ____ VA × 50% ÷ ____ =	____ VA	____ VA	____ VA	____ VA
Continuous duty—220-10(b) ____ VA × 125% ÷ ____ =	____ VA	____ VA	____ VA	____ VA
Multioutlet assembly—220-3(c)				
Heavy duty = ____′ × 180 VA ÷ ____ =	____ VA	____ VA	____ VA	____ VA
Light duty = ____′ × 180 VA ÷ ____ = (5′)	____ VA	____ VA	____ VA	____ VA
3. SPECIAL LOADS—220-10(b) (load)				
____ = ____ VA × ____% ÷ ____ =	____ VA	____ VA	____ VA	____ VA
____ = ____ VA × ____% ÷ ____ =	____ VA	____ VA	____ VA	____ VA
____ = ____ VA × ____% ÷ ____ =	____ VA	____ VA	____ VA	____ VA
____ = ____ VA × ____% ÷ ____ =	____ VA	____ VA	____ VA	____ VA
____ = ____ VA × ____% ÷ ____ =	____ VA	____ VA	____ VA	____ VA
4. HEATING OR A/C LOAD—220-21	____ VA	____ VA	____ VA	____ VA
5. MOTOR LOADS—430-22(a); 430-24; 430-25 (load) [3ϕ motors only: $\times \sqrt{3}$]				
____ = ____ A × ____ V $\times \sqrt{3}$ ÷ ____ =	____ VA	____ VA	____ VA	____ VA
____ = ____ A × ____ V $\times \sqrt{3}$ ÷ ____ =	____ VA	____ VA	____ VA	____ VA
____ = ____ A × ____ V $\times \sqrt{3}$ ÷ ____ =	____ VA	____ VA	____ VA	____ VA
____ = ____ A × ____ V $\times \sqrt{3}$ ÷ ____ =	____ VA	____ VA	____ VA	____ VA
6. LARGEST MOTOR LOAD—220-14				
ϕ ____ A × ____ V × ____% × 25% =	____ VA	____ VA	____ VA	____ VA
N ____ A × ____ V × ____% × 25% =	____ VA	____ VA	____ VA	____ VA
TOTAL =	**VA**	**VA**	**VA**	**VA**

1ϕ service: $I = \frac{VA}{V}$

PHASES $I = \frac{\_\_\_\_ VA}{V}$ = ____ A

NEUTRAL $I = \frac{\_\_\_\_ VA}{V}$ = ____ A

3ϕ service: $I = \frac{VA}{V \times \sqrt{3}}$

$I = \frac{\_\_\_\_ VA}{V \times \sqrt{3}}$ = ____ A

$I = \frac{\_\_\_\_ VA}{V \times \sqrt{3}}$ = ____ A

If over 200 A—220-22; First 200 A × 100% = 200 A

Remaining ____ A × 70% = ____ A

A

STANDARD CALCULATION—COMMERCIAL AND INDUSTRIAL (CARRYING PHASES AND NEUTRAL IN AMPS)

	PHASE	PHASE	PHASE	NEUTRAL
1. LIGHTING LOADS—220, Part B				
General lighting loads—Table 220-3(b); 220-10(b)				
______ sq. ft. × ______ VA × ______% ÷ ______ V =	______ A	______ A	______ A	______ A
Show window loads—220-12				
______′ × 200 VA ÷ ______ V =	______ A	______ A	______ A	______ A
Track lighting loads—410-102				
______′ (2′) × 180 VA ÷ ______ V =	______ A	______ A	______ A	______ A
Outside lighting loads—220-10(b)				
______ VA × ______ units × ______% ÷ ______ V =	______ A	______ A	______ A	______ A
Sign lighting loads—220-3, FPN; 600-6(c)				
______ VA × ______% ÷ ______ V =	______ A	______ A	______ A	______ A
2. RECEPTACLE LOADS—220-3(c)(5)				
Noncontinuous duty—220-10(b)				
______ VA × 100% = ______ VA ÷ ______ V =	______ A	______ A	______ A	______ A
Applying demand factors—Table 220-13				
First 10,000 VA × 100% ÷ ______ V =	______ A	______ A	______ A	______ A
Remaining ______ VA × 50% ÷ ______ V =	______ A	______ A	______ A	______ A
Continuous duty—220-10(b) ______ VA × 125% ÷ ______ V =	______ A	______ A	______ A	______ A
Multioutlet assembly—220-3(c)				
Heavy duty = ______′ × 180 VA ÷ ______ V =	______ A	______ A	______ A	______ A
Light duty = ______′ (5′) × 180 VA ÷ ______ V =	______ A	______ A	______ A	______ A
3. SPECIAL LOADS—220-10(b)				
(load)				
______ = ______ VA × ______% ÷ ______ V =	______ A	______ A	______ A	______ A
______ = ______ VA × ______% ÷ ______ V =	______ A	______ A	______ A	______ A
______ = ______ VA × ______% ÷ ______ V =	______ A	______ A	______ A	______ A
______ = ______ VA × ______% ÷ ______ V =	______ A	______ A	______ A	______ A
______ = ______ VA × ______% ÷ ______ V =	______ A	______ A	______ A	______ A
4. HEATING OR A/C LOAD—220-21	______ A	______ A	______ A	______ A
5. MOTOR LOADS—430-22(a); 430-24; 430-25				
(load)				
______ = ______ A × ______# ÷ ______ =	______ A	______ A	______ A	______ A
______ = ______ A × ______# ÷ ______ =	______ A	______ A	______ A	______ A
______ = ______ A × ______# ÷ ______ =	______ A	______ A	______ A	______ A
______ = ______ A × ______# ÷ ______ =	______ A	______ A	______ A	______ A
6. LARGEST MOTOR LOAD—220-14				
ϕ ______ A × ______ V × ______% × 25% =	______ A	______ A	______ A	______ A
N ______ A × ______ V × ______% × 25% =	______ A	______ A	______ A	______ A
TOTAL =	A	A	A	A

If over 200 A—220-22; First 200 A × 100% = 200 A

Remaining ______ A × 70% = ______ A

A

STANDARD CALCULATION—MOBILE HOME

1. GENERAL LIGHTING LOADS—550-13(a); 550-13(b)(1)

______ sq. ft. × 3 VA = ______ VA

Small appliance loads—550-13(a)

______ VA × ______ circuits = ______ VA

Laundry load—550-13(a)

______ VA × 1 = ______ VA

______ VA

Applying demand factors—550-13(a)

First 3000 VA × 100% = 3 0 0 0 VA

Next ______ VA × 35% = ______ VA

Remaining ______ VA × 25% = ______ VA

		PHASES	NEUTRAL
(total)	______ VA	VA	VA

2. FREE-STANDING RANGE—550-13(b)(5)—(nameplate rating)

		PHASES	NEUTRAL
Range =	______ VA	VA	VA

3. APPLIANCE LOADS—550-13(b)(4)

Dishwasher = ______ VA

Disposal = ______ VA

Water heater = ______ VA

Cooktop = ______ VA

Oven = ______ VA

Dryer = ______ VA

______ = ______ VA

______ = ______ VA

______ = ______ VA

		PHASES	NEUTRAL
(total) ______ VA × 75% =	______ VA	VA	VA

4. HEATING OR A/C LOAD—550-13(b)(2)

Heating unit = ______ VA × 100% = ______ VA

A/C unit = ______ VA × 100% = ______ VA

Heat pump = ______ VA × 100% = ______ VA

		PHASES	NEUTRAL
(largest load)	______ VA	VA	VA

5. MOTOR LOADS—550-13(b)(2)

______ = ______ A × ______ V = ______ VA

______ = ______ A × ______ V = ______ VA

______ = ______ A × ______ V = ______ VA

______ = ______ A × ______ V = ______ VA

		PHASES	NEUTRAL
(total)	______ VA	VA	VA

6. LARGEST MOTOR LOAD—550-13(b)(3)

		PHASES	NEUTRAL
ϕ ______ VA × 25% =	______ VA	VA	
N ______ VA × 25% =	______ VA	______	______ VA
	TOTAL =	VA	VA

1ϕ service: I = $\frac{\textbf{VA}}{\textbf{V}}$

PHASES I = $\frac{\text{\_\_\_\_\_\_ VA}}{\text{V}}$ = ______ **A**

NEUTRAL I = $\frac{\text{\_\_\_\_\_\_ VA}}{\text{V}}$ = ______ **A**

STANDARD CALCULATION—HOTELS AND MOTELS

1. GENERAL LIGHTING LOADS—Table 220-3(b)

________ sq. ft. × 2 VA × ______ units = ________ VA

________ sq. ft. × 2 VA × ______ units = ________ VA

________ sq. ft. × 2 VA × ______ units = ________ VA

________ VA

Applying demand factors—Table 220-11

First 20,000 VA × 50% = 10,000 VA

Next ________ VA × 40% = ________ VA

Remaining ________ VA × 30% = ________ VA

(total) ________ VA ________ VA

Show window loads—220-12

________′ × 200 VA = ________ VA

Track lighting loads—410-102

$\frac{\_\_\_\_\_\_\_\_'}{2'}$ × 180 VA = ________ VA

Outside lighting loads—220-10(b)

________ VA × ______ units × ______% = ________ VA

Sign lighting loads—220-3, FPN; 600-6(c)

________ VA × ________% = ________ VA

		PHASES	NEUTRAL
(total)	________ VA	VA	VA

2. RECEPTACLE LOADS—220-3(c)(5)

Continuous duty—220-10(b); Table 220-13

		PHASES	NEUTRAL
________ VA × 125% = (total)	________ VA	VA	VA

3. SPECIAL LOADS—220-10(b)

________ = ________ VA × ______ units × ______% = ________ VA

________ = ________ VA × ______ units × ______% = ________ VA

________ = ________ VA × ______ units × ______% = ________ VA

________ = ________ VA × ______ units × ______% = ________ VA

________ = ________ VA × ______ units × ______% = ________ VA

________ = ________ VA × ______ units × ______% = ________ VA

________ = ________ VA × ______ units × ______% = ________ VA

		PHASES	NEUTRAL
(total)	________ VA	VA	VA

4. HEATING OR A/C LOAD—220-21

Heating unit = ________ VA × 100% × ______ units = ________ VA

A/C unit = ________ VA × 100% × ______ units = ________ VA

Heat pump = ________ VA × 100% × ______ units = ________ VA

		PHASES	NEUTRAL
(largest load)	________ VA	VA	VA

5. LARGEST MOTOR LOAD—220-14

		PHASES	NEUTRAL
ϕ ________ A × ______% × ________ V × 25% =	________ VA	VA	VA
N ________ A × ______% × ________ V × 25% =	________ VA	VA	VA
	TOTAL =	VA	VA

1ϕ service: $I = \frac{VA}{V}$

PHASES $I = \frac{\_\_\_\_\_\_ VA}{V}$ = ________ A

NEUTRAL $I = \frac{\_\_\_\_\_\_ VA}{V}$ = ________ A

3ϕ service: $I = \frac{VA}{V \times \sqrt{3}}$

$I = \frac{\_\_\_\_\_\_ VA}{V \times \sqrt{3}}$ = ________ A

$I = \frac{\_\_\_\_\_\_ VA}{V \times \sqrt{3}}$ = ________ A

If over 200 A—220-22; First 200 A × 100% = 200 A

Remaining ______ A × 70% = ________ A

A

OPTIONAL CALCULATION—SCHOOL

1. TOTAL VA—Table 220-34

Total connected load = ________ VA

2. TOTAL SQUARE FOOTAGE—Table 220-34

Square footage = ________ sq. ft.

3. VA PER SQ. FT.—Table 220-34

$\frac{\text{\_\_\_\_\_\_ VA}}{\text{sq. ft.}}$ = ________ VA/sq. ft.

4. APPLYING DEMAND FACTORS—Table 220-34

First 3 VA per sq. ft. × 100%

Next 3 VA to 20 VA per sq. ft. × 75%

Remaining 21 VA per sq. ft × 25%

	PHASES
3 VA/sq. ft. × ________ sq. ft. × 100% =	VA
______ VA/sq. ft. × ________ sq. ft. × 75% =	VA
______ VA/sq. ft. × ________ sq. ft. × 25% =	VA
TOTAL =	VA

NEUTRAL

(Loads from Standard Calculation)

1. Lighting loads = ________ VA
2. Receptacle loads = ________ VA
3. Special loads = ________ VA
4. Heating or A/C load = ________ VA
5. Motor loads = ________ VA
6. Largest motor load = ________ VA

TOTAL = ________ **VA**

________ **A**

1φ service: $I = \frac{VA}{V}$

PHASES $I = \frac{\text{\_\_\_\_\_\_ VA}}{V}$ = ________ **A**

NEUTRAL $I = \frac{\text{\_\_\_\_\_\_ VA}}{V}$ = ________ **A**

3φ service: $I = \frac{VA}{V \times \sqrt{3}}$

$I = \frac{\text{\_\_\_\_\_\_ VA}}{V \times \sqrt{3}}$ = ________ **A**

$I = \frac{\text{\_\_\_\_\_\_ VA}}{V \times \sqrt{3}}$ = ________ **A**

If over 200 A—220-22; First 200 A × 100% = 200 A

Remaining ______ A × 70% = ________ A

________ **A**

Index

T

U

V

W

X